T0145319

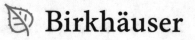

# Operator Theory: Advances and Applications

Volume 274

Founded in 1979 by Israel Gohberg

More information about this series at http://www.springer.com/series/4850

Fabrizio Colombo • Jonathan Gantner

# Quaternionic Closed Operators, Fractional Powers and Fractional Diffusion Processes

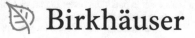 Birkhäuser

Fabrizio Colombo
Dipartimento di Matematica
Politecnico di Milano
Milano, Italy

Jonathan Gantner
Dipartimento di Matematica
Politecnico di Milano
Milano, Italy

ISSN 0255-0156          ISSN 2296-4878  (electronic)
Operator Theory: Advances and Applications
ISBN 978-3-030-16411-9          ISBN 978-3-030-16409-6  (eBook)
https://doi.org/10.1007/978-3-030-16409-6

Mathematics Subject Classification (2010): 47A60

This book is published under the imprint Birkhäuser, www.birkhauser-science.com by the registered company Springer Nature Switzerland AG.
The registered company address is: Gewerbestrasse 11, 6330 Cham, Switzerland

# Preface

The $S$-spectrum of a quaternionic linear operator and the $S$-functional calculus, introduced in 2006, were the starting point to fully understand quaternionic spectral theory and in particular the spectral theory of vector operators. An early version of the $S$-functional calculus for quaternionic operators, for $n$-tuples of non commuting operators and the theory of slice hyperholomorphic functions can be found in the book [93] published in 2011. Since then, the spectral theory based on the $S$-spectrum has grown enormously, thanks to several collaborators. Because of this, a systematic structure, including the most important results obtained after 2011, has been needed. The foundations of quaternionic spectral theory based on the $S$-spectrum have been the topic of the monograph [57]. Precisely, in [57] we study the properties on the $S$-spectrum, the $S$-functional calculus, the $F$-functional calculus (which is a monogenic functional calculus based on the $S$-spectrum), the quaternionic spectral theorem, the theory of spectral integration in quaternionic Banach spaces and quaternionic spectral operators.

This book is the natural continuation of [57], but here we consider mainly unbounded operators and we apply quaternionic spectral theory to fractional diffusion processes. Precisely, we consider the direct formulation of the $S$-functional calculus for unbounded operators, and we show that in the quaternionic setting, the two possible approaches are not fully equivalent, as it is the case for complex operators. Then we introduce the theory of quaternionic groups and semigroups, we study some generation results, and we define the quaternionic Phillips functional calculus for generators of groups of quaternionic operators via the Laplace-Stieltjes-transform. Under suitable conditions, we show that the quaternionic Phillips-functional calculus agrees with the $S$-functional calculus.

Then we dedicate a substantial part of this book to the quaternionic $H^\infty$-functional calculus for sectorial quaternionic operators. This calculus is important to define fractional powers of quaternionic operators and, as a particular case, of vector operators like the gradient operator or its generalizations such as non-constant coefficients first order vector operators. We will also investigate different methods for the definition of the fractional powers of quaternionic operators like the Kato approach.

The possibility to define fractional powers of vector operators has important applications to fractional diffusion problems. In fact, the fractional powers of suitable non-constant coefficients first order vector operators give the associated fractional Fourier's law, this is a non-local law that takes into account global effects in the heat propagation. Our theory applies not only to the heat diffusion process but also, for example, to Fick's law and, more generally, it allows to compute the fractional powers of vector operators that arise in different fields of science and technology. For the convenience of the reader, we summarize in Chapter 1 some of the theoretical aspects and we discuss the applications that will be developed in this book.

**Acknowledgment.** It is a pleasure for the authors to thank Daniel Alpay, Jussi Behrndt, Vladinir Bolotnikov, Paula Cerejeiras, Roman Lávička, Oscar González-Cervantes, Maria Elena Luna-Elizarraras, Uwe Kähler, D. P. Kimsey, Tao Qian, Irene Sabadini, Michael Shapiro, Frank Sommen, Vladimir Souček and Daniele C. Struppa for fruitful collaborations.

# Contents

# Chapter 1

# Introduction

In this chapter we summarize the theoretical aspects of the quaternionic spectral theory that will be developed later in this book and we also show some of the possible applications of this theory to fractional diffusion processes.

We denote the skew-field of quaternions by $\mathbb{H}$. An element $s$ of $\mathbb{H}$ is of the form $s = s_0 + s_1 e_1 + s_2 e_2 + s_3 e_3$, $s_\ell \in \mathbb{R}$, $\ell = 0, 1, 2, 3$, where $e_1$, $e_2$ and $e_3$ are the generating imaginary units of $\mathbb{H}$, which satisfy $e_\ell^2 = -1$ and $e_\ell e_\kappa = -e_\kappa e_\ell$ for $\ell, \kappa = 1, 2, 3$ and $\ell \neq \kappa$. The real part $s_0$ of the quaternion $s$ is also denoted by $\operatorname{Re}(s)$, while its imaginary part is defined as $\operatorname{Im}(s) := s_1 e_1 + s_2 e_2 + s_3 e_3$. We indicate by $\mathbb{S}$ the unit sphere of purely imaginary quaternions, i.e.,

$$\mathbb{S} = \{s = s_1 e_1 + s_1 e_2 + s_3 e_3 : \quad s_1^2 + s_2^2 + s_3^2 = 1\}.$$

Notice that if $j \in \mathbb{S}$, then $j^2 = -1$. For this reason the elements of $\mathbb{S}$ are also called imaginary units. The set $\mathbb{S}$ is a 2-dimensional sphere in $\mathbb{R}^4 \cong \mathbb{H}$. Given a nonreal quaternion $s = \operatorname{Re}(s) + \operatorname{Im}(s)$, we have $s = u + j_s v$ with $u = \operatorname{Re}(s)$, $j_s = \operatorname{Im}(s)/|\operatorname{Im}(s)| \in \mathbb{S}$ and $v = |\operatorname{Im}(s)|$. We can associate to $s$ the 2-dimensional sphere

$$[s] = \{s_0 + j|\operatorname{Im}(s)| : \quad j \in \mathbb{S}\} = \{u + jv : \quad j \in \mathbb{S}\}. \tag{1.1}$$

Before we describe the contents of this book, we recall the problem with the definition of the spectrum of a linear vector operator or, more generally, of a quaternionic linear operator. For more details and for the history of quaternionic spectral theory we refer to the book [57], where we have previously written about the historical development of quaternionic spectral theory and of the related function theories, which can be found in the introduction and in several notes at the end of each chapter. We, however, point out that the main difficulties in developing a mathematically rigorous spectral theory for quaternionic operators was the lack of correct definitions of spectrum and resolvent for such operators. In fact, consider for example, a right linear bounded quaternionic operator $T : X \to X$ acting on a

© Springer Nature Switzerland AG 2019

F. Colombo, J. Gantner, *Quaternionic Closed Operators, Fractional Powers and Fractional Diffusion Processes*, Operator Theory: Advances and Applications 274,
https://doi.org/10.1007/978-3-030-16409-6_1

two-sided quaternionic Banach space $X$, that is,

$$T(x\alpha + y\beta) = T(x)\alpha + T(y)\beta,$$

for all $\alpha, \beta \in \mathbb{H}$ and $x, y \in X$. The symbol $\mathcal{B}(X)$ denotes the Banach space of all bounded right linear operators endowed with the natural norm, and we denote by $\mathcal{I}$ the identity operator. The spectrum of an operator should, in an appropriate way, generalize the set of its eigenvalues. But since the quaternionic multiplication is not commutative, even the notion of eigenvalue in this setting is ambiguous. Indeed, one can either consider left or right eigenvalues, which are determined by the equations

$$Tx = sx \quad \text{and} \quad Tx = xs,$$

respectively. In the classical setting, the spectrum $\sigma(A)$ of a complex linear operator $A$ is defined as the set of all $\lambda \in \mathbb{C}$ such that the operator of the eigenvalue equation $\lambda \mathcal{I} - A$ does not have a bounded inverse. If we try to proceed similarly for the left eigenvalue equation, we obtain the left spectrum $\sigma_L(T)$ of $T$, which is defined as

$$\sigma_L(T) := \{s \in \mathbb{H} : \quad s\mathcal{I} - T \text{ is not invertible in } \mathcal{B}(X)\}, \tag{1.2}$$

where the notation $s\mathcal{I}$ in $\mathcal{B}(X)$ means that $(s\mathcal{I})(x) = sx$. It is associated with the left resolvent operator $(s\mathcal{I} - T)^{-1}$, which is defined on the complement of $\sigma_L(T)$. However, the operator-valued function $s \mapsto (s\mathcal{I} - T)^{-1}$ is not hyperholomorphic in $\mathbb{H} \setminus \sigma_L(T)$ with respect to any known notion of generalized holomorphicity over the quaternions, which limited its usefulness for developing quaternionic spectral theory. Furthermore, the left eigenvalues of an operator (or even a matrix) did not seem to be have any meaningful applications neither in physical applications nor in the mathematical theory.

The notion of right eigenvalues on the other hand seemed to be the more natural notion of eigenvalues, since the considered operators were right linear. Even more, the notion of right eigenvalues had an interpretation in quaternionic quantum mechanics [4] and the spectral theorem for quaternionic matrices is based on the right eigenvalues [114]. The right eigenvalue equation is, however, not quaternionic linear. Indeed, if $Tx = xs$ for some $x \neq 0$ and $a \in \mathbb{H}$ with $as \neq sa$, then

$$T(xa) = T(x)a = (xs)a = x(sa) = (xa)(a^{-1}sa) \neq (xa)s. \tag{1.3}$$

Hence, the operator of the right eigenvalue equation $x \mapsto Tx - xs$ is not linear and its inverse can consequently not be used to define a meaningful notion of quaternionic resolvent. We define the right spectrum $\sigma_R(T)$ of $T$ therefore as the set of right eigenvalues

$$\sigma_R(T) := \{s \in \mathbb{H} : \quad Tx = xs, \text{for some } x \in X \setminus \{0\}\}.$$

Even though the set of right eigenvalues was meaningful both in applications and in the mathematical theory on finite dimensional spaces, it was not clear how to generalize it to a proper notion of spectrum of a right linear operator nor with which resolvent operator this spectrum should be associated. Furthermore, right eigenvalues have two problematic properties that are immediately understood from (1.3). First of all, the set of eigenvectors associated with an individual eigenvalue does not constitute a quaternionic linear space. If $x$ is a right eigenvector of $T$ associated with $s$, then $xa$ is a right eigenvector of $T$ associated with $a^{-1}sa$ instead of $s$. The second problem is that right eigenvalues do not appear individually but in terms of equivalence classes of the form

$$[s] = \{a^{-1}sa : \quad a \in \mathbb{H} \setminus \{0\}\}. \tag{1.4}$$

This set agrees with the symmetry class of $s$ defined in (1.1). Hence, the right spectrum $\sigma_R(T)$ of $T$ is axially symmetric.

The solution of these problems came in 2006, when I. Sabadini and one of the authors, introduced the $S$-spectrum and the $S$-functional calculus for quaternionic linear operators starting from considerations on slice hyperholomorphic functions, see the introduction of the book [57]. This notion is not intuitive because the $S$-spectrum of $T$ is defined for those quaternions $s$ such that the second order operator $T^2 - 2\text{Re}(s)T + |s|^2 \mathcal{I}$ is not invertible, where $\text{Re}(s)$ is the real part of the quaternion $s$ and $|s|^2$ is its squared norm. There exists also a commutative version of the $S$-spectrum and it is very useful in applications. We will denote by $\mathcal{BC}(X)$ the subclass of $\mathcal{B}(X)$ that consists of those quaternionic operators $T$ that can be written as $T = T_0 + e_1 T_1 + e_2 T_2 + e_3 T_3$ where the operators $T_\ell$, $\ell = 0, 1, 2, 3$, commute mutually, and we set $\overline{T} = T_0 - e_1 T_1 - e_2 T_2 - e_3 T_3$. In this case the $S$-spectrum has an equivalent definition that takes into account the commutativity of $T_\ell$, for $\ell = 0, 1, 2, 3$. In the literature the commutative definition of the $S$-spectrum is often called the $F$-spectrum because it is used for the $F$-functional calculus, see [57]. Let $T \in \mathcal{BC}(X)$, we define the commutative version of the $S$-spectrum (or F-spectrum $\sigma_F(T)$) of $T$ as those $s \in \mathbb{H}$ such that the operator $s^2 \mathcal{I} - s(T + \overline{T}) + T\overline{T}$ is not invertible. The S-resolvent set $\rho_S(T)$ is defined as $\rho_S(T) = \mathbb{H} \setminus \sigma_S(T)$.

Since this book is the natural continuation of [57] where the quaternionic spectral theory based on the $S$-spectrum is systematically studied, we summarize in two sections the theoretical aspects and the applications that we develop in this book.

## 1.1 Theoretical aspects

The notion of $S$-spectrum turned out to be the correct notion of spectrum for a quaternionic linear operator $T$, see also the books [57, 93], and it was discovered from considerations on slice hyperholomorphic functions. Moreover, the right eigenvalues $\sigma_R(T)$ are equal to the $S$-eigenvalues of $T$. We limit the discussion to the case of quaternionic operators but the following definition of $S$-spectrum can be adapted to the case of $n$-tuples of non commuting operators. We define

$$\mathcal{Q}_s(T) := T^2 - 2\mathrm{Re}(s)T + |s|^2 \mathcal{I}.$$

If $T$ is a linear quaternionic operator then the $S$-resolvent set is defined as

$$\rho_S(T) = \{s \in \mathbb{H} : \quad \mathcal{Q}_s(T)^{-1} \in \mathcal{B}(X)\},$$

where $\mathcal{Q}_s(T)^{-1}$ is called the pseudo-resolvent operator of $T$ at $s$, while the $S$-spectrum is defined as:

$$\sigma_S(T) := \mathbb{H} \setminus \rho_S(T).$$

Due to the non commutativity of the quaternions, there are two resolvent operators associated with a quaternionic linear operator $T$: when $T$ is bounded, the left $S$-resolvent operator is defined as

$$S_L^{-1}(s,T) := -\mathcal{Q}_s(T)^{-1}(T - \bar{s}\mathcal{I}), \quad s \in \rho_S(T) \tag{1.5}$$

and the right $S$-resolvent operator is

$$S_R^{-1}(s,T) := -(T - \bar{s}\mathcal{I})\mathcal{Q}_s(T)^{-1}, \quad s \in \rho_S(T). \tag{1.6}$$

The first main difference with respect to complex operator theory is the fact that the $S$-resolvent equation involves both the $S$-resolvent operators

$$S_R^{-1}(s,T)S_L^{-1}(p,T) = [[S_R^{-1}(s,T) - S_L^{-1}(p,T)]p$$
$$- \bar{s}[S_R^{-1}(s,T) - S_L^{-1}(p,T)]](p^2 - 2s_0 p + |s|^2)^{-1},$$

for $s,\, p \in \rho_S(T)$ with $s \notin [p]$. A second major difference is the fact that the operator that defines the $S$-spectrum is the pseudo-resolvent operator and not the $S$-resolvent operator, but that the pseudo-resolvent operator $\mathcal{Q}_s(T)^{-1}$ is not slice hyperholomorphic. Only the $S$-resolvent operators are operator-valued slice hyperholomorphic functions.

The $S$-functional calculus (also called quaternionic functional calculus) is the quaternionic version of the Riesz-Dunford functional calculus. It is based on the $S$-spectrum and on the Cauchy formula of slice hyperholomorphic functions. In the next chapter we therefore summarize the main facts on slice hyperholomorphic functions. For more details see [57, 93].

If the operator $T$ is bounded, then its $S$-spectrum $\sigma_S(T)$ is a non-empty compact subset of $\mathbb{H}$ that is bounded by the norm of $T$. We denote by $\mathcal{SH}_L(\sigma_S(T))$ the set of left slice hyperholomorphic functions $f : U \to \mathbb{H}$ where $U$ is a suitable bounded open set that contains $\sigma_S(T)$. Analogously we define $\mathcal{SH}_R(\sigma_S(T))$ for right slice hyperholomorphic functions. The two formulations of the quaternionic

functional calculus for left- and right slice hyperholomorphic functions are then
given by

$$f(T) = \frac{1}{2\pi} \int_{\partial(U \cap \mathbb{C}_j)} S_L^{-1}(s,T) \, ds_j \, f(s), \quad f \in \mathcal{SH}_L(\sigma_S(T)), \qquad (1.7)$$

and

$$f(T) = \frac{1}{2\pi} \int_{\partial(U \cap \mathbb{C}_j)} f(s) \, ds_j \, S_R^{-1}(s,T), \quad f \in \mathcal{SH}_R(\sigma_S(T)), \qquad (1.8)$$

where $ds_j = -dsj$, for $j \in \mathbb{S}$. The $S$-functional calculus is well defined since
the integrals depend neither on the open set $U$ with $\sigma_S(T) \subset U$ nor on the
imaginary unit $j \in \mathbb{S}$. It is important to note that the definition of the quaternionic
functional calculus does not require the linear operator $T$ to be written in terms
of components $T = T_0 + \sum_{\ell=1}^{3} T_\ell e_\ell$ with bounded linear operators $T_\ell$, $\ell = 0, \ldots, 3$,
on a real Banach space. Nor does it require that the components $T_\ell$ commute
mutually as it was the case in earlier developed functional calculi for quaternionic
linear operators that were based on other function theories. If the components $T_\ell$
of the operator $T = T_0 + \sum_{\ell=1}^{3} T_\ell e_\ell$ commute mutually, we set for $s \in \mathbb{H}$

$$\mathcal{Q}_{c,s}(T) := s^2 \mathcal{I} - 2sT_0 + T\overline{T}$$

and we find that the operator $\mathcal{Q}_{c,s}(T)$ is invertible if and only if $\mathcal{Q}_s(T)$ is invertible
and so the $S$-resolvent set of $T$ can also be characterized as

$$\rho_S(T) = \left\{ s \in \mathbb{H} : \quad \mathcal{Q}_{c,s}(T)^{-1} \in \mathcal{B}(X) \right\}. \qquad (1.9)$$

The operator $\mathcal{Q}_{c,s}(T)^{-1}$ is in this case called the commutative pseudo-resolvent
operator. Moreover, for $s \in \rho_S(T)$, the commutative $S$-resolvent operators are

$$S_L^{-1}(s,T) = (s\mathcal{I} - \overline{T})\mathcal{Q}_{c,s}(T)^{-1} \qquad (1.10)$$
$$S_R^{-1}(s,T) = \mathcal{Q}_{c,s}(T)^{-1}(s\mathcal{I} - \overline{T}). \qquad (1.11)$$

The main topics treated in the next chapters are the possible extensions and generalizations of the $S$-functional calculus to unbounded operators while particular
attention is dedicated to sectorial operators.

*Direct approach to the S-functional calculus.* We develop the $S$-functional calculus for closed quaternionic linear operators. The $S$-functional calculus for closed
operators has already been considered in the books [57,93], where the unbounded
operator and the function were suitably transformed so that the $S$-functional calculus for bounded operators could be applied. This strategy is standard in the
complex case, but in the quaternionic case it has the disadvantage that it requires
that $\rho_S(T) \cap \mathbb{R} \neq \emptyset$. Already the most important quaternionic linear operator, the
gradient operator, does not satisfy this condition.

We therefore define the $S$-functional calculus for closed operators in Chapter 3 directly via a slice hyperholomorphic Cauchy integral formula. If $T$ is a closed

operator with nonempty $S$-resolvent set and $f$ is a function that is left slice hyper-holomorphic on a suitable set $U$ with $\sigma_S(T) \subset U$ that contains a neighbourhood of $\infty$, then we can use the Cauchy formula

$$f(x) = f(\infty) + \frac{1}{2\pi} \int_{\partial(U\cap\mathbb{C}_j)} S_L^{-1}(s,x)\, ds_j\, f(s), \quad x \subset U.$$

Formally replacing $x$ by $T$ we define

$$f(T) := f(\infty)\mathcal{I} + \frac{1}{2\pi} \int_{\partial(U\cap\mathbb{C}_j)} S_L^{-1}(s,T)\, ds_j\, f(s), \quad \sigma_S(T) \subset U.$$

This functional calculus is well-defined and its properties agree with those of the Riesz-Dunford-functional calculus for closed complex linear operators. We investigate these properties in detail. In particular, we discuss the product rule and show that this functional calculus is compatible with intrinsic polynomials of $T$ although these polynomials do not belong to the class of admissible functions because they are not slice hyperholomorphic at infinity. Furthermore, we discuss the relation between the $S$-functional calculus for left and the $S$-functional calculus for right slice hyperholomorphic functions, we prove the spectral mapping theorem and we show that the functional calculus is capable of generating Riesz-projectors onto invariant subspaces.

*Generation of groups and semigroups and the Phillips functional calculus.* If $T$ is a bounded right linear operator on a quaternionic Banach space $X$ then, for $s_0$ sufficiently large, the left $S$-resolvent operator can be written as the Laplace transform of $e^{tT}$,

$$S_L^{-1}(s,T) = \int_0^\infty e^{tT} e^{-ts}\, dt,$$

while the right $S$-resolvent operator can be written as:

$$S_R^{-1}(s,T) = \int_0^\infty e^{-ts} e^{tT}\, dt.$$

The above relations hold true also for a class of unbounded linear operators. We investigate the generation of groups and of semigroups. Moreover, we consider the following perturbation problem. Suppose that the closed right linear quaternionic operator $T$ is the infinitesimal generator of the semigroup $\mathcal{U}_T(t)$. Determine a class of closed right linear quaternionic operators $P$ such that $T + P$ is the generator of a quaternionic semigroup $\mathcal{U}_{T+P}(t)$.

In the case the operator $T$ is the generator of a strongly continuous group of quaternionic linear operators, then one can define a slice hyperholomorphic functional calculus via the quaternionic Laplace–Stieltjes transform. The so-called Phillips functional calculus applies to a larger class of functions with respect to the

$S$-functional calculus because it does not require slice hyperholomorphicity at infinity. If $T$ is the infinitesimal generator of a strongly continuous group $\{\mathcal{U}_T(t)\}_{t \in \mathbb{R}}$ with growth bound $\omega > 0$, that is,

$$\sigma_S(T) \subset \{s \in \mathbb{H}: \quad -\omega \leq \mathrm{Re}(s) \leq \omega\} \tag{1.12}$$

and

$$\|\mathcal{U}_T(t)\| \leq M e^{-|t|\omega}, \tag{1.13}$$

for some constant $M > 0$, then we consider the subset $\mathbf{S}(T)$ of all quaternion-valued measures on $\mathbb{R}$ given by

$$\mathbf{S}(T) := \left\{ \mu \in \mathcal{M}(\mathbb{R}, \mathbb{H}): \quad \int_{\mathbb{R}} e^{-|t|(\omega + \varepsilon)} \, d|\mu|(t) < +\infty \right\}, \tag{1.14}$$

where $\varepsilon > 0$ might depend on the measure $\mu$. The quaternionic Laplace–Stieltjes transform

$$f(q) := \int_{\mathbb{R}} d\mu(t) \, e^{-tq}, \quad -(\omega + \varepsilon) < \mathrm{Re}(q) < \omega + \varepsilon$$

is then a right slice hyperholomorphic function and we can define

$$f(T) := \int_{\mathbb{R}} d\mu(t) \, \mathcal{U}_T(-t).$$

We discuss the quaternionic Laplace–Stieltjes transform and show that this functional calculus is well-defined. We study its algebraic properties and show its compatibility with the $S$-functional calculus defined in Chapter 3. Finally, we conclude by showing how to invert the operator $f(T)$ for intrinsic $f$ using an inverting sequence of polynomials.

*The general version of the $H^\infty$-functional calculus.* This is the natural functional calculus for sectorial quaternionic operators and in this book we introduce it in its full generality. Any quaternion can be written as $s = |s| e^{j_s \arg(s)}$ with a unique angle $\arg(s) \in [0, \pi]$. A quaternionic right linear operator is called sectorial if its $S$-spectrum is contained in the closure of a symmetric sector around the positive real axis of the form

$$\Sigma_\omega = \{s \in \mathbb{H}: \quad \arg(s) \in [0, \omega)\}$$

with $\omega \in (0, \pi)$ and for any $\varphi \in (\omega, \pi)$ there exists a constant $C > 0$ such that

$$\|S_L^{-1}(s, T)\| \leq \frac{C}{|s|} \quad \text{and} \quad \|S_R^{-1}(s, T)\| \leq \frac{C}{|s|},$$

for all $s \in \mathbb{H} \setminus \Sigma_\varphi$. If $f$ is left slice hyperholomorphic on a sector $\Sigma_\varphi$ for some $\varphi \in (\omega, \pi)$ and has polynomial limit 0 both at 0 and at infinity, then we can choose $\varphi' \in (\omega, \varphi)$ and define $f(T)$ by a Cauchy integral as

$$f(T) := \frac{1}{2\pi} \int_{\partial(\Sigma_{\varphi'} \cap \mathbb{C}_j)} S_L^{-1}(s, T) \, ds_j \, f(s). \tag{1.15}$$

If $f$ is left slice hyperholomorphic on $\Sigma_\varphi$ and has finite polynomial limits at 0 and infinity, then it is of the form

$$f(q) = \tilde{f}(q) + a + (1+q)^{-1}b \qquad (1.16)$$

with $a, b \in \mathbb{H}$ and $\tilde{f}$ admissible for (1.15). Since $-1 \in \rho_S(T)$, the operator

$$-S_L^{-1}(s,T) = (\mathcal{I} + T)^{-1}$$

exists, and we can define for such functions

$$f(T) := \tilde{f}(T) + \mathcal{I}a + (\mathcal{I} + T)^{-1}b, \qquad (1.17)$$

where $\tilde{f}(T)$ is intended in the sense of (1.15). We denote the class of functions of the form (1.16) by $\mathcal{E}_L(\Sigma_\varphi)$ and the class of intrinsic functions of the form (1.16) by $\mathcal{E}(\Sigma_\varphi)$. Finally, the class of admissible functions can be extended even further, which yields the $H^\infty$-functional calculus. A regulariser for a left slice meromorphic function $f$ on $\Sigma_\varphi$ is a function $e \in \mathcal{E}(\Sigma_\varphi)$ such that $ef \in \mathcal{E}_L(\Sigma_\varphi)$ and such that $e(T)$ is injective. If such a regulariser exists for $f$, then we define

$$f(T) := e(T)^{-1}(ef)(T),$$

where $e(T)$ and $(ef)(T)$ are intended in the sense of (1.17). This operator is not necessarily bounded, because $e(T)^{-1}$ can be unbounded.

We define this functional calculus precisely and discuss its properties. We focus in particular on the composition rule and the spectral mapping theorem. As we will see there are several technical difficulties that have to be overcome when generalising them from the complex to the quaternionic setting.

*Fractional powers of quaternionic linear operators.* We first define fractional powers of sectorial operators with bounded inverse directly by the slice hyperholomorphic Cauchy integral formula

$$T^{-\alpha} := \frac{1}{2\pi} \int_\Gamma s^{-\alpha} \, ds_j \, S_R^{-1}(s,T),$$

where $\Gamma$ is a path that goes from $-\infty e^{j\theta}$ to $\infty e^{-j\theta}$ in the set $\mathbb{C}_j \setminus (\Sigma_\varphi \cup B_\varepsilon(0))$ for sufficiently small $\varepsilon > 0$, sufficiently large $\theta \in (0, \pi)$ and arbitrary $j \in \mathbb{S}$ and avoids the negative real axis. We then discuss the properties of these fractional powers. In particular, we prove several integral representations and the semigroup property. We point out that in the quaternionic setting there exist integral representations that do not exist in the complex setting, for example when $\sigma_S(T) \subset \{s \in \mathbb{H} : \text{Re}(s) > 0\}$ for $\alpha \in (0,1)$, we have

$$T^{-\alpha} = \frac{1}{\pi} \int_0^{+\infty} \tau^{-\alpha} \left( \cos\left(\frac{\alpha\pi}{2}\right) T + \sin\left(\frac{\alpha\pi}{2}\right) \tau\mathcal{I} \right) (T^2 + \tau^2\mathcal{I})^{-1} \, d\tau.$$

The inverse of the quadratic operator $T^2 + \tau^2 \mathcal{I}$, that appears in the above formula, comes from the $S$-resolvent operator and it has no analogue in the complex setting because of noncommutativity. As a second approach, we define the fractional powers of positive exponent via the $H^\infty$-functional calculus. In a third approach we introduce the fractional powers indirectly using an approach of Kato. We first define for $\alpha \in (0,1)$ the operator-valued function

$$F_\alpha(p, T) := \frac{\sin(\alpha\pi)}{\pi} \int_0^{+\infty} t^\alpha \left(p^2 - 2pt^\alpha \cos(\alpha\pi) + t^{2\alpha}\right)^{-1} S_R^{-1}(-t, T)\, dt,$$

which corresponds to an integral representation of $S_R^{-1}(p, T^\alpha)$ of the form (1.15), in which we let $\varphi'$ tend to $\pi$. Then we show that there actually exists a unique closed operator $B_\alpha$ such that $F_\alpha(p, T) = S_R^{-1}(s, B_\alpha)$ and we define $T^\alpha := B_\alpha$.

## 1.2 Applications to fractional diffusion processes

One of the most important facts about quaternionic spectral theory is that it contains, as a particular case, the spectral theory of vector operators like the gradient operator and its generalizations with non constant coefficients. We explain in the following the ideas behind the definition of new fractional diffusion operators that generalize the Fourier law to non-local diffusion processes.

Our strategy for fractional diffusion problems does not apply only to the Fourier law with constant coefficients, but it works for general non constant coefficients Fourier law and it generates the associated non local diffusion operator. Moreover, we can apply our techniques for bounded domains as well as for unbounded domains.

To explain our approach we consider the case of fractional evolution on $\mathbb{R}^3$. We denote by $v : \mathbb{R}^3 \times [0, \infty) \to \mathbb{R}$ the temperature and by $q$ the heat flow. We set the thermal diffusivity equal to 1 and set $x = (x_1, x_2, x_3)$. The heat equation is then deduced from the two laws

$$q(x, t) = -\nabla v(x, t) \quad \text{(Fourier's law)}, \tag{1.18}$$

$$\partial_t v(x, t) + \operatorname{div} q(x, t) = 0 \quad \text{(Conservation of Energy)}. \tag{1.19}$$

Replacing the heat flow in the law of conservation of energy using Fourier's law, we get the classical heat equation:

$$\partial_t v(x, t) - \Delta v(x, t) = 0, \quad (x, t) \in \mathbb{R}^3 \times (0, \infty). \tag{1.20}$$

The fractional heat equation is an alternative model, that takes non-local interactions into account. It is obtained by replacing the negative Laplacian in the heat equation by its fractional power so that one obtains the equation

$$\partial_t v(x, t) + (-\Delta)^\alpha v(x, t) = 0, \quad (x, t) \in \mathbb{R}^3 \times (0, \infty), \quad \alpha \in (0, 1), \tag{1.21}$$

where the fractional Laplacian is given by

$$(-\Delta)^\alpha v(x) = c(\alpha) P.V. \int_{\mathbb{R}^3} \frac{v(x) - v(y)}{|x - y|^{3+2\alpha}} \, dy,$$

and the integral is defined in the sense of the principal value, $c(\alpha)$ is a known constant, and $v : \mathbb{R}^3 \to \mathbb{R}$ must belong to a suitable function space.

*The fractional powers of the gradient operator.* The new approach to fractional diffusion presented in this book consists in replacing the gradient in (1.18) by its fractional power before combining it with (1.19) instead of replacing the negative Laplacian in (1.20) by its fractional power. This is done by interpreting the gradient as a quaternionic linear operator, which allows us to define its fractional powers using the techniques presented in this book.

The following two observations are of crucial importance for defining the new procedure for fractional diffusion processes.

(I) The $S$-spectrum of the gradient operator $\nabla$ on $L^2(\mathbb{R}^3, \mathbb{H})$ is $\sigma_S(\nabla) = \mathbb{R}$. Since the map $s \mapsto s^\alpha$ with $\alpha \in (0,1)$ is not defined on $(-\infty, 0)$, we have to consider the projections of the fractional power $\nabla^\alpha$ to the subspace associated with the subset $[0, +\infty)$ of the $S$-spectrum of $\nabla$. Only for these spectral values, the function $s \mapsto s^\alpha$ is well defined and slice hyperholomorphic. We denote these projections by $P_\alpha(\nabla)$. Precisely, what we call the fractional power $P_\alpha(\nabla)$ of the gradient, is given by the quaternionic Balakrishnan formula (deduced from the $H^\infty$-functional calculus)

$$P_\alpha(\nabla) v = \frac{1}{2\pi} \int_{-j\mathbb{R}} S_L^{-1}(s, \nabla) \, ds_j \, s^{\alpha-1} \nabla v,$$

for $v : \mathbb{R}^3 \to \mathbb{R}$ in $\mathcal{D}(\nabla)$. The path of integration is chosen to take into account just the part of the $S$-spectrum with $\mathrm{Re}(\sigma_S(\nabla)) \geq 0$.

(II) The above procedure gives a quaternionic operator

$$P_\alpha(\nabla) = Z_0 + e_1 Z_1 + e_2 Z_2 + e_3 Z_3,$$

where $Z_\ell$, $\ell = 0, 1, 2, 3$, are real operators obtained by the functional calculus. Finally, we take the vector part $P_\alpha(\nabla)$

$$\mathrm{Vect}(P_\alpha(\nabla)) = e_1 Z_1 + e_2 Z_2 + e_3 Z_3$$

of the quaternionic operator $P_\alpha(\nabla)$ so that we can apply the divergence operator.

With the above definitions and the surprising expression for the left $S$-resolvent operator

$$S_L^{-1}(-jt, \nabla) = (-jt + \nabla) \underbrace{(-t^2 + \Delta)^{-1}}_{=R_{-t^2}(-\Delta)},$$

the fractional powers $P_\alpha(\nabla)$ become

$$P_\alpha(\nabla)v = \underbrace{\frac{1}{2}(-\Delta)^{\frac{\alpha}{2}-1}\nabla^2 v}_{:=\mathrm{Scal}P_\alpha(\nabla)v} + \underbrace{\frac{1}{2}(-\Delta)^{\frac{\alpha-1}{2}}\nabla v}_{:=\mathrm{Vec}P_\alpha(\nabla)v}.$$

Now we observe that

$$\mathrm{div}\,\mathrm{Vec}P_\alpha(\nabla)v = -\frac{1}{2}(-\Delta)^{\frac{\alpha}{2}+1}v.$$

The fractional heat equation for $\alpha \in (1/2, 1)$

$$\partial_t v(t,x) + (-\Delta)^\alpha v(t,x) = 0$$

can hence be written as

$$\partial_t v(t,x) - 2\mathrm{div}\,(\mathrm{Vec}P_\beta(\nabla)v) = 0, \quad \beta = 2\alpha - 1,$$

so this approach coincided with the classical one for the gradient operator.

*The S-spectrum approach to fractional diffusion processes.* We say that $T$ represents a Fourier law with commuting components, when $T$ is a vector operator of the form

$$T = e_1 a_1(x_1)\partial_{x_1} + e_2 a_2(x_2)\partial_{x_2} + e_3 a_3(x_3)\partial_{x_3},$$

where $a_1$, $a_2$, $a_3 : \Omega \to \mathbb{R}$ are suitable real-valued functions that depend on the space variables $x_1$, $x_2$, $x_3$, respectively, where $(x_1, x_2, x_3) \in \Omega$ and $\Omega \subseteq \mathbb{R}^3$. In this case the real operators $a_\ell(x_\ell)\partial_{x_\ell}$, for $\ell = 1, 2, 3$, commute among themselves. Then the $S$-spectrum of $T$ can also be determined using the commutative pseudo-resolvent operator

$$\mathcal{Q}_{c,s}(T) := s^2\mathcal{I} - 2sT_0 + T\overline{T} = a_1^2(x_1)\partial_{x_1}^2 + a_2^2(x_2)\partial_{x_2}^2 + a_3^2(x_3)\partial_{x_3}^2 + s^2\mathcal{I}$$

because $\mathcal{Q}_{c,s}(T)$ is invertible if and only if $\mathcal{Q}_s(T)$ is invertible. The operator $\mathcal{Q}_{c,s}(T)$ is a scalar operator if $s^2$ is a real number. Since $T$ is a vector operator, we have $T_0 = 0$ and $T\overline{T}$ does not contain the imaginary units of the quaternions. Using the non commutative expression of the pseudo-resolvent operator $\mathcal{Q}_s(T)$, we obtain

$$\begin{aligned}\mathcal{Q}_s(T) = &-(a_1(x_1)\partial_{x_1})^2 - (a_2(x_2)\partial_{x_2})^2 - (a_3(x_3)\partial_{x_3})^2 \\ &- 2s_0(e_1 a_1(x_1)\partial_{x_1} + e_2 a_2(x_2)\partial_{x_2} + e_3 a_3(x_3)\partial_{x_3}) + |s|^2\mathcal{I}.\end{aligned}$$

We observe that, according to what we need to show in the commutative case, we have two possibilities to determine the $S$-spectrum: $\mathcal{Q}_{c,s}(T)$ and $\mathcal{Q}_s(T)$.

Now we explicitly describe the procedure of the $S$-spectrum approach to fractional diffusion processes. Suppose that $\Omega \subseteq \mathbb{R}^3$ is a suitable bounded or unbounded domain and let $X$ be a two-sided Banach space. We consider the initial-boundary value problem for non-homogeneous materials. We denote by $T$ the heat flow $q(x, \partial_x)$ and we restrict ourselves to the case of homogeneous boundary conditions (for $\tau > 0$):

$$T(x) = a_1(x_1)\partial_{x_1}e_1 + a_2(x_2)\partial_{x_2}e_2 + a_3(x_3)\partial_{x_3}e_3, \quad x = (x_1, x_2, x_3) \in \Omega,$$
$$\partial_t v(x, t) + \operatorname{div} T(x)v(x, t) = 0, \quad (x, t) \in \Omega \times (0, \tau],$$
$$v(x, 0) = f(x), \quad x \in \Omega,$$
$$v(x, t) = 0, \quad x \in \partial\Omega, \ t \in [0, \tau].$$

Our general procedure consists of the following steps:

(S1)  We study the invertibility of the operator

$$Q_{c,s}(T) := s^2\mathcal{I} - 2sT_0 + T\overline{T} = a_1^2(x_1)\partial_{x_1}^2 + a_2^2(x_2)\partial_{x_2}^2 + a_3^2(x_3)\partial_{x_3}^2 + s^2\mathcal{I},$$

where $\overline{T} = -T$, to get the $S$-resolvent operator. Precisely, let $F : \Omega \to \mathbb{H}$ be a given function with a suitable regularity and denote by $X : \Omega \to \mathbb{H}$ the unknown function of the boundary value problem:

$$\left(a_1^2(x_1)\partial_{x_1}^2 + a_2^2(x_2)\partial_{x_2}^2 + a_3^2(x_3)\partial_{x_3}^2 + s^2\mathcal{I}\right)X(x) = F(x), \quad x \in \Omega,$$
$$X(x) = 0, \quad x \in \partial\Omega.$$

We study under which conditions on the coefficients $a_1$, $a_2$, $a_3 : \mathbb{R}^3 \to \mathbb{R}$ the above equation has a unique solution. We can similarly use the non commutative version of the pseudo-resolvent operator $Q_s(T)$. In the case we deal with an operator $T$ with non-commuting components, then we have to consider $Q_s(T)$, only.

(S2)  From (S1) we get that $s \in \mathbb{H} \setminus \{0\}$ with $\operatorname{Re}(s) = 0$ belongs to $\rho_S(T)$, so we obtain the unique pesudo-resolvent operator $Q_{c,s}(T)^{-1}$ and we define the $S$-resolvent operator

$$S_L^{-1}(s, T) = (s\mathcal{I} - \overline{T})Q_{c,s}(T)^{-1}.$$

Then we prove that, for every $s \in \mathbb{H} \setminus \{0\}$ with $\operatorname{Re}(s) = 0$, the $S$-resolvent operators satisfy the estimates

$$\left\|S_L^{-1}(s, T)\right\| \leq \frac{\Theta}{|s|} \quad \text{and} \quad \left\|S_R^{-1}(s, T)\right\| \leq \frac{\Theta}{|s|} \tag{1.22}$$

with a constant $\Theta > 0$ that does not depend on $s$.

(S3)  Using the Balakrishnan, formula we define $P_\alpha(T)$ as:

$$P_\alpha(T)v = \frac{1}{2\pi} \int_{-j\mathbb{R}} s^{\alpha-1} \, ds_j \, S_R^{-1}(s, T)Tv, \quad \text{for } \alpha \in (0, 1),$$

and $v \in \mathcal{D}(T)$. Analogously, one can use the definition of $P_\alpha(T)$ related to the left $S$-resolvent operator.

(S4) After we define the fractional powers $P_\alpha(T)$ of the vector operator $T$, we consider its vector part $\mathrm{Vec}(P_\alpha(T))$ and we obtain the fractional evolution equation:

$$\partial_t v(t,x) - \mathrm{div}(\mathrm{Vec}(P_\alpha(T)v)(t,x) = 0.$$

As an application of our theory we get Theorems 10.3.1 and 10.3.2 that we summarize in the following result.

*Let $\Omega$ be a bounded domain in $\mathbb{R}^3$ with sufficiently smooth boundary. Let $T = e_1\, a_1(x_1)\partial_{x_1} + e_2\, a_2(x_2)\partial_{x_2} + e_3\, a_3(x_3)\partial_{x_3}$ and assume that the coefficients $a_\ell : \overline{\Omega} \to \mathbb{R}$, for $\ell = 1,2,3$, belong to $C^1(\overline{\Omega},\mathbb{R})$ and $a_\ell(x_\ell) \geq m$ in $\overline{\Omega}$ for some $m > 0$. Moreover, assume that*

$$\inf_{x\in\Omega} \left|a_\ell(x_\ell)^2\right| - \frac{\sqrt{C_\Omega}}{2}\left\|\partial_{x_\ell} a_\ell(x_\ell)^2\right\|_\infty > 0, \quad \ell = 1,2,3,$$

*and*

$$\frac{1}{2} - \frac{1}{2}\|\Phi\|_\infty^2 C_\Omega^2 C_a^2 > 0,$$

*where $C_\Omega$ is the Poincaré constant of $\Omega$ and*

$$\Phi(x) := \sum_{\ell=1}^{3} e_\ell \partial_{x_\ell} a_\ell(x_\ell) \quad and \quad C_a := \sup_{\substack{x\in\Omega \\ \ell=1,2,3}} \frac{1}{|a_\ell(x_\ell)|} = \frac{1}{\inf_{\substack{x\in\Omega \\ \ell=1,2,3}} |a_\ell(x_\ell)|}.$$

*Then any $s \in \mathbb{H}\setminus\{0\}$ with $\mathrm{Re}(s) = 0$ belongs to $\rho_S(T)$ and the S-resolvents satisfy the estimate*

$$\left\|S_L^{-1}(s,T)\right\| \leq \frac{\Theta}{|s|} \quad and \quad \left\|S_R^{-1}(s,T)\right\| \leq \frac{\Theta}{|s|}, \quad if\ \mathrm{Re}(s) = 0, \tag{1.23}$$

*with a constant $\Theta > 0$ that does not depend on $s$. Moreover, for $\alpha \in (0,1)$, and for any $v \in \mathcal{D}(T)$, the integral*

$$P_\alpha(T)v := \frac{1}{2\pi}\int_{-j\mathbb{R}} s^{\alpha-1}\, ds_j\, S_R^{-1}(s,T)Tv$$

*converges absolutely in $L^2(\Omega,\mathbb{H})$.*

The above result in particular holds for $v$ real-valued.

This approach has several advantages:

(I) It modifies the Fourier law but keeps the law of conservation of energy.

(II) It is applicable to a large class of operators that includes the gradient but also operators with non-constant coefficients.

(III) Fractional powers of the operator $T$ provide a more realistic model for non-homogeneous materials.

(IV) The fact that we keep the evolution equation in divergence form allows an immediate definition of the weak solution of the fractional evolution problem.

## 1.3   On quaternionic spectral theories

For the convenience of the reader, we recall in this section some considerations that were already discussed in [57] in order to put the spectral theory on the $S$-spectrum into perspective. The quaternionic spectral theories arise from the Fueter–Sce–Qian mapping theorem that has been widely treated in [57].

In classical complex operator theory, the Cauchy formula of holomorphic functions is a fundamental tool for defining functions of operators. Moreover, the Cauchy–Riemann operator factorizes the Laplace operator, so holomorphic functions also play a crucial role in harmonic analysis and in boundary value problems. In higher dimensions, for quaternion-valued functions or more in general for Clifford-algebra-valued functions, there appear two different notions of hyperholomorphicity. The first one is called slice hyperholomorphicity and the second one is known under different names, depending on the dimension of the algebra and the range of the functions: Cauchy–Fueter regularity for quaternion-valued functions and monogenicity for Clifford algebra-valued functions. The Fueter–Sce–Qian mapping theorem reveals a fundamental relation between the different notions of hyperholomorphicity and it can be illustrated by the following two maps

$$F_1 : \mathrm{Hol}(\Omega) \mapsto \mathcal{N}(U) \quad \text{and} \quad F_2 : \mathcal{N}(U) \mapsto \mathcal{AM}(U).$$

The map $F_1$ transforms holomorphic functions in $\mathrm{Hol}(\Omega)$, where $\Omega$ is a suitable open set $\Omega$ in $\mathbb{C}$, into intrinsic slice hyperholomorphic functions in $\mathcal{N}(U)$ defined on the open set $U$ in $\mathbb{H}$. Applying the second transformation $F_2$ to intrinsic slice hyperholomorphic functions, we get axially Fueter regular (resp. axially monogenic) functions. Roughly speaking the map $F_1$ is defined as follows:

1. We consider a holomorphic function $f(z)$ that depends on a complex variable $z = u + \iota v$ in an open set of the upper complex halfplane. (In order to distinguish the imaginary unit of $\mathbb{C}$ from the quaternionic imaginary units, we denote it by $\iota$). We write

$$f(z) = f_0(u, v) + \iota f_1(u, v),$$

where $f_0$ and $f_1$ are $\mathbb{R}$-valued functions that satisfy the Cauchy–Riemann system.

2. For suitable quaternions $q$, we replace the complex imaginary unit $\iota$ in $f(z) = f_0(u, v) + \iota f_1(u, v)$ by the quaternionic imaginary unit $\frac{\mathrm{Im}(q)}{|\mathrm{Im}(q)|}$ and we set $u = \mathrm{Re}(q) = q_0$ and $v = |\mathrm{Im}(q)|$. We then define

$$f(q) = f_0(q_0, |\mathrm{Im}(q)|) + \frac{\mathrm{Im}(q)}{|\mathrm{Im}(q)|} f_1(q_0, |\mathrm{Im}(q)|).$$

The function $f(q)$ turns out to be slice hyperholomorphic by construction.

When considering quaternion-valued functions, the map $F_2$ is the Laplace operator, i.e., $F_2 = \Delta$. When we work with Clifford algebra-valued functions, then $F_2 = \Delta_{n+1}^{(n-1)/2}$, where $n$ is the number of generating units of the Clifford algebra and $\Delta_{n+1}$ is the Laplace operator in dimension $n+1$. The Fueter–Sce–Qian mapping theorem can be adapted to the more general case in which $\mathcal{N}(U)$ is replaced by slice hyperholomorphic functions and the axially regular (or axially monogenic) functions $\mathcal{AM}(U)$ are replaced by monogenic functions. The generalization of holomorphicity to quaternion- or Clifford algebra-valued functions produces two different notions of hyper-holomorphicity that are useful for different purposes. Precisely, we have that:

(I) The Cauchy formula of slice hyperholomorphic functions leads to the definition of the $S$-spectrum and the $S$-functional calculus for quaternionic linear operators. Moreover, the spectral theorem for quaternionic linear operators is based on the $S$-spectrum. The aim of this book and of the monograph [57] is to give a systematic treatment of this theory and of its applications.

(II) The Cauchy formula associated with Cauchy–Fueter regularity (resp. monogenicity) leads to the notion of monogenic spectrum and produces the Cauchy–Fueter functional calculus for quaternion-valued functions and the monogenic functional calculus for Clifford algebra-valued functions. This theory has applications in harmonic analysis in higher dimensions and in boundary value problems. For an overview on the monogenic functional calculus and its applications see [171] and for applications to boundary values problems see [163] and the references contained in those books.

In this book and in the monograph [57] we treat the quaternionic spectral theory on the $S$-spectrum so, very often, we will refer to it as quaternionic spectral theory because no confusion arises with respect to the monogenic spectral theory.

# Chapter 2

# Preliminary results

This chapter contains two main topics: the theory of slice hyperholomorphic functions and the $S$-functional calculus for bounded quaternionic operators. We limit ourselves to recalling the results we need in this book. For the proofs of the main theorems, we refer the reader to [57], which currently contains the completest version of quaternionic spectral theory.

## 2.1   Slice hyperholomorphic functions

There are three possible ways to define slice hyperholomorphic functions, using the definition in [141], using the global operator of slice hyperholomorphic functions introduced in [62] or using the definition that comes from Fueter–Sce–Qian mapping theorem. This last definition is the most appropriate for operator theory so in this chapter we summarize the properties of slice hyperholomorphic functions that we will use in the sequel. The proofs can be found in Chapter 2 of the monograph [57].

We denote by $\mathbb{H}$ the algebra of quaternions. An element $q$ of $\mathbb{H}$ is of the form

$$q = q_0 + q_1 e_1 + q_2 e_2 + q_3 e_3, \quad q_\ell \in \mathbb{R}, \ \ell = 0, 1, 2, 3,$$

where $e_1$, $e_2$ and $e_3$ are the generating imaginary units of $\mathbb{H}$. They satisfy the relations

$$e_1^2 = e_2^2 = e_3^2 = -1 \tag{2.1}$$

and

$$e_1 e_2 = -e_2 e_1 = e_3, \quad e_2 e_3 = -e_3 e_2 = e_1, \quad e_3 e_1 = -e_1 e_3 = e_2. \tag{2.2}$$

The real part, the imaginary part and the modulus of a quaternion $q = q_0 + q_1 e_1 + q_2 e_2 + q_3 e_3$ are defined as $\mathrm{Re}(q) = q_0$, $\mathrm{Im}(q) = q_1 e_1 + q_2 e_2 + q_3 e_3$ and $|q|^2 = q_0^2 + q_1^2 + q_2^2 + q_3^2$. The conjugate of the quaternion $q$ is

$$\bar{q} = \mathrm{Re}(q) - \mathrm{Im}(q) = q_0 - q_1 e_1 - q_2 e_2 - q_3 e_3$$

© Springer Nature Switzerland AG 2019

F. Colombo, J. Gantner, *Quaternionic Closed Operators, Fractional Powers and Fractional Diffusion Processes*, Operator Theory: Advances and Applications 274,

https://doi.org/10.1007/978-3-030-16409-6_2

and it satisfies

$$|q|^2 = q\bar{q} = \bar{q}q.$$

The inverse of any nonzero element $q$ is hence given by

$$q^{-1} = \frac{\bar{q}}{|q|^2}.$$

We denote by $\mathbb{S}$ the unit sphere of purely imaginary quaternions, i.e.,

$$\mathbb{S} = \{q = q_1 e_1 + q_2 e_2 + q_3 e_3 : q_1^2 + q_2^2 + q_3^2 = 1\}.$$

Notice that if $j \in \mathbb{S}$, then $j^2 = -1$. For this reason the elements of $\mathbb{S}$ are also called imaginary units. The set $\mathbb{S}$ is a 2-dimensional sphere in $\mathbb{R}^4 \cong \mathbb{H}$. Given a nonreal quaternion $q = q_0 + \mathrm{Im}(q)$, we have $q = u + jv$ with $u = \mathrm{Re}(q)$, $v = |\mathrm{Im}(q)|$ and $j = \mathrm{Im}(q)/|\mathrm{Im}(q)| \in \mathbb{S}$. (We will sometimes use the notation $j_q = j = \mathrm{Im}(q)/|\mathrm{Im}(q)|$ when it is necessary to stress the relation between $j$ and $q$.) We can associate to $q$ the 2-dimensional sphere

$$[q] = \{q_0 + j|\mathrm{Im}(q)| : \quad j \in \mathbb{S}\} = \{u + jv : \quad j \in \mathbb{S}\}.$$

This sphere is centered at the real point $q_0 = \mathrm{Re}(q)$ and has radius $|\mathrm{Im}(q)|$. Furthermore, a quaternion $\tilde{q}$ belongs to $[q]$ if and only if there exists $h \in \mathbb{H} \setminus \{0\}$ such that $\tilde{q} = h^{-1}qh$.

If $j \in \mathbb{S}$, then the set

$$\mathbb{C}_j = \{u + jv : \quad u, v \in \mathbb{R}\}$$

is an isomorphic copy of the complex numbers. If moreover $i \in \mathbb{S}$ with $j \perp i$, then $j, i$ and $k := ji$ is a generating basis of $\mathbb{H}$, i.e., this basis also satisfies the relations (2.1) and (2.2). Hence, any quaternion $q \in \mathbb{H}$ can be written as

$$q = z_1 + z_2 i = z_1 + i\bar{z}_2$$

with unique $z_1, z_2 \in \mathbb{C}_j$ and so

$$\mathbb{H} = \mathbb{C}_j + i\mathbb{C}_j \quad \text{and} \quad \mathbb{H} = \mathbb{C}_j + \mathbb{C}_j i. \tag{2.3}$$

Moreover, we observe that

$$\mathbb{H} = \bigcup_{j \in \mathbb{S}} \mathbb{C}_j.$$

Finally, we introduce the notation $\overline{\mathbb{H}} := \mathbb{H} \cup \{\infty\}$.

**Definition 2.1.1.** Let $U \subseteq \mathbb{H}$.

  (i) We say that $U$ is *axially symmetric* if $[q] \subset U$ for any $q \in U$.

  (ii) We say that $U$ is a *slice domain* if $U \cap \mathbb{R} \neq \emptyset$ and if $U \cap \mathbb{C}_j$ is a domain in $\mathbb{C}_j$ for any $j \in \mathbb{S}$.

**Definition 2.1.2** (Slice hyperholomorphic functions). Let $U \subseteq \mathbb{H}$ be an axially symmetric open set and let $\mathcal{U} = \{(u,v) \in \mathbb{R}^2 : u + \mathbb{S}v \subset U\}$. A function $f : U \to \mathbb{H}$ is called left slice function, if it is of the form

$$f(q) = f_0(u,v) + j f_1(u,v), \quad \text{for } q = u + jv \in U$$

with two functions $f_0, f_1 : \mathcal{U} \to \mathbb{H}$ that satisfy the compatibility condition

$$f_0(u,-v) = f_0(u,v), \quad f_1(u,-v) = -f_1(u,v). \tag{2.4}$$

If in addition $f_0$ and $f_1$ satisfy the Cauchy–Riemann equations

$$\frac{\partial}{\partial u} f_0(u,v) - \frac{\partial}{\partial v} f_1(u,v) = 0, \tag{2.5}$$

$$\frac{\partial}{\partial v} f_0(u,v) + \frac{\partial}{\partial u} f_1(u,v) = 0, \tag{2.6}$$

then $f$ is called left slice hyperholomorphic. A function $f : U \to \mathbb{H}$ is called right slice function if it is of the form

$$f(q) = f_0(u,v) + f_1(u,v)j, \quad \text{for } q = u + jv \in U$$

with two functions $f_0, f_1 : \mathcal{U} \to \mathbb{H}$ that satisfy (2.4). If in addition $f_0$ and $f_1$ satisfy the Cauchy–Riemann equations, then $f$ is called right slice hyperholomorphic.

If $f$ is a left (or right) slice function such that $f_0$ and $f_1$ are real-valued, then $f$ is called intrinsic.

We denote the sets of left and right slice hyperholomorphic functions on $U$ by $\mathcal{SH}_L(U)$ and $\mathcal{SH}_R(U)$, respectively. The set of intrinsic slice hyperholomorphic functions on $U$ will be denoted by $\mathcal{N}(U)$.

**Remark 2.1.1.** Any quaternion $q$ can be represented as an element of a complex plane $\mathbb{C}_j$ using at least two different imaginary units $j \in \mathbb{S}$. We have $q = u + jv = u + (-j)(-v)$ and $-j$ also belongs to $\mathbb{S}$. If $q$ is real, then we can use any imaginary unit $j \in \mathbb{S}$ to consider $q$ as an element of $\mathbb{C}_j$. The compatibility condition (2.4) assures that the choice of this imaginary unit is irrelevant. In particular, it forces $f_1(u,v)$ to equal 0 if $v = 0$, that is, if $q \in \mathbb{R}$.

The multiplication and the composition with intrinsic functions preserve slice hyperholomorphicity. This is not true for arbitrary slice hyperholomorphic functions.

**Theorem 2.1.3.** *With the notation above we have the properties:*

(i) *If $f \in \mathcal{N}(U)$ and $g \in \mathcal{SH}_L(U)$, then $fg \in \mathcal{SH}_L(U)$. If $f \in \mathcal{SH}_R(U)$ and $g \in \mathcal{N}(U)$, then $fg \in \mathcal{SH}_R(U)$.*

(ii) *If $g \in \mathcal{N}(U)$ and $f \in \mathcal{SH}_L(g(U))$, then $f \circ g \in \mathcal{SH}_L(U)$. If $g \in \mathcal{N}(U)$ and $f \in \mathcal{SH}_R(g(U))$, then $f \circ g \in \mathcal{SH}_R(U)$.*

**Lemma 2.1.4.** *Let $U \subset \mathbb{H}$ be axially symmetric and let $f$ be a left (or right) slice function on $U$. The following statements are equivalent.*

(i) *The function $f$ is intrinsic.*

(ii) *We have $f(U \cap \mathbb{C}_j) \subset \mathbb{C}_j$ for any $j \in \mathbb{S}$.*

(iii) *We have $f(\overline{q}) = \overline{f(q)}$ for all $q \in U$.*

If we restrict a slice hyperholomorphic function to one of the complex planes $\mathbb{C}_j$, then we obtain a function that is holomorphic in the usual sense.

**Lemma 2.1.5** (The Splitting Lemma). *Let $U \subset \mathbb{H}$ be an axially symmetric open set and let $j, i \in \mathbb{S}$ with $i \perp j$. If $f \in \mathcal{SH}_L(U)$, then the restriction $f_j = f|_{U \cap \mathbb{C}_j}$ satisfies*

$$\frac{1}{2}\left(\frac{\partial}{\partial u}f_j(z) + j\frac{\partial}{\partial v}f_j(z)\right) = 0 \tag{2.7}$$

*for all $z = u + jv \in U \cap \mathbb{C}_j$. Hence*

$$f_j(z) = F_1(z) + F_2(z)i$$

*with holomorphic functions $F_1, F_2 : U \cap \mathbb{C}_j \to \mathbb{C}_j$.*
    *If $f \in \mathcal{SH}_R(U)$, then the restriction $f_j = f|_{U \cap \mathbb{C}_j}$ satisfies*

$$\frac{1}{2}\left(\frac{\partial}{\partial u}f_j(z) + \frac{\partial}{\partial v}f_j(z)j\right) = 0, \tag{2.8}$$

*for all $z = u + jv \in U \cap \mathbb{C}_j$. Hence,*

$$f_j(z) = F_1(z) + iF_2(z)$$

*with holomorphic functions $F_1, F_2 : U \cap \mathbb{C}_j \to \mathbb{C}_j$.*

The splitting lemma states that the restriction of any left slice hyperholomorphic function to a complex plane $\mathbb{C}_j$ is left holomorphic, i.e., it is a holomorphic function with values in the left vector space $\mathbb{H} = \mathbb{C}_j + \mathbb{C}_j i$ over $\mathbb{C}_j$. The restriction of a right slice hyperholomorphic function to a complex plane $\mathbb{C}_j$ is right holomorphic, i.e., it is a holomorphic function with values in the right vector space $\mathbb{H} = \mathbb{C}_j + i\mathbb{C}_j$ over $\mathbb{C}_j$.

**Theorem 2.1.6** (Identity Principle). *Let $U \subset \mathbb{H}$ be an axially symmetric slice domain, let $f, g : U \to \mathbb{H}$ be left (or right) slice hyperholomorphic functions and set $\mathcal{Z} = \{q \in U : f(q) = g(q)\}$. If there exists $j \in \mathbb{S}$ such that $\mathcal{Z} \cap \mathbb{C}_j$ has an accumulation point in $U \cap \mathbb{C}_j$, then $f = g$.*

The most important property of slice functions (and in particular of slice hyperholomorphic functions) is the Structure Formula, which is often also called Representation Formula.

**Theorem 2.1.7** (The Structure (or Representation) Formula). *Let $U \subset \mathbb{H}$ be axially symmetric and let $i \in \mathbb{S}$. A function $f : U \to \mathbb{H}$ is a left slice function on $U$ if and only if for any $q = u + jv \in U$ we have*

$$f(q) = \frac{1}{2}\big[f(\bar{z}) + f(z)\big] + \frac{1}{2}ji\big[f(\bar{z}) - f(z)\big] \tag{2.9}$$

*with $z = u + iv$. A function $f : U \to \mathbb{H}$ is a right slice function on $U$ if and only if for any $q = u + jv \in U$ we have*

$$f(q) = \frac{1}{2}\big[f(\bar{z}) + f(z)\big] + \frac{1}{2}\big[f(\bar{z}) - f(z)\big]ij \tag{2.10}$$

*with $z = u + iv$.*

**Remark 2.1.2.** It is sometimes useful to rewrite (2.9) as

$$f(q) = \frac{1}{2}(1 - ij)f(z) + \frac{1}{2}(1 + ij)f(\bar{z})$$

and (2.10) as

$$f(q) = f(z)(1 - ij)\frac{1}{2} + f(\bar{z})(1 + ij)\frac{1}{2}.$$

The representation formula can be written in a different form that shows that $f(s)$ is determined by the values of $f$ at two arbitrary points in the sphere $[s]$, not necessarily by a point and its conjugate.

**Corollary 2.1.8.** *Let $U \subset \mathbb{H}$ be axially symmetric, let $[s] = u + \mathbb{S}v \subset U$ and let $i$, $j$ and $k$ be three different imaginary units in $\mathbb{S}$. If $f$ is a left slice function on $U$, then*

$$\begin{aligned}
f(u + jv) &= \big((i - k)^{-1}i + j(k - i)^{-1}\big) f(u + iv) \\
&\quad + \big((k - i)^{-1}k + j(k - i)^{-1}\big) f(u + kv).
\end{aligned} \tag{2.11}$$

*Similarly, if $f$ is a right slice function on $U$, then*

$$\begin{aligned}
f(u + jv) &= f(u + iv)\big(i(i - k)^{-1} + (k - i)^{-1}j\big) \\
&\quad + f(u + kv)\big(k(k - i)^{-1} + (k - i)^{-1}j\big).
\end{aligned} \tag{2.12}$$

As a consequence of the Structure Formula, every holomorphic function that is defined on a suitable open set in $\mathbb{C}_j$ has a slice hyperholomorphic extension.

**Lemma 2.1.9.** *Let $O \subset \mathbb{C}_j$ be an open set which is symmetric with respect to the real axis. We call the set $[O] = \bigcup_{z \in O}[z]$ the axially symmetric hull of $O$.*

(i) *Any function $f : O \to \mathbb{H}$ has a unique extension $\mathrm{ext}_L(f)$ to a left slice function on $[O]$ and a unique extension $\mathrm{ext}_R(f)$ to a right slice function on $[O]$.*

(ii) *If $f : O \to \mathbb{H}$ is left holomorphic, i.e., it satisfies (2.7), then $\mathrm{ext}_L(f)$ is left slice hyperholomorphic.*

(iii) *If $f$ is right holomorphic, i.e., it satisfies (2.8), then $\mathrm{ext}_R(f)$ is right slice hyperholomorphic.*

Sometimes in the following we will use the notation $f_j$ instead of $f|_{U \cap \mathbb{C}_j}$ for the restriction $f$ to $U \cap \mathbb{C}_j$ without mentioning it explicitly because it is clear from the context. Slice hyperholomorphic functions admit a special kind of derivative, that yields again a slice hyperholomorphic function.

**Definition 2.1.10.** Let $f : U \subseteq \mathbb{H} \to \mathbb{H}$ and let $q = u + jv \in U$. If $q$ is not real, then we say that $f$ admits left slice derivative in $q$ if

$$\partial_S f(q) := \lim_{p \to q,\, p \in \mathbb{C}_j} (p - q)^{-1}(f_j(p) - f_j(q)) \tag{2.13}$$

exists and is finite. If $q$ is real, then we say that $f$ admits left slice derivative in $q$ if (2.13) exists for any $j \in \mathbb{S}$.

Similarly, we say that $f$ admits right slice derivative at a nonreal point $q = u + jv \in U$ if

$$\partial_S f(q) := \lim_{p \to q,\, p \in \mathbb{C}_j} (f_j(p) - f_j(q))(p - q)^{-1} \tag{2.14}$$

exists and is finite, and we say that $f$ admits right slice derivative at a real point $q \in U$ if (2.14) exists and is finite, for any $j \in \mathbb{S}$.

Observe that $\partial_S f(q)$ is uniquely defined and independent of the choice of $j \in \mathbb{S}$ even if $q$ is real. If $f$ admits slice derivative, then $f_j$ is $\mathbb{C}_j$-complex left (resp. right) differentiable and we find

$$\partial_S f(q) = f_j'(q) = \frac{\partial}{\partial u} f_j(q) = \frac{\partial}{\partial u} f(q), \quad q = u + jv. \tag{2.15}$$

**Proposition 2.1.11.** *Let $U \subseteq \mathbb{H}$ be an axially symmetric open set and let $f : U \to \mathbb{H}$ be a real differentiable function.*

(i) *If $f$ is left (or right) slice hyperholomorphic, it admits left (resp. right) slice derivative and $\partial_S f$ is again left (resp. right) slice hyperholomorphic on $U$.*

(ii) *If $f$ is a left (or right) slice function that admits a left (resp. right) slice derivative, then $f$ is a left (resp. right) slice hyperholomorphic.*

(iii) *If $U$ is a slice domain, then any function that admits left (resp. right) slice derivative is left (resp. right) slice hyperholomorphic.*

Important examples of slice hyperholomorphic functions are power series in the quaternionic variable: power series of the form $\sum_{n=0}^{+\infty} q^n a_n$ with $a_n \in \mathbb{H}$ are left slice hyperholomorphic and power series of the form $\sum_{n=0}^{+\infty} a_n q^n$ are right slice hyperholomorphic. Such a power series is intrinsic if and only if the coefficients $a_n$ are real.

Conversely, any slice hyperholomorphic function can be expanded into a power series, at any real point, due to the splitting lemma.

**Theorem 2.1.12.** *Suppose that $a \in \mathbb{R}$ and $r > 0$. Let $B_r(a) = \{q \in \mathbb{H} : |q - a| < r\}$. If $f \in \mathcal{SH}_L(B_r(a))$, then*

$$f(q) = \sum_{n=0}^{+\infty} (q - a)^n \frac{1}{n!} \partial_S{}^n f(a), \quad \forall q = u + jv \in B_r(a). \qquad (2.16)$$

*If on the other hand $f \in \mathcal{SH}_R(B_r(a))$, then*

$$f(q) = \sum_{n=0}^{+\infty} \frac{1}{n!} \left( \partial_S{}^n f(a) \right) (q - a)^n, \quad \forall q = u + jv \in B_r(a).$$

**Example 2.1.13.** The exponential function is defined by its power series expansion

$$\exp(q) := e^q := \sum_{n=0}^{+\infty} \frac{1}{n!} q^n, \quad q \in \mathbb{H}.$$

Since the power series expansion has real coefficients, the exponential function is both left and right slice hyperholomorphic and even intrinsic. If $p$ and $q$ commute, then it satisfies the usual identity $\exp(p + q) = \exp(p) \exp(q)$. This is, however, not true for arbitrary $p$ and $q$.

**Example 2.1.14.** Any $q = u + jv \in \mathbb{H} \setminus \{0\}$ can be written as $q = |q| \exp(j\theta)$ with a unique angle $\theta \in [0, \pi]$. We define the argument of $q$ as

$$\arg(q) := \theta = \arccos(u/|q|).$$

Observe that, $\arg(q)$ is well defined even though $j$ is not unique if $q$ real: if $q > 0$ then $q = |q| \exp(j0)$ for any $j \in \mathbb{S}$ and hence $\arg(q) = 0$ is independent of the choice of $j \in \mathbb{S}$ and if $q < 0$ then $q = |q| \exp(j\pi)$ for any $j \in \mathbb{S}$ and hence $\arg(q) = \pi$ is independent of the choice of $j \in \mathbb{S}$, too.

The logarithm of a quaternion $q = u + jv \in \mathbb{H} \setminus (-\infty, 0]$ is defined as

$$\log(q) := \ln(|q|) + j \arg(q). \qquad (2.17)$$

It is an intrinsic slice hyperholomorphic function on $\mathbb{H} \setminus (-\infty, 0]$ and satisfies

$$e^{\log(q)} = q \quad \text{for } q \in \mathbb{H},$$

and

$$\log e^q = q, \quad \text{for } q = u + jv \in \mathbb{H} \text{ with } |\mathrm{Im}(q)| = v < \pi.$$

Contrary to the complex setting it is not possible to choose different intrinsic hyperholomorphic branches of the logarithm.

**Remark 2.1.3.** Observe that there exist other definitions of the quaternionic logarithm in the literature. In [162], the logarithm of a quaternion is for instance defined as

$$\log_{k,\ell} x := \begin{cases} \ln|x| + j_x \left( \arccos \frac{x_0}{|x|} + 2k\pi \right), & |\underline{x}| \neq 0 \text{ or } |\underline{x}| = 0, \; x_0 > 0, \\ \ln|x| + e_\ell \pi, & |\underline{x}| = 0, \; x_0 < 0, \end{cases}$$

where $k \in \mathbb{Z}$ and $e_\ell$ is one of the generating units of $\mathbb{H}$. This logarithm is, however, not continuous (and therefore, in particular, not slice hyperholomorphic) on the real line, unless $k = 0$. But for $k = 0$ this definition of the logarithm coincides with the one given above. Even more, the identity principle implies that (2.17) defines the maximal slice hyperholomorphic extension of the natural logarithm on $(0, +\infty)$ to a subset of the quaternions.

**Example 2.1.15.** For $\alpha \in \mathbb{R}$, we define the fractional power $q^\alpha$ of a quaternion $q = u + jv \in \mathbb{H} \setminus (-\infty, 0]$ as

$$q^\alpha := e^{\alpha \log q} = e^{\alpha(\ln|q| + j \arg(q))}. \tag{2.18}$$

The function $q \mapsto q^\alpha$ is an intrinsic slice hyperholomorphic function on its domain $\mathbb{H} \setminus (-\infty, 0]$ by Theorem 2.1.3 and Lemma 2.1.4 as it is the composition of two intrinsic slice hyperholomorphic functions. However, if we try to define fractional powers of non-real components by the formula (2.18), then we do not obtain a slice hyperholomorphic function: the composition of two such functions is only intrinsic if the inner function is slice hyperholomorphic and this is not the case for $\alpha \notin \mathbb{R}$.

As pointed out above, the product of two slice hyperholomorphic functions is not slice hyperholomorphic unless the factor on the appropriate side is intrinsic. However, there exists a regularised product that preserves slice hyperholomorphicity.

**Definition 2.1.16.** For $f = f_0 + jf_1, g = g_0 + jg_1 \in \mathcal{SH}_L(U)$, we define their left slice hyperholomorphic product as

$$f *_L g = (f_0 g_0 - f_1 g_1) + j(f_0 g_1 + f_1 g_0).$$

For $f = f_0 + f_1 j, g = g_0 + g_1 j \in \mathcal{SH}_R(U)$, we define their right slice hyperholomorphic product as

$$f *_R g = (f_0 g_0 - f_1 g_1) + (f_0 g_1 + f_1 g_0)j.$$

The slice hyperholomorphic product is associative and distributive, but it is in general not commutative. If $f$ is intrinsic, then $f *_L g$ coincides with the pointwise product $fg$ and

$$f *_L g = fg = g *_L f. \tag{2.19}$$

Similarly, if $g$ is intrinsic, then $f *_R g$ coincides with the pointwise product $fg$ and

$$f *_R g = fg = g *_R f. \tag{2.20}$$

**Example 2.1.17.** If $f(q) = \sum_{n=0}^{+\infty} q^n a_n$ and $g(q) = \sum_{n=0}^{+\infty} q^n b_n$ are two left slice hyperholomorphic power series, then their slice hyperholomorphic product equals the usual product of formal power series with coefficients in a non-commutative ring

$$\left(\sum_{n=0}^{+\infty} q^n a_n\right) *_L \left(\sum_{n=0}^{+\infty} q^n b_n\right) = (f *_L g)(q) = \sum_{n=0}^{+\infty} q^n \sum_{k=0}^{n} a_k b_{n-k}. \qquad (2.21)$$

Similarly, we have for right slice hyperholomorphic power series that

$$\left(\sum_{n=0}^{+\infty} a_n q^n\right) *_R \left(\sum_{n=0}^{+\infty} b_n q^n\right) = \sum_{n=0}^{+\infty} \left(\sum_{k=0}^{n} a_k b_{n-k}\right) q^n. \qquad (2.22)$$

**Definition 2.1.18.** We define for $f = f_0 + j f_1 \in \mathcal{SH}_L(U)$ its slice hyperholomorphic conjugate $f^c = \overline{f_0} + j\overline{f_1}$ and its symmetrisation $f^s = f *_L f^c = f^c *_L f$. Similarly, we define for $f = f_0 + f_1 j \in \mathcal{SH}_R(U)$ its slice hyperholomorphic conjugate as $f^c = \overline{f_0} + \overline{f_1} j$ and its symmetrisation as $f^s = f *_R f^c = f^c *_R f$.

The symmetrisation of a left slice hyperholomorphic function $f = f_0 + j f_1$ is explicitly given by

$$f^s = |f_0|^2 - |f_1|^2 + j 2\mathrm{Re}\left(f_0 \overline{f_1}\right).$$

Hence, it is an intrinsic function. It is $f^s(q) = 0$ if and only if $f(\tilde{q}) = 0$ for some $\tilde{q} \in [q]$. Furthermore, if $q = u + jv$, one has

$$f^c(q) = \overline{f_0(u,v)} + j\overline{f_1(u,v)} = \overline{f_0(u,v)} + \overline{f_1(u,v)}(-j) = \overline{f(\overline{q})} \qquad (2.23)$$

and an easy computation shows that

$$f *_L g(q) = f(q) g\left(f(q)^{-1} q f(q)\right), \quad \text{if } f(q) \neq 0. \qquad (2.24)$$

For $f(q) \neq 0$, one has

$$\begin{aligned} f^s(q) &= f(q) f^c\left(f(q)^{-1} q f(q)\right) \\ &= f(q) \overline{f\left(\overline{f(q)^{-1} q f(q)}\right)} = f(q) \overline{f\left(f(q)^{-1} \overline{q} f(q)\right)}. \end{aligned} \qquad (2.25)$$

Similar computations hold true in the right slice hyperholomorphic case. Finally, if $f$ is intrinsic, then $f^c(q) = f(q)$ and $f^s(q) = f(q)^2$. Some consequences of the above definitions are collected in the following corollary.

**Corollary 2.1.19.** *The following statements hold true.*

(i) *For $f \in \mathcal{SH}_L(U)$ with $f \not\equiv 0$, its slice hyperholomorphic inverse $f^{-*_L}$, which satisfies $f^{-*_L} *_L f = f *_L f^{-*_L} = 1$, is given by*

$$f^{-*_L} = (f^s)^{-1} *_L f^c = (f^s)^{-1} f^c$$

*and it is defined on $U \setminus [\mathcal{Z}_f]$, where $\mathcal{Z}_f = \{s \in U : f(s) = 0\}$.*

(ii) *For $f \in \mathcal{SH}_R(U)$ with $f \not\equiv 0$, its slice hyperholomorphic inverse $f^{-*R}$, which satisfies $f^{-*R} *_R f = f *_R f^{-*R} = 1$, is given by*

$$f^{-*R} = f^c *_R (f^s)^{-1} = f^c (f^s)^{-1}$$

*and it is defined on $U \setminus [\mathcal{Z}_f]$, where $\mathcal{Z}_f = \{s \in U : f(s) = 0\}$.*

(iii) *If $f \in \mathcal{N}(U)$ with $f \not\equiv 0$, then $f^{-*L} = f^{-*R} = f^{-1}$.*

We observe that the modulus $|f^{-*L}|$ is in a certain sense comparable to $1/|f|$. Since $f^s$ is intrinsic, we have $|f^s(q)| = |f^s(\tilde{q})|$ for any $\tilde{q} \in [q]$. As $f(q)qf(q)^{-1} \in [q]$, we find, for $f(q) \neq 0$ because of (2.25), that

$$\begin{aligned}
|f^s(q)| &= \left| f^s \left( f(q)qf(q)^{-1} \right) \right| \\
&= \left| f \left( f(q)qf(q)^{-1} \right) \overline{f(\bar{q})} \right| \\
&= \left| f \left( f(q)qf(q)^{-1} \right) \right| |f(\bar{q})| .
\end{aligned}$$

Therefore we have, because of (2.23), that

$$\begin{aligned}
\left| f^{-*L}(q) \right| &= \left| f^s(q)^{-1} \right| |f^c(q)| \\
&= \frac{1}{\left| f \left( f(q)qf(q)^{-1} \right) \right| |f(\bar{q})|} |f(\bar{q})| \\
&= \frac{1}{\left| f \left( f(q)\bar{q}f(q)^{-1} \right) \right|}
\end{aligned}$$

and so

$$\left| f^{-*L}(q) \right| = \frac{1}{|f(\tilde{q})|} \quad \text{with } \tilde{q} = f(q)\bar{q}f(q)^{-1} \in [q]. \tag{2.26}$$

An analogous estimate holds for the slice hyperholomorphic inverse of a right slice hyperholomorphic function.

Slice hyperholomorphic functions satisfy a version of Cauchy's integral theorem and a Cauchy formula with a slice hyperholomorphic integral kernel. However, left and right slice hyperholomorphic functions satisfy Cauchy formulas with different kernels. This is contrary to the case of Fueter regular functions, where both left and right Fueter regular functions satisfy a Cauchy formula with the same kernel.

**Theorem 2.1.20** (Cauchy's integral theorem). *Let $U \subset \mathbb{H}$ be open, let $j \in \mathbb{S}$ and let $f \in \mathcal{SH}_L(U)$ and $g \in \mathcal{SH}_R(U)$. Moreover, let $D_j \subset U \cap \mathbb{C}_j$ be an open and bounded subset of the complex plane $\mathbb{C}_j$ with $\overline{D_j} \subset U \cap \mathbb{C}_j$ such that $\partial D_j$ is a finite union of piecewise continuously differentiable Jordan curves. Then*

$$\int_{\partial D_j} g(s) \, ds_j \, f(s) = 0,$$

*where $ds_j = ds(-j)$.*

In order to determine the left and right slice hyperholomorphic Cauchy kernels, we start from an analogy with the classical complex case. We consider the series expansion of the complex Cauchy kernel and determine its closed form under the assumption that $s$ and $q$ are quaternions that do not commute.

**Theorem 2.1.21.** *Let $q, s \in \mathbb{H}$ with $|q| < |s|$. Then,*

$$\sum_{n=0}^{+\infty} q^n s^{-n-1} = -(q^2 - 2\mathrm{Re}(s)q + |s|^2)^{-1}(q - \bar{s}) \tag{2.27}$$

*and*

$$\sum_{n=0}^{+\infty} s^{-n-1} q^n = -(q - \bar{s})(q^2 - 2\mathrm{Re}(s)q + |s|^2)^{-1}. \tag{2.28}$$

**Definition 2.1.22** (Slice hyperholomorphic Cauchy kernels). We define the left slice hyperholomorphic Cauchy kernel as

$$S_L^{-1}(s, q) := -(q^2 - 2\mathrm{Re}(s)q + |s|^2)^{-1}(q - \bar{s}), \quad q \notin [s]$$

and the right slice hyperholomorphic Cauchy kernel as

$$S_R^{-1}(s, q) := -(q - \bar{s})(q^2 - 2\mathrm{Re}(s)q + |s|^2)^{-1}, \quad q \notin [s].$$

The left and right slice hyperholomorphic Cauchy kernels are proper generalizations of the classical Cauchy kernel.

**Lemma 2.1.23.** *If $s$ and $q$ commute, then the left and the right slice hyperholomorphic Cauchy kernel reduce to the complex Cauchy kernel, i.e.,*

$$S_L^{-1}(s, q) = (s - q)^{-1} = S_R^{-1}(s, q) \quad if \ sq = qs.$$

**Remark 2.1.4.** The left and right slice hyperholomorphic Cauchy kernels are the left and right slice hyperholomorphic inverses of the function $q \mapsto s - q$, which is another analogy to the classical case.

As the next proposition shows, the slice hyperholomorphic Cauchy kernels $S_L^{-1}(s, q)$ and $S_R^{-1}(s, q)$ can be written in two different ways, which justifies Definition 2.1.25.

**Proposition 2.1.24.** *If $q, s \in \mathbb{H}$ with $q \notin [s]$, then*

$$-(q^2 - 2q\mathrm{Re}(s) + |s|^2)^{-1}(q - \bar{s}) = (s - \bar{q})(s^2 - 2\mathrm{Re}(q)s + |q|^2)^{-1} \tag{2.29}$$

*and*

$$(s^2 - 2\mathrm{Re}(q)s + |q|^2)^{-1}(s - \bar{q}) = -(q - \bar{s})(q^2 - 2\mathrm{Re}(s)q + |s|^2)^{-1}. \tag{2.30}$$

**Definition 2.1.25.** Let $q, s \in \mathbb{H}$ with $q \notin [s]$.

- We say that $S_L^{-1}(s,q)$ is written in the form I if

$$S_L^{-1}(s,q) := -(q^2 - 2\mathrm{Re}(s)q + |s|^2)^{-1}(q - \bar{s}).$$

- We say that $S_L^{-1}(s,q)$ is written in the form II if

$$S_L^{-1}(s,q) := (s - \bar{q})(s^2 - 2\mathrm{Re}(q)s + |q|^2)^{-1}.$$

- We say that $S_R^{-1}(s,q)$ is written in the form I if

$$S_R^{-1}(s,q) := -(q - \bar{s})(q^2 - 2\mathrm{Re}(s)q + |s|^2)^{-1}.$$

- We say that $S_R^{-1}(s,q)$ is written in the form II if

$$S_R^{-1}(s,q) := (s^2 - 2\mathrm{Re}(q)s + |q|^2)^{-1}(s - \bar{q}).$$

**Corollary 2.1.26.** *For $q, s \in \mathbb{H}$ with $s \notin [q]$, we have*

$$S_L^{-1}(s,q) = -S_R^{-1}(q,s).$$

It is essential for the theory that the left and right slice hyperholomorphic Cauchy kernels are slice hyperholomorphic in both variables. This is what the next lemma shows.

**Lemma 2.1.27.** *Let $q, s \in \mathbb{H}$ with $q \notin [s]$. The left slice hyperholomorphic Cauchy kernel $S_L^{-1}(s,q)$ is left slice hyperholomorphic in $q$ and right slice hyperholomorphic in $s$. The right slice hyperholomorphic Cauchy kernel $S_R^{-1}(s,q)$ is left slice hyperholomorphic in $s$ and right slice hyperholomorphic in $q$.*

As pointed out in Remark 2.1.4, the left and the right Cauchy kernel are the slice hyperholomorphic inverses of the mapping $q \mapsto s - q$. In analogy with the classical case, their slice derivatives are multiples of the $n$-th slice hyperholomorphic inverses of this function.

**Definition 2.1.28.** *For $s, q \in \mathbb{H}$ with $s \notin [q]$ and $n \in \mathbb{N}$, we define*

$$S_L^{-n}(s,q) := (s - q)^{-*_L n} = (q^2 - 2\mathrm{Re}(s)q + |s|^2)^{-n} \sum_{k=0}^{n} \binom{n}{k}(-q)^k \bar{s}^{n-k},$$

$$S_R^{-n}(s,q) := (s - q)^{-*_R n} = \sum_{k=0}^{n} \binom{n}{k}\bar{s}^{n-k}(-q)^k (q^2 - 2\mathrm{Re}(s)q + |s|^2)^{-n}.$$

**Lemma 2.1.29.** *For $s, q \in \mathbb{H}$ with $s \notin [q]$ and $n \in \mathbb{N}$, the slice derivatives in the variable $q$ of the right and left slice hyperholomorphic Cauchy kernels are*

$$\partial_S^n S_L^{-1}(s,q) = (-1)^n \, n! \, S_L^{-(n+1)}(s,q)$$

*and*

$$\partial_S^n S_R^{-1}(s,q) = (-1)^n \, n! \, S_R^{-(n+1)}(s,q).$$

The domains of integration that appear in the slice hyperholomorphic Cauchy formulas as well as in the $S$-functional calculus are slice Cauchy domains. Before we state the Cauchy formulas, we recall some properties of the slice Cauchy domains, the proofs of which can be found in [130].

**Definition 2.1.30** (Slice Cauchy domain). An axially symmetric open set $U \subset \mathbb{H}$ is called a slice Cauchy domain, if $U \cap \mathbb{C}_j$ is a Cauchy domain in $\mathbb{C}_j$ for any $j \in \mathbb{S}$. More precisely, $U$ is a slice Cauchy domain if, for any $j \in \mathbb{S}$, the boundary $\partial(U \cap \mathbb{C}_j)$ of $U \cap \mathbb{C}_j$ is the union of a finite number of non-intersecting piecewise continuously differentiable Jordan curves in $\mathbb{C}_j$.

**Remark 2.1.5.** Any slice Cauchy domain has only finitely many components (i.e., maximal connected subsets). Moreover, at most one of them is unbounded and if there exists an unbounded component, then it contains a neighborhood of $\infty$ in $\mathbb{H}$.

**Theorem 2.1.31.** *Let $C$ be a closed and let $O$ be an open axially symmetric subset of $\mathbb{H}$ such that $C \subset O$ and such that $\partial O$ is nonempty and bounded. Then there exists a slice Cauchy domain $U$ such that $C \subset U$ and $\overline{U} \subset O$. If $O$ is unbounded, then $U$ can be chosen to be unbounded, too.*

The boundary of a slice Cauchy domain in a complex plane $\mathbb{C}_j$ is of course symmetric with respect to the real axis. Hence, it can be fully described by the part that lies in the closed upper half plane

$$\mathbb{C}_j^+ := \{z_0 + jz_1 : \quad z_0 \in \mathbb{R}, \ z_1 \geq 0\}.$$

We specify this idea in the following statements.

**Definition 2.1.32.** For a path $\gamma : [0,1] \to \mathbb{C}_j$, we define the paths $(-\gamma)(t) := \gamma(1-t)$ and $\overline{\gamma}(t) := \overline{\gamma(t)}$.

**Lemma 2.1.33.** *Let $\gamma$ be a Jordan curve in $\mathbb{C}_j$ whose image is symmetric with respect to the real axis. Then $\gamma_+ := \gamma \cap \mathbb{C}_j^+$ consists of a single curve and $\gamma = \gamma_+ \cup \gamma_-$ with $\gamma_- := -\overline{\gamma_+}$.*

Let now $U$ be a slice Cauchy domain and consider any $j \in \mathbb{S}$. The boundary $\partial(U \cap \mathbb{C}_j)$ of $U$ in $\mathbb{C}_j$ consists of a finite union of piecewise continuously differentiable Jordan curves and is symmetric with respect to the real axis. Hence, whenever a curve $\gamma$ belongs to $\partial(U \cap \mathbb{C}_j)$, the curve $-\overline{\gamma}$ belongs to $\partial(U \cap \mathbb{C}_j)$ too. We can therefore decompose $\partial(U \cap \mathbb{C}_j)$ as follows:

- First define $\gamma_{+,1}, \ldots, \gamma_{+,\kappa}$ as those Jordan curves that belong to $\partial(U \cap \mathbb{C}_j)$ and lie entirely in the open upper complex halfplane $\mathbb{C}_j^+$.

  The curves $-\overline{\gamma_{+,1}}, \ldots, -\overline{\gamma_{+,\kappa}}$ are then exactly those Jordan curves that belong to $\partial(U \cap \mathbb{C}_j)$ and lie entirely in the lower complex halfplane $\mathbb{C}_j^-$.

- In a second step, consider the curves $\gamma_{\kappa+1}, \ldots, \gamma_N$ that belong to $\partial(U \cap \mathbb{C}_j)$ and take values both in $\mathbb{C}_j^+$ and $\mathbb{C}_j^-$. Define $\gamma_{+,\ell}$ for $\ell = \kappa + 1, \ldots, N$ as the part of $\gamma_\ell$ that lies in $\mathbb{C}_j^+$ and $\gamma_{-,\ell} = -\overline{\gamma_{+,\ell}}$ as the part of $\gamma_\ell$ that lies in $\mathbb{C}_j^-$, cf. Lemma 2.1.33.

Overall, we obtain the following decomposition of $\partial(U \cap \mathbb{C}_j)$:

$$\partial(U \cap \mathbb{C}_j) = \bigcup_{1 \leq \ell \leq N} \gamma_{+,\ell} \cup -\overline{\gamma_{+,\ell}}.$$

**Definition 2.1.34.** We call the set $\{\gamma_{1,+}, \ldots, \gamma_{N,+}\}$ the part of $\partial(U \cap \mathbb{C}_j)$ that lies in $\mathbb{C}_j^+$.

Finally, we are now able to formulate the Cauchy formulas for slice hyperholomorphic functions. These formulas are also the starting point for the definition of the $S$-functional calculus.

**Theorem 2.1.35** (The Cauchy formulas). *Let $U \subset \mathbb{H}$ be a bounded slice Cauchy domain, let $j \in \mathbb{S}$ and set $ds_j = ds(-j)$. If $f$ is a (left) slice hyperholomorphic function on a set that contains $\overline{U}$, then*

$$f(q) = \frac{1}{2\pi} \int_{\partial(U \cap \mathbb{C}_j)} S_L^{-1}(s,q) \, ds_j \, f(s), \quad \text{for any } q \in U. \tag{2.31}$$

*If $f$ is a right slice hyperholomorphic function on a set that contains $\overline{U}$, then*

$$f(q) = \frac{1}{2\pi} \int_{\partial(U \cap \mathbb{C}_j)} f(s) \, ds_j \, S_R^{-1}(s,q), \quad \text{for any } q \in U. \tag{2.32}$$

*These integrals depend neither on $U$ nor on the imaginary unit $j \in \mathbb{S}$.*

**Theorem 2.1.36** (Cauchy formulas on unbounded slice Cauchy domains). *Let $U \subset \mathbb{H}$ be an unbounded slice Cauchy domain and let $j \in \mathbb{S}$ and set $ds_j = ds(-j)$. If $f \in \mathcal{SH}_L(\overline{U})$ and $f(\infty) := \lim_{|q| \to \infty} f(q)$ exists, then*

$$f(q) = f(\infty) + \frac{1}{2\pi} \int_{\partial(U \cap \mathbb{C}_j)} S_L^{-1}(s,q) \, ds_j \, f(s), \quad \text{for any } q \in U.$$

*If $f \in \mathcal{SH}_R(\overline{U})$ and $f(\infty) := \lim_{|q| \to \infty} f(q)$ exists, then*

$$f(q) = f(\infty) + \frac{1}{2\pi} \int_{\partial(U \cap \mathbb{C}_j)} f(s) \, ds_j \, S_R^{-1}(s,q), \quad \text{for any } q \in U.$$

Finally, just as holomorphic functions, slice hyperholomorphic functions can be approximated by rational functions.

**Definition 2.1.37.** A function $r$ is called left rational if it is of the form $r(q) = P(q)^{-1}Q(q)$ with polynomials $P \in \mathcal{N}(\mathbb{H})$ and $Q \in \mathcal{SH}_L(\mathbb{H})$.

A function $r$ is called right rational if it is of the form $r(q) = Q(q)P(q)^{-1}$ with polynomials $P \in \mathcal{N}(\mathbb{H})$ and $Q \in \mathcal{SH}_R(\mathbb{H})$.

Finally, a function $r$ is called intrinsic rational if it is of the form $r(q) = P(q)^{-1}Q(q)$ with two polynomials $P, Q \in \mathcal{N}(\mathbb{H})$.

**Remark 2.1.6.** The requirement that $P$ is intrinsic is necessary because the function $P^{-1}$ is otherwise not slice hyperholomorphic. This is, however, not a serious restriction since any rational left slice hyperholomorphic function as $f(q) = P(q)^{-*_L} *_L Q(p)$ with left slice hyperholomorphic (but not necessarily intrinsic) polynomials can be represented in the above form $f(q) = \tilde{P}(q)^{-1}\tilde{Q}(q)$ with an intrinsic polynomial $\tilde{P}$ and a left slice hyperholomorphic polynomial $\tilde{Q}$. (Precisely, we have $\tilde{P} = P^s$ and $\tilde{Q} = P^c *_L Q$, cf. Corollary 2.1.19.) An analogous result holds for the right slice hyperholomorphic case.

**Corollary 2.1.38.** *Let $f \in \mathcal{SH}_L(U)$, let $j, i \in \mathbb{S}$ with $i \perp j$ and write $f_j = F_1 + F_2 i$ with holomorphic components $F_1, F_2 : U \cap \mathbb{C}_j \to \mathbb{C}_j$ according to Lemma 2.1.5. Then $f$ is left rational if and only if $F_1$ and $F_2$ are rational functions on $\mathbb{C}_j$.*

*Similarly, if $f \in \mathcal{SH}_R(U)$ and we write $f_j = F_1 + iF_2$ with holomorphic components $F_1$ and $F_2$ according to Lemma 2.1.5, then $f$ is right rational if and only if $F_1, F_2$ are rational functions on $\mathbb{C}_j$.*

**Theorem 2.1.39** (Runge's Theorem). *Let $K \subset \mathbb{H}$ be an axially symmetric compact set and let $A$ be an axially symmetric set such that $A \cap C \neq \emptyset$ for any connected component $C$ of $(\mathbb{H} \cup \{\infty\}) \setminus K$.*

*If $f$ is left slice hyperholomorphic on an axially symmetric open set $U$ with $K \subset U$, then, for any $\varepsilon > 0$, there exists a left rational function $r$ whose poles lie in $A$ such that*

$$\sup\{|f(q) - r(q)| : \quad q \in K\} < \varepsilon. \tag{2.33}$$

*Similarly, if $f$ is right slice hyperholomorphic on an axially symmetric open set $U$ with $K \subset U$, then, for any $\varepsilon > 0$, there exists a right rational function $r$ whose poles lie in $A$ such that (2.33) holds.*

*Finally, if $f \in \mathcal{N}(U)$ for some axially symmetric open set $U$ with $K \subset U$, then, for any $\varepsilon > 0$, there exists a real rational function $r$ whose poles lie in $A$ such that (2.33) holds.*

## 2.2 The $S$-functional calculus for bounded operators

This section is a preliminary to the arguments dealt with in this book. As pointed out in the introduction, the fundamental difficulty in developing a mathematically rigorous theory of quaternionic linear operators was the identification of suitable notions of quaternionic spectrum and of quaternionic resolvent operators. These problems were solved with the introduction of the $S$-spectrum and the $S$-resolvent operators. We summarize these concepts and we define the quaternionic $S$-functional calculus for bounded operators, for the proofs see [57].

Let us start with a precise definition of the various structures of quaternionic vector, Banach and Hilbert spaces.

**Definition 2.2.1.** A quaternionic right vector space is an additive group $(X, +)$ endowed with a quaternionic scalar multiplication from the right such that for all

$x, y \in X$ and all $a, b \in \mathbb{H}$

$$(x + y)a = xa + ya, \quad x(a + b) = xa + xb, \quad y(ab) = (ya)b, \quad y1 = y. \quad (2.34)$$

A quaternionic left vector space is an additive group $(X, +)$ endowed with a quaternionic scalar multiplication from the left such that for all $x, y \in X$ and all $a, b \in \mathbb{H}$

$$a(x + y) = ax + ay, \quad (a + b)y = ay + by, \quad (ab)y = a(by), \quad 1y = y. \quad (2.35)$$

Finally, a two-sided quaternionic vector space is an additive group $(X, +)$ endowed with a quaternionic scalar multiplication from the right and a quaternionic scalar multiplication from the left that satisfy (2.34) (resp. (2.35)) such that in addition $ay = ya$ for all $y \in X$ and all $a \in \mathbb{R}$.

**Remark 2.2.1.** Starting from a real vector space $X_{\mathbb{R}}$, we can easily construct a two-sided quaternionic vector space by setting

$$X_{\mathbb{R}} \otimes \mathbb{H} = \left\{ \sum_{\ell=0}^{3} y_{\ell} \otimes e_{\ell} : \quad y_{\ell} \in X_{\mathbb{R}} \right\},$$

where we denote $e_0 = 1$ for neatness. Together with the componentwise addition $X_{\mathbb{R}} \otimes \mathbb{H}$ forms an additive group. It is a two-sided quaternionic vector space, if we endow it with the right and left scalar multiplications

$$ay = \sum_{\ell,\kappa=0}^{3} (a_{\ell} y_{\kappa}) \otimes (e_{\ell} e_{\kappa}) \quad \text{and} \quad ya = \sum_{\ell,\kappa=0}^{3} (a_{\ell} y_{\kappa}) \otimes (e_{\kappa} e_{\ell})$$

for $a = \sum_{\ell=0}^{3} a_{\ell} e_{\ell} \in \mathbb{H}$ and $y = \sum_{\kappa=0}^{3} y_{\kappa} \otimes e_{\kappa} \in X_{\mathbb{R}} \otimes \mathbb{H}$. Usually one omits the symbol $\otimes$ and simply writes $y = \sum_{\ell=0}^{3} y_{\ell} e_{\ell}$.

Any two-sided quaternionic vector space is essentially of this form. Indeed, we can set

$$X_{\mathbb{R}} = \{y \in X : \quad ay = ya, \ \forall a \in \mathbb{H}\} \quad (2.36)$$

and find that $X$ is isomorphic to $X_{\mathbb{R}} \otimes \mathbb{H}$. If we set $\mathrm{Re}(y) := \frac{1}{4} \sum_{\ell=0}^{3} \overline{e_{\ell}} y e_{\ell}$, then $\mathrm{Re}(y) \in X_{\mathbb{R}}$ and $y = \sum_{\ell=0}^{3} \mathrm{Re}(\overline{e_{\ell}} y) e_{\ell}$.

**Remark 2.2.2.** A quaternionic right or left vector space also carries the structure of a real vector space: if we simply restrict the quaternionic scalar multiplication to $\mathbb{R}$, then we obtain a real vector space. Similarly, if we choose some $j \in \mathbb{S}$ and identify $\mathbb{C}_j$ with the field of complex numbers, then $X$ also carries the structure of a complex vector space over $\mathbb{C}_j$. Again we obtain this structure by restricting the quaternionic scalar multiplication to $\mathbb{C}_j$.

If we consider a two-sided quaternionic vector space, then the left and the right scalar multiplication coincide for real numbers so that we can restrict them to $\mathbb{R}$ in order to obtain again a real vector space. This is, however, not true for the multiplication with scalars in one complex plane $\mathbb{C}_j$. In general, $zy \neq yz$ for $z \in \mathbb{C}_j$ and $y \in X$. Hence, we can only restrict either the left or the right multiplication to $\mathbb{C}_j$ in order to consider $X$ as a complex vector space over $\mathbb{C}_j$, but not both simultaneously.

**Definition 2.2.2.** A norm on a right, left or two-sided quaternionic vector space $X$ is a norm in the sense of real vector spaces (cf. Remark 2.2.2) that is compatible with the quaternionic right, left (resp. two-sided) scalar multiplication. Precisely, this means that $\|ya\| = \|y\||a|$ (or $\|ay\| = |a|\|y\|$ (resp. $\|ay\| = |a|\|y\| = \|ya\|$)) for all $a \in \mathbb{H}$ and all $y \in X$. A quaternionic right, left or two-sided Banach space is a quaternionic right, left or two-sided vector space that is endowed with a norm $\| \cdot \|$ and complete with respect to the topology induced by this norm.

**Remark 2.2.3.** Similar to Remark 2.2.2, we obtain a real Banach space if we restrict the left or right scalar multiplication on a quaternionic Banach space to $\mathbb{R}$ and we obtain a complex Banach space over $\mathbb{C}_j$ if we restrict the left or right scalar multiplication to $\mathbb{C}_j$ for some $j \in \mathbb{S}$.

**Definition 2.2.3.** A quaternionic right Hilbert space $\mathcal{H}$ is a quaternionic right vector space equipped with a scalar product $\langle \cdot, \cdot \rangle : \mathcal{H} \times \mathcal{H} \to \mathbb{H}$ so that for all vectors $x, y, w \in \mathcal{H}$ and all scalars $a \in \mathbb{H}$

(i) $\langle x, x \rangle \geq 0$,

(ii) $\langle x, ya + w \rangle = \langle x, y \rangle a + \langle x, w \rangle$,

(iii) $\langle x, y \rangle = \overline{\langle y, x \rangle}$,

and so that $\mathcal{H}$ is complete with respect to the norm $\|y\| = \sqrt{\langle y, y \rangle}$.

**Remark 2.2.4.** In order to be consistent with our notation, we shall also assume the scalar product of a complex Hilbert space to be sesquilinear in the first and linear in the second variable.

**Remark 2.2.5.** The fundamental concepts that appear in complex Hilbert spaces such as orthogonality, orthonormal bases, etc. can be defined in a similar way in the quaternionic setting and fundamental results such as the Riesz representation theorem hold also in this noncommutative setting.

**Notation 2.2.4.** Since we are working with different number systems and vector space structures, we introduce for a set of vectors $\mathbf{B} := (b_\ell)_{\ell \in \Lambda}$ the quaternionic right-linear span of $\mathbf{B}$

$$\operatorname{span}_{\mathbb{H}} \mathbf{B} := \left\{ \sum_{\ell \in I} b_\ell x_\ell : \quad x_\ell \in \mathbb{H}, \; I \subset \Lambda \text{ finite} \right\}$$

and the $\mathbb{C}_j$-linear span of $\mathbf{B}$

$$\operatorname{span}_{\mathbb{C}_j} \mathbf{B} := \left\{ \sum_{\ell \in I} b_\ell z_\ell : \quad z_\ell \in \mathbb{C}_j, \; I \subset \Lambda \text{ finite} \right\}.$$

**Definition 2.2.5.** A mapping $T : X_R \to W_R$ between two quaternionic right Banach spaces $X_R$ and $W_R$ is called right linear if $T(xa + y) = T(x)a + T(y)$ for all $x, y \in X_R$ and $a \in \mathbb{H}$. It is bounded if $\|T\| := \sup_{\|y\|=1} \|Ty\|$ is finite. We denote the set of all bounded right linear operators $T : X_R \to X_R$ by $\mathcal{B}(X_R)$.

**Remark 2.2.6.** We consider right linear operators by convention. One can also consider left linear operators, which leads to an equivalent theory.

**Remark 2.2.7.** The set $\mathcal{B}(X_R)$ of all bounded right linear operators on a quaternionic right Banach space $X_R$ is a real Banach space with the pointwise addition $(T + U)(y) := T(y) + U(y)$ and the multiplication $(Ta)(y) := T(y)a$ with scalars $a \in \mathbb{R}$. However, if we define $(Ta)(y) := T(ya)$ for $a \in \mathbb{H} \setminus \mathbb{R}$, then we do not obtain a quaternionic right linear operator as

$$(Ta)(y)b = T(y)ab \neq T(yba) = T(yb)a = (Ta)(yb)$$

if $a, b \in \mathbb{H}$ do not belong to the same complex plane. Hence, $Ta \notin \mathcal{B}(X_R)$ for $a \notin \mathbb{R}$ and so $\mathcal{B}(X_R)$ is not a quaternionic linear space.

The space $\mathcal{B}(X)$ of all bounded right linear operators on a two-sided quaternionic Banach space $X$ is on the other hand again a two-sided quaternionic Banach space with the scalar multiplications

$$(aT)(y) = a(T(y)) \quad \text{and} \quad (Ta)(y) = T(ay). \tag{2.37}$$

For this reason, the theory of quaternionic linear operators is usually developed on two-sided quaternionic Banach spaces and we also work on two-sided spaces in this book. An exception is the article [131], which discusses the minimal structure that is necessary to develop quaternionic operator theory and shows that the essential results can also be obtained on one-sided spaces.

If $T$ is a bounded operator on a two-sided quaternionic Banach space $X = X_\mathbb{R} \otimes \mathbb{H}$, then we can write $T$ as

$$T = T_0 + \sum_{\ell=1}^{3} T_\ell e_\ell \tag{2.38}$$

with $\mathbb{R}$-linear components $T_\ell \in \mathcal{B}(X_\mathbb{R})$ for $\ell = 0, \ldots, 3$. If we set $e_0 = 1$ for neatness, then $T$ acts as

$$Ty = \sum_{\ell,\kappa=0}^{3} T_\ell y_\kappa e_\ell e_\kappa \quad \text{for } y = \sum_{\kappa=0}^{3} y_\kappa e_\kappa \in X = X_\mathbb{R} \otimes \mathbb{H}.$$

We thus have

$$\mathcal{B}(X) = \mathcal{B}(X_\mathbb{R}) \otimes \mathbb{H}.$$

**Definition 2.2.6.** Let $X$ be a two-sided quaternionic Banach space. We denote the set of all operators $T = T_0 + \sum_{\ell=1}^{3} T_\ell e_\ell \in \mathcal{B}(X)$ with commuting components by $\mathcal{BC}(X)$. Furthermore, we call a bounded operator $T$ a scalar operator if it is of the form $T = T_0$, that is if $T_1 = T_2 = T_3 = 0$.

We now want to introduce suitable notions of spectrum and resolvent operator for quaternionic operators. In the classical setting, the spectrum $\sigma(A)$ of a complex operator $A$ is the set of complex numbers $\lambda$, for which the resolvent operator $R_\lambda(A) = (\lambda\mathcal{I} - A)^{-1}$ does not exist as a bounded operator. The resolvent on the other hand formally corresponds to the Cauchy kernel $(\lambda - z)^{-1}$, in which the scalar variable $z$ is formally replaced by the operator $A$. This relation is fundamental for the definition of the Riesz–Dunford functional calculus for holomorphic functions.

In order to extend the concepts of spectrum and resolvent operator in the quaternionic setting, we consider the series expansions (2.27) and (2.28) of the slice hyperholomorphic Cauchy kernels, that is the series

$$\sum_{n=0}^{+\infty} q^n s^{-n-1} \quad \text{and} \quad \sum_{n=0}^{+\infty} s^{-n-1} q^n$$

and we give the following definition.

**Definition 2.2.7.** Let $T \in \mathcal{B}(X)$ and $s \in \mathbb{H}$. We call the series

$$\sum_{n=0}^{+\infty} T^n s^{-n-1} \quad \text{and} \quad \sum_{n=0}^{+\infty} s^{-n-1} T^n$$

the left and right Cauchy kernel operator series, respectively.

**Lemma 2.2.8.** *Let $T \in \mathcal{B}(X)$. For $\|T\| < |s|$ the left and the right Cauchy kernel operator series converge in the operator norm.*

Our goal is now to determine the closed form of the Cauchy kernel operator series. We collect in the following theorem some important results.

**Theorem 2.2.9.** *Let $T \in \mathcal{B}(X)$ and let $s \in \mathbb{H}$ with $\|T\| < |s|$. Then the following results hold.*

(i) *We have*

$$(T^2 - 2\mathrm{Re}(s)T + |s|^2\mathcal{I})^{-1} = \sum_{n=0}^{+\infty} T^n \sum_{k=0}^{n} \bar{s}^{-k-1} s^{-n+k-1}. \qquad (2.39)$$

(ii) *The left Cauchy kernel series equals*

$$\sum_{n=0}^{+\infty} T^n s^{-n-1} = -(T^2 - 2\mathrm{Re}(s)T + |s|^2\mathcal{I})^{-1}(T - \bar{s}\mathcal{I}).$$

(iii) *The right Cauchy kernel series equals*

$$\sum_{n=0}^{+\infty} s^{-n-1} T^n = -(T - \bar{s}\mathcal{I})(T^2 - 2\mathrm{Re}(s)T + |s|^2\mathcal{I})^{-1}.$$

The previous result motivates the following definition.

**Definition 2.2.10.** Let $T \in \mathcal{B}(X)$. For $s \in \mathbb{H}$, we set

$$\mathcal{Q}_s(T) := T^2 - 2\mathrm{Re}(s)T + |s|^2 \mathcal{I}.$$

We define the *S-resolvent set* $\rho_S(T)$ of $T$ as

$$\rho_S(T) := \{s \in \mathbb{H} : \quad \mathcal{Q}_s(T) \text{ is invertible in } \mathcal{B}(X)\}$$

and we define the *S-spectrum* $\sigma_S(T)$ of $T$ as

$$\sigma_S(T) := \mathbb{H} \setminus \rho_S(T).$$

For $s \in \rho_S(T)$, the operator $\mathcal{Q}_s(T)^{-1} \in \mathcal{B}(X)$ is called the *pseudo-resolvent operator* of $T$ at $s$.

As the following results show, the $S$-spectrum has a structure that is compatible with the structure of slice hyperholomorphic functions and with the symmetry of the set of right eigenvalues of $T$. Moreover, it generalizes the set of right eigenvalues just as the classical spectrum generalizes the set of eigenvalues of a complex linear operator and has analogous properties.

**Theorem 2.2.11.** *Let $T \in \mathcal{B}(X)$.*

  (i) *The sets $\rho_S(T)$ and $\sigma_S(T)$ are axially symmetric.*

 (ii) *The $S$-spectrum $\sigma_S(T)$ of $T$ is a nonempty, compact set contained in the closed ball $\overline{B_{\|T\|}(0)}$.*

(iii) *Let $T \in \mathcal{B}(X)$. Then $\ker \mathcal{Q}_s(T) \neq \{0\}$ if and only if $s$ is a right eigenvalue of $T$. In particular, any right eigenvalue belongs to $\sigma_S(T)$.*

On the $S$-resolvent set we can now define the slice hyperholomorphic resolvents. As in the complex case, they correspond to the slice hyperholomorphic Cauchy kernel, in which we formally replace the scalar variable $q$ by the operator $T$. Since we distinguish between left and right slice hyperholomorphicity, two different resolvent operators are associated with an operator $T$ in the quaternionic setting.

**Definition 2.2.12.** Let $T \in \mathcal{B}(X)$. For $s \in \rho_S(T)$, we define the *left S-resolvent operator* as

$$S_L^{-1}(s, T) = -\mathcal{Q}_s(T)^{-1}(T - \bar{s}\,\mathcal{I}),$$

and the *right S-resolvent operator* as

$$S_R^{-1}(s, T) = -(T - \bar{s}\mathcal{I})\mathcal{Q}_s(T)^{-1}.$$

To develop the slice hyperholomorphic functional calculus it is necessary that the $S$-resolvents are slice hyperholomorphic in the scalar variable $s$. We observe that the left resolvent operator $(s\mathcal{I} - T)^{-1}$ is not slice hyperholomorphic nor Fueter regular.

**Lemma 2.2.13.** *Let $T \in \mathcal{B}(X)$.*

(i) *The left S-resolvent $S_L^{-1}(s,T)$ is a $\mathcal{B}(X)$-valued right slice hyperholomorphic function of the variable $s$ on $\rho_S(T)$.*

(ii) *The right S-resolvent $S_R^{-1}(s,T)$ is a $\mathcal{B}(X)$-valued left slice hyperholomorphic function of the variable $s$ on $\rho_S(T)$.*

The S-resolvent operators in general do not commute with $T$. However, they satisfy the following relations, that are useful for compensating this fact.

**Theorem 2.2.14.** *Let $T \in \mathcal{B}(X)$ and let $s \in \rho_S(T)$. The left S-resolvent operator satisfies the* left *S-resolvent equation*

$$S_L^{-1}(s,T)s - TS_L^{-1}(s,T) = \mathcal{I} \tag{2.40}$$

*and the right S-resolvent operator satisfies the* right *S-resolvent equation*

$$sS_R^{-1}(s,T) - S_R^{-1}(s,T)T = \mathcal{I}. \tag{2.41}$$

The left and the right S-resolvent equations cannot be considered the generalizations of the classical resolvent equation

$$R_\lambda(A) - R_\mu(A) = (\mu - \lambda)R_\lambda(A)R_\mu(A), \quad \text{for } \lambda, \mu \in \rho(A), \tag{2.42}$$

where $R_\lambda(A) = (\lambda \mathcal{I} - A)^{-1}$ is the resolvent operator of $A$ at $\lambda \in \rho(A)$. This equation provides the possibility to write the product $R_\lambda(A)R_\mu(A)$ in terms of the difference $R_\lambda(A) - R_\mu(A)$. This is not the case for the left and the right S-resolvent equations. The proper generalization of (2.42), which preserves this property, is the S-resolvent equation that we show in the following theorem. It is remarkable that this equation involves both the left and the right S-resolvent operators while a generalization of (2.42), that includes just one of them, has never been found.

**Theorem 2.2.15** (The S-resolvent equation)**.** *Let $T \in \mathcal{B}(X)$ and let $s, q \in \rho_S(T)$ with $q \notin [s]$. Then the equation*

$$S_R^{-1}(s,T)S_L^{-1}(p,T) = \left[ (S_R^{-1}(s,T) - S_L^{-1}(q,T))\, q \right.$$
$$\left. -\bar{s}\left( S_R^{-1}(s,T) - S_L^{-1}(q,T) \right) \right] (q^2 - 2\mathrm{Re}(s)q + |s|^2)^{-1} \tag{2.43}$$

*holds true. Equivalently, it can also be written as*

$$S_R^{-1}(s,T)S_L^{-1}(q,T) = (s^2 - 2\mathrm{Re}(q)s + |q|^2)^{-1}$$
$$\cdot \left[ (S_L^{-1}(q,T) - S_R^{-1}(s,T))\,\bar{q} - s\left( S_L^{-1}(q,T) - S_R^{-1}(s,T) \right) \right]. \tag{2.44}$$

We can now define the S-functional calculus for a bounded quaternionic linear operator $T$ on a two-sided quaternionic Banach space $X$. The S-functional calculus is the quaternionic version of the Riesz–Dunford functional calculus for complex linear operators. We consider a function $f$ that is slice hyperholomorphic on $\sigma_S(T)$ and we use the slice hyperholomorphic Cauchy formula. In order to define $f(T)$, we formally replace the scalar variable $q$ by the operator $T$, in the Cauchy kernels $S_L^{-1}(s,q)$ (resp. $S_R^{-1}(s,q)$). We thus obtain the corresponding $S$-resolvent operators $S_L^{-1}(s,T)$ (resp. $S_R^{-1}(s,T)$). The main references in which the formulations and the properties of $S$-functional calculus for quaternionic operators has been studied are [11, 82, 83].

Before we define the $S$-functional calculus, we show that the procedure described above is actually meaningful. In particular, it must be consistent with functions of $T$ that we can define explicitly, that is with polynomials of $T$.

**Theorem 2.2.16.** *Let $T \in \mathcal{B}(X)$, let $U$ be a bounded slice Cauchy domain that contains $\sigma_S(T)$, let $j \in \mathbb{S}$ and set $ds_j = ds(-j)$. For any left slice hyperholomorphic polynomial $P(q) = \sum_{\ell=0}^{n} q^\ell a_\ell$ with $a_\ell \in \mathbb{H}$, we set $P(T) = \sum_{\ell=0}^{n} T^\ell a_\ell$. Then*

$$P(T) = \frac{1}{2\pi} \int_{\partial(U \cap \mathbb{C}_j)} S_L^{-1}(s,T)\, ds_j\, P(s). \tag{2.45}$$

*Similarly, we set $P(T) = \sum_{\ell=0}^{n} a_\ell T^\ell$ for any right slice hyperholomorphic polynomial $P(q) = \sum_{\ell=0}^{n} a_\ell q^\ell$ with $a_\ell \in \mathbb{H}$. Then*

$$P(T) = \frac{1}{2\pi} \int_{\partial(U \cap \mathbb{C}_j)} P(s)\, ds_j\, S_R^{-1}(s,T). \tag{2.46}$$

*In particular, the operators in (2.45) and (2.46) coincide for any intrinsic polynomial $P(q) = \sum_{\ell=0}^{n} q^\ell a_\ell$ with real coefficients $a_\ell \in \mathbb{R}$.*

The $S$-functional calculus applies to functions that are slice hyperholomorphic on the $S$-spectrum of $T$. We introduce the following notation for this class of functions.

**Definition 2.2.17.** Let $T \in \mathcal{B}(X)$. We denote by $\mathcal{SH}_L(\sigma_S(T))$, $\mathcal{SH}_R(\sigma_S(T))$ and $\mathcal{N}(\sigma_S(T))$ the sets of all left, right and intrinsic slice hyperholomorphic functions $f$ with $\sigma_S(T) \subset \mathcal{D}(f)$, where $\mathcal{D}(f)$ is the domain of the function $f$.

**Remark 2.2.8.** If $f \in \mathcal{SH}_L(\sigma_S(T))$ or $f \in \mathcal{SH}_R(\sigma_S(T))$, then the set $\mathcal{D}(f)$ is an axially symmetric open set that contains the compact axially symmetric set $\sigma_S(T)$. By Theorem 2.1.31 there exists a bounded slice Cauchy domain $U$ such that $\sigma_S(T) \subset U$ and $\overline{U} \subset \mathcal{D}(f)$.

**Definition 2.2.18** (*S-functional calculus*). Let $T \in \mathcal{B}(X)$. For any $f \in \mathcal{SH}_L(\sigma_S(T))$, we define

$$f(T) := \frac{1}{2\pi} \int_{\partial(U \cap \mathbb{C}_j)} S_L^{-1}(s,T)\, ds_j\, f(s), \tag{2.47}$$

where $j$ is an arbitrary imaginary unit in $\mathbb{S}$, $ds_j = ds(-j)$ and $U$ is an arbitrary slice Cauchy domain $U$ as in Remark 2.2.8. For any $f \in \mathcal{SH}_R(\sigma_S(T))$, we define

$$f(T) := \frac{1}{2\pi} \int_{\partial(U \cap \mathbb{C}_j)} f(s) \, ds_j \, S_R^{-1}(s, T), \tag{2.48}$$

where $j$ is again an arbitrary imaginary unit in $\mathbb{S}$, $ds_j = ds(-j)$ and $U$ is an arbitrary slice Cauchy domain as in Remark 2.2.8.

Theorem 2.2.16 shows that the $S$-functional calculus is meaningful because it is consistent with polynomials of $T$. As the next crucial result shows, it is moreover well-defined.

**Theorem 2.2.19.** *Let $T \in \mathcal{B}(X)$. For any $f \in \mathcal{SH}_L(\sigma_S(T))$, the integral in (2.47) that defines the operator $f(T)$ is independent of the choice of the slice Cauchy domain $U$ and of the imaginary unit $j \in \mathbb{S}$. Similarly, for any $f \in \mathcal{SH}_R(\sigma_S(T))$, the integral in (2.48) that defines the operator $f(T)$ is also independent of the choice of $U$ and $j \in \mathbb{S}$.*

Theorem 2.2.19 shows that the $S$-functional calculus is well defined for any left or right slice hyperholomorphic function and Theorem 2.2.16 shows that it is consistent with slice hyperholomorphic polynomials. Even more, it is compatible with any rational slice hyperholomorphic function and with limits of uniformly convergent sequences of functions.

**Lemma 2.2.20.** *Let $T \in \mathcal{B}(X)$. If $P$ is an intrinsic polynomial such that $P^{-1} \in \mathcal{N}(\sigma_S(T))$, then $P^{-1}(T) = P(T)^{-1}$. Moreover, if $r(q) = P(q)^{-1}Q(q)$ is an intrinsic rational function and $P^{-1} \in \mathcal{N}(\sigma_S(T))$, then (2.47) and (2.48) give the same operator $r(T) = P(T)^{-1}Q(T)$.*

**Theorem 2.2.21.** *Let $T \in \mathcal{B}(X)$. Let $f_n, f \in \mathcal{SH}_L(\sigma_S(T))$ or let $f_n, f \in \mathcal{SH}_R(\sigma_S(T))$ for $n \in \mathbb{N}$. If there exists a bounded slice Cauchy domain $U$ with $\sigma_S(T) \subset U$ such that $f_n \to f$ uniformly on $\overline{U}$, then $f_n(T)$ converges to $f(T)$ in the norm topology of $\mathcal{B}(X)$.*

Lemma 2.2.20 implies in particular that the $S$-functional calculus for left slice hyperholomorphic functions and the $S$-functional calculus for right slice hyperholomorphic functions are consistent for intrinsic rational functions. Since, by Theorem 2.1.39, any intrinsic slice hyperholomorphic function can be uniformly approximated by intrinsic rational functions, Theorem 2.2.21 implies that both versions of the $S$-functional calculus are consistent for arbitrary intrinsic slice hyperholomorphic functions.

**Theorem 2.2.22.** *Let $T \in \mathcal{B}(X)$. If $f \in \mathcal{N}(\sigma_S(T))$, then both versions of $S$-functional calculus give the same operator $f(T)$. Precisely, we have*

$$f(T) = \frac{1}{2\pi} \int_{\partial(U \cap \mathbb{C}_j)} S_L^{-1}(s, T) \, ds_j \, f(s) = \frac{1}{2\pi} \int_{\partial(U \cap \mathbb{C}_j)} f(s) \, ds_j \, S_R^{-1}(s, T).$$

**Remark 2.2.9.** We point out that Theorem 2.2.22 is in general not true for arbitrary intrinsic slice hyperholomorphic functions, cf. Example 3.7.9 and the discussion in Section 3.8.

An immediate consequence of Definition 2.2.18 is that the $S$-functional calculus for left slice hyperholomorphic functions is quaternionic right linear and that the $S$-functional calculus for right slice hyperholomorphic functions is quaternionic left linear.

**Lemma 2.2.23.** *Let $T \in \mathcal{B}(X)$.*

(i) *If $f, g \in \mathcal{SH}_L(\sigma_S(T))$ and $a \in \mathbb{H}$, then*

$$(f + g)(T) = f(T) + g(T) \quad and \quad (fa)(T) = f(T)a.$$

(ii) *If $f, g \in \mathcal{SH}_R(\sigma_S(T))$ and $a \in \mathbb{H}$, then*

$$(f + g)(T) = f(T) + g(T) \quad and \quad (af)(T) = af(T).$$

Since the product of two slice hyperholomorphic functions is not necessarily slice hyperholomorphic, we cannot expect to obtain a product rule for arbitrary slice hyperholomorphic functions. However, if $f \in \mathcal{N}(\sigma_S(T))$ and $g \in \mathcal{SH}_L(\sigma_S(T))$ then $fg \in \mathcal{SH}_L(\sigma_S(T))$ and if $f \in \mathcal{SH}_R(\sigma_S(T))$ and $g \in \mathcal{N}(\sigma_S(T)$ then $fg \in \mathcal{SH}_R(\sigma_S(T))$. In order to show that the $S$-functional calculus is at least in these cases compatible with the multiplication of functions, we have used the following lemma.

**Lemma 2.2.24.** *Let $B \in \mathcal{B}(X)$. For any $q, s \in \mathbb{H}$ with $q \notin [s]$, we have*

$$(\bar{s}B - Bq)(q^2 - 2\mathrm{Re}(s)q + |s|^2)^{-1} = (s^2 - 2\mathrm{Re}(q)s + |q|^2)^{-1}(sB - B\bar{q}). \quad (2.49)$$

*If, moreover, $f$ is an intrinsic slice hyperholomorphic function and $U$ is a bounded slice Cauchy domain with $\overline{U} \subset \mathcal{D}(f)$, then*

$$\frac{1}{2\pi} \int_{\partial(U \cap \mathbb{C}_j)} f(s) \, ds_j \, (\bar{s}B - Bq)(q^2 - 2\mathrm{Re}(s)q + |s|^2)^{-1} = Bf(q),$$

*for any $q \in U$ and any $j \in \mathbb{S}$.*

From the above lemma and some computations we get the product rule.

**Theorem 2.2.25** (Product rule)**.** *Let $T \in \mathcal{B}(X)$ and let $f \in \mathcal{N}(\sigma_S(T))$ and $g \in \mathcal{SH}_L(\sigma_S(T))$ or let $f \in \mathcal{SH}_R(\sigma_S(T))$ and $g \in \mathcal{N}(\sigma_S(T))$. Then*

$$(fg)(T) = f(T)g(T).$$

An immediate consequence is the following corollary.

**Corollary 2.2.26.** *Let $T \in \mathcal{B}(X)$ and let $f \in \mathcal{N}(\sigma_S(T))$. If $f^{-1} \in \mathcal{N}(\sigma_S(T))$, then $f(T)$ is invertible and $f(T)^{-1} = f^{-1}(T)$.*

Finally, the $S$-functional calculus has the capability to define the quaternionic Riesz projectors and allows in turn to identify invariant subspaces of $T$ that are associated with sets of spectral values.

**Theorem 2.2.27** (Riesz's projectors). *Let $T \in \mathcal{B}(X)$ and assume that $\sigma_S(T) = \sigma_1 \cup \sigma_2$ with*

$$\mathrm{dist}(\sigma_1, \sigma_2) > 0.$$

*We choose an open axially symmetric set $O$ with $\sigma_1 \subset O$ and $\overline{O} \cap \sigma_2 = \emptyset$ and define $\chi_{\sigma_1}(s) = 1$ for $s \in O$ and $\chi_{\sigma_2}(s) = 0$ for $s \notin O$. Then $\chi_{\sigma_1} \in \mathcal{N}(\sigma_S(T))$ and*

$$P_{\sigma_1} := \chi_{\sigma_1}(T) = \frac{1}{2\pi} \int_{\partial(O \cap \mathbb{C}_j)} S_L^{-1}(s, T)\, ds_j$$

*is a continuous projection that commutes with $T$. Hence, $P_{\sigma_1} X$ is a right linear subspace of $X$ that is invariant under $T$.*

Similar to the product rule, the spectral mapping theorem does not hold for arbitrary slice hyperholomorphic functions. This is not surprising: it is clear that it can only hold true for slice hyperholomorphic functions that preserve the fundamental geometry of the $S$-spectrum, namely its axially symmetry. Again, the class of intrinsic slice hyperholomorphic functions stands out here; in fact this class of functions maps axially symmetric sets to axially symmetric sets.

**Theorem 2.2.28** (The Spectral Mapping Theorem). *Let $T \in \mathcal{B}(X)$ and let $f \in \mathcal{N}(\sigma_S(T))$. Then*

$$\sigma_S(f(T)) = f(\sigma_S(T)) = \{f(s) : \quad s \in \sigma_S(T)\}.$$

The Spectral Mapping Theorem allows us to generalize the Gelfand formula for the spectral radius to quaternionic linear operators.

**Definition 2.2.29.** *Let $T \in \mathcal{B}(X)$. Then the $S$-spectral radius of $T$ is defined to be the nonnegative real number*

$$r_S(T) := \sup\{|s| : \quad s \in \sigma_S(T)\}.$$

**Theorem 2.2.30.** *For $T \in \mathcal{B}(X)$, we have*

$$r_S(T) = \lim_{n \to +\infty} \|T^n\|^{\frac{1}{n}}.$$

Finally, the Spectral Mapping Theorem also allows us to generalize the composition rule.

**Theorem 2.2.31** (Composition rule). *Let $T \in \mathcal{B}(X)$ and let $f \in \mathcal{N}(\sigma_S(T))$. If $g \in \mathcal{SH}_L(\sigma_S(f(T)))$, then $g \circ f \in \mathcal{SH}_L(\sigma_S(T))$ and if $g \in \mathcal{SH}_R(f(\sigma_S(T)))$, then $g \circ f \in \mathcal{SH}_R(\sigma_S(T))$. In both cases,*

$$g(f(T)) = (g \circ f)(T).$$

We recall Theorem 4.7 in [83], or also see Theorem 4.14.14 in [93].

**Theorem 2.2.32** (Perturbation of the $S$-functional calculus). *Let $T, Z \in \mathcal{B}(X)$, $f \in \mathcal{SH}_L(\sigma_S(T))$ and let $\varepsilon > 0$. Then there exists $\delta > 0$ such that, for $\|Z - T\| < \delta$, we have $f \in \mathcal{SH}_L(\sigma_S(Z))$ and*

$$\|f(Z) - f(T)\| < \varepsilon,$$

*where*

$$f(T) = \frac{1}{2\pi} \int_{\partial(U \cap \mathbb{C}_j)} S_L^{-1}(s,T) \, ds_j \, f(s)$$

*and $U \subset \mathbb{H}$ is a Cauchy domain, $ds_j = ds/j$ for $j \in \mathbb{S}$.*

## 2.3   Bounded operators with commuting components

If the components of $T$ commute, then the $S$-spectrum can be characterized by a different operator, which is often easier to handle in the applications. The $S$-resolvent operators, in this case, can be expressed in a form that corresponds to replacing the scalar variable $q$ by the operator $T$ in the slice hyperholomorphic Cauchy kernels when they are written in form II, see Proposition 2.1.24.

We recall that any two-sided quaternionic vector space $X$ is essentially of the form $X = X_\mathbb{R} \otimes \mathbb{H}$, where $X_\mathbb{R}$ is the real vector space consisting of those vectors that commute with all quaternions. If $x = \sum_{\ell=0}^{3} x_\ell e_\ell$ with $x_\ell \in X_\mathbb{R}$, where we set $e_0 = 1$ for neatness, then we can write any operator $T \in \mathcal{B}(X)$ as $T = \sum_{\ell=0}^{3} T_\ell e_\ell$ with components $T_\ell \in \mathcal{B}(X_\mathbb{R})$, where this operator acts as

$$Tx = \left( \sum_{\ell=0}^{3} T_\ell e_\ell \right) \left( \sum_{\kappa=0}^{3} x_\kappa e_\kappa \right) = \sum_{\ell,\kappa=0}^{3} T_\ell(x_\kappa) e_\ell e_\kappa.$$

We find $\mathcal{B}(X) = \mathcal{B}(X_\mathbb{R}) \otimes \mathbb{H}$ and hence we call any operator in $\mathcal{B}(X_\mathbb{R})$ a scalar operator on $X$. Furthermore, we denote the space of all $T = \sum_{\ell=0}^{3} T_\ell e_\ell \in \mathcal{B}(X)$ with mutually commuting components $T_\ell \in \mathcal{B}(X_\mathbb{R}), \ell = 0, \dots, 3$, by $\mathcal{BC}(X)$, cf. Definition 2.2.6.

**Definition 2.3.1.** For $T = T_0 + \sum_{\ell=1}^{3} T_\ell e_\ell \in \mathcal{BC}(X)$, we set

$$\overline{T} := T_0 - \sum_{\ell=1}^{3} T_\ell e_\ell.$$

The following statement shows that for an operator $T \in \mathcal{BC}(X)$ the analogues of the scalar identities $s + \bar{s} = 2\mathrm{Re}(s)$ and $s\bar{s} = \bar{s}s = |s|^2$ hold true. This motivates the idea that we can write the $S$-resolvent operators, for such operators, also by formally replacing $q$ by $T$ in the slice hyperholomorphic Cauchy kernels, when they are written in form II.

**Lemma 2.3.2.** *Let $T = T_0 + \sum_{\ell=1}^{3} T_\ell e_\ell \in \mathcal{BC}(X)$. Then $2T_0 = T + \bar{T}$ and $T\bar{T} = \bar{T}T = \sum_{\ell=0}^{3} T_\ell^2$.*

**Lemma 2.3.3.** *If $T = T_0 + \sum_{\ell=1}^{3} T_\ell e_\ell \in \mathcal{BC}(X)$, then the following statements are equivalent.*

(i) *The operator $T$ is invertible.*

(ii) *The operator $\bar{T}$ is invertible.*

(iii) *The operator $T\bar{T}$ is invertible.*

*In this case we have*

$$\bar{T}^{-1} = \overline{T^{-1}} \quad and \quad T^{-1} = (T\bar{T})^{-1}\bar{T}. \tag{2.50}$$

**Definition 2.3.4.** *Let $T = T_0 + \sum_{\ell=1}^{3} T_\ell e_\ell \in \mathcal{BC}(X)$. For $s \in \mathbb{H}$, we define the operator*

$$\mathcal{Q}_{c,s}(T) := s^2 \mathcal{I} - 2sT_0 + T\bar{T}.$$

**Theorem 2.3.5.** *Let $T = T_0 + \sum_{\ell=1}^{3} T_\ell e_\ell \in \mathcal{BC}(X)$. Then $\mathcal{Q}_{c,s}(T)$ is invertible if and only if $\mathcal{Q}_s(T)^{-1}$ is invertible and so*

$$\rho_S(T) = \left\{ s \in \mathbb{H} : \quad \mathcal{Q}_{c,s}(T)^{-1} \in \mathcal{B}(X) \right\}. \tag{2.51}$$

*Moreover, for $s \in \rho_S(T)$, we have*

$$S_L^{-1}(s,T) = (s\mathcal{I} - \bar{T})\mathcal{Q}_{c,s}(T)^{-1}, \tag{2.52}$$
$$S_R^{-1}(s,T) = \mathcal{Q}_{c,s}(T)^{-1}(s\mathcal{I} - \bar{T}). \tag{2.53}$$

**Definition 2.3.6** (*SC-resolvent operators*)**.** *Let $T \in \mathcal{BC}(X)$. For $s \in \rho_S(T)$, we define the left and right $SC$-resolvent operators of $T$ as*

$$S_{c,L}^{-1}(s,T) = (s\mathcal{I} - \bar{T})\mathcal{Q}_{c,s}(T)^{-1}$$
$$S_{c,R}^{-1}(s,T) = \mathcal{Q}_{c,s}(T)^{-1}(s\mathcal{I} - \bar{T}).$$

So for $T \in \mathcal{BC}(X)$ we have the equivalent definitions of the $S$-functional calculus for operators with commuting components. For $f \in \mathcal{SH}_L(\sigma_S(T))$, we have

$$f(T) = \frac{1}{2\pi} \int_{\partial(U \cap \mathbb{C}_j)} S_{c,L}^{-1}(s,T)\, ds_j\, f(s)$$

and for $f \in \mathcal{SH}_R(\sigma_S(T))$ we have

$$f(T) = \frac{1}{2\pi} \int_{\partial(U \cap \mathbb{C}_j)} f(s) \, ds_j \, S_{c,R}^{-1}(s, T),$$

for any imaginary unit $j \in \mathbb{S}$ and any bounded slice Cauchy domain $U$ with $\sigma_S(T) \subset U$ and $\overline{U} \subset \mathcal{D}(f)$.

The $S$-functional calculus for operators with commuting components defined by the above integrals, that involve the $SC$-resolvents, is often also referred to as the $SC$-functional calculus. Similarly, the $S$-spectrum is sometimes called $F$-spectrum, when it is characterized by the operator $\mathcal{Q}_{c,s}(T)^{-1}$, because it was used in the definition of the $F$-functional calculus.

# Chapter 3

# The direct approach to the $S$-functional calculus

The $S$-functional calculus can also be defined for unbounded operators $T : \mathcal{D}(T) \subset X \to X$, where $X$ is a two-sided quaternionic Banach space $X$. In the papers [69,101] this calculus was defined using suitable transformations in order to reduce the problem to the case of bounded operators. The direct approach has been studied in the more recent paper [130] and it turned out that the two approaches, contrary to the complex setting, are not totally equivalent. In fact, in using the direct approach one can remove the assumption that the $S$-resolvent set contains a real point.

**Definition 3.0.1.** Let $X$ be a two-sided quaternionic Banach space. A right linear operator $T : \mathcal{D}(T) \subset X \to X$ defined on a right-linear subspace $\mathcal{D}(T)$ of $X$ is called closed if its graph is closed in $X \oplus X$. We denote the set of closed right linear operators $T : \mathcal{D}(T) \subset X \to X$ by $\mathcal{K}(X)$.

**Remark 3.0.1.** The notion of a closed right linear operator can also be considered on a right Banach space and does not necessarily require the existence of a left multiplication on $X$. However, for the reasons explained in Remark 2.2.7, one usually works on two-sided Banach spaces.

When we deal with closed operators, we have to pay attention to the domains on which they are defined. The powers of $T$ are defined inductively as $T^0 = \mathcal{I}$ with $\mathcal{D}(T^0) = \mathcal{D}(\mathcal{I}) = X$ and $T^{n+1}v = T(T^n v)$ for $v \in \mathcal{D}(T^{n+1}) := \{v \in \mathcal{D}(T) : T^n \in \mathcal{D}(T)\}$. Polynomials of $T$ with real coefficients are then defined as usual: if $P(s) = \sum_{\ell=0}^{n} a_\ell s^\ell$ with $a_\ell \in \mathbb{R}$, then $P(T)v = \sum_{\ell=0}^{n} a_\ell T^\ell v$ for $v \in \mathcal{D}(T^n)$. However, if the coefficients are not real, then we have to distinguish two cases: for a right slice hyperholomorphic polynomial $P(s) = \sum_{\ell=0}^{n} a_\ell s^\ell$ with $a_\ell \in \mathbb{H}$, we can again set $P(T)v = \sum_{\ell=0}^{n} a_\ell T^\ell v$ for $v \in \mathcal{D}(T^n)$. For a left slice hyperholomorphic polynomial $P(s) = \sum_{\ell=0}^{n} s^\ell a_\ell$ with $a_\ell \in \mathbb{H}$ it is, however, not always possible

© Springer Nature Switzerland AG 2019
F. Colombo, J. Gantner, *Quaternionic Closed Operators, Fractional Powers and Fractional Diffusion Processes*, Operator Theory: Advances and Applications 274,
https://doi.org/10.1007/978-3-030-16409-6_3

to set $P(T)v = \sum_{\ell=0}^{n} T^n a_\ell v$ for $v \in \mathcal{D}(T^n)$. Indeed, since $T$ is right linear, the domain of $\mathcal{D}(T^\ell)$ is a right-linear but not necessarily a left-linear subspace of $X$. Hence, it might happen that $a_\ell v \notin \mathcal{D}(T^\ell)$ even though $v \in \mathcal{D}(T^n)$, so setting $P(T)v = \sum_{\ell=0}^{n} T^n a_\ell v$ is not meaningful.

## 3.1  Properties of the $S$-spectrum of a closed operator

For $T \in \mathcal{K}(X)$, we define

$$\mathcal{Q}_s(T) := T^2 - 2\mathrm{Re}(s)T + |s|^2 \mathcal{I}, \quad \text{for } s \in \mathbb{H},$$

and the operator $\mathcal{Q}_s(T)$ is defined on $\mathcal{D}(T^2)$.

**Definition 3.1.1.** Let $T \in \mathcal{K}(X)$. We define the $S$-resolvent set of $T$ as

$$\rho_S(T) := \big\{ s \in \mathbb{H} : \quad \mathcal{Q}_s(T)^{-1} \in \mathcal{B}(X) \big\}$$

and the $S$-spectrum of $T$ as

$$\sigma_S(T) := \mathbb{H} \setminus \rho_S(T).$$

For $s \in \rho_S(T)$, the operator $\mathcal{Q}_s(T)^{-1}$ is called the pseudo-resolvent of $T$ at $s$. Furthermore, we define the extended $S$-spectrum $\sigma_{SX}(T)$ as

$$\sigma_{SX}(T) := \begin{cases} \sigma_S(T) & \text{if } T \text{ is bounded,} \\ \sigma_S(T) \cup \{\infty\} & \text{if } T \text{ is unbounded.} \end{cases}$$

Before we study the properties of the $S$-spectrum of a closed operator, we need to investigate the differentiability properties of its pseudo-resolvent in detail. The correct tool for studying these properties is a series expansion of $\mathcal{Q}_s(T)^{-1}$, which was found in [52]. An heuristic approach for finding this expansion consists in considering the equation

$$\mathcal{Q}_s(T)^{-1} - \mathcal{Q}_q(T)^{-1} = \mathcal{Q}_s(T)^{-1}(\mathcal{Q}_q(T) - \mathcal{Q}_s(T))\mathcal{Q}_q(T)^{-1} \qquad (3.1)$$

and writing it as

$$\mathcal{Q}_s(T)^{-1} = \mathcal{Q}_q(T)^{-1} + \mathcal{Q}_s(T)^{-1}(\mathcal{Q}_q(T) - \mathcal{Q}_s(T))\mathcal{Q}_q(T)^{-1}.$$

Recursive application of this equation then yields the series expansion proved in the following, where we consider closed axially symmetric neighbourhoods, described by the function $d_S(s, q) = \max\big\{2|s_0 - q_0|, \big||q|^2 - |s|^2\big|\big\}$, which naturally rise from the series expansion of the pseudo-resolvent operator.

**Theorem 3.1.2.** Let $T \in \mathcal{K}(X)$ and $q \in \rho_S(T)$ and let $s \in \mathbb{H}$. If the series

$$\mathcal{J}(s) = \sum_{n=0}^{+\infty} (\mathcal{Q}_q(T) - \mathcal{Q}_s(T))^n \, \mathcal{Q}_q(T)^{-(n+1)} \qquad (3.2)$$

converges absolutely in $\mathcal{B}(X)$, then $s \in \rho_S(T)$ and it equals the pseudo-resolvent $\mathcal{Q}_s(T)^{-1}$ of $T$ at $s$.

The series converges in particular uniformly on any of the closed axially symmetric neighbourhoods

$$C_\varepsilon(q) = \{s \in \mathbb{H} : \quad d_S(s, q) \leq \varepsilon\}$$

of $q$ with

$$d_S(s, q) = \max\left\{2|s_0 - q_0|, \left||q|^2 - |s|^2\right|\right\}$$

and

$$\varepsilon < \frac{1}{\|T\mathcal{Q}_q(T)^{-1}\| + \|\mathcal{Q}_q(T)^{-1}\|}.$$

*Proof.* Let us first consider the question of the convergence of the series. The sets $C_\varepsilon(q)$ are obviously axially symmetric: if $s_j$ belongs to the sphere $[s]$ associated to $s$, then $s_0 = \mathrm{Re}(s) = \mathrm{Re}(s_j)$ and $|s|^2 = |s_j|^2$. Thus, $d_S(s_j, q) = d_S(s, q)$ and in turn $s \in C_\varepsilon(q)$ if and only if $s_j \in C_\varepsilon(q)$. Moreover, since the map $s \mapsto d_S(s, q)$ is continuous, the sets $U_\varepsilon(q) := \{s \in \mathbb{H} : d_S(s, q) < \varepsilon\}$ are open in $\mathbb{H}$. Since $U_\varepsilon(q) \subset C_\varepsilon(q)$, the sets $C_\varepsilon$ are actually neighbourhoods of $q$. In order to simplify the notation, we set

$$\Lambda(q, s) := \mathcal{Q}_q(T) - \mathcal{Q}_s(T) = 2(s_0 - q_0)T + (|q|^2 - |s|^2)\mathcal{I}.$$

Since $\mathcal{Q}_q(T)^{-1}$ maps $X$ to $\mathcal{D}(T^2)$ and $\Lambda(q, s)$ commutes with $\mathcal{Q}_q(T)^{-1}$ on $\mathcal{D}(T^2)$, we have for any $s \in C_\varepsilon(q)$,

$$\sum_{n=0}^{+\infty} \left\| \Lambda(q, s)^n \mathcal{Q}_q(T)^{-(n+1)} \right\|$$

$$= \sum_{n=0}^{+\infty} \left\| \left(\Lambda(q, s)\mathcal{Q}_q(T)^{-1}\right)^n \mathcal{Q}_q(T)^{-1} \right\|$$

$$\leq \sum_{n=0}^{+\infty} \left\| \Lambda(q, s)\mathcal{Q}_q(T)^{-1} \right\|^n \left\| \mathcal{Q}_q(T)^{-1} \right\|.$$

We further have

$$\left\| \Lambda(q, s)\mathcal{Q}_q(T)^{-1} \right\| \leq 2|s_0 - q_0| \left\| T\mathcal{Q}_q(T)^{-1} \right\| + \left| |q|^2 - |s|^2 \right| \left\| \mathcal{Q}_q(T)^{-1} \right\|$$

$$\leq d_S(s, q) \left( \left\| T\mathcal{Q}_q(T)^{-1} \right\| + \left\| \mathcal{Q}_q(T)^{-1} \right\| \right)$$

$$\leq \varepsilon \left( \left\| T\mathcal{Q}_q(T)^{-1} \right\| + \left\| \mathcal{Q}_q(T)^{-1} \right\| \right) =: \varrho.$$

If now $\varepsilon < 1/\left( \left\| T\mathcal{Q}_q(T)^{-1} \right\| + \left\| \mathcal{Q}_q(T)^{-1} \right\| \right)$, then $0 < \varrho < 1$ and thus,

$$\sum_{n=0}^{+\infty} \left\| \Lambda(q, s)^n \mathcal{Q}_q(T)^{-(n+1)} \right\| \leq \left\| \mathcal{Q}_q(T)^{-1} \right\| \sum_{n=0}^{+\infty} \varrho^n < +\infty.$$

and the series converges uniformly in $\mathcal{B}(X)$ on $C_\varepsilon(q)$.

Now assume that the series (3.2) converges and observe that $\mathcal{Q}_s(T)$, $\mathcal{Q}_q(T)$ and $\mathcal{Q}_q(T)^{-1}$ commute on $\mathcal{D}(T^2)$. Hence, we have for $y \in \mathcal{D}(T^2)$ that

$$
\begin{aligned}
\mathcal{J}(s)\mathcal{Q}_s(T)y &= \sum_{n=0}^{+\infty} \Lambda(q,s)^n \mathcal{Q}_q(T)^{-(n+1)} \mathcal{Q}_s(T)y \\
&= \sum_{n=0}^{+\infty} \Lambda(q,s)^n \mathcal{Q}_q(T)^{-(n+1)} \left[ -\Lambda(q,s) + \mathcal{Q}_q(T) \right] y \\
&= -\sum_{n=0}^{+\infty} \Lambda(q,s)^{n+1} \mathcal{Q}_q(T)^{-(n+1)} y \\
&\quad + \sum_{n=0}^{+\infty} \Lambda(q,s)^n \mathcal{Q}_q(T)^{-n} y = y.
\end{aligned}
$$

On the other hand,

$$
y_N := \sum_{n=0}^{N} \Lambda(q,s)^n \mathcal{Q}_q(T)^{-(n+1)} y = \mathcal{Q}_q(T)^{-1} \sum_{n=0}^{N} \Lambda(q,s)^n \mathcal{Q}_q(T)^{-n} y
$$

belongs to $\mathcal{D}(T^2)$ for any $y \in X$ and we have

$$
\begin{aligned}
\mathcal{Q}_s(T)y_N &= (-\Lambda(q,s) + \mathcal{Q}_q(T)) \sum_{n=0}^{N} \Lambda(q,s)^n \mathcal{Q}_q(T)^{-(n+1)} y \\
&= -\sum_{n=0}^{N} \Lambda(q,s)^{n+1} \mathcal{Q}_q(T)^{-(n+1)} y + \sum_{n=0}^{N} \Lambda(q,s)^n \mathcal{Q}_q(T)^{-n} y \\
&= -\Lambda(q,s)^{N+1} \mathcal{Q}_p(T)^{-(n+1)} y + y.
\end{aligned}
$$

Now observe that
$$
\Lambda(q,s) = 2(s_0 - q_0)T + (|q|^2 - |s|^2)\mathcal{I}
$$
is defined on $\mathcal{D}(T)$ and maps $\mathcal{D}(T^2)$ to $\mathcal{D}(T)$. Hence, $\Lambda(q,s)^2 \mathcal{Q}_q(T)^{-1}$ belongs to $\mathcal{B}(X)$ and for $N \geq 1$

$$
\begin{aligned}
&\left\| -\Lambda(q,s)^{N+1} \mathcal{Q}_p(T)^{-(n+1)} y \right\| \\
&= \left\| -\Lambda(q,s)^{N-1} \mathcal{Q}_p(T)^{-N} \Lambda(q,s)^2 \mathcal{Q}_q(T)^{-1} y \right\| \\
&\leq \left\| -\Lambda(q,s)^{N-1} \mathcal{Q}_p(T)^{-N} \right\| \left\| \Lambda(q,s)^2 \mathcal{Q}_q(T)^{-1} y \right\| \overset{N\to\infty}{\longrightarrow} 0
\end{aligned}
$$

because the series (3.2) converges in the norm of $\mathcal{B}(X)$ by assumption. Thus, $\mathcal{Q}_s(T)y_N \to y$ and $y_N \to y_\infty := \mathcal{J}(s)y$ as $N \to \infty$. Since $\mathcal{Q}_s(T)$ is closed, we obtain that

$$\mathcal{J}(s)y \in \mathcal{D}(\mathcal{Q}_s(T)) = \mathcal{D}(T^2) \quad \text{and} \quad \mathcal{Q}_s(T)\mathcal{J}(s)y = y.$$

Hence, $\mathcal{J}(s) = \mathcal{Q}_s(T)^{-1}$ and in turn $s \in \rho_S(T)$. $\qquad\square$

**Lemma 3.1.3.** *Let $T \in \mathcal{K}(X)$. The functions $s \to \mathcal{Q}_s(T)^{-1}$ and $s \to T\mathcal{Q}_s(T)^{-1}$, which are defined on $\rho_S(T)$ and take values in $\mathcal{B}(X)$, are continuous.*

*Proof.* Let $q \in \rho_S(T)$. Then $\mathcal{Q}_s(T)^{-1}$ can be represented by the series (3.2), which converges uniformly on a neighborhood of $q$. Hence, we have

$$\lim_{s \to q} \mathcal{Q}_s(T)^{-1} = \sum_{n=0}^{+\infty} \lim_{s \to q} \left(2(s_0 - q_0)T + \left(|q|^2 - |s|^2\right)\mathcal{I}\right)^n \mathcal{Q}_q(T)^{-(n+1)}$$

$$= \mathcal{Q}_q(T)^{-1},$$

because each term in the sum is a polynomial in $s_0$ and $s_1$ (since $s = s_0 + js_1$ for $j \in \mathbb{S}$) with coefficients in $\mathcal{B}(X)$ and thus, continuous. Indeed

$$\left((s_0 - q_0)T + \left(|q|^2 - |s|^2\right)\mathcal{I}\right)^n \mathcal{Q}_q(T)^{-(n+1)}$$

$$= \sum_{k=0}^{n} \binom{n}{k}(s_0 - q_0)^k \left(|q|^2 - |s|^2\right)^{n-k} T^k \mathcal{Q}_q(T)^{-(n+1)}$$

and the coefficients $T^k \mathcal{Q}_q(T)^{-(n+1)}$ belong to $\mathcal{B}(X)$ because $\mathcal{Q}_q(T)^{-(n+1)}$ maps $X$ to $\mathcal{D}(T^{2(n+1)})$ and $k < 2(n+1)$. The function $s \mapsto T\mathcal{Q}_s(T)^{-1}$ is continuous because the identity (3.1) implies

$$\lim_{h \to 0} \left\| T\mathcal{Q}_{s+h}(T)^{-1} - T\mathcal{Q}_s(T)^{-1} \right\|$$

$$= \lim_{h \to 0} \left\| T\mathcal{Q}_{s+h}(T)^{-1}(\mathcal{Q}_s(T) - \mathcal{Q}_{s+h}(T))\mathcal{Q}_s(T)^{-1} \right\|.$$

The operator $\mathcal{Q}_s(T)^{-1}$ maps $X$ to $\mathcal{D}(T^2)$ and so

$$(\mathcal{Q}_s(T) - \mathcal{Q}_{s+h}(T))\mathcal{Q}_s(T)^{-1} = (2h_0 T + (|s|^2 - |s + h|^2)\mathcal{I})\mathcal{Q}_s(T)^{-1}$$

maps $X$ to $\mathcal{D}(T)$. Since $T$ and $\mathcal{Q}_{s+h}(T)^{-1}$ commute on $\mathcal{D}(T)$ we thus have

$$\lim_{h \to 0} \left\| T\mathcal{Q}_{s+h}(T)^{-1} - T\mathcal{Q}_s(T)^{-1} \right\|$$

$$= \lim_{h \to 0} \left\| \mathcal{Q}_{s+h}(T)^{-1}\left(2h_0 T^2 + \left(|s|^2 - |s + h|^2\right)T\right)\mathcal{Q}_s(T)^{-1} \right\|$$

$$\leq \lim_{h \to 0} \left\| \mathcal{Q}_{s+h}(T)^{-1} \right\| \lim_{h \to 0} 2h_0 \left\| T^2 \mathcal{Q}_s(T)^{-1} \right\|$$

$$+ \lim_{h \to 0} \left\| \mathcal{Q}_{s+h}(T)^{-1} \right\| \lim_{h \to 0} \left(|s|^2 - |s + h|^2\right) \left\| T\mathcal{Q}_s(T)^{-1} \right\| = 0. \qquad\square$$

**Lemma 3.1.4.** *Let $T \in \mathcal{K}(X)$ and $s \in \rho_S(T)$. The pseudo-resolvent $\mathcal{Q}_s(T)^{-1}$ is continuously real differentiable with*

$$\frac{\partial}{\partial s_0} \mathcal{Q}_s(T)^{-1} = (2T - 2s_0 \mathcal{I}) \mathcal{Q}_s(T)^{-2} \quad and \quad \frac{\partial}{\partial s_1} \mathcal{Q}_s(T)^{-1} = -2s_1 \mathcal{Q}_s(T)^{-2}.$$

*Proof.* Let us first compute the partial derivative of $\mathcal{Q}_s(T)^{-1}$ with respect to the real part $s_0$. Applying equation (3.1), we have

$$\frac{\partial}{\partial s_0} \mathcal{Q}_s(T)^{-1} = \lim_{\mathbb{R} \ni h \to 0} \frac{1}{h} \left( \mathcal{Q}_{s+h}(T)^{-1} - \mathcal{Q}_s(T)^{-1} \right)$$

$$= \lim_{\mathbb{R} \ni h \to 0} \frac{1}{h} \mathcal{Q}_{s+h}(T)^{-1} \left( \mathcal{Q}_s(T) - \mathcal{Q}_{s+h}(T) \right) \mathcal{Q}_s(T)^{-1}$$

$$= \lim_{\mathbb{R} \ni h \to 0} \mathcal{Q}_{s+h}(T)^{-1} \left( 2T - 2s_0 \mathcal{I} - h \mathcal{I} \right) \mathcal{Q}_s(T)^{-1},$$

where $\lim_{\mathbb{R} \ni h \to 0} f(h)$ denotes the limit of a function $f$ as $h$ tends to $0$ in $\mathbb{R}$. Since the composition and the multiplication with scalars are continuous operations on $\mathcal{B}(X)$, we further have

$$\frac{\partial}{\partial s_0} \mathcal{Q}_s(T)^{-1} = \lim_{\mathbb{R} \ni h \to 0} \mathcal{Q}_{s+h}(T)^{-1} \lim_{\mathbb{R} \ni h \to 0} \left( (2T - 2s_0 \mathcal{I}) \mathcal{Q}_s(T)^{-1} - h \mathcal{Q}_s(T)^{-1} \right)$$

$$= \mathcal{Q}_s(T)^{-1} (2T - 2s_0 \mathcal{I}) \mathcal{Q}_s(T)^{-1}$$

$$= (2T - 2s_0 \mathcal{I}) \mathcal{Q}_s(T)^{-2},$$

where the last equation holds true because $\mathcal{Q}_s(T)^{-1}$ maps $X$ to $\mathcal{D}(T^2) \subset \mathcal{D}(T)$ and $T$ and $\mathcal{Q}_s(T)^{-1}$ commute on $\mathcal{D}(T)$. Observe that $\frac{\partial}{\partial s_0} \mathcal{Q}_s(T)^{-1}$ is even continuous because it is the sum and product of continuous functions by Lemma 3.1.3.

If we write $s = s_0 + j_s s_1$, then we can argue in a similar way to show that the derivative of $\mathcal{Q}_s(T)^{-1}$ with respect to $s_1$ is

$$\frac{\partial}{\partial s_1} \mathcal{Q}_s(T)^{-1} = \lim_{\mathbb{R} \ni h \to 0} \frac{1}{h} \left( \mathcal{Q}_{s+hj_s}(T)^{-1} - \mathcal{Q}_s(T)^{-1} \right)$$

$$= \lim_{\mathbb{R} \ni h \to 0} \frac{1}{h} \mathcal{Q}_{s+hj_s}(T)^{-1} \left( \mathcal{Q}_s(T) - \mathcal{Q}_{s+hj_s}(T) \right) \mathcal{Q}_s(T)^{-1}$$

$$= \lim_{\mathbb{R} \ni h \to 0} \mathcal{Q}_{s+hj_s}(T)^{-1} \left( -2s_1 - h \right) \mathcal{Q}_s(T)^{-1}$$

$$= \lim_{\mathbb{R} \ni h \to 0} \mathcal{Q}_{s+hj_s}(T)^{-1} \lim_{\mathbb{R} \ni h \to 0} \left( -2s_1 \mathcal{Q}_s(T)^{-1} - h \mathcal{Q}_s(T)^{-1} \right)$$

$$= -2s_1 \mathcal{Q}_s(T)^{-2}.$$

Again this derivative is continuous as it is the product of two continuous functions by Lemma 3.1.3.

Finally, we easily obtain that $\mathcal{Q}_s(T)^{-1}$ is continuously real differentiable from the fact that $\mathcal{Q}_s(T)^{-1}$ is continuously differentiable in the variables $s_0$ and $s_1$. If we write $s$ in terms of its four real coordinates as $s = \xi_0 + \sum_{\ell=1}^{3} \xi_\ell e_\ell$, then the partial derivative with respect to $\xi_0$ corresponds to the partial derivative with respect to $s_0$ and thus, exists and is continuous. The partial derivative with respect to $\xi_\ell$ for $1 \le \ell \le 3$ on the other hand exists and is continuous for $s_1 \ne 0$ because $\mathcal{Q}_s(T)^{-1}$ can be considered as the composition of the continuously differentiable functions $s \mapsto s_1 = \sqrt{\xi_1^2 + \xi_2^2 + \xi_3^2}$ and $s_1 \to \mathcal{Q}_{s+js_1}(T)^{-1}$ with fixed $j \in \mathbb{S}$. We find

$$\frac{\partial}{\partial \xi_\ell} \mathcal{Q}_s(T)^{-1} = -2s_1 \mathcal{Q}_s(T)^{-2} \frac{\partial}{\partial \xi_\ell} s_0 = -2\xi_\ell \mathcal{Q}_s(T)^{-2}.$$

For $s_1 = 0$ (that is for $s \in \mathbb{R}$), we can simply choose $j = e_\ell$ and then the partial derivative with respect to $\xi_\ell$ agrees with the partial derivative with respect to $s_1$. In particular, we see that also the partial derivatives with respect to the real coordinates $\xi_0, \ldots, \xi_3$ are continuous. $\qquad\square$

**Lemma 3.1.5.** *Let $T \in \mathcal{K}(X)$ and $s \in \rho_S(T)$. The function $s \mapsto T\mathcal{Q}_s(T)^{-1}$ is continuously real differentiable with*

$$\frac{\partial}{\partial s_0} T\mathcal{Q}_s(T)^{-1} = (2T^2 - 2s_0 T)\mathcal{Q}_s(T)^{-2}$$

*and*

$$\frac{\partial}{\partial s_1} T\mathcal{Q}_s(T)^{-1} = -2s_1 T\mathcal{Q}_s(T)^{-2}.$$

*Proof.* If $\lim_{\mathbb{R} \ni h \to 0} f(h)$ denotes again the limit of a function $f$ as $h$ tends to 0 in $\mathbb{R}$, then we obtain from (3.1) that

$$\frac{\partial}{\partial s_0} T\mathcal{Q}_s(T)^{-1} = \lim_{\mathbb{R} \ni h \to 0} \frac{1}{h} \left( T\mathcal{Q}_{s+h}(T)^{-1} - T\mathcal{Q}_s(T)^{-1} \right)$$

$$= \lim_{\mathbb{R} \ni h \to 0} \frac{1}{h} T\mathcal{Q}_{s+h}(T)^{-1} \left( \mathcal{Q}_s(T) - \mathcal{Q}_{s+h}(T) \right) \mathcal{Q}_s(T)^{-1}$$

$$= \lim_{\mathbb{R} \ni h \to 0} \frac{1}{h} T\mathcal{Q}_{s+h}(T)^{-1} \left( 2hT - 2hs_0 \mathcal{I} - h^2 \mathcal{I} \right) \mathcal{Q}_s(T)^{-1}$$

$$= \lim_{\mathbb{R} \ni h \to 0} \mathcal{Q}_{s+h}(T)^{-1} \left( 2T^2 - 2s_0 T - hT \right) \mathcal{Q}_s(T)^{-1},$$

because $\left( 2hT - 2hs_0 \mathcal{I} - h^2 \mathcal{I} \right) \mathcal{Q}_s(T)^{-1}$ maps $X$ to $\mathcal{D}(T)$ and $T$ and $\mathcal{Q}_{s+h}(T)^{-1}$ commute on $\mathcal{D}(T)$. Since the composition and the multiplication with scalars are continuous operations on the space $\mathcal{B}(X)$ and since the pseudo-resolvent is continuous by Lemma 3.1.3, we get

$$\frac{\partial}{\partial s_0} T\mathcal{Q}_s(T)^{-1} = \lim_{\mathbb{R} \ni h \to 0} \mathcal{Q}_{s+h}(T)^{-1} \lim_{\mathbb{R} \ni h \to 0} \left( (2T^2 - 2s_0 T) \mathcal{Q}_s(T)^{-1} - hT\mathcal{Q}_s(T)^{-1} \right)$$

$$= \mathcal{Q}_s(T)^{-1} (2T^2 - 2s_0 T)\mathcal{Q}_s(T)^{-1}$$

$$= (2s_0 T - 2T^2)\mathcal{Q}_s(T)^{-2}.$$

This function is continuous because we can write it as the product of functions that are continuous by Lemma 3.1.3.

The derivative with respect to $s_1$ can be computed using similar arguments via

$$\frac{\partial}{\partial s_1} T\mathcal{Q}_s(T)^{-1} = \lim_{\mathbb{R} \ni h \to 0} \frac{1}{h} \left( T\mathcal{Q}_{s+hj_s}(T)^{-1} - T\mathcal{Q}_s(T)^{-1} \right)$$

$$= \lim_{\mathbb{R} \ni h \to 0} \frac{1}{h} T\mathcal{Q}_{s+hj_s}(T)^{-1} \left( \mathcal{Q}_s(T) - \mathcal{Q}_{s+hj_s}(T) \right) \mathcal{Q}_s(T)^{-1}$$

$$= \lim_{\mathbb{R} \ni h \to 0} \frac{1}{h} T\mathcal{Q}_{s+hj_s}(T)^{-1} \left( -2hs_1 - h^2 \right) \mathcal{Q}_s(T)^{-1}$$

$$= \lim_{\mathbb{R} \ni h \to 0} \mathcal{Q}_{s+hj_s}(T)^{-1} \lim_{\mathbb{R} \ni h \to 0} \left( -2s_1 T\mathcal{Q}_s(T)^{-1} - hT\mathcal{Q}_s(T)^{-1} \right)$$

$$= -2s_1 T\mathcal{Q}_s(T)^{-2}.$$

Also this derivative is continuous because

$$\frac{\partial}{\partial s_1} T\mathcal{Q}_s(T)^{-1} = -2s_1 \left( T\mathcal{Q}_s(T)^{-1} \right) \mathcal{Q}_s(T)^{-1}$$

is the product of functions that are continuous by Lemma 3.1.3.

Finally, we see as in the proof of Lemma 3.1.4 that $T\mathcal{Q}_s(T)^{-1}$ is continuously differentiable in the four real coordinates by considering it as the composition of the two continuously real differentiable functions $s \mapsto (s_0, s_1)$ and $(s_0, s_1) \mapsto T\mathcal{Q}_{s+js_1}(T)^{-1}$ choosing $j_s$ appropriately if $s \in \mathbb{R}$.  □

Let us return now to studying the $S$-spectrum of $T$. As we show in the next theorem, it has properties that are analogue to the properties of the usual spectrum of a complex linear operator.

**Theorem 3.1.6.** *Let $T \in \mathcal{K}(X)$.*

(i) *The $S$-spectrum $\sigma_S(T)$ of $T$ is axially symmetric. It contains the set of right eigenvalues $\sigma_R(T)$ of $T$ and if $X$ has finite dimension, then it equals $\sigma_R(T)$.*

(ii) *The $S$-spectrum $\sigma_S(T)$ is a closed subset of $\mathbb{H}$ and the extended $S$-spectrum $\sigma_{SX}(T)$ is a closed and compact subset of $\mathbb{H}_\infty := \mathbb{H} \cup \{\infty\}$.*

(iii) *If $T$ is bounded, then $\sigma_S(T)$ is nonempty and bounded by the norm of $T$.*

*Proof.* We have $q \in [s]$ if and only if $\mathrm{Re}(q) = \mathrm{Re}(s)$ and $|q| = |s|$. In this case

$$\mathcal{Q}_s(T) = T^2 - 2\mathrm{Re}(s)T + |s|^2 \mathcal{I} = T^2 - 2\mathrm{Re}(q) + |q|^2 \mathcal{I} = \mathcal{Q}_q(T)$$

and $s \in \rho_S(T)$ if and only if $q \in \rho_S(T)$. Hence, $\rho_S(T)$ and $\sigma_S(T)$ are both axially symmetric. Furthermore,

$$\mathcal{Q}_s(T)v = T(Tv - vs) - (Tv - vs)\bar{s}. \tag{3.3}$$

If $s \in \sigma_R(T)$, then there exists a right eigenvector $v \in X \setminus \{0\}$ associated with $s$, that is $Tv - vs = 0$, and hence $\mathcal{Q}_s(T)v = 0$ because of (3.3). Therefore $\mathcal{Q}_s(T)v$ is not invertible and so $s \in \sigma_S(T)$.

If furthermore $X$ is finite-dimensional, then $\mathcal{Q}_s(T)$ is invertible if and only if $\ker \mathcal{Q}_s(T) \neq \{0\}$. Hence, if $s \in \sigma_S(T)$, then there exists $v \in X \setminus \{0\}$ with $\mathcal{Q}_s(T)v = 0$. If $Tv = vs$, then we already see that $s \in \sigma_R(T)$. Otherwise, we see from (3.3) that $\tilde{v} := Tv - vs \neq 0$ is a right eigenvector of $T$ associated with $\bar{s}$. If $s = u + jv$ with $j \in \mathbb{S}$, we can choose $i \in \mathbb{S}$ with $i \perp j$. Then $ji = -ij$ and in turn $si = i\bar{s}$ so that

$$T(\tilde{v}i) = T(\tilde{v})i = (\tilde{v}\bar{s})i = (\tilde{v}i)s.$$

Hence, $\tilde{v}$ is a right eigenvector of $T$ associated with $s$ and so $s \in \sigma_R(T)$. Thus, (i) holds true.

If $s \in \rho_S(T)$, then Theorem 3.1.2 shows that there exists an axially symmetric neighborhood of $s$ that also belongs to $\rho_S(T)$. Hence, $\rho_S(T)$ is an open subset of $\mathbb{H}$ and $\sigma_S(T) = \mathbb{H} \setminus \rho_S(T)$ is in turn a closed subset of $\mathbb{H}$. If $\sigma_S(T)$ is bounded in $\mathbb{H}$, then it is also closed in $\mathbb{H}_\infty$. Hence, if $T$ is bounded, then $\sigma_{SX}(T) = \sigma_S(T)$ is closed in $\mathbb{H}_\infty$. Similarly, if $T$ is unbounded and $\sigma_S(T)$ is bounded, then $\sigma_{SX}(T) = \sigma_S(T) \cup \{\infty\}$ is the union of two closed subsets of $\mathbb{H}_\infty$ and hence bounded itself. Finally, if $\sigma_S(T)$ is unbounded, then $T$ must be unbounded and we find that $\sigma_{SX}(T)$ is closed as

$$\sigma_{SX}(T) = \sigma_S(T) \cup \{\infty\} = \overline{\sigma_S(T)}^{\mathbb{H}_\infty}.$$

Hence, (ii) holds true. Finally, (iii) is part of the statement of Theorem 2.2.11. $\square$

**Definition 3.1.7.** Let $T \in \mathcal{K}(X)$.

(i) We call $s \in \mathbb{R}$ an $S$-eigenvalue of $T$ if $(T - s\mathcal{I})x = 0$ for some $x \in X \setminus \{0\}$.

(ii) Let $s \in \mathbb{H} \setminus \mathbb{R}$. We call $[s]$ an eigensphere of $T$ if $\mathcal{Q}_s(T)x = 0$ for some $x \in X \setminus \{0\}$.

In both cases, the respective vector $x$ is called an $S$-eigenvector associated with the $S$-eigenvalue $s$ (resp. the eigensphere $[s]$).

The next theorem clarifies the relation between the $S$-spectrum and the classical spectrum known from the theory of complex linear operators. The quaternionic Banach space $X$ also carries, for any $j \in \mathbb{S}$, the structure of a Banach space over the complex field $\mathbb{C}_j$. We only have to restrict the multiplication of vectors with quaternionic scalars from the right to the complex plane $\mathbb{C}_j$ and obtain a complex Banach space over $\mathbb{C}_j$. We denote this $\mathbb{C}_j$-complex Banach space by $X_j$. (Observe that $\mathbb{C}_j$-complex multiples of the identity $\mathcal{I}_{X_j}$ on $X_j$ act as $(\lambda \mathcal{I}_{X_j})y = y\lambda$ for $\lambda \in \mathbb{C}_j$ and $y \in X_j$.) Any quaternionic right linear operator $T$ on $X$ is then also a $\mathbb{C}_j$-linear operator on $X_j$. We denote the resolvent set and the spectrum of $T$ as a complex linear operator on $X_j$ by $\rho_{\mathbb{C}_j}(T)$ and $\sigma_{\mathbb{C}_j}(T)$.

**Theorem 3.1.8.** *Let $T \in \mathcal{K}(X)$ and choose $j \in \mathbb{S}$. The spectrum $\sigma_{\mathbb{C}_j}(T)$ of $T$ considered as a closed complex linear operator on $X_j$ equals $\sigma_S(T) \cap \mathbb{C}_j$, i.e.,*

$$\sigma_{\mathbb{C}_j}(T) = \sigma_S(T) \cap \mathbb{C}_j. \tag{3.4}$$

*For any $\lambda$ in the resolvent set $\rho_{\mathbb{C}_j}(T)$ of $T$ as a complex linear operator on $X_j$, the $\mathbb{C}_j$-linear resolvent of $T$ is given by $R_\lambda(T) = \left(\overline{\lambda}\mathcal{I}_{X_j} - T\right)\mathcal{Q}_\lambda(T)^{-1}$, i.e.,*

$$R_\lambda(T)y := \mathcal{Q}_\lambda(T)^{-1}y\overline{\lambda} - T\mathcal{Q}_\lambda(T)^{-1}y. \tag{3.5}$$

*For any $i \in \mathbb{S}$ with $j \perp i$, we, moreover, have*

$$R_{\overline{\lambda}}(T)y = -[R_\lambda(T)(yi)]i. \tag{3.6}$$

*Finally, if $s = u + iv \in \rho_S(T)$, we can set $s_j = u + jv$ and find*

$$Q_s(T)^{-1} = R_{s_j}(T)R_{\overline{s_j}}(T). \tag{3.7}$$

*Proof.* Let $\lambda \in \rho_S(T) \cap \mathbb{C}_j$. The resolvent $(\lambda \mathcal{I}_{X_j} - T)^{-1}$ of $T$ as a $\mathbb{C}_j$-linear operator on $X_j$ is then given by (3.5). Indeed, since $T$ and $\mathcal{Q}_\lambda(T)^{-1}$ commute, we have for $y \in \mathcal{D}(T)$ that

$$
\begin{aligned}
&R_\lambda(T)(\lambda \mathcal{I}_{X_j} - T)y \\
&= (\overline{\lambda}\mathcal{I}_{X_j} - T)\mathcal{Q}_\lambda(T)^{-1}(y\lambda - Ty) \\
&= (\overline{\lambda}\mathcal{I}_{X_j} - T)\left(\mathcal{Q}_\lambda(T)^{-1}y\lambda - T\mathcal{Q}_\lambda(T)^{-1}y\right) \\
&= \mathcal{Q}_\lambda(T)^{-1}y\lambda\overline{\lambda} - T\mathcal{Q}_\lambda(T)^{-1}y\overline{\lambda} - T\mathcal{Q}_\lambda(T)^{-1}y\lambda + T^2\mathcal{Q}_\lambda(T)^{-1}y \\
&= (|\lambda|^2\mathcal{I}_{X_j} - 2\lambda_0 T + T^2)\mathcal{Q}_\lambda(T)^{-1}y = y.
\end{aligned}
$$

Similarly, for $y \in X_j$, we have

$$
\begin{aligned}
&(\lambda \mathcal{I}_{X_j} - T)R_\lambda(T)y \\
&= (\lambda \mathcal{I}_{X_j} - T)\left(\mathcal{Q}_\lambda(T)^{-1}y\overline{\lambda} - T\mathcal{Q}_\lambda(T)^{-1}y\right) \\
&= \mathcal{Q}_\lambda(T)^{-1}y\overline{\lambda}\lambda - T\mathcal{Q}_\lambda(T)^{-1}y\lambda - T\mathcal{Q}_\lambda(T)^{-1}y\overline{\lambda} + T^2\mathcal{Q}_\lambda(T)^{-1}y \\
&= (|\lambda|^2\mathcal{I}_{X_j} - 2\lambda_0 T + T^2)\mathcal{Q}_\lambda(T)^{-1}y = y.
\end{aligned}
$$

Since $\mathcal{Q}_\lambda(T)^{-1}$ maps $X_j$ to $\mathcal{D}(T^2) \subset \mathcal{D}(T)$, we find that the operator $R_\lambda(T) = (\lambda \mathcal{I}_{X_j} - T)\mathcal{Q}_\lambda(T)^{-1}$ is bounded and so $\lambda$ belongs to the resolvent set $\rho_{\mathbb{C}_j}(T)$ of $T$ considered as a $\mathbb{C}_j$-linear operator on $X_j$. Hence, $\rho_S(T) \cap \mathbb{C}_j \subset \rho_{\mathbb{C}_j}(T)$ and in turn $\sigma_{\mathbb{C}_j}(T) \subset \sigma_S(T) \cap \mathbb{C}_j$. Together with the axial symmetry of the $S$-spectrum, this further implies

$$\sigma_{\mathbb{C}_j}(T) \cup \overline{\sigma_{\mathbb{C}_j}(T)} \subset (\sigma_S(T) \cap \mathbb{C}_j) \cup (\overline{\sigma_S(T) \cap \mathbb{C}_j}) = \sigma_S(T) \cap \mathbb{C}_j, \tag{3.8}$$

where $\overline{A} = \{\overline{z} : z \in A\}$.

If $\lambda$ and $\overline{\lambda}$ both belong to $\rho_{\mathbb{C}_j}(T)$, then $[\lambda] \subset \rho_S(T)$ because

$$
\begin{aligned}
&(\lambda \mathcal{I}_{X_j} - T)(\overline{\lambda}\mathcal{I}_{X_j} - T)y \\
&= (y\overline{\lambda})\lambda - (Ty)\lambda - T(y\overline{\lambda}) + T^2 y \\
&= (T^2 - 2\lambda_0 T + |\lambda|^2)y
\end{aligned}
$$

and hence $\mathcal{Q}_\lambda(T)^{-1} = R_\lambda(T)R_{\overline{\lambda}}(T) \in \mathcal{B}(X)$. Thus, $\rho_S(T) \cap \mathbb{C}_j \supset \rho_{\mathbb{C}_j}(T) \cap \overline{\rho_{\mathbb{C}_j}(T)}$ and in turn

$$\sigma_S(T) \cap \mathbb{C}_j \subset \sigma_{\mathbb{C}_j}(T) \cup \overline{\sigma_{\mathbb{C}_j}(T)}. \tag{3.9}$$

The two relations (3.8) and (3.9) yield together

$$\sigma_S(T) \cap \mathbb{C}_j = \sigma_{\mathbb{C}_j}(T) \cup \overline{\sigma_{\mathbb{C}_j}(T)}. \tag{3.10}$$

What remains to show is that $\rho_{\mathbb{C}_j}(T)$ and $\sigma_{\mathbb{C}_j}(T)$ are symmetric with respect to the real axis, which then implies

$$\sigma_S(T) \cap \mathbb{C}_j = \sigma_{\mathbb{C}_j}(T) \cup \overline{\sigma_{\mathbb{C}_j}(T)} = \sigma_{\mathbb{C}_j}(T). \tag{3.11}$$

Let $\lambda \in \rho_{\mathbb{C}_j}(T)$ and choose $i \in \mathbb{S}$ with $j \perp i$. We show that $R_{\overline{\lambda}}(T)$ equals the mapping $Ay := -[R_\lambda(T)(yi)]\,i$. As $\lambda i = i\overline{\lambda}$ and $i\lambda = \overline{\lambda}i$, we have for $y \in \mathcal{D}(T)$ that

$$\begin{aligned}
A\left(\overline{\lambda}\mathcal{I}_{X_j} - T\right)y &= A\left(y\overline{\lambda} - Ty\right) \\
&= -\left[R_\lambda(T)\left((y\overline{\lambda})i - (Ty)i\right)\right]i \\
&= -\left[R_\lambda(T)((yi)\lambda - T(yi))\right]i \\
&= -\left[R_\lambda(T)(\lambda\mathcal{I}_{X_j} - T)(yi)\right]i = -yii = y.
\end{aligned}$$

Similarly, for arbitrary $y \in X_j = X$, we have

$$\begin{aligned}
\left(\overline{\lambda}\mathcal{I}_{X_j} - T\right)Ay &= (Ay)\overline{\lambda} - T(Ay) \\
&= -[R_\lambda(T)(yi)]\,i\overline{\lambda} + T\left([R_\lambda(T)(yi)]\,i\right) \\
&= -[R_\lambda(T)(yi)\lambda - T(R_\lambda(T)(yi))]\,i \\
&= -\left[(\lambda\mathcal{I}_{X_j} - T)R_\lambda(T)(yi)\right]i = -yii = y.
\end{aligned}$$

Hence, if $\lambda \in \rho_{\mathbb{C}_j}(T)$, then $R_{\overline{\lambda}}(T) = -[R_\lambda(T)(yi)]\,i$ such that in particular $\overline{\lambda} \in \rho_{\mathbb{C}_j}(T)$. Consequently $\rho_{\mathbb{C}_j}(T)$ and in turn also $\sigma_{\mathbb{C}_j}(T)$ are symmetric with respect to the real axis such that (3.11) holds true. $\qquad\square$

**Remark 3.1.1.** The relations (3.10) and (3.7) had been observed in [159]. Also the relation $R_\lambda(T)R_{\overline{\lambda}}(T) = \mathcal{Q}_\lambda(T)^{-1}$, which is a consequence of (3.5), was understood in that paper. The complete statement, in particular the fact that for a quaternionic linear operator $T$ always $\sigma_{\mathbb{C}_j}(T) = \sigma_{\mathbb{C}_i}(T)$ due to (3.6), was finally established in [131]. For unitary operators, this symmetry was already understood in [196], but the correct notion of spectrum for quaternionic operators had not yet been developed so it was impossible to see the full picture.

## 3.2 The *S*-resolvent of a closed operator

For closed operators, the definition of the *S*-resolvent operators needs a little modification. If we define the left *S*-resolvent operator as in the case of bounded operators, we obtain

$$S_L^{-1}(s,T)x := -\mathcal{Q}_s(T)^{-1}(T - \bar{s}\mathcal{I})x, \tag{3.12}$$

which is only defined for $x \in \mathcal{D}(T)$ and not on all of $X$. However, for $x \in \mathcal{D}(T)$, we have $\mathcal{Q}_s(T)^{-1}Tx = T\mathcal{Q}_s(T)^{-1}x$ and so we can commute $T$ and $\mathcal{Q}_s(T)^{-1}$ in order to obtain an operator that is defined on all of $X$.

**Definition 3.2.1** (The $S$-resolvent operators of a closed operator). Let $T \in \mathcal{K}(X)$. For $s \in \rho_S(T)$, we define the left $S$-resolvent operator of $T$ at $s$ as

$$S_L^{-1}(s,T)x := \mathcal{Q}_s(T)^{-1}\bar{s}x - T\mathcal{Q}_s(T)^{-1}x, \quad \text{for all } x \in X, \tag{3.13}$$

and the right $S$-resolvent operator of $T$ at $s$ as

$$S_R^{-1}(s,T)x := -(T - \mathcal{I}s)\mathcal{Q}_s(T)^{-1}x, \quad \text{for all } x \in X. \tag{3.14}$$

**Remark 3.2.1.** For $s \in \rho_S(T)$, the operator $\mathcal{Q}_s(T)^{-1}$ maps $X$ to $\mathcal{D}(T^2)$. Hence, $T\mathcal{Q}_s(T)^{-1}$ is a bounded operator and so $S_L^{-1}(s,T)$ and $S_R^{-1}(s,T)$ are bounded, too. The converse, however, is not necessarily true. As the next example shows, there might exist points $s \in \mathbb{H}$ that belong to $\sigma_S(T)$ even though $S_L^{-1}(s,T)$ or $S_R^{-1}(s,T)$ are bounded operators. In order to determine the $S$-spectrum of an operator $T$ one therefore always has to work with the operator $\mathcal{Q}_s(T)^{-1}$ even though, as we will see later on, the $S$-resolvent $S_L^{-1}(s,T)$ and $S_R^{-1}(s,T)$ and not the pseudo-resolvent $\mathcal{Q}_s(T)^{-1}$ appear in the $S$-functional calculus.

If a sphere $[s] = u + \mathbb{S}v$ belongs to $\sigma_S(T)$, then the $S$-resolvents can be bounded at most at one point in $[s]$. We will prove in the following that the right $S$-resolvent is left slice hyperholomorphic in $s$. If $S_R^{-1}(s_k,T)$ with $s_k = u + kv$ and $S_R^{-1}(s_i,T)$ with $s_i = u + iv$ are bounded, then (2.11) in Corollary 2.1.8 implies for any $s_j = u + jv$ with $j \in \mathbb{S} \setminus \{i,k\}$ that

$$\left\| S_R^{-1}(s_j,T) \right\| \leq \left| (i-k)^{-1}i + j(k-i)^{-1} \right| \left\| S_R^{-1}(s_i,T) \right\|$$
$$+ \left| (k-i)^{-1}k + j(k-i)^{-1} \right| \left\| S_R^{-1}(s_k,T) \right\| < +\infty.$$

Hence, if $S_R^{-1}(s,T)$ is bounded at two points in $[s]$, then it is bounded at any $s \in [s]$. The estimates that we will show in Lemma 3.2.8 imply then that $[s] \subset \rho_S(T)$. For the left $S$-resolvent, we can argue similarly.

**Example 3.2.2.** Let $\ell^2(\mathbb{H})$ be the quaternionic Hilbert space of all square-summable sequences in $\mathbb{H}$ and let $i \in \mathbb{S}$. On this space, we consider the operator

$$T : \begin{cases} \ell^2(\mathbb{H}) & \to & \ell^2(\mathbb{H}) \\ (a_n)_{n \in \mathbb{N}} & \mapsto & \left(\frac{n-1}{n}ia_n\right)_{n \in \mathbb{N}}. \end{cases}$$

This operator is obviously bounded with $\|T\| = 1$ and if $e_n = (\delta_{n,m})_{m,\in\mathbb{N}}$, where $\delta_{n,m} = 1$ if $m = n$ and $\delta_{n,m} = 0$ if $m \neq n$, then $Te_n = e_n\frac{n-1}{n}i$. Hence, we conclude from Theorem 3.1.6 that

$$\sigma_S(T) \supset \overline{\bigcup_{n \in \mathbb{N}} \frac{n-1}{n}\mathbb{S}} = \mathbb{S} \cup \bigcup_{n \in \mathbb{N}} \frac{n-1}{n}\mathbb{S}. \tag{3.15}$$

Straightforward computations show, that we even have equality in (3.15) since $Q_s(T)^{-1}$ is bounded for any $s \notin \mathbb{S} \cup \bigcup_{n \in \mathbb{N}} \frac{n-1}{n}\mathbb{S}$.

Let us now consider the point $-j$, which obviously belongs to $\sigma_S(T)$. The pseudo-resolvent of $T$ at $-j$ applied to $(a_n)_{n \in \mathbb{N}} \in \ell^2(\mathbb{H})$ is

$$Q_{-j}(T)(a_n)_{n \in \mathbb{N}} = (T^2 + \mathcal{I})^{-1}(a_n)_{n \in \mathbb{N}} = \left(\frac{n^2}{2n-1}a_n\right)_{n \in \mathbb{N}}.$$

As expected, this is an unbounded operator on $\ell^2(\mathbb{H})$, because $\frac{n^2}{2n-1} \to +\infty$. The left $S$-resolvent at $-i$ on the other hand is

$$S_L^{-1}(-i, T)(a_n)_{n \in \mathbb{N}} = Q_{-i}(T)^{-1}(-i\mathcal{I} - T)(a_n)_{n \in \mathbb{N}} = \left(\frac{n}{2n-1}ia_n\right)_{n \in \mathbb{N}},$$

and this is a bounded operator because

$$\|S_L^{-1}(-i, T)\| = \sup_{n \in \mathbb{N}} \left|\frac{n}{2n-1}i\right| < +\infty.$$

Hence, $S_L^{-1}(-i, T)$ is bounded even though $-i \notin \rho_S(T)$.

A second difference between the left and the right $S$-resolvent operators is that the right $S$-resolvent equation only holds true on $\mathcal{D}(T)$.

**Theorem 3.2.3** (The $S$-resolvent equations). *Let $T \in \mathcal{K}(X)$. For $s \in \rho_S(T)$, the left $S$-resolvent operator satisfies the identity*

$$S_L^{-1}(s, T)sx - TS_L^{-1}(s, T)x = x, \quad \text{for all } x \in X. \tag{3.16}$$

*Moreover, the right $S$-resolvent operator satisfies the identity*

$$sS_R^{-1}(s, T)x - S_R^{-1}(s, T)Tx = x, \quad \text{for all } x \in \mathcal{D}(T). \tag{3.17}$$

*Proof.* We have for $x \in \mathcal{D}(T)$ that

$$\begin{aligned}
sS_R^{-1}&(s, T)x - S_R^{-1}(s, T)Tx \\
&= -s(T - \mathcal{I}\bar{s})Q_s(T)^{-1}x + (T - \mathcal{I}\bar{s})Q_s(T)^{-1}Tx \\
&= (-sT + |s|^2\mathcal{I})Q_s(T)^{-1}x + (T^2 - \bar{s}T)Q_s(T)^{-1}x \\
&= (T^2 - 2\mathrm{Re}(s)T + |s|^2\mathcal{I})Q_s(T)^{-1}x = x.
\end{aligned}$$

Similar computations show (3.16). $\square$

**Remark 3.2.2.** We can extend (3.17) to an equation that holds on the entire space $X$, similarly to how we could extend (3.12) to a bounded operator on the entire space $X$. This equation is

$$sS_R^{-1}(s, T)x + (T^2 - \bar{s}T)Q_s(T)^{-1}x = x, \quad \text{for all } x \in X.$$

**Theorem 3.2.4** (*S-resolvent equation*)*. Let* $T \in \mathcal{K}(X)$. *If* $s, q \in \rho_S(T)$ *with* $s \notin [q]$, *then*

$$
\begin{aligned}
S_R^{-1}(s,T)S_L^{-1}(q,T) = [[S_R^{-1}(s,T) &- S_L^{-1}(q,T)]q \\
&- \bar{s}[S_R^{-1}(s,T) - S_L^{-1}(q,T)]](q^2 - 2\mathrm{Re}(s)q + |s|^2)^{-1}.
\end{aligned} \quad (3.18)
$$

*Proof.* As in the case of bounded operators, the $S$-resolvent equation is deduced from the left and the right $S$-resolvent equation. However, we have to pay attention to being consistent with the domains of definition of every operator that appears in the following. We show that, for every $x \in X$, one has

$$
\begin{aligned}
S_R^{-1}(s,T)&S_L^{-1}(q,T)(q^2 - 2s_0 q + |s|^2)x \\
&= [S_R^{-1}(s,T) - S_L^{-1}(q,T)]qx - \bar{s}[S_R^{-1}(s,T) - S_L^{-1}(q,T)]x. \quad (3.19)
\end{aligned}
$$

We then obtain (3.18) by replacing $x$ by $(q^2 - 2s_0 q + |s|^2)^{-1}x$. For $w \in X$, the left $S$-resolvent equation (3.16) implies

$$
S_R^{-1}(s,T)S_L^{-1}(q,T)qw = S_R^{-1}(s,T)TS_L^{-1}(q,T)w + S_R^{-1}(s,T)w.
$$

The pseudo-resolvent $\mathcal{Q}_s(T)^{-1}$ maps $X$ onto $\mathcal{D}(T^2)$. Therefore the left $S$-resolvent operator $S_L^{-1}(s,T) = \mathcal{Q}_s(T)^{-1}\bar{s} - T\mathcal{Q}_s(T)^{-1}$ maps $X$ to $\mathcal{D}(T)$ and so $S_L^{-1}(q,T)w \in \mathcal{D}(T)$. The right $S$-resolvent equation (3.17) yields

$$
\begin{aligned}
S_R^{-1}(s,T)&S_L^{-1}(q,T)qw \\
&= sS_R^{-1}(s,T)S_L^{-1}(q,T)w - S_L^{-1}(q,T)w + S_R^{-1}(s,T)w.
\end{aligned} \quad (3.20)
$$

If we apply this identity with $w = qx$ we get

$$
\begin{aligned}
S_R^{-1}(s,T)&S_L^{-1}(q,T)(q^2 - 2s_0 q + |s|^2)x \\
&= S_R^{-1}(s,T)S_L^{-1}(q,T)q^2 x - 2s_0 S_R^{-1}(s,T)S_L^{-1}(q,T)qx \\
&\quad + |s|^2 S_R^{-1}(s,T)S_L^{-1}(q,T)x \\
&= sS_R^{-1}(s,T)S_L^{-1}(q,T)qx - S_L^{-1}(q,T)qx + S_R^{-1}(s,T)qx \\
&\quad - 2s_0 S_R^{-1}(s,T)S_L^{-1}(q,T)qx + |s|^2 S_R^{-1}(s,T)S_L^{-1}(q,T)x.
\end{aligned}
$$

Applying identity (3.20) again with $w = x$ gives

$$
\begin{aligned}
S_R^{-1}(s,T)&S_L^{-1}(q,T)(q^2 - 2s_0 q + |s|^2)x \\
&= s^2 S_R^{-1}(s,T)S_L^{-1}(q,T)x - sS_L^{-1}(q,T)x + sS_R^{-1}(s,T)x \\
&\quad - S_L^{-1}(q,T)qx + S_R^{-1}(s,T)qx \\
&\quad - 2s_0 sS_R^{-1}(s,T)S_L^{-1}(q,T)x + 2s_0 S_L^{-1}(q,T)x - 2s_0 S_R^{-1}(s,T)x \\
&\quad + |s|^2 S_R^{-1}(s,T)S_L^{-1}(q,T)x \\
&= (s^2 - 2s_0 s + |s|^2)S_R^{-1}(s,T)S_L^{-1}(q,T)x \\
&\quad - (2s_0 - s)[S_R^{-1}(s,T)x - S_L^{-1}(q,T)x] \\
&\quad + [S_R^{-1}(s,T) - S_L^{-1}(q,T)]qx.
\end{aligned}
$$

The identity $2s_0 = s + \bar{s}$ implies $s^2 - 2s_0 s + |s|^2 = 0$ and $2s_0 - s = \bar{s}$ and hence we obtain the desired equation (3.19). □

We want to show now the slice hyperholomorphicity of the $S$-resolvent operators of a closed quaternionic operator. The fact that they are differentiable follows from the series expansion of the pseudo-resolvent that was found in Theorem 3.1.2.

**Lemma 3.2.5.** *Let $T \in \mathcal{K}(X)$ and $s \in \rho_S(T)$. The left and the right S-resolvents of $T$ are continuously real differentiable.*

*Proof.* The $S$-resolvents are sums of functions that are continuously real differentiable by Lemma 3.1.4 and Lemma 3.1.5 and hence continuously real differentiable themselves. □

**Theorem 3.2.6.** *Let $T \in \mathcal{K}(X)$. The left S-resolvent $S_L^{-1}(s, T)$ is right slice hyperholomorphic and the right S-resolvent $S_R^{-1}(s, T)$ is left slice hyperholomorphic in the variable $s$.*

*Proof.* We consider only the case of the left $S$-resolvent, the other one works with analogous arguments. We have

$$S_L^{-1}(s, T) = \alpha(s_0, s_1) + j_s \beta(s_0, s_1)$$

with

$$\alpha(s_0, s_1) = \mathcal{Q}_s(T)^{-1} s_0 - T \mathcal{Q}_s(T) \quad \text{and} \quad \beta(s_0, s_1) = -\mathcal{Q}_s(T)^{-1} s_1.$$

Obviously $\alpha$ and $\beta$ satisfy the compatibility condition (2.4) and hence $S_L^{-1}(s, T)$ is a right slice function in $s$.

Applying Lemma 3.1.4 and Lemma 3.1.5, we have

$$\frac{\partial}{\partial s_0} S_L^{-1}(s, T) = \frac{\partial}{\partial s_0} \mathcal{Q}_s(T)^{-1} \bar{s} - \frac{\partial}{\partial s_0} T \mathcal{Q}_s(T)^{-1}$$

$$= (2T - 2s_0 \mathcal{I}) \mathcal{Q}_s(T)^{-2} \bar{s} + \mathcal{Q}_s(T)^{-1} - \left(2T^2 - 2s_0 T\right) \mathcal{Q}_s(T)^{-2}$$

$$= (2T - 2s_0 \mathcal{I}) \mathcal{Q}_s(T)^{-2} \bar{s} + \left(-T^2 + |s|^2 \mathcal{I}\right) \mathcal{Q}_s(T)^{-2}.$$

Since $s_0$ and $|s|^2$ are real, they commute with $\mathcal{Q}_s(T)^{-2}$. If we apply the identities $2s_0 = s + \bar{s}$ and $|s|^2 = s\bar{s}$, we obtain

$$\frac{\partial}{\partial s_0} S_L^{-1}(s, T) = -T^2 \mathcal{Q}_s(T)^{-2} + 2T \mathcal{Q}_s(T)^{-2} \bar{s} - \mathcal{Q}_s(T)^{-2} \bar{s}^2.$$

For the partial derivative with respect to $s_1$, we obtain

$$\frac{\partial}{\partial s_1} S_L^{-1}(s, T) = \frac{\partial}{\partial s_1} \mathcal{Q}_s(T)^{-1} \bar{s} - \frac{\partial}{\partial s_1} T \mathcal{Q}_s(T)^{-1}$$

$$= -2s_1 \mathcal{Q}_s(T)^{-2} \bar{s} - \mathcal{Q}_s(T)^{-1} j_s + 2s_1 T \mathcal{Q}_s(T)^{-2}$$

$$= -2s_1 \mathcal{Q}_s(T)^{-2} \bar{s} - (T^2 - 2s_0 T + |s|^2 \mathcal{I}) \mathcal{Q}_s(T)^{-2} j_s + 2s_1 T \mathcal{Q}_s(T)^{-2}.$$

We can again commute $2s_0$, $2s_1$ and $|s|^2$ with $\mathcal{Q}_s(T)^{-1}$ because they are real. By exploiting the identities $2s_0 = s + \bar{s}$, $-2s_1 = (s - \bar{s})j_s$ and $|s|^2 = s\bar{s}$, we obtain the formula

$$\frac{\partial}{\partial s_1} S_L^{-1}(s, T) = \left(-T^2 \mathcal{Q}_s(T)^{-2} + 2T\mathcal{Q}_s(T)^{-2}\bar{s} - \mathcal{Q}_s(T)^{-2}(T)\bar{s}^2\right) j_s.$$

So the function $s \mapsto S_L^{-1}(s, T)$ is right slice hyperholomorphic as

$$\frac{1}{2}\left(\frac{\partial}{\partial s_0} S_L^{-1}(s, T) + \frac{\partial}{\partial s_1} S_L^{-1}(s, T)j_s\right) = 0. \qquad \square$$

In Section 8.3 we will need the fact that the $S$-resolvent set is the maximal domain of slice hyperholomorphicity of the $S$-resolvent operators such that they do not have a slice hyperholomorphic continuation. In the complex case this is guaranteed by the well-known estimate

$$\|R(z, A)\| \geq \frac{1}{\operatorname{dist}(z, \sigma(A))}, \tag{3.21}$$

where $R(z, A)$ denotes the resolvent operator and $\sigma(A)$ the spectrum of the complex linear operator $A$. This estimate assures that $\|R(z, A)\| \to +\infty$ as $z$ approaches $\sigma(A)$ and in turn that the resolvent does not have any holomorphic continuation to a larger domain, see [177, 191].

In the quaternionic setting, an estimate similar to (3.21) cannot hold true. We can for example consider the operator $T = \lambda\mathcal{I}$ on a two-sided Banach space $X$ for some $\lambda = \lambda_0 + j_\lambda\lambda_1$ with $\lambda_1 > 0$. Its $S$-spectrum $\sigma_S(T)$ coincides with the sphere $[\lambda]$ associated with $\lambda$ and its left $S$-resolvent is

$$S_L^{-1}(s, T) = (\lambda^2 - 2s_0\lambda + |s|^2)^{-1}(\bar{s} - \lambda)\mathcal{I}.$$

If $s \in \mathbb{C}_{j_\lambda}$, then $\lambda$ and $s$ commute so that the left $S$-resolvent reduces to

$$S_L^{-1}(s, T) = (s - \lambda)^{-1}\mathcal{I}$$

with $\|S_L^{-1}(s, T)\| = 1/|s - \lambda|$. If $s$ tends to $\bar{\lambda}$ in $\mathbb{C}_{j_\lambda}$, then $\operatorname{dist}(s, \sigma_S(T)) \to 0$ because $\bar{\lambda} \in \sigma_S(T)$. But at the same time

$$\|S_L^{-1}(s, T)\| \to 1/|\lambda - \bar{\lambda}| = 1/(2\lambda_1) < +\infty.$$

Nevertheless, although (3.21) does not have a pointwise counterpart in the quaternionic setting, we can show that the norms of the $S$-resolvents explode near the $S$-spectrum. As it happens often in quaternionic operator theory, this requires that we work with spectral spheres of associated quaternions instead of single spectral values.

**Lemma 3.2.7.** *Let $T \in \mathcal{K}(X)$ and $s \in \rho_S(T)$. Then*

$$\left\| \mathcal{Q}_s(T)^{-1} \right\| + \left\| T\mathcal{Q}_s(T)^{-1} \right\| \geq \frac{1}{d_S(s, \sigma_S(T))}, \tag{3.22}$$

*where*

$$d_S(s, \sigma_S(T)) = \inf_{q \in \sigma_S(T)} d_S(s, q)$$

*and $d_S(s, x)$ is defined as in Lemma 3.1.2.*

*Proof.* Set $C_s := \left\| \mathcal{Q}_s(T)^{-1} \right\| + \left\| T\mathcal{Q}_s(T)^{-1} \right\|$. If $d_S(s, q) < 1/C_s$, then $x \in \rho_S(T)$ by Lemma 3.1.2. Thus, $d_S(s, q) \geq 1/C_s$ for any $q \in \sigma_S(T)$. If we take the infimum over all $q \in \sigma_S(T)$, this inequality still holds true and we obtain $d_S(s, \sigma_S(T)) \geq 1/C_s$, which is equivalent to (3.22). $\qquad\square$

**Lemma 3.2.8.** *Let $T \in \mathcal{K}(X)$ and $s \in \rho_S(T)$. Then*

$$\sqrt{2} \left\| \mathcal{Q}_s(T)^{-1} \right\| \leq \left\| S_L^{-1}(s, T) \right\| + \left\| S_L^{-1}(\bar{s}, T) \right\|$$

*and in turn*

$$\sqrt{\left\| \mathcal{Q}_s(T)^{-1} \right\|} \leq \sqrt{2} \sup_{s_j \in [s]} \left\| S_L^{-1}(s_j, T) \right\|.$$

*Analogous estimates hold for the right S-resolvent operator.*

*Proof.* Observe that $\mathcal{Q}_s(T)^{-1} = \mathcal{Q}_{\bar{s}}(T)^{-1}$ for $s \in \rho_S(T)$. Because of $2s_0 = s + \bar{s}$, we have

$$\begin{aligned}
&S_L^{-1}(s, T)S_L^{-1}(s, T) + S_L^{-1}(s, T)S_L^{-1}(\bar{s}, T) \\
&= \left( \mathcal{Q}_s(T)^{-1}\bar{s} - T\mathcal{Q}_s(T)^{-1} \right) \left( \mathcal{Q}_s(T)^{-1}\bar{s} - T\mathcal{Q}_s(T)^{-1} \right) \\
&\quad + \left( \mathcal{Q}_s(T)^{-1}\bar{s} - T\mathcal{Q}_s(T)^{-1} \right) \left( \mathcal{Q}_s(T)^{-1}s - T\mathcal{Q}_s(T)^{-1} \right) \\
&= \left( \mathcal{Q}_s(T)^{-1}\bar{s} - T\mathcal{Q}_s(T)^{-1} \right) 2\left( s_0\mathcal{I} - T \right) \mathcal{Q}_s(T)^{-1}
\end{aligned}$$

and similarly

$$\begin{aligned}
&S_L^{-1}(\bar{s}, T)S_L^{-1}(s, T) + S_L^{-1}(\bar{s}, T)S_L^{-1}(\bar{s}, T) \\
&= \left( \mathcal{Q}_s(T)^{-1}s - T\mathcal{Q}_s(T)^{-1} \right) 2\left( s_0\mathcal{I} - T \right) \mathcal{Q}_s(T)^{-1}.
\end{aligned}$$

Therefore

$$\begin{aligned}
&S_L^{-1}(s, T)S_L^{-1}(s, T) + S_L^{-1}(s, T)S_L^{-1}(\bar{s}, T) \\
&\quad + S_L^{-1}(\bar{s}, T)S_L^{-1}(s, T) + S_L^{-1}(\bar{s}, T)S_L^{-1}(\bar{s}, T) \\
&= \left( \mathcal{Q}_s(T)^{-1}\bar{s} - T\mathcal{Q}_s(T)^{-1} \right) 2\left( s_0\mathcal{I} - T \right) \mathcal{Q}_s(T)^{-1} \\
&\quad + \left( \mathcal{Q}_s(T)^{-1}s - T\mathcal{Q}_s(T)^{-1} \right) 2\left( s_0\mathcal{I} - T \right) \mathcal{Q}_s(T)^{-1}
\end{aligned}$$

$$= 2\,(s_0\mathcal{I} - T)\,\mathcal{Q}_s(T)^{-1}2\,(s_0\mathcal{I} - T)\,\mathcal{Q}_s(T)^{-1}$$
$$= 4(T^2 - 2s_0T + s_0^2\mathcal{I})\mathcal{Q}_s(T)^{-2} = 4\mathcal{Q}_s(T)^{-1} - 4s_1^2\mathcal{Q}_s(T)^{-2},$$

which can be rewritten as

$$4\mathcal{Q}_s(T)^{-1} = S_L^{-1}(s,T)S_L^{-1}(s,T) + S_L^{-1}(s,T)S_L^{-1}(\bar{s},T)$$
$$+ S_L^{-1}(\bar{s},T)S_L^{-1}(s,T) + S_L^{-1}(\bar{s},T)S_L^{-1}(\bar{s},T) + 4s_1^2\mathcal{Q}_s(T)^{-2}.$$

Thus, we can estimate

$$4\left\|\mathcal{Q}_s(T)^{-1}\right\|$$
$$= \left\|S_L^{-1}(s,T)\right\|\left\|S_L^{-1}(s,T)\right\| + \left\|S_L^{-1}(s,T)\right\|\left\|S_L^{-1}(\bar{s},T)\right\|$$
$$+ \left\|S_L^{-1}(\bar{s},T)\right\|\left\|S_L^{-1}(s,T)\right\| + \left\|S_L^{-1}(\bar{s},T)\right\|\left\|S_L^{-1}(\bar{s},T)\right\|$$
$$+ 4\left\|s_1^2\mathcal{Q}_s(T)^{-2}\right\|$$
$$= \left(\left\|S_L^{-1}(s,T)\right\| + \left\|S_L^{-1}(\bar{s},T)\right\|\right)^2 + \left\|2s_1\mathcal{Q}_s(T)^{-1}\right\|\left\|2s_1\mathcal{Q}_s(T)^{-1}\right\|. \tag{3.23}$$

Finally observe that

$$2\mathcal{Q}_s(T)^{-1}s_1j_s = T\mathcal{Q}_s(T)^{-1} - \mathcal{Q}_s(T)^{-1}(s_0 - j_ss_1)$$
$$- \left(T\mathcal{Q}_s(T)^{-1} - \mathcal{Q}_s(T)^{-1}(s_0 + j_ss_1)\right) = S_L^{-1}(s,T) - S_L^{-1}(\bar{s},T)$$

and hence

$$\left\|2s_1\mathcal{Q}_s(T)^{-1}\right\| = \left\|2\mathcal{Q}_s(T)^{-1}s_1j_s\right\| \leq \left\|S_L^{-1}(s,T)\right\| + \left\|S_L^{-1}(\bar{s},T)\right\|.$$

Combining this estimate with (3.23), we finally obtain

$$2\left\|\mathcal{Q}_s(T)^{-1}\right\| \leq \left(\left\|S_L^{-1}(s,T)\right\| + \left\|S_L^{-1}(\bar{s},T)\right\|\right)^2$$

and hence the statement for the left $S$-resolvent operator. The estimates for the right $S$-resolvent operator can be shown with similar computations. $\quad\square$

From the above results we get:

**Lemma 3.2.9.** *Let $T \in \mathcal{K}(X)$. If $(s_n)_{n\in\mathbb{N}}$ is a bounded sequence in $\rho_S(T)$ with*

$$\lim_{n\to\infty} \mathrm{dist}(s_n, \sigma_S(T)) = 0,$$

*then*

$$\lim_{n\to\infty} \sup_{s\in[s_n]} \left\|S_L^{-1}(s,T)\right\| = +\infty \quad \text{and} \quad \lim_{n\to\infty} \sup_{s\in[s_n]} \left\|S_R^{-1}(s,T)\right\| = +\infty.$$

*Proof.* First of all observe that $\text{dist}(s_n, \sigma_S(T)) \to 0$ if and only if $d_S(s_n, \sigma_S(T)) \to 0$ because $\sigma_S(T)$ is axially symmetric. Indeed, for any $n \in \mathbb{N}$, there exits $x_n \in \sigma_S(T)$ such that

$$|s_n - x_n| < \text{dist}(s_n, \sigma_S(T)) + 1/n.$$

If $\text{dist}(s_n, \sigma_S(T)) \to 0$, then $|s_n - x_n| \to 0$ and hence $|s_{n,0} - x_{n,0}| \to 0$. Since the sequence $s_n$ is bounded, the sequence $x_n$ is bounded too and we also have

$$\left| |s_n|^2 - |x_n|^2 \right| \le |s_n||\overline{s_n} - \overline{x_n}| + |s_n - x_n||\overline{x_n}| \to 0$$

and in turn

$$0 < d_S(s_n, \sigma_S(T)) \le d_S(s_n, x_n) = \max\left\{ |s_{n,0} - x_{n,0}|, \left||s_n|^2 - |x_n|^2\right| \right\} \longrightarrow 0.$$

If on the other hand $d_S(s_n, \sigma_S(T))$ tends to zero, then there exists a sequence $(x_n)_{n\in\mathbb{N}}$ in $\sigma_S(T)$ such that

$$d_S(s_n, x_n) < d_S(s_n, \sigma_S(T)) + 1/n$$

and in turn $d_S(s_n, x_n) \to 0$. Since $\sigma_S(T)$ is axially symmetric and $d(s_n, x_{n,j}) = d(s_n, x_n)$ for any $x_{n,j} \in [x_n]$, we can, moreover, assume that $j_{x_n} = j_{s_n}$. Then

$$0 \le |s_{n,0} - x_{n,0}| \le d_S(s_n, x_n) \to 0.$$

Since $s_n$ and in turn also $x_n$ are bounded, this implies $|s_{n,0}^2 - x_{n,0}^2| \to 0$, from which we deduce that also $|s_{n,1}^2 - x_{n,1}^2| \to 0$ because

$$0 \le \left| s_{n,0}^2 - x_{n,0}^2 + s_{n,1}^2 - x_{n,1}^2 \right| = \left||s_n|^2 - |x_n|^2\right| \le d_S(s_n, x_n) \to 0.$$

Since $s_{n,1} \ge 0$ and $x_{n,1} \ge 0$, we conclude that $s_{n,1} - x_{n,1} \to 0$ and, since $j_s = j_x$, also

$$0 < \text{dist}(s_n, \sigma_S(T)) \le |s_n - x_n| = \sqrt{(s_{n,0} - x_{n,0})^2 + (s_{n,1} - x_{n,1})^2} \to 0.$$

Now assume that $s_n \in \rho_S(T)$ with $\text{dist}(s_n, \sigma_S(T)) \to 0$. By the above considerations and (3.22), we have

$$\|\mathcal{Q}_{s_n}(T)^{-1}\| + \|T\mathcal{Q}_{s_n}(T)^{-1}\| \to +\infty. \tag{3.24}$$

We show now that every subsequence $(s_{n_k})_{k\in\mathbb{N}}$ has a subsequence $(s_{n_{k_j}})_{j\in\mathbb{N}}$ such that

$$\lim_{j \to +\infty} \sup_{s \in [s_{n_{k_j}}]} \|S_L^{-1}(s, T)\| = +\infty, \tag{3.25}$$

which implies $\lim_{n\to+\infty} \sup_{s\in[s_n]} \|S_L^{-1}(s, T)\| = +\infty$. We consider an arbitrary subsequence $(s_{n_k})_{k\in\mathbb{N}}$ of $(s_n)_{n\in\mathbb{N}}$. If this subsequence has a subsequence $(s_{n_{k_j}})_{j\in\mathbb{N}}$ such that $\|\mathcal{Q}_{s_{n_{k_j}}}(T)\| \to +\infty$, then Lemma 3.2.8 implies (3.25). Otherwise,

$$\|\mathcal{Q}_{s_{n_k}}(T)^{-1}\| \le C$$

for some constant $C > 0$ and we deduce from (3.24) that $\|TQ_{s_{n_k}}(T)^{-1}\| \to +\infty$. Observe that

$$TQ_{s_{n_k}}(T)^{-1} = -\frac{1}{2}S_L^{-1}(s_{n_k}, T) - \frac{1}{2}S_L^{-1}(\overline{s_{n_k}}, T) + s_{n_k,0}Q_{s_{n_k}}(T)^{-1},$$

from which we obtain the estimate

$$\left\|TQ_{s_{n_k}}(T)^{-1}\right\| \leq \sup_{s \in [s_{n_k}]} \left\|S_L^{-1}(s_{n_k}, T)\right\| + |s_{n_k,0}|\left\|Q_{s_{n_k}}(T)^{-1}\right\|$$

$$\leq \sup_{s \in [s_{n_k}]} \left\|S_L^{-1}(s_{n_k}, T)\right\| + CM$$

with $M = \sup_{n \in \mathbb{N}} |s_n| < +\infty$. Since the left-hand side tends to infinity as $k \to +\infty$, we obtain that also

$$\sup_{s \in [s_{n_k}]} \left\|S_L^{-1}(s_{n_k}, T)\right\| \to +\infty$$

and thus, the statement holds true. The case of the right $S$-resolvent can be shown with analogous arguments. $\qquad\square$

**Definition 3.2.10** (Slice hyperholomorphic continuation). Let $f$ be a left (or right) slice hyperholomorphic function defined on an axially symmetric open set $U$. A left (or right) slice hyperholomorphic function $g$ defined on an axially symmetric open set $U'$ with $U \subsetneq U'$ is called a slice hyperholomorphic continuation of $f$ if $f(s) = g(s)$ for all $s \in U$. It is called nontrivial if $V = U' \setminus U$ cannot be separated from $U$, i.e. if $U' \neq U \cup V$ for some open set $V$ with $\overline{V} \cap \overline{U} = \emptyset$.

**Theorem 3.2.11.** *Let $T \in \mathcal{K}(X)$. There does not exist any nontrivial slice hyperholomorphic continuation of the left or of the right $S$-resolvent operators.*

*Proof.* Assume that there exists a nontrivial continuation $f$ of $S_L^{-1}(s, T)$ to an axially symmetric open set $U$ with $\rho_S(T) \subsetneq U$. Then there exists a point $s \in U \cap \partial(\rho_S(T))$ and a sequence $s_n \in \rho_S(T)$ with $\lim_{n \to +\infty} s_n = s$ such that

$$\lim_{n \to +\infty} \left\|S_L^{-1}(s_n, T)\right\| = \lim_{n \to +\infty} \|f(s_n)\| = \|f(s)\| < +\infty.$$

Moreover, $\overline{s_n} \to \overline{s}$ as $n \to +\infty$ and in turn

$$\lim_{n \to +\infty} \left\|S_L^{-1}(\overline{s_n}, T)\right\| = \lim_{n \to +\infty} \|f(\overline{s_n})\| = \|f(\overline{s})\| < +\infty.$$

From the representation formula (2.1.7) we then deduce

$$\lim_{n \to +\infty} \sup_{s \in [s_n]} \left\|S_L^{-1}(s, T)\right\| \leq \lim_{n \to +\infty} \left\|S_L^{-1}(s_n, T)\right\| + \left\|S_L^{-1}(\overline{s_n}, T)\right\| < +\infty.$$

On the other hand the sequence $s_n$ is bounded and

$$\mathrm{dist}(s_n, \sigma_S(T)) \leq |s_n - s| \to 0.$$

Lemma 3.2.9 therefore implies

$$\lim_{n \to +\infty} \sup_{s \in [s_n]} \left\| S_L^{-1}(s, T) \right\| = +\infty,$$

which is a contradiction. Thus, the analytic continuation $(f, U)$ cannot exist. For the right $S$-resolvent, we argue analogously. □

One could suspect that it might be possible to improve the above results by finding an estimate of the form (3.21) for the pseudo-resolvent $\mathcal{Q}_s(T)^{-1}$ instead of the $S$-resolvents. This is, however, not possible as the following example shows.

**Example 3.2.12.** We consider for $p \in [1, +\infty)$ the space $\ell^p(\mathbb{N})$ of $p$-summable sequences with quaternionic entries. Any sequence $(\lambda_n)_{n \in \mathbb{N}}$ with $\lambda_n \in \mathbb{H}$ obviously defines a right linear, densely defined and closed operator on $\ell^p(\mathbb{N})$ via $T(y) = (\lambda_n v_n)_{n \in \mathbb{N}}$ for $y = (v_n)_{n \in \mathbb{N}} \in \ell^p(\mathbb{N})$. If $(\lambda_n)_{n \in \mathbb{N}}$ is unbounded, then $T$ is unbounded. Otherwise, we have

$$\|T\| = \sup_{n \in \mathbb{N}} |\lambda_n| = \|(\lambda_n)_{n \in \mathbb{N}}\|_\infty.$$

Indeed,

$$\|T(y)\|_p = \sqrt[p]{\sum_{n \in \mathbb{N}} |\lambda_n v_n|^p} \leq \|(\lambda_n)_{n \in \mathbb{N}}\|_\infty \sqrt[p]{\sum_{n \in \mathbb{N}} |v_n|^p} = \|(\lambda_n)_{n \in \mathbb{N}}\|_\infty \|y\|_p$$

such that $\|T\| \leq \|(\lambda_n)_{n \in \mathbb{N}}\|_\infty$. On the other hand, with $e_m = (\delta_{m,n})_{n \in \mathbb{N}}$ where $\delta_{m,n}$ is the Kronecker delta,

$$|\lambda_m| = \sqrt[p]{\sum_{n \in \mathbb{N}} |\lambda_n \delta_{m,n}\|^p} = \|T(e_m)\| \leq \|T\|,$$

for any $m \in \mathbb{N}$ such that also $\|(\lambda_n)_{n \in \mathbb{N}}\|_\infty \leq \|T\|$. The $S$-spectrum of $T$ is

$$\sigma_S(T) = \overline{\bigcup_{n \in \mathbb{N}} [\lambda_n]} \tag{3.26}$$

as one can see easily: any $\lambda_n$ is a right eigenvalue of $T$ since $T(e_n) = e_n \lambda_n$ and hence the relation $\supset$ in (3.26) holds true by Theorem 3.1.6. If on the other hand $s$ does not belong to the right hand side of (3.26), then

$$\delta_s = \inf_{n \in \mathbb{N}} \operatorname{dist}(s, [\lambda_n]) = \inf_{n \in \mathbb{N}} |s_{j_{\lambda_n}} - \lambda_n| > 0,$$

where $s_{j_{\lambda_n}} = s_0 + j_{\lambda_n} s_1$. As

$$\mathcal{Q}_s(T)y = \left( (\lambda_n - s_{j_{\lambda_n}})(\lambda_n - \overline{s_{j_{\lambda_n}}}) v_n \right)_{n \in \mathbb{N}}$$

and in turn

$$\mathcal{Q}_s(T)^{-1} y = \left( (\lambda_n - s_{j_{\lambda_n}})^{-1} (\lambda_n - \overline{s_{j_{\lambda_n}}})^{-1} v_n \right)_{n \in \mathbb{N}},$$

we have $\|Q_s(T)^{-1}\| \le 1/\delta_s^2 < +\infty$ such that $s \in \rho_S(T)$. Thus, the relation $\subset$ in (3.26) also holds true.

Now choose a sequence $(\lambda_n)_{n\in\mathbb{N}}$ such that $\lambda_{n,1} \to +\infty$ as $n \to +\infty$ and consider the respective operator $T$ on $\ell^p(\mathbb{N})$. For simplicity, consider for instance $\lambda_n = jn$ with $j \in \mathbb{S}$. By the above considerations, the sequence $s_N = j(N + 1/N)$ with $N = 2, 3, \ldots$ does then satisfy $\mathrm{dist}(s_N, \sigma_S(T)) \to 0$ as $N \to +\infty$ and

$$\|Q_{s_n}(T)^{-1}\| = \sup_{n\in\mathbb{N}} \frac{1}{|\lambda_n - s_N||\lambda_n - \overline{s_N}|}$$
$$= \frac{1}{|\lambda_N - s_N||\lambda_N - \overline{s_N}|} = \frac{1}{2 + \frac{1}{N^2}}. \tag{3.27}$$

Indeed, if $n < N$, then some simple computations show that the inequality

$$\frac{1}{|\lambda_n - s_N||\lambda_n - \overline{s_N}|} = \frac{1}{N + \frac{1}{N} - n} \frac{1}{n + N + \frac{1}{N}}$$
$$< \frac{1}{2 + \frac{1}{N^2}} = \frac{1}{|\lambda_N - s_N||\lambda_N - \overline{s_N}|}$$

is equivalent to $0 < N^2 - n^2$, which is obviously true. Similarly, in the case $n > N$, the inequality

$$\frac{1}{|\lambda_n - s_N||\lambda_n - \overline{s_N}|} = \frac{1}{n - N - \frac{1}{N}} \frac{1}{n + N + \frac{1}{N}}$$
$$< \frac{1}{2 + \frac{1}{N^2}} = \frac{1}{|\lambda_N - s_N||\lambda_N - \overline{s_N}|}$$

is equivalent to $4 + 1/N^2 < n^2 - N^2$, which holds true since $2 \le N < n$.

From (3.27), we see that $\|Q_{s_n}(T)^{-1}\| \le 2$ although $\mathrm{dist}(s_N, \sigma_S(T)) \to 0$. Consequently, the pseudo-resolvent cannot satisfy an estimate analogue to (3.21).

Also controlling the norm of $TQ_s(T)^{-1}$ by the norm of $Q_s(T)^{-1}$ in order to improve (3.22) is not possible: if we consider the operator $TQ_{s_n}(T)^{-1}$ in the above example, then

$$TQ_{s_n}(T)^{-1}y = \left( \frac{n}{n - N - \frac{1}{N}} \frac{1}{j\left(n + N + \frac{1}{N}\right)} v_n \right)_{n\in\mathbb{N}}$$

and

$$\|TQ_{s_n}(T)^{-1}\| \ge \|TQ_{s_n}(T)^{-1}(e_N)\| = \frac{N^2}{2N + \frac{1}{N}} \to +\infty$$

shows that $\|TQ_{s_n}(T)^{-1}\|$ tends to infinity although $\|Q_{s_n}(T)^{-1}\|$ stays bounded.

## 3.3 Closed operators with commuting components

Closed right linear operators cannot always be decomposed into components as it is the case for bounded operators, cf. (2.38).

However, this is possible if $\mathcal{D}(T)$ is a two-sided subspace of $X$, that is if it is of the form $\mathcal{D}(T) = X_0 \otimes \mathbb{H}$ for some subspace $X_0$ of $X_\mathbb{R}$.

If on the other hand $T_0, \ldots, T_3$ are operators on $X_\mathbb{R}$, then we can define the operator

$$T = T_0 + \sum_{\ell=1}^{3} T_\ell e_\ell \quad \text{with} \quad \mathcal{D}(T) = \left( \bigcap_{\ell=0}^{4} \mathcal{D}(T_\ell) \right) \otimes \mathbb{H}.$$

**Definition 3.3.1.** Let $X$ be a two-sided quaternionic Banach space. We define $\mathcal{KC}(X)$ as the set of all operators $T \in \mathcal{K}(X)$ that admit a decomposition of the form $T = T_0 + \sum_{\ell=1}^{3} T_\ell e_\ell$ with closed operators $T_\ell \in \mathcal{K}(X_\mathbb{R})$ such that

(i) $\mathcal{D}(T^2) = \bigcap_{\ell,\kappa=0}^{3} \mathcal{D}(T_\ell T_\kappa) = \bigcap_{\ell=0}^{3} \mathcal{D}(T_\ell^2)$,

(ii) $\mathcal{D}(T_\ell T_\kappa) = \mathcal{D}(T_\kappa T_\ell)$, for $\ell, \kappa \in \{0, \ldots, 3\}$,

(iii) $T_\ell T_\kappa y = T_\kappa T_\ell y$, for all $y \in \mathcal{D}(T^2)$ for $\ell, \kappa \in \{0, \ldots, 3\}$.

Furthermore, we call a closed operator $T$ a scalar operator if it is of the form $T = T_0$, that is if $T_1 = T_2 = T_3 = 0$ or equivalently if $T$ is the extension of a closed operator on $X_\mathbb{R}$ to $X$.

**Remark 3.3.1.** A scalar operator $T \in \mathcal{K}(X)$ commutes with any $a \in \mathbb{H}$.

The $S$-spectrum $\sigma_S(T)$ of any operator $T \in \mathcal{KC}(X)$ can be characterized in a different way that takes the commutativity of the components into account. The corresponding characterization for bounded operators has been presented in Section 2.3.

**Definition 3.3.2.** Let $X$ be a two-sided quaternionic Banach space. For a closed operator $T = T_0 + \sum_{\ell=1}^{3} T_\ell e_\ell \in \mathcal{KC}(X)$ with commuting components, we define $\overline{T} = T - \sum_{\ell=1}^{3} T_\ell e_\ell$ with $\mathcal{D}\left(\overline{T}\right) = \bigcap_{\ell=0}^{3} \mathcal{D}(T_\ell) = \mathcal{D}(T)$.

McIntosh and Pryde showed in [179, Theorem 3.3] that an operator $T \in \mathcal{B}(X)$ with commuting components is invertible if and only if $T\overline{T} = \overline{T}T = \sum_{\ell=0}^{3} T_\ell^2$ is invertible. This holds true also for an unbounded operator with commuting components as the next lemma shows.

**Lemma 3.3.3.** Let $T \in \mathcal{KC}(X)$. Then the following statements are equivalent.

(i) The operator $T$ has a bounded inverse.

(ii) The operator $\overline{T}$ has a bounded inverse.

(iii) The operator $\overline{T}T$ has a bounded inverse.

*Proof.* First of all, we observe that, due to $\mathcal{D}(T) = \mathcal{D}\left(\overline{T}\right)$, we have

$$\mathcal{D}\left(\overline{T}T\right) = \{y \in X : Ty \in \mathcal{D}(T)\} = \mathcal{D}\left(T^2\right).$$

Since $\mathcal{D}\left(T^2\right) = \bigcap_{\ell,\kappa=0}^3 \mathcal{D}\left(T_\ell T_\kappa\right) = \bigcap_{\ell=0}^3 \mathcal{D}(T_\ell^2)$ and

$$\overline{T}Ty = T_0^2 y + \sum_{\ell=1}^3 e_\ell T_0 T_\ell y - \sum_{\ell=1}^3 e_\ell T_\ell T_0 y - \sum_{\ell,\kappa=1}^3 e_\ell e_\kappa T_\ell T_\kappa y = \sum_{\ell=0}^3 T_\ell^2 y$$

because $e_\ell e_\kappa = -e_\kappa e_\ell$ and $e_\ell^2 = -1$ for $1 \leq \ell, \kappa \leq 3$ with $\ell \neq \kappa$, we thus have $\overline{T}T = \sum_{\ell=0}^3 T_\ell^2$. In particular, $\overline{T}T$ is a scalar operator and hence commutes with any quaternion.

If $T\overline{T}$ is invertible, then $\left(T\overline{T}\right)^{-1} = \left(\sum_{\ell=0}^3 T_\ell^2\right)^{-1}$ commutes with each of the components $T_\ell$ and it also commutes with the imaginary units $e_\ell$. Hence, it commutes with $T$ and so the inverse $T^{-1}$ is given by $T^{-1} = \overline{T}\left(T\overline{T}\right)^{-1}$ because

$$\left(\overline{T}\left(T\overline{T}\right)^{-1}\right)Ty = \overline{T}T\left(T\overline{T}\right)^{-1}y, \quad \forall y \in \mathcal{D}(T)$$

and

$$T\left(\overline{T}\left(T\overline{T}\right)^{-1}\right)y = \left(T\overline{T}\right)\left(T\overline{T}\right)^{-1}y = y, \quad \forall y \in X.$$

Consequently, the invertibility of $T\overline{T}$ implies the invertibility of $T$.

If on the other hand $T$ is invertible and $T^{-1} = S_0 + \sum_{\kappa=1}^3 S_\kappa e_\kappa \in \mathcal{B}(X)$, then

$$\mathcal{I}|_{\mathcal{D}(T)} = T^{-1}T = \left(S_0 + \sum_{\kappa=1}^3 S_\kappa e_\kappa\right)\left(T_0 + \sum_{\ell=1}^3 T_\ell e_\ell\right)$$

$$= S_0 T_0 - \sum_{\ell=1}^3 S_\ell T_\ell + (S_2 T_3 - S_3 T_2)e_1$$

$$+ (S_3 T_1 - S_1 T_3)e_2 + (S_1 T_2 - S_2 T_1)e_3,$$

from which we conclude that

$$\mathcal{I}|_{\mathcal{D}(T)} = S_0 T_0 - \sum_{\ell=1}^3 S_\ell T_\ell \quad \text{and} \quad S_\ell T_\kappa - S_\kappa T_\ell = 0, \quad 1 \leq \ell < \kappa \leq 3.$$

Therefore

$$\overline{S}\,\overline{T} = \left(S_0 - \sum_{\ell=1}^3 S_\ell e_\ell\right)\left(T_0 - \sum_{\ell=1}^3 T_\ell e_\ell\right)$$

$$= S_0 T_0 - \sum_{\ell=1}^3 S_\ell T_\ell + (S_2 T_3 - S_3 T_2)e_1$$

$$+ (S_3 T_1 - S_1 T_3)e_2 + (S_1 T_2 - S_2 T_1)e_3 = \mathcal{I}|_{\mathcal{D}(T)}.$$

Similarly, we see that $TS = \mathcal{I}$ also implies $\overline{T}\,\overline{S} = \mathcal{I}$. Hence, the invertibility of $T$ implies the invertibility of $\overline{T}$ and $\overline{T}^{-1} = \overline{T^{-1}}$. Thus, if $T$ is invertible, we have $(T\overline{T})^{-1} = \overline{T}^{-1}T^{-1} \in \mathcal{B}(X)$. Altogether, we find that $T$ is invertible if and only if $T\overline{T} = \overline{T}T$ is invertible. $\qquad\square$

**Theorem 3.3.4.** *Let* $T = T_0 + \sum_{\ell=1}^{3} T_\ell e_\ell \in \mathcal{KC}(X)$ *with dense domain. If we set*

$$\mathcal{Q}_{c,s}(T) = s^2\mathcal{I} - 2sT_0 + T\overline{T},$$

*then*

$$\rho_S(T) = \left\{ s \in \mathbb{H} : \quad \mathcal{Q}_{c,s}(T)^{-1} \in \mathcal{B}(X) \right\} \tag{3.28}$$

*and*

$$S_L^{-1}(s, T) = (s\mathcal{I} - \overline{T})\mathcal{Q}_{c,s}(T)$$
$$S_R^{-1}(s, T) = \mathcal{Q}_{c,s}(T)^{-1}s - \sum_{\ell=0}^{3} T_\ell \mathcal{Q}_{c,s}(T)^{-1}e_\ell. \tag{3.29}$$

*Proof.* Since $T$ and $\overline{T}$ commute, we have $\overline{\mathcal{Q}_s(T)} = \mathcal{Q}_s(\overline{T})$ and $\overline{\mathcal{Q}_{c,s}(T)} = \mathcal{Q}_{c,\overline{s}}(T)$. For $y \in \mathcal{D}(T^4) = \mathcal{D}\left(\mathcal{Q}_{c,s}(T)\mathcal{Q}_{c,\overline{s}}(T)\right)$, we thus find

$$\begin{aligned}
\mathcal{Q}_{c,s}(T)\overline{\mathcal{Q}_{c,s}(T)}y &= (s^2\mathcal{I} - 2sT_0 + T\overline{T})(\overline{s}^2\mathcal{I} - 2\overline{s}T_0 + T\overline{T})y \\
&= |s|^4\mathcal{I}y - 2s|s|^2T_0y + s^2T\overline{T}y \\
&\quad - 2|s|^2T_0\overline{s}y + 4|s|^2T_0^2y - 2sT_0T\overline{T}y \\
&\quad + \overline{s}^2T\overline{T}y - 2\overline{s}T_0T\overline{T}y + (T\overline{T})^2y \\
&= |s|^4\mathcal{I}y - 2s_0|s|^2Ty - 2s_0|s|^2\overline{T}y + 2\mathrm{Re}(s^2)T\overline{T}y \\
&\quad + 4|s|^2T_0^2y - 2s_0T^2\overline{T}y - 2s_0T\overline{T}^2y + T^2\overline{T}^2y,
\end{aligned}$$

where we used in the last identity that $2s_0 = s + \overline{s}$, that $|s|^2 = s\overline{s}$, and that $2T_0y = Ty + \overline{T}y$. As

$$2\mathrm{Re}(s^2)T\overline{T}y = 2s_0^2T\overline{T}y - 2s_1^2T\overline{T}y$$

and

$$4|s|^2T_0^2y = |s|^2(T + \overline{T})^2y = |s|^2T^2y + 2s_0^2T\overline{T}y + s_1^2T\overline{T}y + |s|^2\overline{T}^2y,$$

we further find

$$\begin{aligned}
\mathcal{Q}_{c,s}(T)\overline{\mathcal{Q}_{c,s}(T)}y &= |s|^2(|s|^2\mathcal{I} - 2s_0T + T^2)y \\
&\quad - 2s_0\overline{T}(|s|^2\mathcal{I} - 2s_0T + T^2)y \\
&\quad + \overline{T}^2(|s|^2\mathcal{I} - 2s_0T + T^2)y = \mathcal{Q}_s(T)\overline{\mathcal{Q}_s(T)}y.
\end{aligned}$$

By the above arguments, we hence have

$$\mathcal{Q}_{c,s}(T)^{-1} \in \mathcal{B}(X) \iff \left(\mathcal{Q}_{c,s}(T)\overline{\mathcal{Q}_{c,s}(T)}\right)^{-1} \in \mathcal{B}(X)$$

$$\iff \left(\mathcal{Q}_s(T)\overline{\mathcal{Q}_s(T)}\right)^{-1} \in \mathcal{B}(X) \iff \mathcal{Q}_s(T)^{-1} \in \mathcal{B}(X)$$

and hence (3.28) holds true.

If $y \in \mathcal{D}(T^2) = \mathcal{D}(\mathcal{Q}_{c,s}(T))$ with $\mathcal{Q}_{c,s}(T) \in \mathcal{D}(T)$, we have

$$
\begin{aligned}
&(s\mathcal{I} - T)\mathcal{Q}_{c,s}(T)y \\
&= (\bar{s}\mathcal{I} - T)\left(s^2\mathcal{I} - 2sT_0 + T\overline{T}\right)y \\
&= |s|^2 s\mathcal{I}y - Ts^2 y - 2|s|^2 T_0 y + 2TT_0 sy + \bar{s}T\overline{T}y - T^2\overline{T}y \\
&= |s|^2 s\mathcal{I}y - Ts^2 y - |s|^2 Ty - |s|^2 \overline{T}y + T^2 sy + T\overline{T}sy + \bar{s}T\overline{T}y - T^2\overline{T}y \\
&= |s|^2 \left(s\mathcal{I} - \overline{T}\right)y - 2s_0 T \left(s\mathcal{I} - \overline{T}\right)y + T^2 \left(s\mathcal{I} - \overline{T}\right)y \\
&= \left(T^2 - 2s_0 T + |s|^2 \mathcal{I}\right)\left(s\mathcal{I} - \overline{T}\right)y = \mathcal{Q}_s(T)\left(s\mathcal{I} - \overline{T}\right)y.
\end{aligned}
$$

For any $x \in \mathcal{D}(T)$, we can set $y = \mathcal{Q}_{c,s}(T)^{-1}x \in \mathcal{D}(T^2)$. If we apply the operator $\mathcal{Q}_s(T)^{-1}$ to the above identity from the right, we then obtain

$$S_L^{-1}(s,T)x = \mathcal{Q}_s(T)^{-1}(s\mathcal{I} - T)x = \left(s\mathcal{I} - \overline{T}\right)\mathcal{Q}_{c,s}(T)^{-1}x$$

and a density argument shows that (3.29) holds true for the left $S$-resolvent operator. Similar computations show also the identity for the right $S$-resolvent equation. $\qquad\square$

## 3.4   The $S$-functional calculus and its properties

We want to define the $S$-functional calculus for an arbitrary operator in $\mathcal{K}(X)$ with nonempty $S$-resolvent set via the slice hyperholomorphic Cauchy integral. The domain of integration is thereby the boundary of a suitable slice Cauchy domain $U$ in one of the complex planes $\mathbb{C}_j$, for $j \in \mathbb{S}$. In order for the $S$-functional calculus to be well-defined, we have to show that these integrals are independent of the choice of the slice Cauchy domain $U$ and of the complex plane $\mathbb{C}_j$. We follow the strategy known from the bounded case.

**Theorem 3.4.1.** *Let $T \in \mathcal{K}(X)$ with $\rho_S(T) \neq \emptyset$. If $f \in \mathcal{SH}_L(\sigma_S(T) \cup \{\infty\})$, then there exists an unbounded slice Cauchy domain $U$ with $\sigma_S(T) \subset U$ and $\overline{U} \subset \mathcal{D}(f)$. The integral*

$$\frac{1}{2\pi} \int_{\partial(U \cap \mathbb{C}_j)} S_L^{-1}(s,T)\, ds_j\, f(s) \tag{3.30}$$

*defines an operator in $\mathcal{B}(X)$ and this operator is the same for any choice of the imaginary unit $j \in \mathbb{S}$ and for any choice of the slice Cauchy domain $U$ that satisfies the above conditions.*

*Similarly, if $f \in \mathcal{SH}_R(\sigma_S(T) \cup \{\infty\})$, then there exists an unbounded slice Cauchy domain $U$ such that $\sigma_S(T) \subset U$ and $\overline{U} \subset \mathcal{D}(f)$. Again, the integral*

$$\frac{1}{2\pi} \int_{\partial(U \cap \mathbb{C}_j)} f(s) \, ds_j \, S_R^{-1}(s, T)$$

*defines an operator in $\mathcal{B}(X)$ and this operator is the same for any choice of the imaginary unit $j \in \mathbb{S}$ and for any choice of the slice Cauchy domain $U$ that satisfies the above conditions.*

*Proof.* Let $f \in \mathcal{SH}_L(\sigma_S(T) \cup \{\infty\})$ and $q \in \rho_S(T)$. Since $\rho_S(T)$ is open, there exists a closed ball $B_\varepsilon(q) \subset \rho_S(T)$ and since $\rho_S(T)$ is axially symmetric we have

$$\left[\overline{B_\varepsilon(q)}\right] = \{s = s_0 + j_s s_1 \in \mathbb{H} : \quad (s_0 - q_0)^2 + (s_1 - q_1)^2 \leq \varepsilon\} \subset \rho_S(T).$$

The existence of the slice Cauchy domain $U$ follows from Theorem 2.1.31 applied with $C = \sigma_S(T)$ and $O = \mathcal{D}(f) \cap \left(\mathbb{H} \setminus \overline{B_\varepsilon(q)}\right)$.

The boundary of $U$ in $\mathbb{C}_j$ consists of a finite set of closed piecewise differentiable Jordan curves and so it is compact. Hence, (3.30) is the integral of a bounded integrand over a compact domain. Thus, it converges in $\mathcal{B}(X)$ and defines an operator in $\mathcal{B}(X)$.

We now show the independence of the slice Cauchy domain. Consider first the case of another unbounded slice Cauchy domain $U'$ such that $\sigma_S(T) \subset U'$ and $\overline{U'} \subset \mathcal{D}(f)$. Let us for the moment furthermore assume that $\overline{U'} \subset U$. Then the set $W = U \setminus \overline{U'}$ is a bounded slice Cauchy domain and

$$\partial(W \cap \mathbb{C}_j) = \partial(U \cap \mathbb{C}_j) \cup \left(-\partial(U' \cap \mathbb{C}_j)\right),$$

where $-\partial(U' \cap \mathbb{C}_j)$ denotes the inversely orientated boundary of $U'$ in $\mathbb{C}_j$. Moreover, the function $s \mapsto S_L^{-1}(s, T)$ is right and the function $s \mapsto f(s)$ is left slice hyperholomorphic on $\overline{W}$. Thus, Theorem 2.1.20 implies

$$0 = \frac{1}{2\pi} \int_{\partial(W \cap \mathbb{C}_j)} S_L^{-1}(s, T) \, ds_j \, f(s)$$

$$= \frac{1}{2\pi} \int_{\partial(U \cap \mathbb{C}_j)} S_L^{-1}(s, T) \, ds_j \, f(s) - \frac{1}{2\pi} \int_{\partial(U' \cap \mathbb{C}_j)} S_L^{-1}(s, T) \, ds_j \, f(s).$$

If $\overline{U'}$ is not contained in $U$, then $U \cap U'$ is an axially symmetric open set that contains $\sigma_S(T)$ such that $\partial(U \cap U')$ is nonempty and bounded. Theorem 2.1.31 implies the existence of a third slice Cauchy domain $W$ such that $\sigma_S(T) \subset W$ and $\overline{W} \subset U \cap U'$. By the above arguments, the choice of any of them yields the same operator in (3.30).

Finally, we consider another imaginary unit $i \in \mathbb{S}$ and choose another unbounded slice Cauchy domain $W$ with $\sigma_S(T) \subset W$ and $\overline{W} \subset U$. By the above arguments and the Cauchy formulae, we have

$$\frac{1}{2\pi} \int_{\partial(U\cap\mathbb{C}_j)} S_L^{-1}(s,T)\, ds_j\, f(s) = \frac{1}{2\pi} \int_{\partial(W\cap\mathbb{C}_j)} S_L^{-1}(s,T)\, ds_j\, f(s)$$

$$= \frac{1}{(2\pi)^2} \int_{\partial(W\cap\mathbb{C}_j)} S_L^{-1}(s,T)\, ds_j \left( f(\infty) + \int_{\partial(U\cap\mathbb{C}_i)} S_L^{-1}(q,s)\, dq_i\, f(q) \right)$$

$$= \frac{1}{(2\pi)^2} \int_{\partial(W\cap\mathbb{C}_j)} S_L^{-1}(s,T)\, ds_j\, f(\infty)$$

$$- \frac{1}{(2\pi)^2} \int_{\partial(U\cap\mathbb{C}_i)} \int_{\partial(W^c\cap\mathbb{C}_j)} S_L^{-1}(s,T)\, ds_j\, S_L^{-1}(q,s)\, dq_i\, f(q),$$

where Fubini's theorem allows us to exchange the order of integration in the last equation because we integrate a bounded function over a finite domain. The set $W^c$ is a bounded slice Cauchy domain and the left $S$-resolvent is right slice hyperholomorphic in $s$ on $\overline{W^c}$. Theorem 2.1.20 implies

$$\frac{1}{(2\pi)^2} \int_{\partial(W\cap\mathbb{C}_j)} S_L^{-1}(s,T)\, ds_j\, f(\infty)$$

$$= -\frac{1}{(2\pi)^2} \int_{\partial(W^c\cap\mathbb{C}_j)} S_L^{-1}(s,T)\, ds_j\, f(\infty) = 0.$$

Since any $q \in \partial(U\cap\mathbb{C}_j)$ belongs to $W^c$ by our choices of $U$ and $W$ and since $S_L^{-1}(q,s) = -S_R^{-1}(s,q)$, we deduce from the Cauchy formulae

$$\frac{1}{2\pi} \int_{\partial(U\cap\mathbb{C}_j)} S_L^{-1}(s,T)\, ds_j\, f(s)$$

$$= \frac{1}{2\pi} \int_{\partial(U\cap\mathbb{C}_i)} \left( \frac{1}{2\pi} \int_{\partial(W^c\cap\mathbb{C}_j)} S_L^{-1}(s,T)\, ds_j\, S_R^{-1}(s,q) \right) dq_i\, f(q)$$

$$= \frac{1}{2\pi} \int_{\partial(U\cap\mathbb{C}_i)} S_L^{-1}(q,T)\, dq_i\, f(q). \qquad \square$$

**Definition 3.4.2.** Let $T \in \mathcal{K}(X)$ with $\rho_S(T) \neq \emptyset$. For any $f \in \mathcal{SH}_L(\sigma_S(T) \cup \{\infty\})$, we define

$$f(T) := f(\infty)\mathcal{I} + \frac{1}{2\pi} \int_{\partial(U\cap\mathbb{C}_j)} S_L^{-1}(s,T)\, ds_j\, f(s), \qquad (3.31)$$

and for $f \in \mathcal{SH}_R(\sigma_S(T) \cup \{\infty\})$, we define

$$f(T) := f(\infty)\mathcal{I} + \frac{1}{2\pi} \int_{\partial(U\cap\mathbb{C}_j)} f(s)\, ds_j\, S_R^{-1}(s,T), \qquad (3.32)$$

where $j \in \mathbb{S}$ is arbitrary and $U$ is any slice Cauchy domain as in Theorem 3.4.1.

**Remark 3.4.1.** If $\rho_S(T) \cap \mathbb{R} \neq \emptyset$, then our approach is consistent with the approach that defines the $S$-functional calculus of an unbounded operator by suitably transforming both the function and the operator and then applying the $S$-functional calculus for bounded operators. Precisely, one chooses $\alpha \in \rho_S(T) \cap \mathbb{R}$ and sets $\Phi_\alpha(s) = (s - \alpha)^{-1}$. Then $A := (T - \alpha\mathcal{I})^{-1} = S_R^{-1}(\alpha, T)$ is a bounded operator and formally corresponds to $\Phi_\alpha(T)$. Furthermore, a function $f$ belongs to $\mathcal{SH}_L(\sigma_S(T) \cup \{\infty\})$ or $\mathcal{SH}_R(\sigma_S(T) \cup \{\infty\})$ if and only if $f \circ \Phi_\alpha^{-1}$ belongs to $\mathcal{SH}_L(\sigma_S(A))$ (resp. $\mathcal{SH}_R(\sigma_S(A))$). One then defines

$$f(T) := f \circ \Phi_\alpha^{-1}(A).$$

This approach was presented in [57, 93]. In the complex setting, it is equivalent to the direct approach via a Cauchy integral, which was developed above. In the quaternionic setting it, however, requires that $\rho_S(T) \cap \mathbb{R} \neq \emptyset$, which is not always true.

The $S$-functional calculus for closed operators is furthermore consistent with the $S$-functional calculus for bounded operators. Since we do not require connectedness of $\mathcal{D}(f)$ in Definition 3.4.2, we might extend $f \in \mathcal{SH}_L(\sigma_S(T))$ for bounded $T$ to a function in $\mathcal{SH}_L(\sigma_S(T) \cup \{\infty\})$, for instance by setting $f(s) = c$ with $c \in \mathbb{H}$ on $\mathbb{H} \backslash B_r(0)$. We can then use the unbounded slice Cauchy domain $(\mathbb{H} \backslash B_r(0)) \cup U$ in (3.31). Since the left $S$-resolvent is then right slice hyperholomorphic on $\mathbb{H} \backslash B_r(0)$ and $f(s)$ is left slice hyperholomorphic on this set, we obtain

$$f(T) = f(\infty)\mathcal{I} + \frac{1}{2\pi} \int_{-\partial(B_r(0) \cap \mathbb{C}_j)} f(s) \, ds_j \, S_L^{-1}(s, T)$$

$$+ \frac{1}{2\pi} \int_{\partial(U \cap \mathbb{C}_j)} f(s) \, ds_j \, S_L^{-1}(s, T)$$

$$= \frac{1}{2\pi} \int_{\partial(U \cap \mathbb{C}_j)} f(s) \, ds_j \, S_L^{-1}(s, T)$$

because Theorem 2.1.20 implies that the sum of $f(\infty)\mathcal{I}$ and the integral over the boundary of $B_r(0)$ vanishes.

**Example 3.4.3.** Let $T \in \mathcal{K}(X)$ with $\rho_S(T) \neq \emptyset$. Consider the left slice hyperholomorphic function $f(s) = a$ for some $a \in \mathbb{H}$ and choose an arbitrary unbounded slice Cauchy domain $U$ with $\sigma_S(T) \subset U$ and an imaginary unit $j \in \mathbb{S}$. Then

$$f(T) = f(\infty)\mathcal{I} + \frac{1}{2\pi} \int_{\partial(U \cap \mathbb{C}_j)} S_L^{-1}(s, T) \, ds_j \, f(s) = a\mathcal{I}, \qquad (3.33)$$

because $f(\infty) = a$ and the integral vanishes by Theorem 2.1.20 as the left $S$-resolvent is right slice hyperholomorphic in $s$ on a superset of $\mathbb{H} \backslash U$ and vanishes at infinity. An analogue argument shows that also $f(T) = \mathcal{I}a$ if $f$ is considered right slice hyperholomorphic.

The following algebraic properties of the $S$-functional calculus follow immediately from the left and right linearity of the integral.

**Corollary 3.4.4.** *Let $T \in \mathcal{K}(X)$ with $\rho_S(T) \neq \emptyset$.*

(i) *If $f, g \in \mathcal{SH}_L(\sigma_S(T) \cup \{\infty\})$ and $a \in \mathbb{H}$, then*

$$(f + g)(T) = f(T) + g(T) \quad \text{and} \quad (fa)(T) = f(T)a.$$

(ii) *If $f, g \in \mathcal{SH}_R(\sigma_S(T) \cup \{\infty\})$ and $a \in \mathbb{H}$, then*

$$(f + g)(T) = f(T) + g(T) \quad \text{and} \quad (af)(T) = af(T).$$

Theorem 3.4.1 ensures that the $S$-functional calculus for left slice hyperholomorphic functions and the $S$-functional calculus for right slice hyperholomorphic functions are well-defined in the sense that they are independent of the choices of the imaginary unit $j \in \mathbb{S}$ and the slice Cauchy domain $U$. Another important question is whether they are consistent. We show now that this is the case, if the function $f$ is intrinsic.

**Lemma 3.4.5.** *Let $T \in \mathcal{K}(X)$ with $\rho_S(T) \neq \emptyset$ and let $f \in \mathcal{N}(\sigma_S(T) \cup \{\infty\})$. Furthermore, consider a slice Cauchy domain $U$ such that $\sigma_S(T) \subset U$ and $\overline{U} \subset \mathcal{D}(f)$ and some imaginary unit $j \in \mathbb{S}$. If $\gamma_1, \ldots, \gamma_N$ is the part of $\partial(U \cap \mathbb{C}_j)$ that lies in $\mathbb{C}_j^+$ as in Definition 2.1.34, then*

$$\int_{\partial(U \cap \mathbb{C}_j)} f(s) \, ds_j \, S_R^{-1}(s, T)$$

$$= \sum_{\ell=1}^{N} \int_0^1 2\mathrm{Re}\left( f(\gamma_\ell(t))(-j)\gamma_\ell'(t)\overline{\gamma_\ell(t)} \right) \mathcal{Q}_{\gamma_\ell(t)}(T)^{-1} \, dt \qquad (3.34)$$

$$- \sum_{\ell=1}^{N} \int_0^1 2\mathrm{Re}\left( f(\gamma_\ell(t))(-j)\gamma_\ell'(t) \right) T\mathcal{Q}_{\gamma_\ell(t)}(T)^{-1} \, dt.$$

*Proof.* We have

$$\int_{\partial(U \cap \mathbb{C}_j)} f(s) \, ds_j \, S_R^{-1}(s, T)$$

$$= \sum_{\ell=1}^{N} \int_{\gamma_\ell} f(s) \, ds_j \, S_R^{-1}(s, T) + \sum_{\ell=1}^{N} \int_{-\overline{\gamma_\ell}} f(s) \, ds_j \, S_R^{-1}(s, T)$$

$$= \sum_{\ell=1}^{N} \int_0^1 f(\gamma_\ell(t))(-j)\gamma_\ell'(t) \left( \overline{\gamma_\ell(t)} - T \right) \mathcal{Q}_{\gamma_\ell(t)}(T)^{-1} \, dt$$

$$+ \sum_{\ell=1}^{N} \int_0^1 f\left( \overline{\gamma_\ell(1-t)} \right) j\overline{\gamma_\ell'(1-t)}(\gamma_\ell(1-t) - T)\mathcal{Q}_{\overline{\gamma_\ell(1-t)}}(T)^{-1} \, dt.$$

Since $f(\bar{s}) = \overline{f(s)}$ as $f$ is intrinsic and $\mathcal{Q}_{\bar{s}}(T)^{-1} = \mathcal{Q}_s(T)^{-1}$ for $s \in \rho_S(T)$, we get, after a change of variables in the integrals of the second sum,

$$\int_{\partial(U \cap \mathbb{C}_j)} f(s)\, ds_j\, S_R^{-1}(s,T)$$

$$= \sum_{\ell=1}^{N} \int_0^1 f(\gamma_\ell(t))(-j)\gamma_\ell'(t)\left(\overline{\gamma_\ell(t)} - T\right)\mathcal{Q}_{\gamma_\ell(t)}(T)^{-1}\, dt$$

$$+ \sum_{\ell=1}^{N} \int_0^1 \overline{f(\gamma_\ell(t))}(-j)\overline{\gamma_\ell'(t)}(\gamma_\ell(t) - T)\mathcal{Q}_{\gamma(t)}(T)^{-1}\, dt$$

$$= \sum_{\ell=1}^{N} \int_0^1 2\mathrm{Re}\left(f(\gamma_\ell(t))(-j)\gamma_\ell'(t)\overline{\gamma_\ell(t)}\right)\mathcal{Q}_{\gamma_\ell(t)}(T)^{-1}\, dt$$

$$- \sum_{\ell=1}^{N} \int_0^1 2\mathrm{Re}\left(f(\gamma_\ell(t))(-j)\gamma_\ell'(t)\right)T\mathcal{Q}_{\gamma_\ell(t)}(T)^{-1}\, dt. \qquad \square$$

**Theorem 3.4.6.** *Let $T \in \mathcal{K}(X)$ with $\rho_S(T) \neq \emptyset$. If $f \in \mathcal{N}(\sigma_S(T) \cup \{\infty\})$, then*

$$\frac{1}{2\pi} \int_{\partial(U \cap \mathbb{C}_j)} S_L^{-1}(s,T)\, ds_j\, f(s) = \frac{1}{2\pi} \int_{\partial(U \cap \mathbb{C}_j)} f(s)\, ds_j\, S_R^{-1}(s,T),$$

*for any $j \in \mathbb{S}$ and any slice Cauchy domain as in Theorem 3.4.1.*

*Proof.* Fix $U$ and $j \in \mathbb{S}$, let $\gamma_1, \ldots \gamma_N$ be the part of $\partial(U \cap \mathbb{C}_j)$ that lies in $\mathbb{C}_j^+$ and write the integral involving the right $S$-resolvent as an integral over these paths as in (3.34). Any operator commutes with real numbers and $f(\gamma_\ell(t))$, $\gamma_\ell'(t)$ and $\overline{\gamma_\ell(t)}$ commute mutually since they all belong to the same complex plane $\mathbb{C}_j$. Hence,

$$\int_{\partial(U \cap \mathbb{C}_j)} f(s)\, ds_j\, S_R^{-1}(s,T)$$

$$= \sum_{\ell=1}^{N} \int_0^1 \mathcal{Q}_{\gamma(t)}(T)^{-1} 2\mathrm{Re}\left(\overline{\gamma_\ell(t)}\gamma_\ell'(t)(-j)f(\gamma_\ell(t))\right)\, dt$$

$$- \sum_{\ell=1}^{N} \int_0^1 T\mathcal{Q}_{\gamma_\ell(t)}(T)^{-1} 2\mathrm{Re}\left(\gamma_\ell'(t)(-j)f(\gamma_\ell(t))\right)\, dt$$

$$= \sum_{\ell=1}^{N} \int_0^1 \left(T\mathcal{Q}_{\gamma_\ell(t)}(T)^{-1} - \mathcal{Q}_{\gamma_\ell(t)}(T)^{-1}\overline{\gamma_\ell(t)}\right)\gamma_\ell'(t)(-j)f(\gamma_\ell(t))\, dt$$

$$+ \sum_{\ell=1}^{N} \int_0^1 \left(T\mathcal{Q}_{\overline{\gamma_\ell(t)}}(T)^{-1} - \mathcal{Q}_{\overline{\gamma_\ell(t)}}(T)^{-1}\gamma_\ell(t)\right)\overline{\gamma_\ell'(t)}j\overline{f(\gamma_\ell(t))}\, dt$$

$$= \sum_{\ell=1}^{N} \int_{\gamma_\ell} S_L^{-1}(s,T)\, ds_j\, f(s) + \sum_{\ell=1}^{N} \int_{-\overline{\gamma_\ell}} S_L^{-1}(s,T)\, ds_j\, f(s)$$

$$= \int_{\partial(U \cap \mathbb{C}_j)} S_L^{-1}(s,T)\, ds_j\, f(s). \qquad\qquad \Box$$

Since $f(\infty) = \lim_{s \to \infty} f(s) \in \mathbb{R}$ as $f(s) \in \mathbb{R}$ for $s \in \mathbb{R}$ if $f$ is intrinsic, we can rephrase the above result as,

**Corollary 3.4.7.** *Let $T \in \mathcal{K}(X)$ with $\rho_S(T) \neq \emptyset$. The S-functional calculus for left slice hyperholomorphic functions and the S-functional calculus for right slice hyperholomorphic functions agree for intrinsic functions: if $f \in \mathcal{N}(\sigma_S(T) \cup \{\infty\})$, then (3.31) and (3.32) give the same operator.*

**Remark 3.4.2.** For intrinsic functions, slice hyperholomorphic Cauchy integrals of the form (3.31) and (3.32) are always equivalent. We have shown this only for the $S$-functional calculus, but with the same technique one can show this equivalence also for the $H^\infty$-functional calculus or for fractional powers of quaternionic linear operators. Since the technique for showing this equivalence is the same in any situation, we will use it without proving it explicitly at every occurrence.

We have shown that the two versions of the $S$-functional calculus are consistent for intrinsic functions. However, there exist functions that are both left and right slice hyperholomorphic, but not intrinsic. We want to clarify the relation between the versions of the $S$-functional calculus for such functions and we start by characterising functions of this type.

Recall that a function $f$ on $U$ is called locally constant if every point $q \in U$ has a neighborhood $B_q \subset U$ such that $f$ is constant on $U$. A locally constant function $f$ is constant on every connected subset of its domain. Thus, since every sphere $[q]$ is connected, the function $f$ is constant on every sphere if its domain $U$ is axially symmetric, i.e., it is of the form $f(q) = c(u,v)$ for $q = u + jv$, where $c$ is locally constant on an appropriate subset of $\mathbb{R}^2$. Therefore, $f$ can be considered a left and a right slice function and it is even left and right slice hyperholomorphic because the partial derivatives of a locally constant function vanish.

**Lemma 3.4.8.** *A function $f$ is both left and right slice hyperholomorphic if and only if $f = c + \tilde{f}$, where $c$ is a locally constant slice function and $\tilde{f}$ is intrinsic slice hyperholomorphic.*

*Proof.* Obviously any function that admits a decomposition of this type is both left and right slice hyperholomorphic. Assume on the other hand that $f$ is left and right slice hyperholomorphic such that for $q = u + jv$

$$f(q) = f_0(u,v) + j f_1(u,v)$$

and

$$f(q) = \hat{f}_0(u,v) + \hat{f}_1(u,v) j.$$

The compatibility condition (2.4) implies

$$f_0(u, v) = \frac{1}{2}\left(f(q) + f(\overline{q})\right) = \hat{f}_0(u, v),$$

from which we deduce $jf_1(u, v) = f(q_j) - f_0(u, v) = \hat{f}_1(u, v)j$ with $q_j = u + jv$ for any $j \in \mathbb{S}$. Hence, we have

$$jf_1(u, v)j^{-1} = \hat{f}_1(u, v).$$

If we choose $j$ such that $f_1(u, v) \in \mathbb{C}_j$, then $j$ and $f_1(u, v)$ commute and we obtain $f_1(u, v) = \hat{f}_1(u, v)$. We further conclude that $f_1(u, v)$ commutes with every $j \in \mathbb{S}$ because

$$jf_1(u, v) = \hat{f}_1(u, v)j = f_1(u, v)j.$$

This implies that $f_1(u, v)$ is real.

Since $f_1$ takes real values, its partial derivatives $\frac{\partial}{\partial u}f_1(u, v)$ and $\frac{\partial}{\partial v}f_1(u, v)$ are real-valued too. Thus, since $f_0$ and $f_1$ satisfy the Cauchy–Riemann equations (2.5), the partial derivatives of $f_0$ also take real-values.

Now define $\tilde{f}_0(u, v) = \text{Re}(f_0(u, v))$ and $\tilde{f}_1(u, v) = f_1(u, v)$ and set $\tilde{f}(q) = \tilde{f}_0(u, v) + j\tilde{f}_1(u, v)$ and $c(q) = f(q) - \tilde{f}(q) = \text{Im}(f_0(u, v))$ for $q = u + jv$. Obviously, $\tilde{f}_0$ and $\tilde{f}_1$ satisfy the compatibility condition (2.4). Moreover, the partial derivatives of $\tilde{f}_0$ and $\tilde{f}_1$ coincide with the partial derivatives of $f_0$ (resp. $f_1$). For $\tilde{f}_1 = \hat{f}_1$ this is obvious and for $\tilde{f}_0$ this follows from

$$\frac{\partial}{\partial\nu}\tilde{f}_0(u, v) = \frac{\partial}{\partial\nu}\text{Re}(f_0(u, v)) = \text{Re}\left(\frac{\partial}{\partial\nu}f_0(u, v)\right) = \frac{\partial}{\partial\nu}f_0(u, v)$$

for $\nu \in \{u, v\}$ since $\frac{\partial}{\partial\nu}f_0(u, v)$ is real-valued by the above arguments. We conclude that $\tilde{f}_0$ and $\tilde{f}_1$ satisfy the Cauchy–Riemann equations (2.5) because $f_0$ and $f_1$ satisfy them. Therefore, $\tilde{f}$ is a left slice hyperholomorphic function with real-valued components, thus intrinsic.

It remains to show that $c$ is locally constant. Since $c(q) = c(u + jv) = \text{Im}(f_0(u, v))$ depends only on $u$ and $v$ but not on the imaginary unit $j$, it is constant on every sphere $[q] \subset U$. Moreover, as the sum of two left slice hyperholomorphic functions, it is left slice hyperholomorphic and thus, its restriction $c_j$ to any complex plane $\mathbb{C}_j$ is a $\mathbb{H}$-valued left holomorphic function. But

$$c_j'(q) = \frac{\partial}{\partial q_0}c_j(q) = \frac{\partial}{\partial q_0}f(q) - \frac{\partial}{\partial q_0}\tilde{f}(q) = 0, \quad q \in U \cap \mathbb{C}_j$$

and hence $c$ is locally constant on $U \cap \mathbb{C}_j$. If $q = u + jv \in U$, we can therefore find a neighborhood $B_j$ of $q$ in $U \cap \mathbb{C}_j$ such that $c_j$ is constant on $B_j$. Since $c$ is constant on every sphere, it is even constant on the axially symmetric hull $B = [B_j]$ of $B_j$, which is a neighborhood of $q$ in $U$. $\qquad\square$

**Corollary 3.4.9.** *Let $T \in \mathcal{K}(X)$ with $\rho_S(T) \neq \emptyset$ and let $f$ be both left and right slice hyperholomorphic on $\sigma_S(T)$ and at infinity. If $\mathcal{D}(f)$ is connected, then (3.31) and (3.32) give the same operator.*

*Proof.* By applying Lemma 3.4.8 we obtain a decomposition $f = c + \tilde{f}$ of $f$ into the sum of a locally constant function $c$ and an intrinsic function $\tilde{f}$. Since $\mathcal{D}(f)$ is connected, $c$ is a constant function. Thus, Corollary 3.4.7 and Example 3.4.3 imply

$$f(\infty)\mathcal{I} + \frac{1}{2\pi} \int_{\partial(U \cap \mathbb{C}_j)} f(s)\, ds_j\, S_R^{-1}(s, T)$$

$$= c\left(\mathcal{I} + \frac{1}{2\pi} \int_{\partial(U \cap \mathbb{C}_j)} ds_j\, S_R^{-1}(s, T)\right)$$

$$+ \tilde{f}(\infty)\mathcal{I} + \frac{1}{2\pi} \int_{\partial(U \cap \mathbb{C}_j)} \tilde{f}(s)\, ds_j\, S_R^{-1}(s, T)$$

$$= c\mathcal{I} + \tilde{f}(T) = \mathcal{I}c + \tilde{f}(T)$$

$$= \left(\mathcal{I} + \frac{1}{2\pi} \int_{\partial(U \cap \mathbb{C}_j)} S_L^{-1}(s, T)\, ds_j\right) c$$

$$+ \tilde{f}(\infty)\mathcal{I} + \frac{1}{2\pi} \int_{\partial(U \cap \mathbb{C}_j)} S_L^{-1}(s, T)\, ds_j\, \tilde{f}(s)$$

$$= f(\infty)\mathcal{I} + \frac{1}{2\pi} \int_{\partial(U \cap \mathbb{C}_j)} S_L^{-1}(s, T)\, ds_j\, f(s),$$

where $U$ and $j \in \mathbb{S}$ are chosen as in Definition 3.4.2.  $\square$

**Remark 3.4.3.** As we have shown, the two versions of the $S$-functional calculus are consistent for intrinsic slice hyperholomorphic functions and for functions defined on connected sets. However, in general, this is not true. If $\mathcal{D}(f)$ is not connected, then $c$ is only locally constant, i.e., it is of the form $c(s) = \sum_{\ell} \chi_{\Delta_\ell}(s) c_\ell$ with $c_\ell \in \mathbb{H}$, where the $\Delta_\ell$ are disjoint axially symmetric sets. The function $\chi_{\Delta_\ell}(s)$ is the characteristic function of $\Delta_\ell$, which is obviously intrinsic. The functional calculi for left and right slice hyperholomorphic functions yield then $c(T) = \sum_{\ell} \chi_{\Delta_\ell}(T) c_\ell$ and $c(T) = \sum_{\ell} c_\ell \chi_{\Delta_\ell}(T)$, respectively. These two operators coincide only if the operators $\chi_{\Delta_\ell}(T)$ commute with the scalars $c_\ell$. As we will see in Section 3.7, the operators $\chi_{\Delta_\ell}(T)$ are projections onto invariant subspaces of the operator $T$. Since the operator $T$ is right linear, its invariant subspaces are right subspaces of $X$. But if a projection $\chi_{\Delta_\ell}(T)$ commutes with any scalar, then

$$ay = a\chi_{\Delta_\ell}(T)y = \chi_\Delta(T)ay \in \chi_{\Delta_\ell}(T)X,$$

for any $y \in \chi_{\Delta_\ell}(T)X$ and any $a \in \mathbb{H}$. Thus, $\chi_{\Delta_\ell}(T)X$ is also a left-sided and therefore, even a two-sided subspace of $X$. In general, this is not true: the invariant subspaces obtained from spectral projections are only right-sided. Hence, the projections $\chi_{\Delta_\ell}(T)$ do not necessarily commute with any scalar and it might happen that

$$\sum_\ell \chi_{\Delta_\ell}(T)c_\ell \neq \sum_\ell c_\ell \chi_{\Delta_\ell}(T),$$

i.e., the two functional calculi give different operators for the same function. An explicit example for this situation is given in Example 3.7.9.

Finally, we show that the $S$-functional calculus admits, for intrinsic functions, a representation that only depends on the right linear structure of the space. In particular, this representation also shows the compatibility of the $S$-functional calculus and its classical counterpart form the theory of complex linear operators, the Riesz–Dunford functional calculus for holomorphic functions.

**Definition 3.4.10.** Let $T \in \mathcal{K}(X)$. We define the $X$-valued function

$$\mathcal{R}_s(T; y) = \mathcal{Q}_s(T)^{-1} y \overline{s} - T \mathcal{Q}_s(T)^{-1} y \quad \forall y \in X, \ s \in \rho_S(T).$$

**Remark 3.4.4.** By Theorem 3.1.8, the mapping $y \mapsto \mathcal{R}_s(T; y)$ coincides with the resolvent of $T$ at $s$ applied to $y$ if $T$ is considered a $\mathbb{C}_{j_s}$-linear operator on $X_{j_s} = X$.

**Theorem 3.4.11.** *Let $T \in \mathcal{K}(X)$ be a closed operator on a two-sided quaternionic Banach space $X$ with $\rho_S(T) \neq \emptyset$ and let $f \in \mathcal{N}(\sigma_{SX}(T))$. For any $j \in \mathbb{S}$ and any unbounded slice Cauchy domain $U$ with $\sigma_S(T) \subset U$ and $\overline{U} \subset \mathcal{D}(F)$, the operator $f(T)$ obtained via the $S$-functional calculus satisfies*

$$f(T)y = y f(\infty) + \int_{\partial(U \cap \mathbb{C}_j)} \mathcal{R}_z(T; y) f(z) \, dz \frac{-j}{2\pi} \quad \forall y \in X. \qquad (3.35)$$

*Proof.* Let $U$ be a slice Cauchy domain such that $\sigma_S(T) \subset U$ and $\overline{U} \subset \mathcal{D}(f)$. We then have for any $j \in \mathbb{S}$ and any $y \in X$ that

$$f(T)y = f(\infty)y + \frac{1}{2\pi} \int_{\partial(U \cap \mathbb{C}_j)} f(s) \, ds_j \, S_R^{-1}(s, T) y. \qquad (3.36)$$

If $\gamma_\ell : [0, 1] \to \mathbb{C}_j^+$, $\ell = 1, \ldots N$, is the part of $\partial(U \cap \mathbb{C}_j)$ that lies in $\mathbb{C}_j^+$ as in Definition 2.1.34, then we have by Lemma 3.4.5 that

$$\int_{\partial(U \cap \mathbb{C}_j)} f(s) \, ds_j \, S_R^{-1}(s, T) y$$

$$= \sum_{\ell=1}^N \int_0^1 2\mathrm{Re}\left( f(\gamma_\ell(t))(-j) \gamma_\ell'(t) \overline{\gamma_\ell(t)} y \right) \mathcal{Q}_{\gamma_\ell(t)}(T)^{-1} y \, dt \qquad (3.37)$$

$$- \sum_{\ell=1}^N \int_0^1 2\mathrm{Re}\left( f(\gamma_\ell(t))(-j) \gamma_\ell'(t) \right) T \mathcal{Q}_{\gamma_\ell(t)}(T)^{-1} y \, dt.$$

Since $\mathcal{Q}_{\gamma_\ell(t)}(T)^{-1}y$ and $T\mathcal{Q}_{\gamma_\ell(t)}(T)^{-1}y$ commute with real numbers, we furthermore have

$$\int_{\partial(U\cap\mathbb{C}_j)} f(s)\,ds_j\,S_R^{-1}(s,T)y$$

$$=\sum_{\ell=1}^{N}\int_0^1 \mathcal{Q}_{\gamma_\ell(t)}(T)^{-1}y\,2\mathrm{Re}\left(f(\gamma_\ell(t))(-j)\gamma_\ell'(t)\overline{\gamma_\ell(t)}\right)dt$$

$$-\sum_{\ell=1}^{N}\int_0^1 T\mathcal{Q}_{\gamma_\ell(t)}(T)^{-1}y\,2\mathrm{Re}\left(f(\gamma_\ell(t))(-j)\gamma_\ell'(t)\right)dt$$

$$=\sum_{\ell=1}^{N}\int_0^1\left(\mathcal{Q}_{\gamma_\ell(t)}(T)^{-1}y\overline{\gamma_\ell(t)}-T\mathcal{Q}_{\gamma_\ell(t)}(T)^{-1}y\right)f(\gamma_\ell(t))\gamma_\ell'(t)\,dt(-j)$$

$$-\sum_{\ell=1}^{N}\int_0^1\left(\mathcal{Q}_{\gamma_\ell(t)}(T)^{-1}y\gamma_\ell(t)-T\mathcal{Q}_{\gamma_\ell(t)}(T)^{-1}y\right)\overline{f(\gamma_\ell(t))\gamma_\ell'(t)}\,dt(-j).$$

Recalling that $f(\overline{x})=\overline{f(x)}$ because $f$ is intrinsic, that $\mathcal{Q}_{\overline{s}}(T)^{-1}=\mathcal{Q}_s(T)^{-1}$ for any $s\in\rho_S(T)$ and that $(-\overline{\gamma_\ell})(t)=-\gamma_\ell'(1-t)$, we thus find

$$\int_{\partial(U\cap\mathbb{C}_j)} f(s)\,ds_j\,S_R^{-1}(s,T)y$$

$$=\sum_{\ell=1}^{N}\int_{\gamma_\ell}\left(\mathcal{Q}_z(T)^{-1}y\overline{z}-T\mathcal{Q}_z(T)^{-1}y\right)f(z)\,dz(-j)$$

$$+\sum_{\ell=1}^{N}\int_{-\overline{\gamma_\ell}}\left(\mathcal{Q}_z(T)^{-1}y\overline{z}-T\mathcal{Q}_z(T)^{-1}y\right)f(z)\,dt(-j)$$

$$=\int_{\partial(U\cap\mathbb{C}_j)}\left(\mathcal{Q}_z(T)^{-1}y\overline{z}-T\mathcal{Q}_z(T)^{-1}y\right)f(z)\,dz(-j)$$

$$=\int_{\partial(U\cap\mathbb{C}_j)}\mathcal{R}_z(T;y)f(z)\,dz(-j).$$

Finally, observe that $f(\infty)=\lim_{s\to\infty}f(s)\in\mathbb{R}$ because, as an intrinsic function, $f$ takes only real values on the real line. Since any vector commutes with real numbers, we can hence rewrite (3.36) as

$$f(T)y=yf(\infty)+\int_{\partial(U\cap\mathbb{C}_j)}\mathcal{R}_z(T;y)f(z)\,dz\frac{(-j)}{2\pi}. \qquad\square$$

**Remark 3.4.5.** We point out that (3.35) contains neither the multiplication of vectors with non-real scalars from the left nor the multiplication of any operator with a non-real scalar. Hence, this expression is independent from the left multiplication defined on the $X$, cf. Remark 2.2.7. Instead, it shows that the operator $f(T)$

can be expressed in terms of only the right linear structure on the space $X$ if $f$ is intrinsic.

**Remark 3.4.6.** Theorem 3.4.11 shows that complex and quaternionic operator theory are consistent. Indeed, we can also obtain $f(T)$ by the following procedure: we choose $j \in \mathbb{S}$ and consider the complex numbers as embedded into the quaternions by identifying them with the plane $\mathbb{C}_j$ determined by $j$. The quaternionic Banach space $X$ is then also a complex Banach space over $\mathbb{C}_j$ and we denote the space $X$ considered as a complex Banach space over $\mathbb{C}_j$ by $X_j$. Any operator $T \in \mathcal{K}(X)$ is then also a complex linear operator on $X_j$. We have $\sigma_{\mathbb{C}_j}(T) = \sigma_S(T) \cap \mathbb{C}_j$ and $R_s(T; y)$ is for $s \in \rho_{\mathbb{C}_j} = \rho_S(T) \cap \mathbb{C}_j$ exactly the resolvent of $T$ as a complex linear operator on $X_j$, cf. Theorem 3.1.8. If $f \in \mathcal{N}(\sigma_S(T))$, then $f_j = f|_{\mathbb{C}_j}$ is a holomorphic function on $\sigma_{\mathbb{C}_j}(T)$ and the right-hand side of (3.35) is hence the formula that determines $f_j(T)$ in terms of the Riesz-Dunford functional calculus for $T$ on $X_j$. (A similar relation also holds for other functional calculi such as the Phillips functional calculus or the continuous functional calculus.) The converse is however not true: if $f_j$ is an arbitrary holomorphic function on a neighborhood of $\sigma_{\mathbb{C}_j}(T)$ in $\mathbb{C}_j$, then $f_j(T)$ obtained by the Riesz–Dunford functional calculus does not coincide with the operator $f(T)$ obtained by applying the $S$-functional calculus to $f = \mathrm{ext}_L(f_j)$. This is only true if $f$ is an intrinsic function. Indeed, $f_j(T)$ is otherwise only $\mathbb{C}_j$-linear, but not necessarily quaternionic linear.

## 3.5    The product rule and polynomials in $T$

One of the most important properties of the $S$-functional calculus is the product rule.

**Theorem 3.5.1** (Product Rule). *Let* $T \in \mathcal{K}(X)$ *with* $\rho_S(T) \neq \emptyset$. *If* $f \in \mathcal{N}(\sigma_S(T) \cup \{\infty\})$ *and* $g \in \mathcal{SH}_L(\sigma_S(T) \cup \{\infty\})$, *then*

$$(fg)(T) = f(T)g(T). \tag{3.38}$$

*Similarly, if* $f \in \mathcal{SH}_R(\sigma_S(T) \cup \{\infty\})$ *and* $g \in \mathcal{N}(\sigma_S(T) \cup \{\infty\})$, *then the product rule* (3.38) *also holds true.*

*Proof.* Let $f \in \mathcal{N}(\sigma_S(\sigma_S(T) \cup \{\infty\})$ and let $g \in \mathcal{SH}_L(\sigma_S(T) \cup \{\infty\})$. By Theorem 3.4.1, there exist unbounded slice Cauchy domains $U_p$ and $U_s$ such that $\sigma_S(T) \subset U_p$ and $\overline{U_p} \subset U_s$ and $\overline{U_s} \subset \mathcal{D}(f) \cap \mathcal{D}(g)$. The subscripts $s$ and $p$ indicate the respective variable of integration in the following computation. Moreover, we use the notation $[\partial O]_j := \partial(O \cap \mathbb{C}_j)$ for an axially symmetric set $O$ in order to obtain compacter formulas.

Recall that the operator $f(T)$ can, by Theorem 3.4.6, also be represented using the right $S$-resolvent operator and hence

$$f(T)g(T) = \left( f(\infty)\mathcal{I} + \frac{1}{2\pi} \int_{[\partial U_s]_j} f(s)\, ds_j\, S_R^{-1}(s,T) \right)$$

$$\cdot \left( g(\infty)\mathcal{I} + \frac{1}{2\pi} \int_{[\partial U_p]_j} S_L^{-1}(p,T)\, dp_j\, g(p) \right).$$

For the product of the integrals, the $S$-resolvent equation gives us that

$$\int_{[\partial U_s]_j} f(s)\, ds_j\, S_R^{-1}(s,T) \int_{[\partial U_p]_j} S_L^{-1}(p,T)\, dp_j\, g(p)$$

$$= \int_{[\partial U_s]_j} \int_{[\partial U_p]_j} f(s)\, ds_j\, S_R^{-1}(s,T) S_L^{-1}(p,T)\, dp_j\, g(p)$$

$$= \int_{[\partial U_s]_j} \int_{[\partial U_p]_j} f(s)\, ds_j\, S_R^{-1}(s,T) p(p^2 - 2s_0 p + |s|^2)^{-1}\, dp_j\, g(p)$$

$$- \int_{[\partial U_s]_j} \int_{[\partial U_p]_j} f(s)\, ds_j\, S_L^{-1}(p,T) p(p^2 - 2s_0 p + |s|^2)^{-1}\, dp_j\, g(p)$$

$$- \int_{[\partial U_s]_j} \int_{[\partial U_p]_j} f(s)\, ds_j\, \bar{s} S_R^{-1}(s,T)(p^2 - 2s_0 p + |s|^2)^{-1}\, dp_j\, g(p)$$

$$+ \int_{[\partial U_s]_j} \int_{[\partial U_p]_j} f(s)\, ds_j\, \bar{s} S_L^{-1}(p,T)(p^2 - 2s_0 p + |s|^2)^{-1}\, dp_j\, g(p).$$

For the sake of readability, let us denote these last four integrals by $I_1, \ldots I_4$.

If $r > 0$ is large enough, then $\mathbb{H} \setminus U_s$ is entirely contained in $B_r(0)$. In particular, $W := B_r(0) \cap U_p$ is then a bounded slice Cauchy domain with boundary

$$\partial(W \cap \mathbb{C}_j) = \partial(U_p \cap \mathbb{C}_j) \cup \partial(B_r(0) \cap \mathbb{C}_j).$$

From Lemma 2.2.24, we deduce

$$I_1 = \int_{[\partial U_s]_j} f(s)\, ds_j\, S_R^{-1}(s,T) \int_{[\partial U_p]_j} p(p^2 - 2s_0 p + |s|^2)^{-1}\, dp_j\, g(p)$$

$$= \int_{[\partial U_s]_j} f(s)\, ds_j\, S_R^{-1}(s,T) \int_{[\partial W]_j} p(p^2 - 2s_0 p + |s|^2)^{-1}\, dp_j\, g(p)$$

$$- \int_{[\partial U_s]_j} f(s)\, ds_j\, S_R^{-1}(s,T) \int_{[\partial B_r(0)]_j} p(p^2 - 2s_0 p + |s|^2)^{-1}\, dp_j\, g(p)$$

$$= - \int_{[\partial U_s]_j} f(s)\, ds_j\, S_R^{-1}(s,T) \int_{[\partial B_r(0)]_j} p(p^2 - 2s_0 p + |s|^2)^{-1}\, dp_j\, g(p),$$

where the last equality follows from the Cauchy integral theorem since, by our choice of $U_s$ and $U_p$, the function $p \mapsto p(p^2 - 2s_0 p + |s|^2)^{-1}$ is left slice hyperholomorphic and the function $p \mapsto g(p)$ is right slice hyperholomorphic on $\overline{W}$. If we let $r$ tend to $+\infty$ and apply Lebesgue's theorem in order to exchange limit and integration, the inner integral tends to $2\pi g(\infty)$ and hence

$$I_1 = -2\pi \left( \int_{[\partial U_s]_j} f(s)\, ds_j\, S_R^{-1}(s,T) \right) g(\infty).$$

We also have

$$-I_2 + I_4 = \int_{[\partial U_s]_j} \int_{[\partial U_p]_j} f(s)\, ds_j\, \left( \bar{s} S_L^{-1}(p,T) - p S_L^{-1}(p,T) \right)$$
$$\cdot (p^2 - 2s_0 p + |s|^2)^{-1}\, dp_j\, g(p)$$

and applying Fubini's theorem allows us to change the order of integration. If we now set $W = B_r(0) \cap U_s$ with $r$ sufficiently large, we obtain, as before, a bounded slice Cauchy domain with $\partial(W \cap \mathbb{C}_j) = \partial(U_s \cap \mathbb{C}_j) \cup \partial(B_r(0) \cap \mathbb{C}_j)$. Applying Lemma 2.2.24 with $B = S_L^{-1}(p,T)$, we find

$$-I_2 + I_4 = \int_{[\partial U_p]_j} \int_{[\partial W]_j} f(s)\, ds_j\, \left( \bar{s} S_L^{-1}(p,T) - p S_L^{-1}(p,T) \right) \cdot$$
$$\cdot (p^2 - 2s_0 p + |s|^2)^{-1}\, dp_j\, g(p)$$
$$- \int_{[\partial U_p]_j} \int_{[\partial B_r(0)]_j} f(s)\, ds_j\, \left( \bar{s} S_L^{-1}(p,T) - p S_L^{-1}(p,T) \right) \cdot$$
$$\cdot (p^2 - 2s_0 p + |s|^2)^{-1}\, dp_j\, g(p)$$
$$= 2\pi \int_{[\partial U_p]_j} S_L^{-1}(p,T) f(p)\, dp_j\, g(p)$$
$$- \int_{[\partial U_p]_j} \int_{[\partial B_r(0)]_j} f(s)\, ds_j\, \bar{s} S_L^{-1}(p,T)(p^2 - 2s_0 p + |s|^2)^{-1}\, dp_j\, g(p)$$
$$- \int_{[\partial U_p]_j} \int_{[\partial B_r(0)]_j} f(s)\, ds_j\, p S_L^{-1}(p,T)(p^2 - 2s_0 p + |s|^2)^{-1}\, dp_j\, g(p).$$

Observe that the third integral tends to zero as $r \to +\infty$. For the second one, by applying Lebesgue's theorem, we obtain

$$\int_{[\partial U_p]_j} \int_{[\partial B_r(0)]_j} f(s)\, ds_j\, \bar{s} S_L^{-1}(p,T)(p^2 - 2s_0 p + |s|^2)^{-1}\, dp_j\, g(p)$$
$$= \int_{[\partial U_p]_j} \left( \int_0^{2\pi} f(re^{i\phi}) r^2 S_L^{-1}(p,T)(p^2 - 2r\cos(\phi)p + r^2)^{-1}\, d\phi \right) dp_j\, g(p)$$
$$\xrightarrow{r \to +\infty} 2\pi f(\infty) \int_{[\partial U_p]_j} S_L^{-1}(p,T)\, dp_j\, g(p).$$

Since $f$ is intrinsic, $f(p)$ commutes with $dp_j$, and hence

$$-I_2 + I_4 = 2\pi \int_{[\partial U_p]_j} S_L^{-1}(p,T)\, dp_j\, f(p)g(p)$$

$$- 2\pi f(\infty) \int_{[\partial U_p]_j} S_L^{-1}(p,T)\, dp_j\, g(p).$$

Finally, we consider the integral $I_3$. If we set again $W = B_r(0) \cap U_p$ with $r$ sufficiently large, then

$$-I_3 = -\int_{[\partial U_s]_j} \int_{[\partial W]_j} f(s)\, ds_j\, \bar{s} S_R^{-1}(s,T)(p^2 - 2s_0 p + |s|^2)^{-1}\, dp_j\, g(p)$$

$$+ \int_{[\partial U_s]_j} \int_{[\partial B_r(0)]_j} f(s)\, ds_j\, \bar{s} S_R^{-1}(s,T)(p^2 - 2s_0 p + |s|^2)^{-1}\, dp_j\, g(p).$$

By our choice of $U_s$ and $U_p$, the functions $p \mapsto (p^2 - 2s_0 p + |s|^2)^{-1}$ and $p \mapsto g(p)$ are left (resp. right) slice hyperholomorphic on $\overline{W}$. Hence, Cauchy's integral theorem implies that the first integral equals zero. Letting $r$ tend to infinity, we can apply Lebesgue's theorem in order to exchange limit and integration and we see that

$$-I_3 = \int_{[\partial U_s]_j} \int_{[\partial B_r(0)]_j} f(s)\, ds_j\, \bar{s} S_R^{-1}(s,T)(p^2 - 2s_0 p + |s|^2)^{-1}\, dp_j\, g(p) \to 0.$$

Altogether, we obtain

$$\frac{1}{(2\pi)^2} \int_{[\partial U_s]_j} f(s)\, ds_j\, S_R^{-1}(s,T) \int_{[\partial U_p]_j} S_L^{-1}(p,T)\, dp_j\, g(p)$$

$$= -\frac{1}{2\pi} \left( \int_{[\partial U_s]_j} f(s)\, ds_j\, S_R^{-1}(s,T) \right) g(\infty) + \frac{1}{2\pi} \int_{[\partial U_p]_j} S_L^{-1}(p,T)\, dp_j\, f(p)g(p)$$

$$- f(\infty) \frac{1}{2\pi} \int_{[\partial U_p]_j} S_L^{-1}(p,T)\, dp_j\, g(p).$$

We thus have

$$f(T)g(T) = f(\infty)g(\infty)\mathcal{I} + f(\infty)\frac{1}{2\pi} \int_{[\partial U_p]_j} S_L^{-1}(p,T)\, dp_j\, g(p)$$

$$+ \left( \frac{1}{2\pi} \int_{[\partial U_s]_j} f(s)\, ds_j\, S_R^{-1}(s,T) \right) g(\infty)$$

$$+ \frac{1}{(2\pi)^2} \int_{[\partial U_s]_j} f(s)\, ds_j\, S_R^{-1}(s,T) \int_{[\partial U_p]_j} S_L^{-1}(p,T)\, dp_j\, g(p)$$

$$= f(\infty)g(\infty)\mathcal{I} + \frac{1}{2\pi} \int_{[\partial U_p]_j} S_L^{-1}(p,T)\, dp_j\, f(p)g(p) = (fg)(T). \qquad \square$$

If the operator $T$ is bounded, then slice hyperholomorphic polynomials of $T$ belong to the class of functions that are admissible within $S$-functional calculus. In the unbounded case, this is not true, but the $S$-functional calculus is in some sense still compatible, at least with intrinsic polynomials. For such polynomial $P(s) = \sum_{k=0}^{n} a_k s^k$ with $a_k \in \mathbb{R}$, the operator $P(T)$ is as usual defined as the operator

$$P(T)y := \sum_{k=0}^{n} a_k T^k y, \quad y \in \mathcal{D}(T^n).$$

**Lemma 3.5.2.** *Let $T \in \mathcal{K}(X)$ with $\rho_S(T) \neq \emptyset$, let $f \in \mathcal{N}(\sigma_S(T) \cup \{\infty\})$ and let $P$ be an intrinsic polynomial of degree $n \in \mathbb{N}_0$. If $y \in \mathcal{D}(T^n)$, then $f(T)y \in \mathcal{D}(T^n)$ and*

$$f(T)P(T)y = P(T)f(T)y.$$

*Proof.* We consider first the special case $P(s) = s$. Let $U$ be an unbounded slice Cauchy domain with $\sigma_S(T) \subset U$, let $j \in \mathbb{S}$ and let $\{\gamma_1, \ldots, \gamma_n\}$ be the part of $\partial(U \cap \mathbb{C}_j)$ in $\mathbb{C}_j^+$ as in Definition 2.1.34. We apply Lemma 3.4.5 and write

$$\int_{\partial(U \cap \mathbb{C}_j)} f(s)\, ds_j\, S_R^{-1}(s,T)$$
$$= \sum_{\ell=1}^{N} \int_0^1 2\mathrm{Re}\left(f(\gamma_\ell(t))(-j)\gamma_\ell'(t)\overline{\gamma_\ell(t)}\right) \mathcal{Q}_{\gamma_\ell(t)}(T)^{-1}\, dt$$
$$- \sum_{\ell=1}^{N} \int_0^1 2\mathrm{Re}\left(f(\gamma_\ell(t))(-j)\gamma_\ell'(t)\right) T\mathcal{Q}_{\gamma_\ell(t)}(T)^{-1}\, dt.$$

Observe that $\mathcal{Q}_{\gamma_\ell(t)}(T)^{-1}Ty = T\mathcal{Q}_{\gamma_\ell(t)}(T)^{-1}y$ for $y \in \mathcal{D}(T)$ and that $T$ also commutes with real numbers. By applying Hille's theorem for the Bochner integral, Theorem 20 in [110, Chapter III.6], we can move $T$ in front of the integral and find

$$\frac{1}{2\pi}\int_{\partial(U \cap \mathbb{C}_j)} f(s)\, ds_j\, S_R^{-1}(s,T)Ty$$
$$= \sum_{\ell=1}^{n} T\frac{1}{2\pi}\int_0^1 2\mathrm{Re}\left(f(\gamma_\ell(t))(-j)\gamma_\ell'(t)\overline{\gamma_\ell(t)}\right) \mathcal{Q}_{\gamma_\ell(t)}(T)^{-1}y$$
$$- \sum_{\ell=1}^{n} T\frac{1}{2\pi}\int_0^1 2\mathrm{Re}\left(f(\gamma_\ell(t))(-j)\gamma_\ell'(t)\right) T\mathcal{Q}_{\gamma_\ell(t)}(T)^{-1}y$$
$$= T\frac{1}{2\pi}\int_{\partial(U \cap \mathbb{C}_j)} f(s)\, ds_j\, S_R^{-1}(s,T)y,$$

where the last equation follows again from Lemma 3.4.5. Finally, observe that $f(\infty) = \lim_{s \to \infty} f(s)$ is real since $f(s) \in \mathbb{R}$ for any $s \in \mathbb{R}$ because $f$ is intrinsic.

Hence, we find

$$f(T)Ty = f(\infty)Ty + \frac{1}{2\pi} \int_{\partial(U \cap \mathbb{C}_j)} f(s) \, ds_j \, S_R^{-1}(s,T)Ty$$

$$= Tf(\infty)y + T\frac{1}{2\pi} \int_{\partial(U \cap \mathbb{C}_j)} f(s) \, ds_j \, S_R^{-1}(s,T)y = Tf(T)y.$$

In particular, this implies $f(T)y \in \mathcal{D}(T)$.

We show the general statement by induction with respect to the degree $n$ of the polynomial. If $n = 0$ then the statement follows immediately from Example 3.4.3. Now assume that it is true for $n-1$ and consider $P(s) = a_k s^n + P_{n-1}(s)$, where $a_n \in \mathbb{R}$ and $P_{n-1}(s)$ is an intrinsic polynomial of degree lower or equal to $n-1$. For $y \in \mathcal{D}(T^n)$ the above argumentation implies then $f(T)T^{n-1}y \in \mathcal{D}(T)$ and

$$f(T)P(T)y = f(T)a_n T^n y + f(T)P_{n-1}(T)y$$

$$= a_n T f(T) T^{n-1} y + f(T) P_{n-1}(T)y.$$

From the induction hypothesis, we further deduce $f(T)T^{n-1}y = T^{n-1}f(T)y$ and $f(T)P_{n-1}(T)y = P_{n-1}(T)f(T)y$ and hence

$$f(T)P(T)y = a_n T^n f(T)y + P_{n-1}(T)f(T)y = P(T)f(T)y.$$

In particular, we see that $f(T)y$ belongs to $\mathcal{D}(T^n)$ and we obtain that the statement is true. $\qquad\square$

**Remark 3.5.1.** We only considered intrinsic polynomials in Lemma 3.5.2 because only multiplying with such functions yields again a slice hyperholomorphic function. However, even the definition of $P(T)$ is not straightforward if $P$ does not have real coefficients. Indeed, if $T$ is unbounded and $P(s) = \sum_{k=0}^n s^k a_k$ with $a_k \notin \mathbb{R}$, then setting $P(T)v = \sum_{k=0}^n T^k a_k v$ might not be meaningful for all $v \in \mathcal{D}(T^n)$. Unless $\mathcal{D}(T^n)$ is a two-sided subspace of $X$, it is not clear that $a_k v \in \mathcal{D}(T^k)$ even if $v \in \mathcal{D}(T)$.

As in the complex case, we say that $f$ has a zero of order $n$ at $\infty$ if the first $n-1$ coefficients in the Taylor series expansion of $s \mapsto f(s^{-1})$ at 0 vanish and the $n$-th coefficient does not. Equivalently, $f$ has a zero of order $n$ if $\lim_{s \to \infty} f(s)s^n$ is bounded and nonzero. We say that $f$ has a zero of infinite order at infinity, if it vanishes on a neighborhood of $\infty$.

**Lemma 3.5.3.** *Let $T \in \mathcal{K}(X)$ with $\rho_S(T) \neq \emptyset$ and assume that $f \in \mathcal{N}(\sigma_S(T) \cup \{\infty\})$ has a zero of order $n \in \mathbb{N}_0 \cup \{+\infty\}$ at infinity.*

(i) *For any intrinsic polynomial $P$ of degree lower than or equal to $n$, we have*
$$P(T)f(T) = (Pf)(T).$$

(ii) *If $y \in \mathcal{D}(T^m)$ for some $m \in \mathbb{N}_0 \cup \{\infty\}$, then $f(T)y \in \mathcal{D}(T^{m+n})$.*

*Proof.* Assume first that $f$ has a zero of order greater than or equal to one at infinity and consider $P(s) = s$. Then $Pf \in \mathcal{N}(\sigma_S(T) \cup \{\infty\})$ and for $y \in X$

$$(Pf)(T)y = \lim_{s \to \infty} sf(s)y + \frac{1}{2\pi} \int_{\partial(U \cap \mathbb{C}_j)} S_L^{-1}(s,T) \, ds_j \, sf(s)y,$$

with an appropriate slice Cauchy domain $U$ and any imaginary unit $j \in \mathbb{S}$. Since $s$ and $ds_j$ commute, we deduce from the left $S$-resolvent equation that

$$\frac{1}{2\pi} \int_{\partial(U \cap \mathbb{C}_j)} S_L^{-1}(s,T) \, ds_j \, sf(s)y$$

$$= \frac{1}{2\pi} \int_{\partial(U \cap \mathbb{C}_j)} TS_L^{-1}(s,T) \, ds_j \, f(s)y + \frac{1}{2\pi} \int_{\partial(U \cap \mathbb{C}_j)} ds_j \, f(s)y$$

Any sufficiently large Ball $\overline{B_r(0)}$ contains $\partial U$. The function $f(s)y$ is then right slice hyperholomorphic on $\overline{B_r(0)} \cap U$ and Cauchy's integral theorem implies

$$\frac{1}{2\pi} \int_{\partial(U \cap \mathbb{C}_j)} ds_j \, f(s)y = \lim_{r \to +\infty} -\frac{1}{2\pi} \int_{\partial(B_r(0) \cap \mathbb{C}_j)} ds_j \, f(s)y$$

$$= \lim_{r \to +\infty} -\frac{1}{2\pi} \int_0^{2\pi} re^{j\varphi} f(re^{j\varphi})y \, d\varphi = -\lim_{s \to +\infty} sf(s)y.$$

Thus, after applying Hille's theorem for the Bochner integral, Theorem 20 in [110, Chapter III.6], in order to write the operator $T$ in front of the integral, we obtain

$$(Pf)(T)y = T\frac{1}{2\pi} \int_{\partial(U \cap \mathbb{C}_j)} S_L^{-1}(s,T) \, ds_j \, f(s)y = P(T)f(T)y.$$

In particular, we see that $f(T)y \in \mathcal{D}(T)$.

We show (i) for monomials by induction and assume that it is true for $P(s) = s^{n-1}$ if $f$ has a zero of order greater than or equal to $n-1$ at infinity. If the order of $f$ at infinity is even greater than or equal to $n$, then $g(s) = s^{n-1}f(s)$ has a zero of order at least 1 at infinity and, from the above argumentation and the induction hypothesis, we conclude for $P(s) = s^n$

$$(Pf)(T)y = Tg(T)y = TT^{n-1}f(T)y = T^n f(T)y,$$

which implies also $f(T)y \in \mathcal{D}(T^n)$. For arbitrary intrinsic polynomials the statement finally follows from the linearity of the $S$-functional calculus.

In order to show (ii) assume first $y \in \mathcal{D}(T^m)$ for $m \in \mathbb{N}$. If $f$ has a zero of order $n \in \mathbb{N}$ at infinity, then (i) with $P(s) = s^n$ and Lemma 3.5.2 imply

$$(Pf)(T)T^m y = T^n f(T)T^m y = T^n T^m f(T)y = T^{m+n} f(T)y$$

and hence $f(T)y \in \mathcal{D}(T^{m+n})$. Finally, if $m = +\infty$ then $y \in \mathcal{D}(T^k)$ and hence $f(T)y \in \mathcal{D}(T^{k+n})$ for any $k \in \mathbb{N}$. Thus, $y \in \mathcal{D}(T^\infty)$. $\qquad \square$

**Corollary 3.5.4.** *Let $T \in \mathcal{K}(X)$ with $\rho_S(T) \neq \emptyset$. For any intrinsic polynomial $P$, the operator $P(T)$ is closed.*

*Proof.* We choose $s \in \rho_S(T)$ and $n \in \mathbb{N}$ such that $m \leq 2n$, where $m$ is the degree of $P$. Then $f(p) = P(p)\mathcal{Q}_s(p)^{-n}$ belongs to $\mathcal{N}(\sigma_S(T) \cup \{\infty\})$ and has a zero of order $2n - m$ at infinity. Applying Lemma 3.5.3, we see that

$$P(T)y = P(T)\mathcal{Q}_s(T)^n \mathcal{Q}_s(T)^{-n}y = \mathcal{Q}_s(T)^n P(T)\mathcal{Q}_s(T)^{-n}y = \mathcal{Q}_s(T)^n f(T)y$$

for $y \in \mathcal{D}(T^m)$. Since its inverse is bounded, the operator $\mathcal{Q}_s(T)^n$ is closed and in turn $P(T)$ is closed as it is the composition of a closed and a bounded operator. $\quad\square$

**Corollary 3.5.5.** *Let $T \in \mathcal{K}(X)$ with $\rho_S(T) \neq \emptyset$. If $f \in \mathcal{N}(\sigma_S(T) \cup \{\infty\})$ does not have any zeros on $\sigma_S(T)$ and a zero of even order $n$ at infinity, then $\mathrm{ran}(f(T)) = \mathcal{D}(T^n)$ and $f(T)$ is invertible in the sense of closed operators. If $\rho_S(T) \cap \mathbb{R} \neq \emptyset$, this holds true for any order $n \in \mathbb{N}$.*

*Proof.* Let $p \in \rho_S(T)$ and set $k = n/2$. The function $h(s) = f(s)\mathcal{Q}_p(s)^k$ with $\mathcal{Q}_p(s) = s^2 - 2p_0s + |p|^2$ belongs to $\mathcal{N}(\sigma_S(T) \cup \{\infty\})$ and does not have any zeros in $\sigma_S(T)$. Furthermore, $h(\infty) = \lim_{s \to \infty} h(s)$ is finite and nonzero. Hence, the function $s \mapsto h(s)^{-1}$ belongs to $\mathcal{N}(\sigma_S(T) \cup \{\infty\})$ and we deduce from Theorem 3.5.1 that $h(T)$ is invertible in $\mathcal{B}(X)$ with $h(T)^{-1} = h^{-1}(T)$. Theorem 3.5.1 moreover implies $f(T) = \mathcal{Q}_p(T)^{-k}h(T)$. Now observe that $h(T)$ maps $X$ bijectively onto $X$ and that $\mathcal{Q}_p(T)^{-k}$ maps $X$ onto $\mathcal{D}(T^{2k}) = \mathcal{D}(T^n)$. Thus $\mathrm{ran}(f(T)) = \mathcal{D}(T^n)$.

Finally, $f(T)^{-1} := h^{-1}(T)\mathcal{Q}_p(T)^k$ is a closed operator because $h$ is bijective and continuous and $\mathcal{Q}_p(T)^k$ is closed by Corollary 3.5.4. So it satisfies $f(T)^{-1}f(T)y = y$ for $y \in X$ and $f(T)f(T)^{-1}y = y$ for $y \in \mathcal{D}(T^n)$. Thus, it is the inverse of $f(T)$.

In the case there exists a point $a \in \rho_S(T) \cap \mathbb{R}$, similar arguments hold with $P(s) = (s-a)^n$ instead of $\mathcal{Q}_p(s)^k$. In particular, this allows us to include functions with a zero of odd order at infinity too. $\quad\square$

We conclude this section by determining the slice derivatives of the left and right $S$-resolvents of $T$ as an application of the above theorems.

**Definition 3.5.6.** Let $T \in \mathcal{B}(X)$ and let $s \in \rho_S(T)$. For $n \geq 0$, we define

$$S_L^{-n}(s, T) := \sum_{k=0}^{n} (-1)^k \binom{n}{k} T^k \mathcal{Q}_s(T)^{-n} \overline{s}^{n-k}$$

and, similarly, we define

$$S_R^{-n}(s, T) := \sum_{k=0}^{n} (-1)^k \binom{n}{k} \overline{s}^{n-k} T^k \mathcal{Q}_s(T)^{-n}.$$

**Remark 3.5.2.** Since the function $\mathcal{Q}_s(q)^{-n}$ is intrinsic, the above definitions are due to the product rule compatibility with the $S$-functional calculus, that is,

$$\left[S_L^{-n}(s, \cdot)\right](T) = S_L^{-n}(s, T) \quad \text{and} \quad \left[S_R^{-n}(s, \cdot)\right](T) = S_R^{-n}(s, T).$$

**Proposition 3.5.7.** *Let $T \in \mathcal{B}(X)$ and let $s \in \rho_S(T)$. Then*

$$\partial_S{}^m S_L^{-1}(s,T) = (-1)^m m! \, S_L^{-(m+1)}(s,T) \qquad (3.39)$$

*and*

$$\partial_S{}^m S_R^{-1}(s,T) = (-1)^m m! \, S_R^{-(m+1)}(s,T), \qquad (3.40)$$

*for any $m \geq 0$.*

*Proof.* Recall that the slice derivative coincides with the partial derivative with respect to the real part of $s$. We show only (3.39), since (3.40) follows by analogous computations.

We prove the statement by induction. For $m = 0$, the identity (3.39) is obvious. We assume that $\partial_S{}^{m-1} S_L^{-1}(s,T) = (-1)^{m-1}(m-1)! \, S_L^{-m}(s,T)$ and we compute $\partial_S{}^m S_L^{-1}(s,T)$. We represent $S_L^{-m}(s,T)$ using the $S$-functional calculus. If we choose the path of integration $\partial(U \cap \mathbb{C}_j)$ in the complex plane $\mathbb{C}_j$ that contains $s$, then $S_L^{-m}(s,p) = (s-p)^{-m}$ for any $p \in \partial(U \cap \mathbb{C}_j)$ so that

$$\begin{aligned}
\partial_S S_L^{-m}(s,T) &= \partial_S \frac{1}{2\pi} \int_{\partial(U \cap \mathbb{C}_j)} S_L^{-1}(p,T) \, dp_j \, S_L^{-m}(s,p) \\
&= \frac{1}{2\pi} \int_{\partial(U \cap \mathbb{C}_j)} S_L^{-1}(p,T) \, dp_j \, \frac{\partial}{\partial s_0}(s-p)^{-m} \\
&= -m \frac{1}{2\pi} \int_{\partial(U \cap \mathbb{C}_j)} S_L^{-1}(p,T) \, dp_j \, (s-p)^{-(m+1)} \\
&= -m \, S_L^{-(m+1)}(s,T),
\end{aligned}$$

and in turn,

$$\begin{aligned}
\partial_S{}^m S_L^{-1}(s,T) &= \partial_S \left( \partial_S{}^{m-1} S_L^{-1}(s,T) \right) \\
&= (-1)^{m-1}(m-1)! \partial_S S_L^{-m}(s,T) = (-1)^m m! \, S_L^{-(m+1)}(s,T). \quad \square
\end{aligned}$$

## 3.6 The spectral mapping theorem

We recall that the extended $S$-spectrum $\sigma_{SX}(T)$ equals $\sigma_S(T)$ if $T$ is bounded and it equals $\sigma_S(T) \cup \{\infty\}$ if $T$ is unbounded.

**Theorem 3.6.1** (Spectral Mapping Theorem). *Let $T \in \mathcal{K}(X)$ with $\rho_S(T) \neq \emptyset$. For any function $f \in \mathcal{N}(\sigma_S(T) \cup \{\infty\})$, we have $\sigma_S(f(T)) = f(\sigma_{SX}(T))$.*

*Proof.* Let us first show the relation $\sigma_S(f(T)) \supset f(\sigma_{SX}(T))$. For $p \in \sigma_S(T)$ consider the function

$$g(s) := (f(s)^2 - 2\mathrm{Re}(f(p))f(s) - |f(p)|^2)(s^2 - 2\mathrm{Re}(p)s + |p|^2)^{-1},$$

which is defined on $\mathcal{D}(f) \setminus [p]$. If we set $p_{j_s} = p_0 + j_s p_1$, then $p_{j_s}$ and $s$ commute. Since $f$ is intrinsic, it maps $\mathbb{C}_j$ into $\mathbb{C}_j$ and hence $f(p_{j_s})$ and $f(s)$ commute, too. Thus

$$g(s) = \frac{(f(s) - f(p_{j_s}))(f(s) - \overline{f(p_{j_s})})}{(s - p_{j_s})(s - \overline{p_{j_s}})}$$

and we can extend $g$ to all of $\mathcal{D}(f)$ by setting

$$g(s) = \begin{cases} \partial_S f(s) \left( f(p) p^{-1} \right), & s \in [p] \text{ if } p \notin \mathbb{R}, \\ (\partial_S f(s))^2, & s = p, \text{ if } p \in \mathbb{R}. \end{cases} \tag{3.41}$$

Now observe that

$$(s^2 - 2\operatorname{Re}(p)s + |p|^2)g(s) = f(s)^2 + 2\operatorname{Re}(f(p))f(s) + |f(p)|^2$$

and that $g$ has a zero of order greater or equal to 2 at infinity. Hence, we can apply the $S$-functional calculus to deduce from Lemma 3.5.3, Theorem 3.5.1 and Example 3.4.3 that

$$(T^2 - 2\operatorname{Re}(p)T + |p|^2 \mathcal{I})g(T)y = (f(T)^2 + 2\operatorname{Re}(f(p))f(T) + |f(p)|\mathcal{I})y,$$

for any $y \in X$ and

$$g(T)(T^2 - 2\operatorname{Re}(p)T + |p|^2 \mathcal{I})y = (f(T)^2 + 2\operatorname{Re}(f(p))f(T) + |f(p)|\mathcal{I})y,$$

for $y \in \mathcal{D}(T^2)$. If $f(p) \in \rho_S(T)$, then

$$\mathcal{Q}_{f(p)}(f(T)) = f(T)^2 - 2\operatorname{Re}(f(p))f(T) + |f(p)|\mathcal{I}$$

is invertible and

$$\mathcal{Q}_{f(p)}(f(T))^{-1}g(T) = g(T)\mathcal{Q}_{f(p)}(f(T))^{-1}$$

is the inverse of the operator $\mathcal{Q}_p(T) = T^2 - 2\operatorname{Re}(p)T + |p|^2 \mathcal{I}$. Hence, $f(p) \notin \sigma_S(f(T))$ implies $p \notin \sigma_S(T)$ and as a consequence $p \in \sigma_S(T)$ implies $f(p) \in \sigma_S(f(T))$, that is $f(\sigma_S(T)) \subset \sigma_S(f(T))$.

Finally, observe that $f(\infty) = \lim_{p \to \infty} f(p)$ is real because $f$ is intrinsic and thus takes real values on the real line. If $T$ is unbounded and $f(\infty) \neq f(p)$ for any point $p \in \sigma_S(T)$ (otherwise we already have $f(\infty) \in f(\sigma_S(T)) \subset \sigma_S(f(T))$), then the function $h(s) = (f(s) - f(\infty))^2$ belongs to $\mathcal{N}(\sigma_S(T) \cup \{\infty\})$ and has a zero of even order $n$ at infinity but no zero in $\sigma_S(T)$. By Corollary 3.5.5, the range of $h(T) = \mathcal{Q}_{f(\infty)}(f(T))$ is $\mathcal{D}(T^n)$. Thus, it does not admit a bounded inverse and we obtain $f(\infty) \in \sigma_S(f(T))$. Altogether, we have $f(\sigma_{SX}(T)) \subset \sigma_S(f(T))$.

In order to show the relation $\sigma_S(f(T)) \subset f(\sigma_{SX}(T))$, we first consider a point $c \in \sigma_S(f(T))$ such that $c \neq f(\infty)$. We want to show $c \in f(\sigma_S(T))$ and assume the converse, i.e., $f(s) - c$ has no zeros on $\sigma_S(T)$.

If $c$ is real, then the function $h(s) = f(s) - c$ is intrinsic, has no zeros on $\sigma_S(T)$ and $\lim_{s \to \infty} h(s) = f(\infty) - c \neq 0$. Hence, $h^{-1}(s) = (f(s) - c)^{-1}$ belongs to $\mathcal{N}(\sigma_S(T) \cup \{\infty\})$. Applying the $S$-functional calculus, we deduce from Theorem 3.5.1 that $h^{-1}(T)$ is the inverse of $f(T) - c\mathcal{I}$ and hence $\mathcal{Q}_c(f(T))^{-1} = (h^{-1}(T))^2$, which is a contradiction as $c \in \sigma_S(f(T))$. Thus, $c = f(p)$ for some $p \in \sigma_S(T)$.

If on the other hand $c = c_0 + ic_1$ is not real, then $f - c_j \neq 0$ for any $c_j = c_0 + jc_1 \in [c]$. Indeed, $f(p) = f_0(u, v) + kf_1(u, v) = c_0 + jc_1$ for $p = u + kv$ would imply $k = j$ and $f_0(u, v) = c_0$ and $f_1(u, v) = c_1$ as $f_0$ and $f_1$ are real-valued because $f$ is intrinsic. This would in turn imply $f(p_i) = f(u + iv) = f_0(u, v) + if_1(u, v) = c$ for $p = u + iv$, which would contradict our assumption. Therefore, the function

$$h(s) = (f(s)^2 - 2\mathrm{Re}(c)f(s) + |c|^2) = (f(s) - c_{j_s})(f(s) - \overline{c_{j_s}})$$

with $c_{j_s} = c_0 + jc_2$ for $s = u + jv$ does not have any zeros on $\sigma_S(T)$. Moreover, since $f(\infty)$ is real, we have

$$h(\infty) = (f(\infty) - c)\overline{(f(\infty) - c)} = |f(\infty) - c|^2 \neq 0$$

and hence $h^{-1}(s) = (f(s)^2 - 2\mathrm{Re}(c)f(s) + |c|^2)^{-1}$ belongs to $\mathcal{N}(\sigma_S(T) \cup \{\infty\})$. Applying the $S$-functional calculus, we deduce again from Theorem 3.5.1 that the operator $h^{-1}(T)$ is the inverse of $\mathcal{Q}_c(T)$, which contradicts $c \in \sigma_S(f(T))$. Hence, there must exist some $p \in \sigma_S(T)$ such that $c = f(p)$.

Altogether, we obtain $\sigma_S(f(T)) \setminus \{f(\infty)\}$ is contained in $f(\sigma_S(T))$.

Finally, let us consider the case that the point $c = f(\infty)$ belongs to $\sigma_S(f(T))$. If $T$ is unbounded, then $\infty \in \sigma_{SX}(T)$ and hence $c \in f(\sigma_{SX}(T))$. If on the other hand $T$ is bounded, then there exists a function $g \in \mathcal{N}(\sigma_S(T) \cup \{\infty\})$ that coincides on an axially symmetric neighborhood $\sigma_S(T)$ with $f$ but satisfies $c \neq g(\infty)$. In this case $f(T) = g(T)$, as pointed out in Remark 3.4.1, and we can apply the above argumentation with $g$ instead of $f$ to see that $c \in g(\sigma_S(T)) = f(\sigma_S(T))$. $\square$

**Theorem 3.6.2.** *If $T \in \mathcal{K}(X)$ with $\sigma_S(T) \neq \emptyset$, then $P(\sigma_S(T)) = \sigma_S(P(T))$ for any intrinsic polynomial $P$.*

*Proof.* The arguments are similar to those in the proof of Theorem 3.6.1: in order to show $P(\sigma_S(T)) \subset \sigma_S(P(T))$, we consider the polynomial $\mathcal{Q}_{P(p)}(P(s))$, which is given by $\mathcal{Q}_{P(p)}(P(s)) = P(s)^2 - 2\mathrm{Re}(P(p))P(s) + |P(p)|^2$ for any $p \in \sigma_S(T)$. As $p$ and $\overline{p}$ are both zeros of $\mathcal{Q}_{P(p)}(P(s))$ (resp. as $p$ is a zero of even order of $\mathcal{Q}_{P(p)}(P(s)) = (P(s) - P(p))^2$ if $p$ is real), there exists an intrinsic polynomial $R(s)$ such that

$$\mathcal{Q}_{P(p)}(P(s)) = \mathcal{Q}_p(s)R(s).$$

If $P(p) \notin \sigma_S(P(T))$, then $\mathcal{Q}_{P(p)}(P(T))$ is invertible and Lemma 3.5.3 and Example 3.4.3 imply that $\mathcal{Q}_{P(p)}(P(T))^{-1}R(T)$ is the inverse of $\mathcal{Q}_p(T)$, which is a contradiction because we assumed $p \in \sigma_S(T)$. Therefore $P(p) \in \sigma_S(P(T))$.

Conversely assume that $p \notin P(\sigma_S(T))$. Then the function

$$\mathcal{Q}_p(P(s)) = P(s)^2 - 2\mathrm{Re}(p)P(s) + |p|^2$$

does not take any zero on $\sigma_S(T)$ and we conclude from Corollary 3.5.5 that $\mathcal{Q}_p(P(T))$ has a bounded inverse. Thus $p \notin \sigma_S(P(T))$ and in turn $\sigma_S(P(T)) \subset P(\sigma_S(T))$. □

**Theorem 3.6.3** (Composition rule). *Let* $T \in \mathcal{K}(X)$ *with* $\rho_S(T) \neq \emptyset$. *If* $f \in \mathcal{N}(\sigma_S(T) \cup \{\infty\})$ *and* $g \in \mathcal{SH}_L(f(\sigma_{SX}(T))$ *or* $g \in \mathcal{SH}_R(f(\sigma_{SX}(T))$, *then*

$$(g \circ f)(T) = g(f(T)).$$

*Proof.* Because of Remark 3.4.1, we can assume that $f(\infty)$ belongs to $f(\sigma_{SX}(T))$. We apply Theorem 2.1.31 in order to choose an unbounded slice Cauchy domain $U_p$ such that $\sigma_S(f(T)) = f(\sigma_{SX}(T)) \subset U_p$ and $\overline{U}_p \subset \mathcal{D}(g)$ and a second unbounded slice Cauchy domain $U_s$ such that $\sigma_S(T) \subset U_s$ and $\overline{U}_s \subset f^{-1}(U_p) \cap \mathcal{D}(f)$. The subscripts are chosen in order to indicate the respective variable of integration in the following computation.

After choosing an imaginary unit $j \in \mathbb{S}$, we deduce from Cauchy's integral formula, that

$$
\begin{aligned}
&(g \circ f)(T) - (g \circ f)(\infty)\mathcal{I} \\
&= \frac{1}{2\pi} \int_{\partial(U_s \cap \mathbb{C}_j)} S_L^{-1}(s, T) \, ds_j \, (g \circ f)(s) \\
&= \frac{1}{2\pi} \int_{\partial(U_s \cap \mathbb{C}_j)} S_L^{-1}(s, T) \, ds_j \left( \frac{1}{2\pi} \int_{\partial(U_p \cap \mathbb{C}_j)} S_L^{-1}(p, f(s)) \, dp_j \, g(p) \right).
\end{aligned}
$$

Changing the order of integration by applying Fubini's theorem, we obtain

$$
\begin{aligned}
&(g \circ f)(T) - (g \circ f)(\infty)\mathcal{I} \\
&= \frac{1}{2\pi} \int_{\partial(U_p \cap \mathbb{C}_j)} \left( \frac{1}{2\pi} \int_{\partial(U_s \cap \mathbb{C}_j)} S_L^{-1}(s, T) \, ds_j \, S_L^{-1}(p, f(s)) \right) dp_j \, g(p) \\
&= \frac{1}{2\pi} \int_{\partial(U_p \cap \mathbb{C}_j)} S_L^{-1}(p, f(T)) \, dp_j \, g(p) \\
&\quad - \frac{1}{2\pi} \int_{\partial(U_p \cap \mathbb{C}_j)} S_L^{-1}(p, f(\infty)) \, dp_j \, g(p)\mathcal{I} \\
&= g(f(T)) - g(f(\infty))\mathcal{I}
\end{aligned}
$$

and hence $(g \circ f)(T) = g(f(T))$. □

## 3.7 Spectral sets and projections onto invariant subspaces

As in the complex case, the $S$-functional calculus allows to associate subspaces of $X$ that are invariant under $T$ to certain subsets of $\sigma_S(T)$.

**Definition 3.7.1** (Spectral set). A subset $\sigma$ of $\sigma_{SX}(T)$ is called a spectral set if it is open and closed in $\sigma_{SX}(T)$.

Just as $\sigma_S(T)$ and $\sigma_{SX}(T)$, every spectral set is axially symmetric: if $s \in \sigma$ then the entire sphere $[s]$ is contained in $\sigma$. Indeed, the set $\sigma \cap [s]$ is then a nonempty, open and closed subset of $\sigma_{SX}(T) \cap [s] = [s]$. Since $[s]$ is connected this implies $\sigma \cap [s] = [s]$. Moreover, if $\sigma$ is a spectral set, then $\sigma' = \sigma_{SX}(T) \setminus \sigma$ is a spectral set, too.

If $\sigma$ is a spectral set of $T$, then $\sigma$ and $\sigma'$ can be separated in $\mathbb{H}_\infty$ by axially symmetric open sets and hence Theorem 2.1.31 implies the existence of two slice Cauchy domains $U_\sigma$ and $U_{\sigma'}$ containing $\sigma$ and $\sigma'$, respectively, such that one of them is unbounded and $\overline{U} \cap \overline{U_{\sigma'}} = \emptyset$. We define

$$\chi_\sigma(x) := \begin{cases} 1 & \text{if } x \in U_\sigma, \\ 0 & \text{if } x \in U_{\sigma'}. \end{cases}$$

The function $\chi_\sigma(x)$ obviously belongs to $\mathcal{N}(\sigma_S(T) \cup \{\infty\})$.

**Definition 3.7.2** (Spectral projection). Let $T \in \mathcal{K}(X)$ with $\rho_S(T) \neq \emptyset$ and let $\sigma \subset \sigma_S(T)$ be a spectral set of $T$. The spectral projection associated with $\sigma$ is the operator $E_\sigma := \chi_\sigma(T)$ obtained by applying the $S$-functional calculus to the function $\chi_\sigma$. Furthermore, we define $X_\sigma := E_\sigma X$ and $T_\sigma = T|_{\mathcal{D}(T_\sigma)}$ with $\mathcal{D}(T_\sigma) = \mathcal{D}(T) \cap X_\sigma$.

Explicit formulas for the operator $E_\sigma$ are for bounded $\sigma$ are given by

$$E_\sigma = \frac{1}{2\pi} \int_{\partial(U_\sigma \cap \mathbb{C}_j)} S_L^{-1}(s,T)\, ds_j = \frac{1}{2\pi} \int_{\partial(U_\sigma \cap \mathbb{C}_j)} ds_j\, S_R^{-1}(s,T)$$

and for unbounded $\sigma$

$$E_\sigma = \mathcal{I} + \frac{1}{2\pi} \int_{\partial(U_\sigma \cap \mathbb{C}_j)} S_L^{-1}(s,T)\, ds_j = \mathcal{I} + \frac{1}{2\pi} \int_{\partial(U_\sigma \cap \mathbb{C}_j)} ds_j\, S_R^{-1}(s,T),$$

where the imaginary unit $j \in \mathbb{S}$ can be chosen arbitrarily.

**Corollary 3.7.3.** *Let $T \in \mathcal{K}(X)$ such that $\rho_S(T) \neq \emptyset$ and let $\sigma$ be a spectral set of $T$.*

(i) *The operator $E_\sigma$ is a projection, i.e., $E_\sigma^2 = E_\sigma$.*

(ii) *Set $\sigma' = \sigma_{SX}(T) \setminus \sigma$. Then $E_\sigma + E_{\sigma'} = \mathcal{I}$ and $E_\sigma E_{\sigma'} = E_{\sigma'} E_\sigma = 0$.*

*Proof.* This follows immediately from the algebraic properties of the $S$-functional calculus shown in Corollary 3.4.4 and Theorem 3.5.1 as $\chi_\sigma^2 = \chi_\sigma$ and $\chi_\sigma + \chi_{\sigma'} = 1$ and $\chi_\sigma \chi_{\sigma'} = \chi_{\sigma'} \chi_\sigma = 0$. $\qquad\square$

The following Lemma 3.7.4 is a special case of [47, Chapter II §1.9, Proposition 14] and Lemma 3.7.5 is an immediate consequence of the fact that any projection with closed range is continuous.

**Lemma 3.7.4.** *Let $A$, $B$, $M$ and $N$ be right linear subspaces of a quaternionic right vector space $X_R$ such that $A \subset M$ and $B \subset M$. If $A \oplus B = M \oplus N$, then $A = M$ and $B = N$.*

**Lemma 3.7.5.** *Let $A$, $B$, $M$ and $N$ be right linear subspaces of a quaternionic Banach vector space $X_R$ such that $A \subset M$, $B \subset N$ and such that $M$, $N$ and $M \oplus N$ are closed. Then $A \oplus B$ is dense in $M \oplus N$ if and only if $A$ is dense in $M$ and $B$ is dense in $N$.*

**Definition 3.7.6.** *Let $T : D(T) \to X$. We split the $S$-spectrum into the three disjoint sets:*

(P) *The point $S$-spectrum of $T$:*

$$\sigma_{Sp}(T) = \{s \in \mathbb{H} : \quad \ker(\mathcal{Q}_s(T)) \neq \{0\}\}.$$

(R) *The residual $S$-spectrum of $T$:*

$$\sigma_{Sr}(T) = \left\{s \in \mathbb{H} : \quad \ker(\mathcal{Q}_s(T)) = \{0\}, \ \overline{\operatorname{ran}(\mathcal{Q}_s(T))} \neq X\right\}.$$

(C) *The continuous $S$-spectrum of $T$:*

$$\sigma_{Sc}(T) = \left\{s \in \mathbb{H} : \quad \ker(\mathcal{Q}_s(T)) = \{0\}, \ \overline{\operatorname{ran}(\mathcal{Q}_s(T))} = X, \ \mathcal{Q}_s(T)^{-1} \notin \mathcal{B}(X)\right\}.$$

There are different possible ways to split the $S$-spectrum. We refer to Section 9.2 in [57] for more details and comments.

**Theorem 3.7.7.** *Let $T \in \mathcal{K}(X)$ with $\rho_S(T) \neq \emptyset$ and let $E_1, E_2 \in \mathcal{B}(X)$ be projections such that $E_1 + E_2 = \mathcal{I}$ (and hence $E_1 E_2 = E_2 E_1 = 0$). Denote $X_\ell := E_\ell(X)$ and $D(T_\ell) := E_\ell(D(T))$ and assume that $T(D(T_\ell)) \subset X_\ell$ such that $T_\ell := T|_{D(T_\ell)}$ is a closed operator on the right Banach space $X_\ell$ for $\ell = 1, 2$. Then*

(i) $E_\ell T y = T E_\ell y$ *for $y \in D(T)$,*

(ii) $D(T_\ell^2) = E_\ell(D(T^2))$ *for $\ell = 1, 2$,*

(iii) $\operatorname{ran}(\mathcal{Q}_s(T)) = \operatorname{ran}(\mathcal{Q}_s(T_1)) \oplus \operatorname{ran}(\mathcal{Q}_s(T_2))$, *for any $s \in \mathbb{H}$,*

(iv) $\sigma_S(T) = \sigma_S(T_1) \cup \sigma_S(T_2)$ *and*

(v) $\sigma_{Sp}(T) = \sigma_{Sp}(T_1) \cup \sigma_{Sp}(T_2)$.

*If moreover $\sigma_S(T_1) \cap \sigma_S(T_2) = \emptyset$, then*

*(vi)* $\sigma_{Sc}(T) = \sigma_{Sc}(T_1) \cup \sigma_{Sc}(T_2)$ *and*

*(vii)* $\sigma_{Sr}(T) = \sigma_{Sr}(T_1) \cup \sigma_{Sr}(T_2)$.

*Proof.* The assertions (i) to (iii) are obvious. Now assume that $s \in \rho_S(T)$. Then $\operatorname{ran}(\mathcal{Q}_s(T)) = X$ and from (iii) we deduce

$$X_1 \oplus X_2 = X = \operatorname{ran}(\mathcal{Q}_s(T)) = \operatorname{ran}(\mathcal{Q}_s(T_1)) \oplus \operatorname{ran}(\mathcal{Q}_s(T_2)).$$

As $\operatorname{ran}(\mathcal{Q}_s(T_\ell)) \subset X_\ell$, Lemma 3.7.4 implies $\operatorname{ran}(\mathcal{Q}_s(T_\ell)) = X_\ell$ and hence $\mathcal{Q}_s(T_\ell)^{-1} = \mathcal{Q}_s(T)^{-1}|_{X_\ell}$ as $\mathcal{Q}_s(T_\ell) = \mathcal{Q}_s(T)|_{\mathcal{D}(T_\ell^2)}$. Indeed, we have

$$\mathcal{Q}_s(T)^{-1}\mathcal{Q}_s(T_\ell)y = \mathcal{Q}_s(T)^{-1}\mathcal{Q}_s(T)y = y \quad \text{for } y \in \mathcal{D}(T_\ell^2)$$

and, since $\mathcal{Q}_s(T)^{-1}y \in \mathcal{D}(T_\ell^2)$ for $y \in X_\ell$, also

$$\mathcal{Q}_s(T_\ell)\mathcal{Q}_s(T)^{-1}y = \mathcal{Q}_s(T)\mathcal{Q}_s(T)^{-1}y = y \quad \text{for } y \in X_\ell.$$

Thus, $s \in \rho_S(T_1) \cap \rho_S(T_2)$. Conversely, if $s \in \rho_S(T_1) \cap \rho_S(T_2)$, then the operator $\mathcal{Q}_s(T_1)^{-1}E_1 + \mathcal{Q}_s(T_2)^{-1}E_2$ is the inverse of $\mathcal{Q}_s(T)$ and hence $s \in \rho_S(T)$. Altogether, $\rho_S(T) = \rho_S(T_1) \cap \rho_S(T_2)$, which is equivalent to $\sigma_S(T) = \sigma_S(T_1) \cup \sigma_S(T_2)$ and hence (iv) holds true.

Obviously, $\sigma_{Sp}(T_\ell) \subset \sigma_{Sp}(T)$ as any $S$-eigenvector of $T_\ell$ is also an $S$-eigenvector of $T$ associated with the same eigensphere. Conversely, if $y \neq 0$ is an $S$-eigenvector of $T$ associated with the eigensphere $[s] = s_0 + \mathbb{S}s_1$, then set $y_\ell = E_\ell y$ and we observe that

$$0 = \mathcal{Q}_s(T)y = \mathcal{Q}_s(T_1)y_1 + \mathcal{Q}_s(T_2)y_2.$$

As $\mathcal{Q}_s(T_\ell)y_\ell \in X_\ell$ and $X_1 \cap X_2 = \{0\}$, this implies $\mathcal{Q}_s(T_\ell)y_\ell = 0$ for $\ell = 1, 2$. As $y \neq 0$, at least one of the vectors $y_\ell$ is nonzero and therefore an $S$-eigenvalue of $T_\ell$ associated with the eigensphere $[s]$. Thus $[s] \subset \sigma_{Sp}(T_1) \cup \sigma_{Sp}(T_2)$ and in turn $\sigma_{Sp}(T) = \sigma_{Sp}(T_1) \cup \sigma_{Sp}(T_2)$ so that (v) holds true.

We assume now that $\sigma_S(T_1) \cap \sigma_S(T_2) = \emptyset$. Then assertions (iv) and (v) imply that $s \in \sigma_{Sc}(T) \cup \sigma_{Sr}(T)$ if and only if $s \in \sigma_{Sc}(T_\ell) \cup \sigma_{Sr}(T_\ell)$ for either $\ell = 1$ or $\ell = 2$. We assume without loss of generality $s \in \sigma_{Sc}(T_1) \cup \sigma_{Sr}(T_1)$ and thus $s \in \rho_S(T_2)$. As $\operatorname{ran}(\mathcal{Q}_s(T_2)) = X_2$, we deduce from (iii) and Lemma 3.7.5 that $\operatorname{ran}(\mathcal{Q}_s(T))$ is dense in $X = X_1 \oplus X_2$ if and only if $\operatorname{ran}(\mathcal{Q}_s(T_1))$ is dense in $X$. In other words, $s \in \sigma_{Sc}(T)$ if and only if $s \in \sigma_{Sc}(T_1)$ and in turn $s \in \sigma_{Sr}(T)$ if and only if $s \in \sigma_{Sr}(T_1)$. $\qquad \square$

**Theorem 3.7.8.** *Let $T \in \mathcal{K}(X)$ with $\rho_S(T) \neq \emptyset$ and let $\sigma \subset \sigma_S(T)$ be a spectral set of $T$. Then*

(i) $E_\sigma(\mathcal{D}(T)) \subset \mathcal{D}(T)$,

(ii) $T(\mathcal{D}(T) \cap X_\sigma) \subset X_\sigma$,

(iii) $\sigma = \sigma_{SX}(T_\sigma)$,

(iv) $\sigma \cap \sigma_{Sp}(T) = \sigma_{Sp}(T_\sigma)$,

(v) $\sigma \cap \sigma_{Sc}(T) = \sigma_{Sc}(T_\sigma)$,

(vi) $\sigma \cap \sigma_{Sr}(T) = \sigma_{Sr}(T_\sigma)$.

*If the spectral set $\sigma$ is bounded, then we further have:*

*(vii) $X_\sigma \subset \mathcal{D}(T^\infty)$ and*

*(viii) $T_\sigma$ is a bounded operator on $X_\sigma$.*

*Proof.* Assertion (i) follows from the definition of $E_\sigma$ and Lemma 3.5.2. In order to prove (ii), we observe that if $y \in \mathcal{D}(T) \cap X_\sigma$, then $E_\sigma y = y$. Hence, we deduce from Lemma 3.5.2 that $E_\sigma T y = T E_\sigma y = T y$, which implies $T y \in X_\sigma$.

If $\sigma$ is bounded, then we can choose $U_\sigma$ bounded and hence $\chi_\sigma$ has a zero of infinite order at infinity. We conclude from Lemma 3.5.3 that $y = E_\sigma y = \chi_\sigma(T) y \in \mathcal{D}(T^\infty)$ for any $y \in X_\sigma$ and hence (vii) holds true. In particular, $X_\sigma \subset \mathcal{D}(T)$. Therefore, $T_\sigma$ is a bounded operator on $X_\sigma$ as it is closed and everywhere defined.

We show now assertion (iii) and consider first a point $s \in \mathbb{H} \setminus \sigma$. We show that $s$ belongs to $\rho_S(T_\sigma)$. For an appropriately chosen slice Cauchy domain $U_\sigma$, the function $f(s) := \mathcal{Q}_s(p)^{-1}\chi_{U_\sigma}(s)$ belongs to $\mathcal{N}(\sigma_S(T) \cup \{\infty\})$. By Lemma 3.5.3 and Lemma 3.5.2, we have

$$f(T)\mathcal{Q}_s(T)y = \chi_{U_\sigma}(T)y = E_\sigma y, \quad \text{for } y \in \mathcal{D}(T^2) \cap X_\sigma$$

and

$$\mathcal{Q}_s(T)f(T)y = \chi_{U_\sigma}(T)y = E_\sigma y = y \quad \text{for } y \in X_\sigma.$$

Hence, $\mathcal{Q}_s(T_\sigma) = \mathcal{Q}_s(T)|_{X_\sigma \cap \mathcal{D}(T^2)}$ has the inverse $f(T)|_{X_\sigma} \in \mathcal{B}(X_\sigma)$. Thus, we find $s \in \rho_S(T_\sigma)$ and in turn $\sigma_S(T_\sigma) \subset \sigma \cap \mathbb{H} =: \sigma_1$. The same arguments applied to $T_{\sigma'}$ with $\sigma' = \sigma_{SX}(T) \setminus \sigma$ show that $\sigma_S(T_{\sigma'}) \subset \sigma' \cap \mathbb{H} := \sigma_2$. But by (iv) in Theorem 3.7.7, we have

$$\sigma_S(T_\sigma) \cup \sigma_S(T_\sigma) = \sigma_S(T) = \sigma_1 \cup \sigma_2$$

and hence $\sigma_S(T_\sigma) = \sigma_1 = \sigma \cap \mathbb{H}$ and $\sigma_S(T_{\sigma'}) = \sigma_2 = \sigma' \cap \mathbb{H}$. If $\sigma$ is bounded, then this is equivalent to (iii) because of (viii). If $\sigma$ is not bounded, then $\infty \in \sigma$ and $T$ is not bounded on $X$. However, in this case $\sigma'$ is bounded and hence $T_{\sigma'} \in \mathcal{B}(X_{\sigma'})$. But as $T = T_\sigma E_\sigma + T_{\sigma'} E_{\sigma'}$, we conclude that $T_\sigma$ is unbounded as $T$ is unbounded. Hence, $\infty \in \sigma_{SX}(T_\sigma)$ and (viii) holds true also in this case.

Finally, (iv) to (vi) are direct consequences of (v) to (vii) in Theorem 3.7.7 as we know now that $\sigma_S(T_\sigma)$ and $\sigma_S(T_{\sigma'})$ are disjoint.                      $\square$

**Example 3.7.9.** We choose a generating basis $j, i$ and $k := ji$ of $\mathbb{H}$ and consider the quaternionic right-linear operator $T$ on $X = \mathbb{H}^2$ that is defined by its action on the two right linearly independent right eigenvectors $y_1 = (1, j)^T$ and $y_2 = (i, -k)^T$, namely

$$\begin{pmatrix} 1 \\ j \end{pmatrix} \mapsto \begin{pmatrix} 0 \\ 0 \end{pmatrix} \quad \text{and} \quad \begin{pmatrix} i \\ -k \end{pmatrix} \mapsto \begin{pmatrix} -k \\ -i \end{pmatrix} = \begin{pmatrix} i \\ -k \end{pmatrix} j.$$

Its matrix representation is

$$T = \frac{1}{2} \begin{pmatrix} -j & 1 \\ -1 & -j \end{pmatrix}.$$

Since, for operators on finite-dimensional spaces, the $S$-spectrum coincides with the set of right-eigenvalues by Theorem 3.1.6, we have $\sigma_S(T) = \sigma_R(T) = \{0\} \cup \mathbb{S}$. Indeed, we have

$$Q_s(T) = \frac{1}{2} \begin{pmatrix} -1 & -j \\ j & -1 \end{pmatrix} - s_0 \begin{pmatrix} -j & 1 \\ -1 & -j \end{pmatrix} + |s|^2 \begin{pmatrix} 1 & 0 \\ 0 & 1 \end{pmatrix}$$

$$= \begin{pmatrix} -\frac{1}{2} + |s|^2 + s_0 j & -s_0 - \frac{1}{2}j \\ s_0 + \frac{1}{2}j & -\frac{1}{2} + |s|^2 + s_0 j \end{pmatrix}$$

and hence

$$Q_s(T)^{-1} = |s|^{-2}(-1 + 2js_0 + |s|^2)^{-1} \begin{pmatrix} -\frac{1}{2} + |s|^2 + js_0 & \frac{1}{2}j + s_0 \\ -\frac{1}{2}j - s_0 & -\frac{1}{2} + |s|^2 + js_0 \end{pmatrix},$$

which is defined for any $s \notin \{0\} \cup \mathbb{S}$. For any $s \in \rho_S(T)$, the left $S$-resolvent is therefore given by

$$S_L^{-1}(s, T) = \frac{1}{2}|s|^{-2}(-1 + |s|^2 + 2js_0)^{-1}$$

$$\cdot \begin{pmatrix} |s|^2(j + 2\bar{s}) + \bar{s}(-1 + 2js_0) & -|s|^2 + \bar{s}(j + 2s_0) \\ |s|^2 - \bar{s}(j + 2s_0) & |s|^2(j + 2\bar{s}) + \bar{s}(-1 + 2js_0) \end{pmatrix}.$$

Since $\sigma_S(T) \cap \mathbb{C}_j = \{0, j, -j\}$, we choose $U_{\{0\}} = B_{1/2}(0)$ and set $U_{\mathbb{S}} = B_2(0) \setminus B_{2/3}(0)$. For $s = \frac{1}{2}e^{j\varphi} \in \partial U_{\{0\}}(0) \cap \mathbb{C}_j$, we have

$$S_L^{-1}(s, T) = 2e^{-j\varphi} \left( 3j + 4\mathrm{Re}\left(e^{j\varphi}\right) \right)^{-1}$$

$$\cdot \begin{pmatrix} j + e^{j\varphi} + 2\cos(\varphi) & 2 + je^{j\varphi} + 2j\cos\varphi \\ -2 - je^{j\varphi} + 2j\cos\varphi & j + e^{j\varphi} + 2\cos\varphi \end{pmatrix}$$

and so

$$E_{\{0\}} = \frac{1}{2\pi} \int_{\partial(U_{\{0\}} \cap \mathbb{C}_j)} S_L^{-1}(s,T)\, ds_j$$

$$= \frac{1}{2\pi} \int_0^{2\pi} 2e^{-j\varphi} \left(3j + 4\mathrm{Re}\left(e^{j\varphi}\right)\right)^{-1} \cdot$$

$$\cdot \begin{pmatrix} j + e^{j\varphi} + 2\cos(\varphi) & 2 + je^{j\varphi} - 2j\cos\varphi \\ -2 - je^{j\varphi} + 2j\cos\varphi & j + e^{j\varphi} + 2\cos\varphi \end{pmatrix} \frac{1}{2} e^{j\varphi} j(-j)\, d\varphi$$

$$= \frac{1}{2} \begin{pmatrix} 1 & -j \\ j & 1 \end{pmatrix}.$$

A similar computation shows that

$$E_{\mathbb{S}} = \frac{1}{2\pi} \int_{\partial(U_{\mathbb{S}} \cap \mathbb{C}_j)} S_L^{-1}(s,T)\, ds_j = \frac{1}{2} \begin{pmatrix} 1 & j \\ -j & 1 \end{pmatrix}.$$

Straightforward calculations show that these matrices actually define projections on $\mathbb{H}^2$ with $E_{\{0\}} + E_{\mathbb{S}} = \mathcal{I}$. Moreover, we have $E_{\{0\}} y_1 = y_1$ and $E_{\mathbb{S}} y_1 = 0$ as well as $E_{\{0\}} y_2 = 0$ and $E_{\mathbb{S}} y_2 = y_2$. Thus, the invariant subspace $E_{\{0\}} X$ associated with the spectral set $\{0\}$ is the right linear span of $y_1$, which consist of all eigenvectors with respect to the real eigenvalue 0 as $T(y_1 a) = T(y_1)a = 0$ for all $a \in \mathbb{H}$. The invariant subspace $E_{\mathbb{S}}$ associated with the spectral set $\mathbb{S}$ consists of the right linear span of $y_2$. For $a \in \mathbb{H} \setminus \{0\}$, we have $T(y_2 a) = T(y_2)a = y_2 j a = (y_2 a)(a^{-1} j a)$. Thus, as $a^{-1} j a \in \mathbb{S}$, the subspace $E_{\mathbb{S}}$ consists of all right eigenvectors associated with eigenvalues in $\mathbb{S}$. (This is true only because the associated subspace is one-dimensional! Otherwise, the subspace would consist of sums of eigenvectors associated with possibly different eigenvalues in the sphere $\mathbb{S}$. Such vectors are in general not right eigenvectors, but they are $S$-eigenvectors associated with the eigensphere $\mathbb{S}$.)

Finally, we can construct functions, which are left and right slice hyperholomorphic on $\sigma_S(T)$, but for which the $S$-functional calculi for left and right slice hyperholomorphic functions yield different operators: consider the function

$$f(s) = c_1 \chi_{U_{\{0\}}}(s) + c_2 \chi_{U_{\mathbb{S}}}(s)$$

such that $c_1$ or $c_2$ does not belong to $\mathbb{C}_j$. Choose for instance $c_1 = i$ and $c_2 = 0$ for the sake of simplicity. This function is a locally constant slice function on $U = U_{\{0\}} \cup U_{\mathbb{S}}$ and thus left and right slice hyperholomorphic by Lemma 3.4.8. Then

$$\frac{1}{2\pi} \int_{\partial(U \cap \mathbb{C}_j)} S_L^{-1}(s,T)\, ds_j\, f(s) = \left( \frac{1}{2\pi} \int_{\partial(B_{1/2}(0) \cap \mathbb{C}_j)} S_L^{-1}(s,T)\, ds_j \right) i$$

$$= \frac{1}{2} \begin{pmatrix} 1 & -j \\ j & 1 \end{pmatrix} i = \frac{1}{2} \begin{pmatrix} i & -k \\ k & i \end{pmatrix},$$

but

$$\frac{1}{2\pi} \int_{\partial(U \cap \mathbb{C}_j)} f(s)\, ds_j\, S_R^{-1}(s,T) = i \left( \frac{1}{2\pi} \int_{\partial(B_{1/2}(0) \cap \mathbb{C}_j)} ds_j\, S_R^{-1}(s,T) \right)$$

$$= \frac{1}{2} i \begin{pmatrix} 1 & -j \\ j & 1 \end{pmatrix} = \frac{1}{2} \begin{pmatrix} i & k \\ -k & i \end{pmatrix}.$$

The reason for why we obtain different operators is that the spectral projections $E_{\mathbb{S}}$ and $E_{\{0\}}$ cannot commute with arbitrary scalars because the respective invariant subspaces are not two-sided. Indeed, $-iy_2 = (1,j) = y_1$, which obviously does not belong to $E_{\mathbb{S}}X$.

# 3.8 The special roles of intrinsic functions and the left multiplication

As we saw in this chapter, the role of intrinsic slice hyperholomorphic functions stands out in quaternionic operator theory. Important results such as the product rule, the spectral mapping theorem and the composition rule only hold for these functions. This is not surprising since, on the level of functions, slice hyperholomorphicity is only compatible with multiplication and composition of intrinsic functions, not of arbitrary slice hyperholomorphic functions. There exists, however, a deeper, more fundamental reason for this special role of intrinsic functions that we want to explain in the following.

A functional calculus for an operator $T$ is a mathematical method that allows to define an operator $f(T)$ such that $f(T)$ generalizes the mapping behavior of $T$ for each $f$ in a certain class of functions on the spectrum of $T$ (for instance the class of holomorphic, continuous, measurable or slice hyperholomorphic functions on the spectrum of $T$). This is useful for generating new operators, and it is also useful for understanding the operator $T$ itself. The way $f(T)$ changes as $f$ varies in the corresponding class of functions gives information about $T$ and allows to identify, for instance, eigenspaces, invariant subspaces, or, if $T$ is a normal operator on a Hilbert space, even its spectral resolution. This is, however, only possible if the mapping behavior of $T$ and $f(T)$ are related in a suitable way. Intuitively, the operator $f(T)$ should be obtained by letting $f$ act on spectral values of $T$. In particular, if $v \in X \setminus \{0\}$ is an eigenvector of $T$ associated with $s$, i.e.,

$$Tv = vs, \tag{3.42}$$

then $v$ should be an eigenvector of $f(T)$ associated with $f(s)$, i.e.,

$$f(T)v = vf(s). \tag{3.43}$$

One of the fundamental peculiarities of operator theory in the quaternionic setting is the axial symmetry of the set of eigenvalues and the $S$-spectrum of an operator. In particular, if (3.42) holds and $h \in \mathbb{H} \setminus \{0\}$, then

$$T(vh) = (Tv)h = vsh = (vh)(h^{-1}sh). \tag{3.44}$$

Consequently, if (3.42) implies (3.43), then $vh$ is an eigenvector of $f(T)$ associated with $f(h^{-1}sh)$, that is

$$f(T)(vh) = (vh)f(h^{-1}sh). \tag{3.45}$$

On the other hand, (3.43) implies

$$f(T)(vh) = (f(T)v)h = vf(s)h = (vh)(h^{-1}f(s)h). \tag{3.46}$$

Combining (3.45) and (3.46), we find that $f$ must satisfy

$$f(h^{-1}sh) = h^{-1}f(s)h \quad \forall h \in \mathbb{H} \setminus \{0\}. \tag{3.47}$$

Now assume that $s = u + jv \in \mathbb{H}$ and choose $h = j$. Since $s$ and $j$ commute, we conclude from (3.47) that

$$jf(s) = jf(j^{-1}sj) = jj^{-1}f(s)j = f(s)j.$$

A quaternion commutes with $j$ if and only it belongs to $\mathbb{C}_j$. Hence, $f(s) \in \mathbb{C}_j$ and so $f(s)$ belongs to the same complex plane as $s$. Let $\alpha, \beta \in \mathbb{R}$ such that $f(s) = \alpha + j\beta$. If $\tilde{s} = u + iv \in [s]$ with $i \in \mathbb{S}$ arbitrary, then there exists $h \in \mathbb{H} \setminus \{0\}$ such that $\tilde{s} = h^{-1}sh$. Furthermore, we conclude from

$$u + iv = \tilde{s} = h^{-1}sh = u + h^{-1}jhv$$

that $i = h^{-1}jh$. The identity (3.47) then implies

$$f(u + iv) = f(\tilde{s}) = f(h^{-1}sh) = h^{-1}f(s)h = \alpha + (h^{-1}jh)\beta = \alpha + i\beta.$$

Setting $f_0(u, v) := \alpha \in \mathbb{R}$ and $f_1(u, v) := \beta \in \mathbb{R}$, we find

$$f(u + iv) = f_0(u, v) + if_1(u, v), \quad \forall i \in \mathbb{S}.$$

Thus, $f$ is an intrinsic slice function. In the quaternionic setting, any proper functional calculus must therefore necessarily apply to a class of intrinsic slice functions—otherwise it does not follow the most fundamental intuition of such calculus, namely that (3.42) implies (3.43), and the mapping behavior of $f(T)$ is not related with the mapping behavior of $T$.

    This explains why the $S$-functional calculus shows undesirable properties when non-intrinsic functions are considered, such as the voidness of the product rule and the spectral mapping theorem or such as the inconsistencies between the $S$-functional calculi for left- and right slice hyperholomorphic functions. (These phenomena are not restricted to the $S$-functional calculus but appear, due to the reasons explained above, in any quaternionic functional calculus, for instance, in the continuous functional calculus for normal quaternionic operators [57, 148].)

We make another important observation: intrinsic slice hyperholomorphic functions of an operator can be expressed in terms of only the right linear structure on the space, cf. Theorem 3.4.11. Hence, they do not depend on the left multiplication. A right linear operator is, via the linearity condition, only related with the right multiplication, not with the left multiplication on the space. It is therefore plausible that, as a general principle, only the right linear structure should be important for the spectral properties of such operator. Indeed, we assume the existence of a left multiplication on $X$ only because the space $\mathcal{B}(X)$ of right linear operators on $X$ is otherwise only a real, not a quaternionic Banach space.

We can show the independence of intrinsic slice hyperholomorphic functions of an operator of the left multiplication with a different argument, which applies in other situations, too. If $f$ is an intrinsic slice hyperholomorphic function on $\sigma_{SX}(T)$, then $f$ can be approximated uniformly on $\sigma_{SX}(T)$ by intrinsic rational functions $R_n$ due to Runge's theorem. Intrinsic rational functions are rational functions with real coefficients. Hence, they are precisely those rational functions of $T$ that can be defined even if $\mathcal{B}(X)$ is considered only as a real Banach space, that is, if only the right linear structure on $X$ is considered. For any $R_n$, the operator $R_n(T)$ therefore does not depend on the left multiplication. Instead, $R_n(T)$ is fully determined by the right linear structure on $X$. Furthermore, the operator norm $\|T\| = \sup_{\|v\|=1} \|Tv\|$, and in turn also the topology on $\mathcal{B}(X)$, is independent of whether we consider $X$ as a quaternionic two-sided Banach space or a quaternionic right Banach space. We have

$$f = \lim_{n \to +\infty} R_n$$

uniformly on $\sigma_S(T)$. As the $S$-functional calculus is compatible with uniform limits, we find

$$f(T) = \lim_{n \to +\infty} R_n(T).$$

Since the operators $R_n(T)$ and the topology on $\mathcal{B}(X)$ are determined by the right linear structure on $X$ and do not depend on the left multiplication, this is also true for the operator $f(T)$.

Similarly, the continuous functional calculus for a bounded normal quaternionic operator $T$ is defined by approximating a continuous intrinsic slice function $f$ on $\sigma_S(T)$ uniformly by intrinsic polynomials in $s$ and $\overline{s}$, that is by polynomials of the form

$$P_n(s) = \sum_{0 \leq \ell, k \leq n} a_{\ell,k} s^\ell \overline{s}^k \quad \text{with } a_{\ell,k} \in \mathbb{R}.$$

The operator

$$P_n(T) := \sum_{0 \leq \ell, k \leq n} a_{\ell,k} T^\ell (T^*)^k,$$

where $T^*$ denotes the adjoint of $T$, is then again fully determined by the right linear structure on the space since it contains only real coefficients. Consequently, also the operator $f(T) = \lim_{n \to +\infty} P_n(T)$ depends only on the right linear structure and not on any left multiplication [57].

Other important functional calculi such as the $H^\infty$-functional calculus or the measurable functional calculus are extensions of these two calculi. Hence, they inherit the independence from the left linear structure on $X$ (as long as only intrinsic slice functions are considered).

The fact that functional calculi for quaternionic right linear operators are determined by the right linear structure on the space brings up the question of clarifying the role that the left multiplication plays in this theory. In particular, we have to ask whether it has any influence on the spectral properties of an operator or not. The spectral properties of a quaternionic operator $T$ must be independent of the concrete model of this operator that is considered a change of basis for instance, must not effect these properties. More general, let $X$ be a two-sided quaternionic Banach space and let $T \in \mathcal{K}(X)$. If $Y$ is another two-sided quaternionic Banach space and $U : X \to Y$ is a norm-preserving and bijective right-linear mapping, then

$$S := UTU^{-1}$$

is a model for $T$ in $Y$. The spectral properties of $S$ should correspond to the spectral properties of $T$ and, indeed, we have

$$\mathcal{Q}_s(S) = \mathcal{Q}_s\left(UTU^{-1}\right) = U\mathcal{Q}_s(T)U^{-1}, \quad \forall s \in \mathbb{H}.$$

Hence, we find

$$\rho_S(T) = \rho_S(S) \quad \text{and} \quad \sigma_S(T) = \sigma_S(S),$$

and

$$\mathcal{Q}_s(S)^{-1} = U\mathcal{Q}_s(T)^{-1}U^{-1}, \quad \forall s \in \rho_S(T).$$

If $P(s) = \sum_{k=0}^{n} a_k s^k$ with $a_n \in \mathbb{R}$ is an intrinsic polynomial, then

$$P(S) = \sum_{k=0}^{n} a_k S^k = U\sum_{k=0}^{n} a_k T^k U^{-1} = UP(T)U^{-1}.$$

For any intrinsic rational function $R(s) = P(s)Q(s)^{-1}$ with intrinsic polynomials $P$ and $Q$ such that the zeros of $Q$ (resp. the poles of $R$) lie in $\rho_S(T) = \rho_S(S)$, we therefore find that

$$R(S) = P(S)Q(S)^{-1} = UP(T)Q(T)^{-1}U^{-1} = UR(T)U^{-1}.$$

If $f \in \mathcal{N}(\sigma_{SX}(T)) = \mathcal{N}(\sigma_{SX}(S))$, then Runge's theorem implies the existence of a sequence of intrinsic rational functions $R_n, n \in \mathbb{N}$, the poles of which lie in $\rho_S(T) = \rho_S(S)$ such that $f(s) = \lim_{n \to +\infty} R_n(s)$ uniformly on $\sigma_{SX}(T)$. Since the $S$-functional calculus is compatible with uniform limits on $\sigma_{SX}(T)$, we obtain that

$$f(S) = \lim_{n \to +\infty} R_n(S) = \lim_{n \to +\infty} R_n(S) = \lim_{n \to +\infty} UR_n(T)U^{-1} = Uf(T)U^{-1}.$$

Similarly, it also follows that $f(S) = Uf(T)U^{-1}$ for any continuous intrinsic slice function $f$ on $\sigma_S(T)$ if $T$ is a normal operator on a quaternionic Hilbert space and $U$ is a unitary right linear bijection. This correspondence is inherited by the extensions of these functional calculi such as the $H^\infty$-functional calculus or the measurable functional calculus for normal operators. (For the $S$-functional calculus the identity $f(S) = Uf(T)U^{-1}$ can also be deduced directly from the integral representation (3.35). However, for the intrinsic functional calculus such integral representation does not exist and one has to follow the strategy described above.)

Objects and techniques that depend on the left multiplication or that apply to functions other than intrinsic functions are, on the other hand, not invariant under the transformation $U$. Consider for instance the constant function $f(s) = a$ with $a \in \mathbb{H} \setminus \mathbb{R}$. This function is both left and right slice hyperholomorphic on $\sigma_{SX}(T) = \sigma_{SX}(S)$, but not intrinsic. If we apply the $S$-functional calculus, we find $f(S) = a\mathcal{I}_Y$ and $f(T) = a\mathcal{I}_X$. However, unless $aU = Ua$, we have

$$f(S) = a\mathcal{I}_Y \neq U(a\mathcal{I}_X)U^{-1} = Uf(T)U^{-1}.$$

Actually, even for the $S$-resolvents, in general, we have

$$S_L^{-1}(s, S) \neq US_L^{-1}(s, T)U^{-1} \quad \text{and} \quad S_R^{-1}(s, S) \neq US_R^{-1}(s, T)U^{-1}.$$

Indeed, unless $U\overline{s} = \overline{s}U$, it is

$$
\begin{aligned}
S_L^{-1}(s, S) &= \mathcal{Q}_s(S)^{-1}\overline{s} - S\mathcal{Q}_s(S)^{-1} \\
&= U\mathcal{Q}_s(T)^{-1}U^{-1}\overline{s} - UT\mathcal{Q}_s(T)^{-1}U^{-1} \\
&\neq U\left(\mathcal{Q}_s(T)^{-1}\overline{s} - T\mathcal{Q}_s(T)^{-1}\right)U^{-1} = US_L^{-1}(s, T)U^{-1}.
\end{aligned}
$$

Due to the symmetry of the path of integration in the $S$-functional calculus, the $S$-resolvents are always simultaneously evaluated at $s$ and $\overline{s}$ and it is this fact that ensures the independence of the $S$-functional calculus for intrinsic functions from the left multiplication.

One could argue that, since we are working on two-sided quaternionic Banach spaces, only transformations $U$ that are compatible with the entire structure of $X$, that is with both the left and the right multiplication, should be considered in the arguments above. Hence, one should assume that $U$ is both left and right linear. Such a transformation would satisfy $aU = Ua$ for all $a \in \mathbb{H}$ and the problems described above would not occur. The transformations of this type can be characterized easily: if

$$X_\mathbb{R} = \{v \in X : \quad av = va, \ \forall a \in \mathbb{H}\} \quad \text{and} \quad Y_\mathbb{R} = \{v \in X : \quad av = va, \ \forall a \in \mathbb{H}\}$$

such that

$$X = X_\mathbb{R} \otimes \mathbb{H} \quad \text{and} \quad Y = Y_\mathbb{R} \otimes \mathbb{H},$$

then an operator $U : X \to Y$ is both left and right linear if and only if it is the quaternionic right linear extension of an $\mathbb{R}$-linear operator $U_{\mathbb{R}} : X_{\mathbb{R}} \to Y_{\mathbb{R}}$. (In the terminology introduced in Definitions 2.2.6 and 3.3.1, this is equivalent to $U$ being a scalar operator.)

However, restricting ourselves to such transformations is not feasible. We can consider for example $X = \mathbb{H}^n$ with its natural left and right multiplication and endowed with an arbitrary norm. This yields a two-sided quaternionic Banach space of dimension $n$. A right linear operator $T$ on $\mathbb{H}^n$ can be represented by an $n \times n$-matrix with quaternionic entries and an operator is both left and right linear if it is represented by a matrix with only real coefficients. This matrix can be put in Jordan normal form, i.e., there exists an invertible matrix $U$ such that $T = USU^{-1}$, where $S$ is a block diagonal matrix with the diagonal that consists of Jordan blocks [190]. The matrix $U$ however does not necessarily only have real entries. It is in general a matrix with quaternionic entries and hence it represents an operator that is only right, but not necessarily left linear. If we require that the spectral properties are only invariant under transformations that are both left and right linear, this would imply that the spectral properties of $T$ are not necessarily invariant under the transformation $U$. Hence, $T$ and its model in Jordan normal form $S$ might have different spectral properties, which is absurd.

Spectral properties of an operator must therefore be invariant under norm-preserving and bijective right linear transformations. Since the left multiplication is not invariant under such transformations, the spectral properties of an operator cannot depend on it. *We conclude that right linear quaternionic operators have to be understood in terms of the right linear structure only. The left multiplication is a useful auxiliary tool, but spectral properties of the operator cannot depend on it.* However, the left multiplication is necessary in order to consider $\mathcal{B}(X)$ as a quaternionic linear space. Without it, it is not possible to apply quaternionic techniques to elements of $\mathcal{B}(X)$ and to give intuitive integral representations for the $S$-functional calculus in $\mathcal{B}(X)$. Furthermore, without assuming the existence of a left multiplication, it would not have been possible to develop the fundamental concepts of quaternionic operator theory. In particular, the $S$-spectrum could not have been found as its definition was understood by finding the closed form of the Cauchy kernel operator series $\sum_{n=0}^{+\infty} T^n s^{-(n+1)}$, cf. Theorem 2.2.9. Giving meaning to this series requires $\mathcal{B}(X)$ to be a quaternionic linear space.

Finally, in certain situations, the left multiplication of $X$ is particularly useful for simplifying computations, since this left multiplication might allow us to write $T$ in terms of components as $T = T_0 + \sum_{\ell=1}^{3} T_\ell e_\ell$, cf. Definition 2.2.6 and Definition 3.3.1. If there exists a model of the operator $T$ in a space with a left multiplication, such that $T$ has commuting components, then this model can be used to significantly simplify computations for investigating $T$, cf. Theorem 3.3.4.

# Chapter 4

# The Quaternionic Evolution Operator

In this chapter we study the theory of semigroups and groups of quaternionic linear operators, the development of which started in [79]. We show that the theory of Hille, Phillips and Yosida can be restored in the quaternionic setting. We consider the problem of determining when an unbounded closed operator $T$ is the infinitesimal generator of a strongly continuous semigroup $\mathcal{U}_T$. This allows us to consider the abstract Cauchy problem: *given a closed unbounded quaternionic operator $T$ on a quaternionic Banach space $X$ and an initial datum $q_0 \in X$, determine the continuous $X$-valued function $q(t)$ defined for $t \geq 0$, such that $q(t)$ belongs to the domain of $T$ for any $t > 0$ and such that*

$$\begin{cases} \frac{d}{dt}q(t) = T\,q(t), & t > 0 \\ q(0) = q_0. \end{cases}$$

In this chapter we generalize the following classical results, see for example [110], to the quaternionic setting: *If a semigroup $E(t)$ is continuous in the uniform operator topology, then there exists a bounded linear operator $B$, acting on a Banach space, such that $E(t) = e^{tB}$. Conversely, for any bounded linear operator $B$, the operator $e^{tB}$ is a uniformly continuous semigroup. The operator $B$ is called the infinitesimal generator of the semigroup. Moreover, the infinitesimal generator $B$ can be obtained as the limit $\lim_{h \to 0}(E(h)-I)/h$ and the Laplace transform of the semigroup $E(t)$ gives, for $\mathrm{Re}(\lambda)$ suitable large, the resolvent operator of $B$:*

$$(\lambda I - B)^{-1} = \int_0^\infty e^{-\lambda t}\, E(t)dt.$$

If $E(t)$ is merely continuous in the strong operator topology, then the problem becomes more difficult to study. In an appropriate sense $E(t)$ still equals $e^{tB}$, but

F. Colombo, J. Gantner, *Quaternionic Closed Operators, Fractional Powers and Fractional Diffusion Processes*, Operator Theory: Advances and Applications 274, https://doi.org/10.1007/978-3-030-16409-6_4

now $B$ is a closed, but possibly unbounded, linear operator with dense domain. In fact, the characterization is in this case given by: *A necessary and sufficient condition in order that a closed linear operator $B$ with dense domain be the infinitesimal generator of a strongly continuous semigroup is that there exist real numbers $M > 0$ and $\omega$ such that every real number $\lambda > \omega$ belongs to the resolvent set of $B$ and*

$$\|(\lambda I - B)^{-n}\| \leq M(\lambda - \omega)^{-n}, \quad \text{for } n \in \mathbb{N}.$$

In the following we will consider right linear quaternionic operators, but the theory can be developed also for left linear quaternionic operators. We will show that it is possible to use the left S-resolvent operator, and also the right S-resolvent operator to prove our results.

## 4.1 Uniformly continuous quaternionic semigroups

We begin by extending the following classical result to the quaternionic setting: a semigroup has a bounded infinitesimal generator if and only if it is uniformly continuous. We recall the definition of uniformly continuous and of strongly continuous semigroups and show some preliminary results which will be useful in the sequel. We will always consider a two-sided quaternionic Banach space $X$, even if we do not specify this explicitly. Moreover, we prove some results that will be used in the next section where we study strongly continuous semigroups.

**Definition 4.1.1.** Let $X$ be a two-sided quaternionic Banach space. A family of bounded linear quaternionic operators $\{\mathcal{U}(t)\}_{t\geq 0}$ in $X$ is called a *strongly continuous quaternionic semigroup* if

(1) $\mathcal{U}(t + \tau) = \mathcal{U}(t)\mathcal{U}(\tau)$, $t, \tau \geq 0$,

(2) $\mathcal{U}(0) = \mathcal{I}$,

(3) for every $v \in X$, the mapping $t \mapsto \mathcal{U}(t)v$ is continuous on $t \in [0, \infty)$.

The family $\{\mathcal{U}(t)\}_{t\geq 0}$ is called a *uniformly continuous quaternionic semigroup* in $\mathcal{B}(X)$ if in addition

(4) the map $t \to \mathcal{U}(t)$ is continuous in the uniform operator topology for $t \in [0, +\infty)$.

Finally, a *strongly* (resp. *uniformly*) *continuous group* of bounded linear quaternionic operators is a family $\{\mathcal{U}(t)\}_{t\in\mathbb{R}}$ of operators in $\mathcal{B}(X)$ such that (1), (2) and (3) (resp. (4)) holds for all $t, \tau \in \mathbb{R}$.

If $T \in \mathcal{B}(X)$, then the S-functional calculus for bounded operators implies that the operator $e^{tT}$ is a uniformly continuous quaternionic semigroup in $\mathcal{B}(X)$. The following theorem shows that the converse is also true, i.e., every uniformly continuous quaternionic semigroup is of this form.

**Theorem 4.1.2.** *Let* $\{\mathcal{U}(t)\}_{t\geq 0}$ *be a uniformly continuous quaternionic semigroup in* $\mathcal{B}(X)$. *Then:*

(1) *there exists a bounded linear quaternionic operator* $T$ *such that*

$$\mathcal{U}(t) = e^{tT},$$

(2) *the quaternionic operator* $T$ *is given by the formula*

$$T = \lim_{h^+ \to 0} \frac{\mathcal{U}(h) - \mathcal{U}(0)}{h},$$

(3) *we have the relation*

$$\frac{d}{dt}e^{tT} = T\,e^{tT} = e^{tT}\,T.$$

*Proof.* Point (1). We recall that the logarithm of a quaternion $q = u + jv \in \mathbb{H} \setminus (-\infty, 0]$ is defined as $\log q = \ln|q| + j\arg(q)$ (see Example 2.1.14). Since the $S$-spectrum of $\mathcal{U}(0) = \mathcal{I}$ consists only of the real point 1, it follows from Theorem 2.2.32 that there exists $\varepsilon > 0$ such that $P(t) = \ln \mathcal{U}(t)$ is defined and continuous for $t \in [0, \varepsilon]$. If $nt \leq \varepsilon$ then we have, due to the semigroup properties, that

$$P(nt) = \ln \mathcal{U}(nt) = \ln \left(\mathcal{U}(t)\right)^n = n\,P(t)$$

and so

$$P(t) = n\,P(t/n) \quad \text{for every } t \in [0, \varepsilon].$$

As a consequence, for each rational number $m/n$ such that $m/n \in [0, 1]$ and for each $t \in [0, \varepsilon]$, we have

$$\frac{m}{n}\,P(t) = m\,P(t/n) = P(mt/n). \tag{4.1}$$

In particular, we have

$$\frac{m}{n}\,P(\varepsilon) = P(m\varepsilon/n). \tag{4.2}$$

Due to the continuity of $P$, we conclude from (4.2) that even

$$tP(\varepsilon) = P(t\varepsilon) \quad \text{for every } t \in [0, 1],$$

and

$$P(t) = \frac{t}{\varepsilon}P(\varepsilon) \quad \text{for every } t \in [0, \varepsilon].$$

If we set

$$T := \frac{1}{\varepsilon}P(\varepsilon),$$

we obtain, due to the composition rule of the $S$-functional calculus, that

$$\mathcal{U}(t) = e^{P(t)} = e^{tT} \quad \text{for every } t \in [0, \varepsilon].$$

If $t > 0$ is arbitrary, then $t/n < \varepsilon$ for sufficiently large $n$, and so we obtain

$$e^{tT} = (e^{(t/n)T})^n = [\mathcal{U}(t/n)]^n = \mathcal{U}(t).$$

Point (2). Since $\lim_{h \to 0+}(e^{hq} - 1)/h = q$ uniformly in any bounded set of $\mathbb{H}$, Theorem 2.2.21 implies that $\lim_{h \to 0+} \frac{1}{h}(e^{hT} - \mathcal{I})$ converges to $T$ in $\mathcal{B}(X)$.

Point (3). We can deduce it from the $S$-functional calculus. Let

$$e^{tT} = \frac{1}{2\pi} \int_{\partial(U \cap \mathbb{C}_j)} e^{ts}\, ds_j\, S_R^{-1}(s,T)$$

where $U$ is a bounded slice Cauchy domain that contains the $S$-spectrum of the bounded operator $T$. Let $h > 0$ and consider

$$\frac{e^{(t+h)T} - e^{tT}}{h} = \frac{1}{2\pi} \int_{\partial(U \cap \mathbb{C}_j)} (e^{(t+h)s} - e^{ts})\,(hs)^{-1}\, ds_j\, S_R^{-1}(s,T).$$

By taking the limit as $h \to 0$ we get

$$\frac{d}{dt} e^{tT} = \lim_{h \to 0} \frac{e^{(t+h)T} - e^{tT}}{h} = e^{tT}T.$$

Similarly, we can also write

$$\frac{e^{(t+h)T} - e^{tT}}{h} = \frac{1}{2\pi} \int_{\partial(U \cap \mathbb{C}_j)} (hs)^{-1}(e^{(t+h)s} - e^{ts})\, ds_j\, S_R^{-1}(s,T)$$

and taking the limit as $h \to 0$ we find that

$$\frac{d}{dt} e^{tT} = \lim_{h \to 0} \frac{e^{(t+h)T} - e^{tT}}{h} = T e^{tT}. \qquad \square$$

We now want to generalize the important result that the Laplace transform of a semigroup $e^{tB}$ that is generated by a bounded complex linear operator $B$ is the resolvent operator $(\lambda I - B)^{-1}$ of $B$. Both the left and the right S-resolvent operators, $S_L^{-1}(s,T)$ and $S_R^{-1}(s,T)$, can be represented as the Laplace transform of the semigroup if we use the two different possible definitions of the Laplace transform based on left or on right slice hyperholomorphic functions, as it is shown in the following theorem.

**Theorem 4.1.3.** *Let* $T \in \mathcal{B}(X)$ *and let* $s \in \mathbb{H}$ *with* $s_0 > \|T\|$. *Then the left S-resolvent operator* $S_L^{-1}(s,T)$ *is given by*

$$S_L^{-1}(s,T) = \int_0^{+\infty} e^{tT}\, e^{-ts}\, dt, \qquad (4.3)$$

*and the right S-resolvent operator* $S_R^{-1}(s,T)$ *is given by*

$$S_R^{-1}(s,T) = \int_0^{+\infty} e^{-ts}\, e^{tT}\, dt. \qquad (4.4)$$

*Proof.* We will prove just (4.4) since (4.3) can be shown with similar arguments. Since $|\bar{s}| = |s| > \|T\|$, the point $\bar{s}$ does not belong to $\sigma_L(T)$. Indeed, the inverse of $\bar{s}\mathcal{I} - T$ is given by

$$(\bar{s} - \mathcal{I})^{-1} = \sum_{n=0}^{+\infty} \bar{s}^{-1}(T\bar{s}^{-1})^n.$$

Therefore, the operator

$$e^{-ts}\, e^{t\,T} S_R(s, T) = -e^{-ts}\, e^{t\,T}(T^2 - 2s_0 T + |s|^2 \mathcal{I})(T - \bar{s}\mathcal{I})^{-1}$$

is well defined. Since $T^2 - 2s_0 T + |s|^2 \mathcal{I} = -s(T - \bar{s}\mathcal{I}) + (T - \bar{s}\mathcal{I})T$ and since $T$ and $e^{t\,T}$ commute, we can write

$$
\begin{aligned}
e^{-ts}\, e^{t\,T} S_R(s, T) &= e^{-ts}\, s(T - \bar{s}\mathcal{I})e^{t\,T}(T - \bar{s}\mathcal{I})^{-1} \\
&\quad - e^{-ts}\,(T - \bar{s}\mathcal{I})e^{t\,T}T(T - \bar{s}\mathcal{I})^{-1} \\
&= -\frac{d}{dt}[e^{-ts}(T - \bar{s}\mathcal{I})e^{t\,T}(T - \bar{s}\mathcal{I})^{-1}].
\end{aligned}
$$

Integrating the last equality over $[0, \theta]$, for $\theta > 0$, gives

$$
\begin{aligned}
\int_0^\theta e^{-ts}\, e^{t\,T} S_R(s, T) dt &= -\int_0^\theta \frac{d}{dt}[e^{-ts}(T - \bar{s}\mathcal{I})e^{t\,T}(T - \bar{s}\mathcal{I})^{-1}] dt \\
&= \mathcal{I} - e^{-\theta s}(T - \bar{s}\mathcal{I})e^{\theta\,T}(T - \bar{s}\mathcal{I})^{-1}.
\end{aligned}
$$

Passing to the limit for $\theta \to +\infty$, due to the assumption $s_0 > \|T\|$, we have

$$\|e^{-\theta\,s}(T - \bar{s}\mathcal{I})e^{\theta\,T}(T - \bar{s}\mathcal{I})^{-1}\| \le e^{-\theta\,s_0}e^{\theta\|T\|}\|(T - \bar{s}\mathcal{I})\|\,\|(T - \bar{s}\mathcal{I})^{-1}\| \to 0.$$

From the latter equality we get

$$\left[\int_0^{+\infty} e^{-ts}\, e^{t\,T}\, dt\right] S_R(s, T) = \int_0^{+\infty} e^{-ts}\, e^{t\,T} S_R(s, T)\, dt = \mathcal{I},$$

so we finally obtain (4.4).                                                                    $\square$

Now we prove a very important result on quaternionic semigroups. We follow the line of the classical result in [110, p. 616]. We will use several results from integration theory that are usually formulated for functions with values in real Banach spaces. Their proofs are, however, identical for functions with values in a quaternionic Banach space so we simply cite them without repeating their proofs explicitly.

**Proposition 4.1.4.** *Let $X$ be a two-sided quaternionic Banach space.*

(1) *Let $\{\mathcal{U}(t)\}$ be a family of bounded linear quaternionic operators defined on a finite closed interval $[a, b]$ such that $\mathcal{U}(t)v$ is continuous in $t$ for each $v \in X$. Then $\|\mathcal{U}(\cdot)\|$ is measurable and bounded on $[a, b]$.*

(2) *Conversely, if $\{\mathcal{U}(t)\}_{t\geq 0}$ is a semigroup of bounded linear quaternionic operators in $X$ and if $\mathcal{U}(\cdot)v$ is measurable on $(0,\infty)$ for each $v \in X$, then $\mathcal{U}(\cdot)v$ is continuous at every point in $(0,\infty)$.*

*Proof.* We first show (1). For each $v \in X$, the function $t \mapsto \mathcal{U}(t)v$ is continuous and so it is bounded on $[a,b]$. The boundedness of $\|\mathcal{U}(\cdot)\|$ follows from the Uniform Boundedness Principle in Theorem 12.0.9. In order to prove that $\|\mathcal{U}(\cdot)\|$ is measurable let $\delta > 0$ and set

$$U = \{t \in \mathbb{R} : \quad \|\mathcal{U}(t)\| > \delta\}.$$

If we pick $t_0 \in U$, then there exists $v$ with $\|v\| = 1$ such that $\|\mathcal{U}(t_0)v\| > \delta$. Due to the continuity of $\mathcal{U}(t)v$, there exists an interval $(-\varepsilon + t_0, t_0 + \varepsilon)$ such that $\|\mathcal{U}(t)v\| > \delta$ and in turn also $\|\mathcal{U}(t)\| > \delta$ for any $t \in (-\varepsilon + t_0, t_0 + \varepsilon)$. So $U$ is open in $[a,b]$ and the statement follows from Theorem III.6.10 in [110], which holds also in the quaternionic setting.

Now we show (2). First we show that if $\{\mathcal{U}(t)\}_{t\geq 0}$ is a quaternionic semigroup such that

(i) $\mathcal{U}(\cdot)v$ is measurable on $(0,\infty)$ for every $v \in X$,

(ii) $\|\mathcal{U}(\cdot)\|$ is bounded over each interval of the form $[\delta, 1/\delta]$, $\delta > 0$,

then $\mathcal{U}(\cdot)v$ is continuous at each point $t_0 > 0$ for any $v \in X$. Therefore, let $0 < a < b < t_0$ and choose $\delta > 0$ such that

$$2\delta < \max\left\{1, a, t_0 - b, (1 + t_0)^{-1}\right\}.$$

By the semigroup property we have

$$\mathcal{U}(t_0)v = \mathcal{U}(t)[\mathcal{U}(t_0 - t)v], \quad \text{for } t_0 - t > 0 \tag{4.5}$$

and since the right-hand side of (4.5) is independent of $t$, we can integrate it on $[a,b]$. If $|\varepsilon| < \delta$, we obtain

$$(b - a)[\mathcal{U}(t_0 + \varepsilon) - \mathcal{U}(t_0)]v = \int_a^b \mathcal{U}(t)[\mathcal{U}(t_0 + \varepsilon - t) - \mathcal{U}(t_0 - t)]v \, dt.$$

By assumption there exists a positive constant $M$ such that $\|\mathcal{U}(t)\| < M$ for $t \in [a,b]$. Moreover, the function

$$t \mapsto [\mathcal{U}(t_0 + \varepsilon - t) - \mathcal{U}(t_0 - t)]v$$

is bounded and measurable for $t \in [a, b]$. So we obtain the chain of inequalities

$$(b - a)\|[\mathcal{U}(t_0 + \varepsilon) - \mathcal{U}(t_0)]v\|$$

$$\leq \int_a^b \|\mathcal{U}(t)\| \|[\mathcal{U}(t_0 + \varepsilon - t) - \mathcal{U}(t_0 - t)]v\| \, dt$$

$$\leq M \int_a^b \|[\mathcal{U}(t_0 + \varepsilon - t) - \mathcal{U}(t_0 - t)]v\| \, dt$$

$$\leq M \int_{t_0 - b}^{t_0 - a} \|[\mathcal{U}(\tau + \varepsilon) - \mathcal{U}(\tau)]v\| \, d\tau \xrightarrow{\varepsilon \to 0} 0$$

and hence $\mathcal{U}(t)v$ is continuous at $t_0$.

What remains to show is that if $\mathcal{U}(t)v$ is measurable on $(0, \infty)$ for each $v \in X$, then $\|\mathcal{U}(\cdot)\|$ is bounded on the interval $[\delta, 1/\delta]$ for each $\delta > 0$. We split the proof of this statement into two steps.

Step (i). For any $v_0 \in X$, there exist a separable closed right linear subspace $X_0$ of $X$ with $v_0 \in X_0$ and a null set $\mathcal{E}_0$ of $(0, \infty)$ such that $\mathcal{U}(t)X_0 \subset X_0$ for any $t \notin \mathcal{E}_0$.

By Lemma III.6.9 in [110] there exists a null set $F_0$ such that $\{\mathcal{U}(t)v_0 \mid t \notin F_0\}$ is separable. Thus,

$$X_0 := \overline{\text{span} \{\{v_0\} \cup \{\mathcal{U}(t)v_0 : \quad t \notin F_0\}\}}$$

is a closed separable rightlinear subspace of $X$. Furthermore there exists a sequence $\{t_n\}_{n \in \mathbb{N}}$ with $t_n \notin F_0$ such that the set $\{v_0\} \cup \{\mathcal{U}(t_n)v_0 : n \in \mathbb{N}\}$ is fundamental in $X_0$, i.e.,

$$X_0 = \overline{\text{span} \{\{v_0\} \cup \{\mathcal{U}(t_n)v_0 : \quad n \in \mathbb{N}\}\}}. \tag{4.6}$$

If $t \in (0, +\infty)$ is now such that $t \notin F_0$ and for any $n \in \mathbb{N}$ also $t + t_n \notin F_0$, then $T(t)v_0 \in X_0$ and $T(t)T(t_n)v_0 = T(t + t_n)v_0 \in X_0$ for any $n \in \mathbb{N}$. Since $\mathcal{U}(t)$ is right linear and continuous, we conclude from (4.6) that even $T(t)v \in X_0$ for any $v \in X_0$, that is $T(t)X_0 \subset X_0$. We define now $\mathcal{E}_0$ as the set of all $t \in (0, +\infty)$ for which this is not true, i.e.,

$$\mathcal{E}_0 = F_0 \cup \bigcup_{n=1}^{\infty} (F_0 - t_n) \cap (0, +\infty),$$

which is a null set as it is the countable union of null sets. By the above arguments, we have $\mathcal{U}(t)X_0 \subset X_0$ for any $t \notin \mathcal{E}_0$.

Step (ii). Suppose now that there exists an interval $[\delta, 1/\delta]$ with $\delta > 0$, on which $\|\mathcal{U}(t)\|$ is not bounded. Then there exists a sequence of real numbers $t_n \in [\delta, 1/\delta]$ and of vectors $v_n \in X$ with $\|v_n\| = 1$, such that $\|\mathcal{U}(t_n)v_n\| > n$ for $n \in \mathbb{N}$. By Step (i), for each $n \in \mathbb{N}$ there exists a separable subspace $X_n$ with

$v_n \in X_n$ and a null set $\mathcal{E}_n$ such that $\mathcal{U}(t)X_n \subseteq X_n$ for any $t \notin \mathcal{E}_n$. We define

$$X_\infty = \overline{\mathrm{span}\{X_n : \quad n \in \mathbb{N}\}} \quad \text{and} \quad \mathcal{E}_\infty = \bigcup_{n=1}^{\infty} \mathcal{E}_n.$$

It is clear that $X_\infty$ is separable, that $\mathcal{E}_\infty$ is a null set and that $\mathcal{U}(t)X_\infty \subseteq X_\infty$ for $t \notin \mathcal{E}_\infty$. For each $t$ set

$$\|\mathcal{U}(t)\|' = \sup\{\|\mathcal{U}(t)v\| : \quad v \in X_\infty : \quad \|v\| \leq 1\}.$$

Since $v_n \in X_\infty$, it is $\|\mathcal{U}(t_n)\|' > n$ for $n \in \mathbb{N}$.

In the unit sphere of $X_\infty$, we consider a dense and countable subset $\{w_n\}_{n \in \mathbb{N}}$. Since $\|\mathcal{U}(\cdot)w_n\|$ is a measurable real function by Theorem III.6.10 in [110], we have that

$$\|\mathcal{U}(\cdot)\|' = \sup\{\|\mathcal{U}(\cdot)w_n\| : \quad n \in \mathbb{N}\}$$

is also measurable. Moreover, if $\tau_2 \notin \mathcal{E}_\infty$, then for any $v \in X_\infty$ it is $\mathcal{U}(\tau_2)v \in X_\infty$ and

$$\|\mathcal{U}(\tau_2)v\| \leq \|\mathcal{U}(\tau_2)\|'\|v\|.$$

So we observe that

$$\begin{aligned}
\|\mathcal{U}(\tau_1 + \tau_2)\|' &= \sup\{\|\mathcal{U}(\tau_1)[\mathcal{U}(\tau_2)v]\| : \quad v \in X_\infty : \quad \|v\| \leq 1\} \\
&\leq \sup\{\|\mathcal{U}(\tau_1)z\| : \quad z \in X_\infty : \quad \|z\| \leq \|\mathcal{U}(\tau_1)\|'\} \\
&\leq \|\mathcal{U}(\tau_1)\|'\|\mathcal{U}(\tau_2)\|',
\end{aligned}$$

provided that $\tau_2 \notin \mathcal{E}_\infty$. Define now $\omega \in (0, \infty)$ by

$$\omega(t) := \ln \|\mathcal{U}(t)\|'.$$

From the above discussion we know that $\omega$ is a measurable function that never takes the value $+\infty$ and it is $\omega(\tau_1 + \tau_2) \leq \omega(\tau_1) + \omega(\tau_2)$ provided at least one of the points $\tau_1$ and $\tau_2$ does not belong to $\mathcal{E}_\infty$. Furthermore, $\omega(t_n) > \ln n$ where the $t_n$ are points in the interval $[\delta, 1/\delta]$. That this is a contradiction follows from the following statement: if $\omega$ is a measurable extended real-valued function on $(0, \infty)$ and if $\omega(t) < +\infty$ for all $t > 0$ and $\omega(t_1 + t_2) \leq \omega(t_1) + \omega(t_2)$ whenever one of the $t_i$ is not in a the null set $\mathcal{E}$, then $\omega$ is bounded above in each finite closed interval.

To see this fact, let $a > 0$ and let $A = \omega(a)$. Then if $t_1 + t_2 = a$ and $t_2 \notin \mathcal{E}$, we have that $A = \omega(a) \leq \omega(t_1) + \omega(t_2)$. So if

$$F = \{t \in \mathbb{R} : \quad 0 < t < a, \ \omega(t) \geq A/2\}$$

and if $a - F := \{a - t : t \in F\}$, then $\mathcal{E} \cup F \cup (a - F) \supseteq [0, a]$. As a consequence, if we denote by $\mu$ the Lebesgue measure, we have $\mu(F) + \mu(a - F) \geq a$, but $\mu(F) = \mu(a - F)$ and so $\mu(F) \geq a/2$. Assume now that $\omega$ is not bounded above on a finite interval $[a, b] \subset (0, +\infty)$, i.e., there exists a sequence of points $t_n \in [a, b]$ such that $\omega(t_n) > 2n$. The Lebesgue measure of $F_n := \{t \in [0, b] : \omega(t) \geq n\}$ can then be estimated by $\mu(F_n) > t_n/2 > a/2$. We conclude that the set $\{t \in \mathbb{R} : \omega(t) = +\infty\}$ has at least measure $a/2$. Conversely, if $\omega$ does not take the value $+\infty$, then it must be bounded above on any finite interval. $\square$

**Proposition 4.1.5.** *Let $\{\mathcal{U}(t)\}_{t \geq 0}$ be a strongly continuous semigroup of bounded linear quaternionic operators on a quaternionic Banach space $X$. If*

$$\omega(t) := \ln \|\mathcal{U}(t)\|$$

*is bounded above on the interval $(0, a)$ for every positive $a \in \mathbb{R}$, then*

$$\lim_{t \to +\infty} \frac{1}{t} \ln \|\mathcal{U}(t)\| = \inf_{t > 0} \frac{1}{t} \ln \|\mathcal{U}(t)\| \in [-\infty, +\infty).$$

*Proof.* The function $\omega(t) := \ln \|\mathcal{U}(t)\|$ is subadditive in $[0, \infty)$ and is bounded on each finite subinterval of $[0, \infty)$ for Proposition 4.1.4. We show that

$$\omega_0 := \inf_{t > 0} \frac{1}{t} \omega(t)$$

is finite or equals $-\infty$ and that

$$\omega_0 = \lim_{t \to +\infty} \frac{\omega(t)}{t}.$$

Indeed, for every number $\eta > \omega_0$ there exists a point $t_0$ such that

$$\frac{\omega(t_0)}{t_0} < \eta.$$

Any $t$ can be written as $t = n(t)t_0 + r$, where $n(t)$ is an integer and $r \in [0, t_0)$. Then we have

$$
\begin{aligned}
\frac{\omega(t)}{t} &\leq \frac{\omega(n(t)t_0)}{t} + \frac{\omega(r)}{t} \\
&\leq \frac{n(t)\omega(t_0)}{t} + \frac{\omega(r)}{t} \\
&= \frac{\omega(t_0)}{t_0 + r/n(t)} + \frac{\omega(r)}{t}.
\end{aligned}
$$

So we get

$$\omega_0 \leq \lim_{t \to +\infty} \inf \frac{1}{t} \omega(t) \leq \limsup_{t \to +\infty} \frac{\omega(t)}{t} \leq \frac{\omega(t_0)}{t_0} < \eta.$$

Since $\mu > \omega_0$ was arbitrary, we conclude that

$$\omega_0 = \lim_{t \to +\infty} \inf \frac{1}{t} \omega(t) = \limsup_{t \to +\infty} \frac{\omega(t)}{t}$$

and hence, in particular, $\omega_0 = \lim_{t \to +\infty} \inf \frac{1}{t} \omega(t)$. $\square$

As a direct consequence of Proposition 4.1.5, we obtain the following proposition.

**Proposition 4.1.6.** *Let $\{\mathcal{U}(t)\}_{t \geq 0}$ be a family of bounded linear quaternionic operators on a two-sided quaternionic Banach space $X$. Then:*

(1) *the limit $\omega_0 := \lim_{t \to +\infty} \frac{1}{t} \ln \|\mathcal{U}(t)\|$ exists,*

(2) *for each $\delta > \omega_0$ there exists a positive constant $M_\delta$ such that*

$$\|\mathcal{U}(t)\| \leq M_\delta e^{\delta t}, \quad \forall t \geq 0.$$

**Definition 4.1.7.** Let $\{\mathcal{U}(t)\}_{t \geq 0}$ be a semi-group of bounded linear quaternionic operators on a quaternionic Banach space $X$.

(1) For each $h > 0$, we define the linear quaternionic operator

$$T_h v = \frac{\mathcal{U}(h)v - v}{h}, \quad v \in X.$$

(2) We define the set

$$\mathcal{D}(T) := \left\{ v \in X : \lim_{h \to 0^+} T_h v \text{ exists in } X \right\}$$

and define the quaternionic operator $T$ with domain $\mathcal{D}(T)$ by the formula

$$Tv = \lim_{h \to 0^+} T_h v, \quad v \in \mathcal{D}(T).$$

The operator $T$, with domain $\mathcal{D}(T)$, is called the infinitesimal quaternionic generator of the quaternionic semigroup $\mathcal{U}(t)$.

**Proposition 4.1.8.** *Let $T$ be the infinitesimal quaternionic generator of the strongly continuous quaternionic semigroup $\{\mathcal{U}(t)\}_{t \geq 0}$ and let $\mathcal{D}(T)$ be its domain. Then*

(1) *The set $\mathcal{D}(T)$ is a linear subspace of $X$ and $T$ is linear on $\mathcal{D}(T)$.*

(2) *If $v \in \mathcal{D}(T)$, then $\mathcal{U}(t)v \in \mathcal{D}(T)$ for $t \geq 0$ and*

$$\frac{d}{dt}\mathcal{U}(t)v = T\mathcal{U}(t)v = \mathcal{U}(t)Tv, \quad v \in \mathcal{D}(T).$$

(3) *If $v \in \mathcal{D}(T)$, then*

$$\mathcal{U}(t)v - \mathcal{U}(\tau)v = \int_{\tau}^{t} \mathcal{U}(\theta)\, T v \, d\theta, \quad 0 \leq \tau < t < \infty. \tag{4.7}$$

(4) *Let $g : [0, +\infty) \to \mathbb{H}$ be a Lebesgue integrable function. If $g$ is continuous at point $t \in [0, +\infty)$, then*

$$\lim_{h \to 0^+} \frac{1}{h} \int_t^{t+h} \mathcal{U}(\theta) \, g(\theta) \, v \, d\theta = \mathcal{U}(t) \, g(t) \, v.$$

*Proof.* Point (1) follows immediately from the definition. Let us show Point (2). Let $h > 0$, let $t \geq 0$ and consider $v \in \mathcal{D}(T)$. Then we can write

$$\mathcal{U}(t)T_h v = T_h \mathcal{U}(t)v$$

and passing to the limit we find that

$$\lim_{h \to 0^+} \mathcal{U}(t)T_h v = \lim_{h \to 0^+} T_h \mathcal{U}(t)v.$$

We have that $\mathcal{U}(t)v \in \mathcal{D}(T)$ if $v \in \mathcal{D}(T)$ and

$$\mathcal{U}(t)Tv = T\mathcal{U}(t)v, \quad v \in \mathcal{D}(T).$$

By the semigroup properties and the definition of $T_h$, we furthermore have for $v \in \mathcal{D}(T)$ that

$$\frac{\mathcal{U}(t)v - \mathcal{U}(t-h)v}{h} - \mathcal{U}(t)Tv$$

$$= \mathcal{U}(t-h)\frac{\mathcal{U}(h)v - v}{h} - \mathcal{U}(t)Tv$$

$$= \mathcal{U}(t-h)\frac{\mathcal{U}(h)v - v}{h} - \mathcal{U}(t-h)Tv + \mathcal{U}(t-h)Tv - \mathcal{U}(t)Tv$$

$$= \mathcal{U}(t-h)(T_h v - Tv) + [\mathcal{U}(t-h) - \mathcal{U}(t)]Tv.$$

Since the semigroup is strongly continuous and because of Proposition 4.1.4, we thus find that

$$\lim_{h \to 0^+} \frac{\mathcal{U}(t)v - \mathcal{U}(t-h)v}{h} - \mathcal{U}(t)Tv = 0.$$

On the other hand, we have

$$\lim_{h \to 0^+} \frac{\mathcal{U}(t+h)v - \mathcal{U}(t)v}{h} = \lim_{h \to 0^+} \mathcal{U}(t)T_h v = U(t)Tv$$

and so

$$\frac{d}{dt}\mathcal{U}(t)v = T\mathcal{U}(t)v = \mathcal{U}(t)Tv, \quad \text{for all } v \in \mathcal{D}(T).$$

In order to prove point (3), we observe that point (2) implies that

$$\frac{d}{d\tau}\langle \varphi, \mathcal{U}(\tau)v \rangle = \left\langle \varphi, \frac{d}{d\tau}\mathcal{U}(\tau)v \right\rangle = \langle \varphi, \mathcal{U}(\tau)Tv \rangle,$$

for all linear and continuous functionals $\varphi \in X^*$. Hence, we have

$$\langle \varphi, \mathcal{U}(t)v \rangle - \langle \varphi, \mathcal{U}(s)v \rangle = \int_s^t \frac{d}{d\tau} \langle \varphi, \mathcal{U}(\tau) \, v \rangle \, d\tau$$

$$= \int_s^t \langle \varphi, \mathcal{U}(\tau)Tv \rangle \, d\tau = \left\langle \varphi, \int_s^t \mathcal{U}(\tau) \, Tv \, d\tau \right\rangle,$$

for all $\varphi \in X^*$, which implies (4.7). Finally, point (4) follows from Theorem III.12.8 in [110] which holds also in this setting, with obvious modifications. □

**Lemma 4.1.9.** *The linear subspace $\mathcal{D}(T)$ in Definition 4.1.7 is dense in $X$ and $T$ is closed on $\mathcal{D}(T)$.*

*Proof.* Let $T_h$ be as in Definition 4.1.7. Take $v \in X$ and for $h > 0$ and $t > 0$ consider

$$T_h \int_0^t \mathcal{U}(\tau)v d\tau = \frac{1}{h} \int_0^t [\mathcal{U}(h + \tau)v - \mathcal{U}(\tau)v] d\tau$$

$$= \frac{1}{h} \int_h^{h+t} \mathcal{U}(\tau)v \, d\tau - \frac{1}{h} \int_0^t \mathcal{U}(\tau)v \, d\tau$$

$$= \frac{1}{h} \int_t^h \mathcal{U}(\tau)v \, d\tau + \frac{1}{h} \int_h^{h+t} \mathcal{U}(\tau)v \, d\tau$$

$$- \frac{1}{h} \int_0^t \mathcal{U}(\tau)v \, d\tau - \frac{1}{h} \int_t^h \mathcal{U}(\tau)v \, d\tau$$

$$= \frac{1}{h} \int_t^{h+t} \mathcal{U}(\tau)v \, d\tau - \frac{1}{h} \int_0^h \mathcal{U}(\tau)v \, d\tau.$$

By Proposition 4.1.8 (4) we get

$$\lim_{h \to 0^+} T_h \int_0^t \mathcal{U}(\tau)v \, d\tau = \mathcal{U}(t)v - v,$$

and so $\int_0^t \mathcal{U}(\tau)v \, d\tau \in \mathcal{D}(T)$. Since

$$v = \lim_{t \to 0^+} \frac{1}{t} \int_0^t \mathcal{U}(\tau)v d\tau$$

we conclude that $\mathcal{D}(T)$ is dense in $X$.

We now prove that $T$ is closed. Let us take a sequence $\{v_n\}_{n \in \mathbb{N}} \subset \mathcal{D}(T)$ such that $\lim_{n \to \infty} v_n = v_0$ and $\lim_{n \to \infty} Tv_n = y_0$. Thanks to Proposition 4.1.8 (3) we have

$$\mathcal{U}(t)v_0 - v_0 = \lim_{n \to \infty} [\mathcal{U}(t)v_n - v_n]$$

$$= \lim_{n \to \infty} \int_0^t \mathcal{U}(\tau)Tv_n \, d\tau = \int_0^t \mathcal{U}(\tau)y_0 \, d\tau$$

where we have used the fact that

$$\lim_{n\to\infty} \mathcal{U}(\tau)Tv_n = \mathcal{U}(\tau)y_0, \quad \text{uniformly in } [0,t].$$

Proposition 4.1.8 point (4) yields

$$\lim_{t\to 0^+} T_t v_0 = \lim_{t\to 0^+} \frac{1}{t} \int_0^t \mathcal{U}(\tau)y_0\, d\tau = y_0.$$

This implies that $v_0 \in \mathcal{D}(T)$ and $Tv_0 = y_0$, i.e., $T$ is closed. □

We can now prove the following characterization result.

**Theorem 4.1.10.** *Let $\{\mathcal{U}(t)\}_{t\geq 0}$ be a quaternionic semigroup on a quaternionic Banach space $X$. Then $\{\mathcal{U}(t)\}_{t\geq 0}$ has a bounded infinitesimal quaternionic generator if and only if it is uniformly continuous.*

*Proof.* If $\{\mathcal{U}(t)\}_{t\geq 0}$ is a uniformly continuous semigroup then by Theorem 4.1.2 it has a bounded infinitesimal quaternionic generator. To prove the converse, we suppose that $\{\mathcal{U}(t)\}_{t\geq 0}$ has a bounded infinitesimal quaternionic generator $T$. Lemma 4.1.9 implies that $T$ is defined everywhere. By Proposition 4.1.4 and the uniform boundedness principle, Theorem 12.0.9, we have

$$K := \sup_{h\in[0,1]} \|T_h\| < +\infty.$$

From Proposition 4.1.4, we also have that for every $\tau \geq 0$ there exists a positive constant $C(\tau)$ such that

$$\|\mathcal{U}(t)\| \leq C(\tau), \quad \text{for } t \geq 0 \text{ with } |t - \tau| \leq 1.$$

The semigroup properties give

$$\mathcal{U}(t) - \mathcal{U}(\tau) = \mathcal{U}(\tau)[\mathcal{U}(t - \tau) - \mathcal{I}], \quad \text{for } t > \tau \tag{4.8}$$

and

$$\mathcal{U}(t) - \mathcal{U}(\tau) = -\mathcal{U}(t)[\mathcal{U}(\tau - t) - \mathcal{I}], \quad \text{for } \tau > t. \tag{4.9}$$

By taking the norm of (4.8) and (4.9), we get

$$\|\mathcal{U}(t) - \mathcal{U}(\tau)\| \leq C(\tau)K|t - \tau|, \quad \text{for } t \geq 0, \ |t - \tau| \leq 1,$$

which proves that $\{\mathcal{U}(t)\}_{t\geq 0}$ is a uniformly continuous quaternionic semigroup. □

## 4.2 Strongly continuous quaternionic semigroups

We now show that it is possible to define the Laplace transform of a strongly continuous quaternionic semigroup, and we prove that the Hille–Yosida–Phillips theorem can be extended to the quaternionic setting.

**Theorem 4.2.1.** Let $\{\mathcal{U}(t)\}_{t\geq 0}$ be a strongly continuous quaternionic semigroup and let $T_h$ be as in Definition 4.1.7. Then

$$\mathcal{U}(t)v = \lim_{h\to 0^+} e^{t\,T_h}\,v, \quad v \in X$$

uniformly for $t$ in any finite interval.

*Proof.* First of all, we observe that since $t$ and $h$ are real numbers, the operators $\frac{t}{h}I$ and $\frac{t}{h}\mathcal{U}(h)$ commute so the quaternionic operator $e^{tT_h}$ can be written as

$$e^{tT_h} = e^{-\frac{t}{h}I}\, e^{\frac{t}{h}\mathcal{U}(h)}.$$

Since $\mathcal{U}(t)$ is a bounded operator, we can use the power series expansion of the exponential and write

$$e^{t\,T_h} = e^{-\frac{t}{h}I} \sum_{n=0}^{+\infty} \frac{t^n}{n!h^n}\mathcal{U}(h)^n = e^{-\frac{t}{h}I} \sum_{n=0}^{\infty} \frac{t^n}{n!\,h^n}\,\mathcal{U}(nh). \tag{4.10}$$

By taking the norm of (4.10) and using point (2) in Proposition 4.1.6, we have that, for every $\delta > \omega_0$, there exists a positive constant $M_\delta$ such that

$$\|e^{t\,T_h}\| \leq e^{-\frac{t}{h}}\, M_\delta \sum_{n=0}^{\infty} \frac{t^n}{n!\,h^n}\, e^{nh\delta} = M_\delta\, e^{m(t,\delta,h)},$$

where

$$m(t,\delta,h) := \frac{t}{h}(e^{h\delta} - 1).$$

So there exists a positive constant $K_t$ for which

$$\|e^{\tau\,T_h}\| \leq K_t, \quad \text{for } \tau \in [0,t],\ h \in (0,1].$$

Now if $v \in \mathcal{D}(T)$ and $t \leq t_0$ then, by point (3) in Proposition 4.1.8, we obtain

$$\mathcal{U}(t)v - e^{t\,T_h}v = \int_0^t \frac{d}{d\tau}(e^{(t-\tau)T_h}\mathcal{U}(\tau)v)\,d\tau$$

$$= \int_0^t e^{(t-\tau)T_h}\mathcal{U}(\tau)(Tv - T_hv)\,d\tau.$$

Taking the norm, we have

$$\|\mathcal{U}(t)v - e^{t\,T_h}v\| \leq t_0 K_{t_0} M_\delta e^{\delta t_0}\|Tv - T_hv\|,$$

so

$$\|\mathcal{U}(t)v - e^{tT_h}v\| \to 0 \quad \text{as } h \to 0.$$

Lemma 4.1.9 implies that $\mathcal{D}(T)$ is dense in $X$. The statement follows by Theorem II.3.6 in [110], (see Theorem 12.0.16 in the appendix) which holds also in the quaternionic setting, with obvious modifications. $\qquad\square$

Now we introduce the Laplace transform for strongly continuous quaternionic semigroups.

**Theorem 4.2.2** (Laplace transform and the S-resolvent operators). *Let $\{\mathcal{U}(t)\}_{t\geq 0}$ be a strongly continuous quaternionic semigroup and set*

$$\omega_0 := \lim_{t\to+\infty} \frac{1}{t} \ln \|\mathcal{U}(t)\|.$$

*Assume that $\{\mathcal{U}(t)\}_{t\geq 0}$ is generated by a linear quaternionic operator $T$ and take $s \in \mathbb{H}$ such that $\mathrm{Re}(s) > \omega_0$. Then $s \in \rho_S(T)$ and*

$$S_L^{-1}(s,T) = \int_0^\infty \mathcal{U}(t)\, e^{-ts}\, dt, \tag{4.11}$$

$$S_R^{-1}(s,T) = \int_0^\infty e^{-ts}\, \mathcal{U}(t)\, dt. \tag{4.12}$$

*Proof.* We choose $\delta$ with $\omega_0 < \delta < \mathrm{Re}(s)$. By Proposition 4.1.6, there exists a positive constant $M_\delta$ such that

$$\|\mathcal{U}(t)\| \leq M_\delta\, e^{\delta t}, \quad \text{for } t \geq 0. \tag{4.13}$$

As a consequence, the integrals (4.11) and (4.12) exist for $\mathrm{Re}(s) > \omega_0$ and define bounded linear operators. Let us set, for $\mathrm{Re}(s) > \omega_0$,

$$F(s) := \int_0^\infty \mathcal{U}(t)\, e^{-ts} \quad \text{and} \quad G(s) := \int_0^\infty e^{-ts}\, \mathcal{U}(t).$$

We have to prove that $s \in \rho_S(T)$ and that $F(s) = S_L^{-1}(s,T)$ and $G(s) = S_R^{-1}(s,T)$.

Step 1. We show that $F(s)$ satisfies the left $S$-resolvent equation and that $F(s)v \in \mathcal{D}(T)$ for any $v \in X$.

We apply the operator $T_h$ to $F(s)v$ and observe that

$$(T_h F)(s)v = \int_0^\infty \frac{\mathcal{U}(h) - \mathcal{I}}{h}\, \mathcal{U}(t)\, e^{-ts}\, v\, dt$$

$$= \frac{1}{h} \int_0^\infty \mathcal{U}(t+h)\, e^{-ts}\, v\, dt$$

$$- \frac{1}{h} \int_0^\infty \mathcal{U}(t)\, e^{-ts}\, v\, dt.$$

With a change of variable we get

$$(T_h F)(s)v = \frac{1}{h} \int_h^\infty \mathcal{U}(\tau)\, e^{-s(\tau-h)}\, v\, d\tau - \frac{1}{h} \int_0^\infty \mathcal{U}(t)\, e^{-ts}\, v\, dt.$$

By denoting again the variable $\tau$ with $t$ we have

$$(T_h F)(s)v = \frac{1}{h} \int_0^h \mathcal{U}(t)\, e^{-s(t-h)}\, v\, dt$$

$$+ \frac{1}{h} \int_h^\infty \mathcal{U}(t)\, e^{-s(t-h)}\, v\, dt$$

$$- \frac{1}{h} \int_0^\infty \mathcal{U}(t)\, e^{-ts}\, v\, dt$$

$$- \frac{1}{h} \int_0^h \mathcal{U}(t)\, e^{-s(t-h)}\, v\, dt$$

$$= \int_0^\infty \mathcal{U}(t)\, e^{-st}\, \frac{(e^{sh} - 1)}{h}\, v\, dt$$

$$- \frac{1}{h} \int_0^h \mathcal{U}(t)\, e^{-s(t-h)}\, v\, dt.$$

So we finally obtain

$$T_h F(s)v = F(s) \frac{1}{h}(e^{sh} - 1)\, v - \frac{1}{h} \int_0^h \mathcal{U}(t)\, e^{-st}\, e^{sh}\, v\, dt. \qquad (4.14)$$

Taking the limit for $h \to 0$ in (4.14), and using point (4) in Proposition 4.1.8, we obtain

$$TF(s)v = F(s)sv - \mathcal{I}v, \quad \text{for every } v \in X. \qquad (4.15)$$

Hence, we find that $F(s)v \in \mathcal{D}(T)$ and also that $F(s)$ satisfies the left S-resolvent equation.

  Step 2. The operator $G(s)$ satisfies the right $S$-resolvent equation for $v \in \mathcal{D}(T)$, i.e.,

$$sG(s)v - G(s)Tv = \mathcal{I}v, \quad v \in \mathcal{D}(T).$$

Moreover the operator $G(s)T$ extends to a bounded linear operator, denoted by $G_T(s)$ defined on all of $X$. So the right $S$-resolvent equation holds for $v \in X$.

  Now, consider

$$G(s)T_h v = \int_0^\infty e^{-ts}\mathcal{U}(t)\, \frac{\mathcal{U}(h) - \mathcal{I}}{h} v\, dt$$

$$= \frac{1}{h} \int_0^\infty e^{-ts}\, \mathcal{U}(t+h)\, v\, dt$$

$$- \frac{1}{h} \int_0^\infty e^{-ts}\mathcal{U}(t)\, v\, dt.$$

With a change of variable we get

$$G(s)T_h v = \int_0^\infty \frac{(e^{sh} - 1)}{h}\, e^{-st}\, \mathcal{U}(t)\, v\, dt \; - \frac{1}{h} \int_0^h e^{-s(t-h)}\mathcal{U}(t)\, v\, dt$$

and taking the limit we get

$$sG(s)v - G(s)Tv = \mathcal{I}v, \quad \text{for every } v \in \mathcal{D}(T). \tag{4.16}$$

Finally observe that the right-hand side of the equation

$$G(s)Tv = sG(s)v - \mathcal{I}v$$

is defined on $X$ so that $G(s)Tv$ extends to an operator that is defined on all of $X$. We denote this operator by $G_T(s)$. We have

$$sG(s)v - G_T(s)v = \mathcal{I}v, \quad \text{for every } v \in X. \tag{4.17}$$

Step 3. We show the slice hyperholomorphicity of $F(s)$ and of $G(s)$. If $s = u + jv$, then we have

$$F(s) = \int_0^{+\infty} \mathcal{U}(t)e^{-ts}\, dt = \int_0^{+\infty} \mathcal{U}(t)e^{-tu}e^{-tjv}\, dt$$

$$= \int_0^{+\infty} \mathcal{U}(t)e^{-tu}\cos(-tv)\, dt + \int_0^{+\infty} \mathcal{U}(t)e^{-tu}\sin(-tv)\, dt j.$$

It is immediate that

$$F_0(u,v) := \int_0^{+\infty} \mathcal{U}(t)e^{-tu}\cos(-tv)\, dt$$

and

$$F_1(u,v) := \int_0^{+\infty} \mathcal{U}(t)e^{-tu}\sin(-tv)\, dt$$

satisfy the compatibility condition (2.4) and hence $F$ is a right slice function. Furthermore, we can, due to (4.13), exchange differentiation and integration and obtain

$$\frac{1}{2}\left(\frac{\partial}{\partial u}F(s) + \frac{\partial}{\partial v}F(s)j\right) = \int_0^{+\infty} \mathcal{U}(t)\frac{1}{2}\left(\frac{\partial}{\partial u}e^{-ts} + \frac{\partial}{\partial v}e^{-ts}j\right) dt = 0,$$

which is equivalent to $F_0$ and $F_1$ satisfying the Cauchy–Riemann equations (2.5). Hence, $F(s)$ is right slice hyperholomorphic. The left slice hyperholomorphicity of $G(s)$ can be shown with analogous arguments.

Step 4. If we consider $s \in \mathbb{R}$, i.e., $s = s_0 \in \mathbb{R}$ with $s_0 > \omega_0$, then $F(s_0) = G(s_0)$ as in this case

$$F(s_0) = \int_0^{\infty} \mathcal{U}(t)\, e^{-ts_0}\, dt = \int_0^{\infty} e^{-ts_0}\mathcal{U}(t)\, dt = G(s_0).$$

Because of (4.15), we have that

$$(s_0 \mathcal{I} - T)F(s_0)v = v \quad \text{for every } v \in X.$$

From (4.16) on the other hand, we obtain that

$$G(s_0)(s_0 \mathcal{I} - T)v = v \quad \text{for every } v \in \mathcal{D}(T)$$

and hence

$$F(s_0) = G(s_0) = (s_0 \mathcal{I} - T)^{-1} = S_R^{-1}(s_0, T) = S_L^{-1}(s_0, T).$$

In particular, we find that $s = s_0 \in \rho_S(T)$ as

$$\mathcal{Q}_s(T)^{-1} = (T^2 - 2s_0 T + s_0^2 \mathcal{I})^{-1} = (T - s_0 \mathcal{I})^{-2} \in \mathcal{B}(X).$$

We know that $F(s)$ and $G(s)$ are the left (resp. right) $S$-resolvents of $T$ for $s = s_0 \in \mathbb{R}$ with $s_0 > w_0$. Furthermore, $F(s)$ is a right slice hyperholomorphic continuation of $S_L^{-1}(s, T)$ and $G(s)$ is a left slice hyperholomorphic continuation of $S_R^{-1}(s, T)$ to the set $\Omega_{w_0} := \{s \in \mathbb{H} : \operatorname{Re}(s) > w_0\}$. From Theorem 3.2.11, we conclude that $\Omega_{w_0} \subset \rho_S(T)$ and

$$F(s) = S_L^{-1}(s, T) \quad \text{and} \quad G(s) = S_R^{-1}(s, T) \quad \text{for } s \in \Omega_{w_0}. \qquad \square$$

**Remark 4.2.1.** The operators $F(s)$ and $G(s)$ satisfy the $S$-resolvent equation, as one can verify also directly. Indeed, this follows from the relations (4.15) and (4.17), which were shown in the above proof,

$$F(q)qv = TF(q)v + \mathcal{I}v, \quad \text{for every } v \in X \tag{4.18}$$

and

$$G_T(s)v = sG(s)v - \mathcal{I}v, \quad \text{for every } v \in X. \tag{4.19}$$

Assuming that $s \notin [q]$, we can apply the operator $G$ to both sides of (4.18) and obtain

$$G(s)F(q)qv = (G(s)T)F(q)v + G(s)v.$$

Using (4.18), this becomes

$$G(s)F(q)qv = sG(s)F(q)v - F(q)v + G(s)v. \tag{4.20}$$

Now we can iteratively use (4.20). Indeed, we replace $v$ by $qv$ in (4.20) and get

$$G(s)F(q)q^2 v = sG(s)F(q)qv - F(q)qv + G(s)qv. \tag{4.21}$$

Then we multiply (4.20) by $-2s_0$ and we obtain

$$-2s_0 G(s)F(q)qv = -2s_0 sG(s)F(q)v + 2s_0 F(q)v - 2s_0 G(s)v. \tag{4.22}$$

If we sum the equations (4.21) and (4.22) and add the identity

$$|s|^2 G(s)F(q)v = |s|^2 G(s)F(q)v,$$

then we get

$$G(s)F(q)(q^2 - 2s_0 q + |s|^2)v = (s[G(s)F(q)qv] - F(q)qv + G(s)qv$$
$$- 2s_0 sG(s)F(q)v + 2s_0 F(q)v - 2s_0 G(s)v$$
$$+ |s|^2 G(s)F(q)v.$$

Now we replace the term $[G(s)F(q)qv]$ in the right side of the above equation, by the right side of (4.20), and we have

$$G(s)F(q)(q^2 - 2s_0 q + |s|^2)v = (s[sG(s)F(q)v - F(q)v + G(s)v] - F(q)qv + G(s)qv$$
$$- 2s_0 sG(s)F(q)v + 2s_0 F(q)v - 2s_0 G(s)v$$
$$+ |s|^2 G(s)F(q)v.$$

Finally, with simple computations, we get

$$G(s)F(q)(q^2 - 2s_0 q + |s|^2)v = (s^2 - 2s_0 s + |s|^2)G(s)F(q)v$$
$$\cdot \overline{s}(F(q) - G(s))v - (F(q) - G(s))qv$$

but since $s^2 - 2s_0 s + |s|^2 = 0$, if we replace $v$ by $(q^2 - 2s_0 q + |s|^2)^{-1}v$, we get that $G(s)$ and $F(q)$ satisfy the $S$-resolvent equation.

**Theorem 4.2.3** (Hille–Yosida–Phillips: necessary condition). *Let $T$ be a closed linear quaternionic operator with dense domain whose $S$-spectrum lies in the half space of quaternions $s$ with $\mathrm{Re}(s) \leq \omega$ for some $\omega \in \mathbb{R}$. Let $\{\mathcal{U}(t)\}_{t\geq 0}$ be a strongly continuous semigroup and assume that there exists $M > 0$ such that*

$$\|\mathcal{U}(t)\| \leq Me^{\omega t}, \quad t \geq 0,$$

*and*

$$S_L^{-1}(s,T) = \int_0^\infty \mathcal{U}(t)\,e^{-st}\,dt, \quad \mathrm{Re}(s) > \omega. \tag{4.23}$$

*Then for any $s \in \mathbb{H}$ with $\mathrm{Re}(s) > \omega$ and any $n \in \mathbb{N}$, we have*

$$\|S_L^{-n}(s,T)\| \leq \frac{M}{(\mathrm{Re}(s) - \omega)^n} \quad \text{and} \quad \|S_R^{-n}(s,T)\| \leq \frac{M}{(\mathrm{Re}(s) - \omega)^n}, \tag{4.24}$$

*where $S_L^{-n}(s,T)$ and $S_R^{-n}(s,T)$ denote the $n$-th slice power of $S_L^{-1}(s,T)$ (resp. $S_R^{-1}(s,T)$), which were defined in Definition 3.5.6, namely*

$$S_L^{-n}(s,T) = \sum_{k=0}^n (-1)^k \binom{n}{k} T^k \mathcal{Q}_s(T)^{-n}\overline{s}^{n-k}, \tag{4.25}$$

*and*

$$S_R^{-n}(s,T) = \sum_{k=0}^n (-1)^k \binom{n}{k} \overline{s}^{n-k} T^k \mathcal{Q}_s(T)^{-n}. \tag{4.26}$$

*Proof.* We observe that we can exchange the slice derivative with the integral in (4.23): if $q, s \in \mathbb{H}$ belong to the same complex plane $\mathbb{C}_j$ and satisfy $\omega < \delta < \mathrm{Re}(s) < \mathrm{Re}(q)$, we set

$$f(q, s, t) := \left(e^{-qt} - e^{-st}\right)(q - s)^{-1}.$$

Then

$$|f(q, s, t)| = \left| e^{-st} \int_0^t e^{-(q-s)\tau} \, d\tau \right| \le e^{-\mathrm{Re}(s)t} \int_0^t e^{-\mathrm{Re}(q-s)\tau} \, d\tau < t e^{-\delta t},$$

and so

$$\|f(q, s, t) t^n \mathcal{U}(t)\| \le M t^{n+1} e^{-(\delta - \omega)t}.$$

This estimate allows us to exchange the slice derivative with the integral in (4.23) and we find

$$\partial_S{}^n S_L^{-1}(s, T) = \int_0^\infty \mathcal{U}(t) \, \partial_S{}^n e^{-st} \, dt = \int_0^\infty \mathcal{U}(t)(-t)^n e^{-st} \, dt.$$

We recall that by Proposition 3.5.7, we have for every $n \in \mathbb{N}$

$$\partial_S{}^n S_L^{-1}(s, T) = (-1)^n \, n! \, S_L^{-(n+1)}(s, T), \quad s \in \rho_S(T) \tag{4.27}$$

and so

$$S_L^{-(n+1)}(s, T) = \frac{1}{n!} \int_0^\infty \mathcal{U}(t) t^n e^{-st} \, dt \quad \text{for} \ \ \mathrm{Re}(s) > \omega.$$

Taking the norm, we obtain the desired estimate

$$\left\| S_L^{-(n+1)}(s, T) \right\| \le \frac{M}{(n-1)!} \int_0^\infty e^{-t(\mathrm{Re}(s)-\omega)} \, t^{n-1} \, dt = \frac{M}{(\mathrm{Re}(s) - \omega)^n}.$$

The estimate for the right $\|S_R^{-1}(s, T)\|$ can be shown with analogous arguments.
□

**Definition 4.2.4** (Yosida approximations). Let $T \in \mathcal{K}(X)$ and $s \in \rho_S(T)$. We define the left and right Yosida approximations of $T$ as

$$\mathcal{Y}_L(s) := -[\mathcal{I} - S_L^{-1}(s, T)s]s \tag{4.28}$$

and

$$\mathcal{Y}_R(s) := -s[\mathcal{I} - s S_R^{-1}(s, T)], \tag{4.29}$$

respectively.

**Remark 4.2.2.** We observe that, if $s \in \mathbb{R}$, we obviously have $\mathcal{Y}_L(s) = \mathcal{Y}_R(s)$.

**Lemma 4.2.5.** *Let $T \in \mathcal{K}(X)$ have dense domain and assume that there exist constants $\omega \in \mathbb{R}$ and $M > 0$ such that*

(i) *the S-spectrum of $T$ lies in the half space of quaternions $s$ with $\mathrm{Re}(s) \leq \omega$ and*

(ii) *for any $n \in \mathbb{N}$ and $s_0 > \omega$, we have*

$$\left\| S_L^{-n}(s_0, T) \right\| \leq \frac{M}{(s_0 - \omega)^n}.$$

*Then*

$$\lim_{\mathbb{R} \ni s_0 \to +\infty} \mathcal{Y}_L(s_0)v = Tv, \quad \text{for } v \in \mathcal{D}(T). \tag{4.30}$$

*Proof.* For any real $s_0 > \omega$, we have $S_L^{-1}(s_0, T) = (s_0 \mathcal{I} - T)^{-1}$ and so

$$\begin{aligned}
\mathcal{Y}_L(s_0)v &= -[\mathcal{I} - S_L^{-1}(s_0, T)s_0]s_0 v \\
&= -[\mathcal{I} - s_0(s_0\mathcal{I} - T)^{-1}]s_0 v = s_0(s_0\mathcal{I} - T)^{-1}Tv
\end{aligned} \tag{4.31}$$

for $v \in \mathcal{D}(T)$. For $w \in X$, we set

$$\mathcal{A}_{s_0} w := s_0(s_0\mathcal{I} - T)^{-1}w$$

and observe that, for sufficiently large $s_0$,

$$\|\mathcal{A}_{s_0} w\| = \|s_0(s_0\mathcal{I} - T)^{-1}w\| \leq \frac{s_0 M}{s_0 - \omega} \leq 2M.$$

If $w \in \mathcal{D}(T)$, we furthermore have

$$\begin{aligned}
\|\mathcal{A}_{s_0} w - w\| &= \|s_0(s_0\mathcal{I} - T)^{-1}w - w\| \\
&= \|(s_0\mathcal{I} - T)^{-1}Tw\| \leq \|Tw\| \frac{M}{s_0 - \omega} \xrightarrow{s_0 \to +\infty} 0.
\end{aligned}$$

So by Theorem 12.0.16, we find that $\mathcal{A}_{s_0}$ converges strongly to a bounded linear operator, namely

$$\lim_{s_0 \to \infty} s_0(s_0\mathcal{I} - T)^{-1}w = w.$$

From (4.31), we conclude

$$\lim_{s_0 \to +\infty} \mathcal{Y}_L(s_0)v = \lim_{s_0 \to +\infty} s_0(s_0\mathcal{I} - T)^{-1}Tv = Tv \quad \text{for } v \in \mathcal{D}(T). \qquad \square$$

**Theorem 4.2.6** (Hille–Yosida–Phillips: sufficient condition)*. Let $T \in \mathcal{K}(X)$ with dense domain. If there exist $M > 0$ and $\omega \in \mathbb{R}$ such that for every real number $s_0 > \omega$ we have $s_0 \in \rho_S(T)$ and*

$$\|(s_0\mathcal{I} - T)^{-n}\| \leq \frac{M}{(s_0 - \omega)^n}, \quad n \in \mathbb{N}, \tag{4.32}$$

*then $T$ is the infinitesimal generator of a strongly continuous semigroup.*

*Proof.* If $s_0 > \omega$, then $s_0 \in \rho_s(T)$ and hence

$$\mathcal{Y}_L(s_0) = -s_0 \mathcal{I} + s_0^2(s_0 \mathcal{I} - T)^{-1}$$

is a bounded operator. Since $s_0$ is real, $-s_0 \mathcal{I}$ and $s_0^2(s_0 \mathcal{I} - T)^{-1}$ commute and so

$$e^{t\,\mathcal{Y}_L(s_0)} = e^{-t\,s_0}\, e^{t\,s_0^2(s_0\mathcal{I}-T)^{-1}}$$

$$= e^{-t\,s_0} \sum_{n=0}^{\infty} \frac{1}{n!}\,(t\,s_0^2)^n\,(s_0\mathcal{I} - T)^{-n}.$$

Taking the norm we have

$$\|e^{t\,\mathcal{Y}_L(s_0)}\| \le e^{-t\,s_0} \sum_{n=0}^{\infty} \frac{1}{n!}\,(t\,s_0^2)^n\,\|(s_0\mathcal{I} - T)^{-n}\|$$

$$\le M\,e^{-t\,s_0} \sum_{n=0}^{\infty} \frac{(t\,s_0^2)^n}{n!\,(s_0 - \omega)^n}$$

$$= M\,\exp(-t\,s_0)\,\exp\left(\frac{t\,s_0^2}{s_0 - \omega}\right) = M\exp\left(\frac{t\,s_0\,\omega}{s_0 - \omega}\right).$$

If $\omega_1 > \omega$, then we have for $s_0$ sufficiently large

$$\frac{s_0\,\omega}{s_0 - \omega} < \omega_1,$$

and for all $t \ge 0$ we obtain

$$\|e^{t\,\mathcal{Y}_L(s_0)}\| \le M e^{t\,\omega_1}. \tag{4.33}$$

Lemma 4.2.5 implies that

$$\lim_{s_0 \to \infty} \mathcal{Y}_L(s_0)v = Tv, \quad \text{for } v \in \mathcal{D}(T).$$

We observe that, for any $s_0$, $p_0 \in \mathbb{R}$, we have

$$\mathcal{Y}_L(s_0)\mathcal{Y}_L(p_0) = [-s_0\mathcal{I} + s_0^2(s_0\mathcal{I} - T)^{-1}][-p_0\mathcal{I} + p_0^2(p_0\mathcal{I} - T)^{-1}]$$
$$= \mathcal{Y}_L(p_0)\mathcal{Y}_L(s_0)$$

and set

$$\mathcal{U}_{s_0}(t) := e^{t\,\mathcal{Y}_L(s_0)}.$$

From the power series expansion

$$\mathcal{U}_{s_0}(t) := \sum_{n=0}^{\infty} \frac{t^n}{n!}\,(\mathcal{Y}_L(s_0))^n,$$

we see that

$$\mathcal{Y}_L(p_0)\mathcal{U}_{s_0}(t) := \sum_{n=0}^{\infty} \frac{t^n}{n!} (\mathcal{Y}_L(s_0))^n \mathcal{Y}_L(p_0) = \mathcal{U}_{s_0}(t)\mathcal{Y}_L(p_0).$$

Take $v \in \mathcal{D}(T)$ and apply point (3) in Proposition 4.1.8 to get

$$\mathcal{U}_{s_0}(t)v - \mathcal{U}_{p_0}(t)v = \int_0^t \frac{d}{d\tau} [\mathcal{U}_{p_0}(t-\tau)\mathcal{U}_{s_0}(\tau)v] \, d\tau$$

$$= \int_0^t \mathcal{U}_{p_0}(t-\tau)[\mathcal{Y}_L(s_0) - \mathcal{Y}_L(p_0)]\mathcal{U}_{s_0}(\tau)v \, d\tau$$

$$= \int_0^t \mathcal{U}_{p_0}(t-\tau)\mathcal{U}_{s_0}(\tau)[\mathcal{Y}_L(s_0) - \mathcal{Y}_L(p_0)]v \, d\tau.$$

By taking the norm and considering $s_0$ and $p_0$ sufficiently large, we have thanks to (4.33)

$$\|\mathcal{U}_{s_0}(t)v - \mathcal{U}_{p_0}(t)v\| \leq M^2 t e^{t\omega_1} \|[\mathcal{Y}_L(s_0) - \mathcal{Y}_L(p_0)]v\|.$$

Hence, $\mathcal{U}_{s_0}(t)v$ converges to a limit uniformly in each finite interval. By assumption $\mathcal{D}(T)$ is dense in $X$. Thanks to estimate (4.33) and to Theorem 12.0.16, there exists a linear quaternionic bounded operator $\mathcal{U}(t)$ such that

$$\lim_{s_0 \to \infty} \mathcal{U}_{s_0}(t)v = \mathcal{U}(t)v, \quad v \in X,$$

and moreover

$$\|\mathcal{U}(t)v\| \leq \liminf_{s_0 \to \infty} \|\mathcal{U}_{s_0}(t)v\| \leq e^{t\omega_1}.$$

We observe that the map $t \to \mathcal{U}(t)v$ is continuous because of the uniform convergence of $\mathcal{U}_{s_0}(t)$ as $s_0$ tends to infinity and that $\mathcal{U}(t)$ is a semigroup because $\mathcal{U}_{s_0}(t)$ is a semigroup for each $s_0$.

What remains to show is that $T$ is the infinitesimal generator of $\mathcal{U}(t)$. We observe that the following estimates hold

$$\|\mathcal{U}_{s_0}(t)\mathcal{Y}_L(s_0)v - \mathcal{U}(t)Tv\|$$
$$\leq \|\mathcal{U}_{s_0}(t)[\mathcal{Y}_L(s_0)v - Tv]\| + \|[\mathcal{U}_{s_0}(t) - \mathcal{U}(t)]Tv\| \qquad (4.34)$$
$$\leq M e^{t\omega_0} \|\mathcal{Y}_L(s_0)v - Tv\| + 2M e^{t\omega_0} \|Tv\|,$$

for any $v \in \mathcal{D}(T)$. By estimate (4.34) and the Lebesgue Dominated Convergence Theorem, we can take the limit as $s_0 \to \infty$ in both sides of

$$\mathcal{U}_{s_0}(t)v - v = \int_0^t \mathcal{U}_{s_0}(\tau)\mathcal{Y}_L(s_0)v \, d\tau$$

and we get

$$\mathcal{U}(t)v - v = \int_0^t \mathcal{U}(\tau)Tv \, d\tau.$$

Now if we denote by $Z$ the infinitesimal quaternionic generator of $\mathcal{U}(t)$, we have

$$Zv = \lim_{t \to 0} \frac{\mathcal{U}(t)v - v}{t} = \lim_{t \to 0} \frac{1}{t} \int_0^t \mathcal{U}(\tau)Tv \, d\tau = Tv, \quad \text{for } v \in \mathcal{D}(T).$$

This means that $\mathcal{D}(T) \subseteq \mathcal{D}(Z)$ and $Z$ is an extension of $T$. But for $s_0$ sufficiently large, we find that $s_0 \in \rho_S(T) \bigcap \rho_S(Z)$ and the relations

$$(Z - s_0\mathcal{I})\mathcal{D}(T) = (T - s_0\mathcal{I})\mathcal{D}(T) = X \quad \text{and} \quad (Z - s_0\mathcal{I})\mathcal{D}(Z) = X$$

imply $\mathcal{D}(T) = \mathcal{D}(Z)$. Hence, $Z = T$.                                         □

The Hille–Yosida–Phillips theorem has several consequences.

**Corollary 4.2.7.** *An operator $T \in \mathcal{K}(X)$ with dense domain generates a strongly continuous quaternionic semigroup $\{\mathcal{U}(t)\}_{t \geq 0}$ of bounded quaternionic operators such that*

$$\|\mathcal{U}(t)\| \leq e^{t\omega}, \quad \text{for some real number } \omega$$

*if and only if the estimate*

$$\|S_L^{-1}(s_0, T)\| \leq \frac{1}{s_0 - \omega}, \quad \text{for } s_0 > \omega \tag{4.35}$$

*holds (where $S_L^{-1}(s_0, T) = S_R^{-1}(s_0, T)$ since $s_0 \in \mathbb{R}$).*

*Proof.* The Laplace transform of the semigroup

$$S_L^{-1}(s_0, T) = \int_0^\infty \mathcal{U}(t)e^{-s_0 t}dt$$

and Theorem 4.2.3 imply the necessity of the estimate (4.35). Now consider $s \in \rho_S(T) \cap \mathbb{R}$. We have

$$S_L^{-1}(s_0, T) = (s_0\mathcal{I} - T)^{-1} = S_R^{-1}(s_0, T).$$

Estimate (4.35) implies that

$$\|S_L^{-1}(s_0, T)^n\| \leq \|S_L^{-1}(s_0, T)\|^n \leq \frac{1}{(s_0 - \omega)^n}, \quad s_0 > \omega,$$

and by Theorem 4.2.6, the operator $T$ is the generator of a semigroup $\{\mathcal{U}(t)\}_{t \geq 0}$. From the proof of Theorem 4.2.6 we also have that $\|\mathcal{U}(t)\| \leq e^{t\omega}$, for $M = 1$.     □

To prove the next corollary we need a technical lemma. The result is well known for real functions (see, for example, XIII.1.15 in [110]). The proof of the quaternionic version follows the same lines.

**Lemma 4.2.8.** *Let $u \in L^1((0,\infty); \mathbb{H})$, $s \in \mathbb{H}$ and*

$$\int_0^\infty e^{-st} u(t) dt = 0$$

*for $\operatorname{Re}(s)$ sufficiently large. Then $u(t) = 0$ almost everywhere.*

**Corollary 4.2.9.** *Let $T$ be a linear closed quaternionic operator with dense domain. Then $T$ is the quaternionic infinitesimal generator of a strongly continuous quaternionic semigroup if and only if there exists a strongly continuous family $\mathcal{W}(t)$, $t \geq 0$, of bounded linear quaternionic operators, satisfying, for some real numbers $M > 0$ and $\omega$, the conditions*

- $\mathcal{W}(0) = \mathcal{I}$,
- $\|\mathcal{W}(t)\| \leq M e^{t\omega}$, and
- $S_L^{-1}(s_0, T) = S_R^{-1}(s_0, T) = \int_0^\infty e^{-s_0 t} \mathcal{W}(t) \, dt, \quad s_0 > \omega.$

*Then $\mathcal{W}(t)$ is the quaternionic semigroup with infinitesimal generator $T$.*

*Proof.* Theorem 4.2.3 implies

$$\|(T - s_0 \mathcal{I})^{-n}\| \leq \frac{M}{(s_0 - \omega)^n}, \quad \text{for } s_0 > \omega, \ n \in \mathbb{N},$$

and by Theorem 4.2.6 the operator $T$ is the infinitesimal generator of a strongly continuous semigroup $\{\mathcal{U}(t)\}_{t \geq 0}$ with $\|\mathcal{U}(t)\| \leq M e^{t\omega}$. Theorem 4.2.2 yields

$$S_R^{-1}(s_0, T) v = (T - s_0 \mathcal{I})^{-1} v = \int_0^\infty e^{-t s_0} \, \mathcal{U}(t) \, v \, dt, \quad v \in X, \ s_0 > \omega.$$

We now reason by duality. Let $\varphi$ be an element in the dual space $X^*$ and let $v \in X$. Then

$$\int_0^\infty e^{-s_0 t} \langle \varphi, \mathcal{U}(t) \, v - \mathcal{W}(t) \, v \rangle \, dt = 0, \quad \text{for } s_0 > \omega.$$

If we define the function

$$u(t) := e^{-(\omega+1)t} \langle \varphi, \mathcal{U}(t) \, v - \mathcal{W}(t) \, v \rangle, \quad \varphi \in X^*,$$

then

$$\int_0^\infty e^{-s_0 t} \, u(t) \, dt = 0, \quad \text{for } s_0 \geq 0.$$

From Lemma 4.2.8, we get that $\langle \varphi, \mathcal{U}(t) \, v - \mathcal{W}(t) \, v \rangle = 0$ for almost all $t \geq 0$. By continuity, such equation holds for all $t \geq 0$ and thus as a consequence of the quaternionic version of the Hahn–Banach theorem, we get $\mathcal{U}(t) = \mathcal{W}(t)$ for all $t \geq 0$. $\square$

## 4.3   Strongly continuous groups

Now we consider the problem of characterizing when a strongly continuous semi-group of operators defined on $[0, \infty)$ can be extended to a group of operators $\mathcal{U}(t)$ defined for $t \in \mathbb{R}$. This extension is unique if it exists and the family $\mathcal{Z}(t) = \mathcal{U}(-t)$, for $t \geq 0$, is a strongly continuous semigroup. Consider the identity

$$\frac{1}{t}[\mathcal{Z}(t)v - v] = \frac{1}{-t}[-\mathcal{U}(-2)[\mathcal{U}(2 - t)v - \mathcal{U}(2)v]], \quad \text{for } t \in (0, 1).$$

By taking the limit for $t \to 0$, we see that the infinitesimal generator of $\mathcal{Z}(t)$ is $-T$ with $\mathcal{D}(-T) = \mathcal{D}(T)$. In this case $T$ is called the quaternionic infinitesimal generator of the group $\mathcal{U}(t)$.

**Theorem 4.3.1.** *Let $T \in \mathcal{K}(X)$ with dense domain. Then $T$ is the quaternionic infinitesimal generator of a strongly continuous quaternionic group of bounded operators if and only if there exist real numbers $M > 0$ and $\omega \geq 0$ such that*

$$\|(S_L^{-1}(s_0, T))^n\| \leq \frac{M}{(|s_0| - \omega)^n}, \quad s_0 > \omega \text{ or } s_0 < -\omega. \tag{4.36}$$

*If $T$ generates the group $\{\mathcal{Z}(t)\}_{t \in \mathbb{R}}$, then $\|\mathcal{Z}(t)\| \leq Me^{\omega|t|}$.*

*Proof.* The necessity of estimate 4.36 follows from the above considerations, Theorem 4.2.6 as well as the relations $\mathcal{Q}_s(-T)^{-1} = \mathcal{Q}_{-s}(T)^{-1}$ and $\sigma_S(-T) = -\sigma_S(T)$.

On the other hand, we observe that if (4.36) holds for both $T$ and $-T$, then (4.36) satisfies the condition of Theorem 4.2.6 so $T$ and $-T$ generate strongly continuous semigroups $\mathcal{U}^+(t)$ and $\mathcal{U}^-(t)$, respectively. Observe that the approximations

$$\mathcal{U}_{s_0}^+(t) := \sum_{n=0}^{\infty} \frac{1}{n!} t^n \left(\mathcal{Y}_L^+(s_0)\right)^n \quad \text{and}$$

$$\mathcal{U}_{s_0}^-(t) := \sum_{n=0}^{\infty} \frac{1}{n!} t^n \left(\mathcal{Y}_L^-(s_0)\right)^n$$

commute and as a consequence $\mathcal{U}^+(t)$ and $\mathcal{U}^-(t)$ commute. Thus

$$\mathcal{Z}(t) = \mathcal{U}^+(t)\mathcal{U}^-(t)$$

is a semigroup defined on $[0, \infty)$. For $v \in \mathcal{D}(T) = \mathcal{D}(-T)$, we have

$$\frac{1}{t}[\mathcal{Z}(t)v - v] = \mathcal{U}^-(t) \frac{\mathcal{U}^+(t)v - v}{t} + \frac{\mathcal{U}^-(t)v - v}{t}$$

and taking the limit for $t \to 0$ we get

$$\lim_{t \to 0} \frac{1}{t}[\mathcal{Z}(t)v - v] = Tv - Tv = 0$$

that is,

$$\frac{d}{dt}\mathcal{Z}(t)v = 0,$$

which implies $\mathcal{Z}(t)v = v$ for all $v \in \mathcal{D}(T)$. By assumption, $\mathcal{D}(T)$ is dense in $X$ so $\mathcal{U}^-(t) = (\mathcal{U}^+(t))^{-1}$. Therefore, we define

$$\mathcal{Z}(t) = \begin{cases} \mathcal{U}^+(t) & \text{if } t \geq 0, \\ \mathcal{U}^-(t) & \text{if } t < 0. \end{cases}$$

Then $\mathcal{Z}(t)$ is a strongly continuous group whose infinitesimal generator is $T$ and the estimate $\|\mathcal{Z}(t)\| \leq Me^{\omega|t|}$ holds. □

# Chapter 5

# Perturbations of the generator of a group

In the applications, it is in general not trivial to verify the conditions of the Hille–Phillips–Yosida theorem. So an other aspect that we will investigate is the generation by perturbation. Precisely, given a closed operator $T$ that generates the evolution operator $\mathcal{U}_T(t)$, we are interested in finding under which conditions a closed operator $P$ is such that $T + P$ generates the evolution operator $\mathcal{U}_{T+P}(t)$. In the sequel, we will consider only right linear quaternionic operators even though the theory can be developed for left linear quaternionic operators following similar lines.

## 5.1   A series expansion of the S-resolvent operator

For right linear operators, we are also in need of the notion of left resolvent set and of left spectrum.

**Definition 5.1.1.** Let $X$ be a two-sided quaternionic Banach space and let $T$ be a right linear closed quaternionic operator.

We define the left resolvent set of $T$ and denote it by $\rho_L(T)$ as

$$\rho_L(T) = \{\lambda \in \mathbb{H} : \quad (\lambda \mathcal{I} - T)^{-1} \in \mathcal{B}(X)\},$$

where the notation $\lambda \mathcal{I}$ in $\mathcal{B}(X)$ means that $(\lambda \mathcal{I})(v) = \lambda v$.

The operator $(\lambda \mathcal{I} - T)^{-1}$ is called the left resolvent operator.

We define the left spectrum of $T$ as

$$\sigma_L(T) = \mathbb{H} \setminus \rho_L(T).$$

Since the operator $T$ is assumed to be right linear, then the left resolvent operator $(\lambda \mathcal{I} - T)^{-1}$ in Definition 5.1.1 is right linear. The S-spectrum and the

© Springer Nature Switzerland AG 2019

F. Colombo, J. Gantner, *Quaternionic Closed Operators, Fractional Powers and Fractional Diffusion Processes*, Operator Theory: Advances and Applications 274, https://doi.org/10.1007/978-3-030-16409-6_5

left spectrum are not, in general, related and we point out that there is no notion of holomorphicity over the quaternions such that $(\lambda \mathcal{I} - T)^{-1}$ turns out to be hyper-holomorphic on the resolvent set $\rho_L(T)$. In the sequel, we will need the following expansion of the S-resolvent operator.

**Proposition 5.1.2** (Expansion of the S-resolvent operator). *Let $X$ be a two-sided quaternionic Banach space. Let $T : \mathcal{D}(T) \subset X \to X$ and $P : \mathcal{D}(P) \subset X \to X$ be right linear closed quaternionic operators and assume that*

(a) $\lambda \in \rho_S(T)$,

(b) $\mathcal{D}(T) \subset \mathcal{D}(P)$,

(c) $B_\lambda \mathcal{Q}_\lambda^{-1}(T) : X \mapsto X, \quad$ *for all* $\lambda \in \rho_S(T)$,

(d) $\|B_\lambda \mathcal{Q}_\lambda^{-1}(T)\| < 1, \quad$ *for some* $\lambda \in \rho_S(T)$,

*where*

$$\mathcal{Q}_\lambda^{-1}(T) := (T^2 - 2\lambda_0 T + |\lambda|^2)^{-1}$$

*and*

$$B_\lambda := 2\lambda_0 P - P^2 - TP - PT. \tag{5.1}$$

*Then $\lambda \in \rho_S(T + P)$ and*

$$\mathcal{Q}_\lambda^{-1}(T + P) = \sum_{m=0}^{\infty} \mathcal{Q}_\lambda^{-1}(T)(B_\lambda \mathcal{Q}_\lambda^{-1}(T))^m; \tag{5.2}$$

*moreover, $S_R^{-1}(\lambda, T + P)$ is given by*

$$S_R^{-1}(\lambda, T + P)v = (\overline{\lambda}\mathcal{I} - T - P) \sum_{m=0}^{\infty} \mathcal{Q}_\lambda^{-1}(T)(B_\lambda \mathcal{Q}_\lambda^{-1}(T))^m v, \quad v \in X. \tag{5.3}$$

*Proof.* Let $\lambda \in \rho_S(T)$, so we have

$$\begin{aligned}
\mathcal{Q}_\lambda^{-1}(T + P) &= [(T + P)^2 - 2\lambda_0(T + P) + |\lambda|^2]^{-1} \\
&= [T^2 - 2\lambda_0 T + |\lambda|^2 - (2\lambda_0 P - P^2 - TP - PT)]^{-1} \\
&= \sum_{m=0}^{\infty} (\mathcal{Q}_\lambda^{-1}(T)(2\lambda_0 P - P^2 - TP - PT))^m \mathcal{Q}_\lambda^{-1}(T) \\
&= \sum_{m=0}^{\infty} \mathcal{Q}_\lambda^{-1}(T)((2\lambda_0 P - P^2 - TP - PT)\mathcal{Q}_\lambda^{-1}(T))^m.
\end{aligned}$$

Recalling (5.1), this concludes the proof. Finally (5.3) follows from the definition of $S_R^{-1}(\lambda, T + P)$ and from (5.2). $\qquad \square$

**Remark 5.1.1.** In the case the operators $T$ and $P$ anti-commute, then the operator $B_\lambda$ depends only on $P$, in fact it is

$$B_\lambda = 2\lambda_0 P - P^2.$$

In this case, the operator $B_\lambda$ depends just on the perturbation $P$.

**Proposition 5.1.3.** *If* $\overline{\lambda} \in \rho_L(T)$ *and* $\mathrm{Re}(\lambda) > \omega_0$, *where*

$$\omega_0 := \lim_{t \to +\infty} \frac{1}{t} \ln \|\mathcal{U}_T(t)\|, \tag{5.4}$$

*then we have*

$$\mathcal{Q}_\lambda^{-1}(T)v = (\overline{\lambda}\mathcal{I} - T)^{-1} \int_0^\infty e^{-t\lambda}\, \mathcal{U}_T(t)\, v\, dt, \quad v \in X. \tag{5.5}$$

*Proof.* Since $\overline{\lambda} \in \rho_L(T)$, then $(\overline{\lambda}\mathcal{I} - T)^{-1}$ is a bounded linear operator. From Theorem 4.2.2, we can write $S_R^{-1}(\lambda, T)$ as the Laplace transform of the evolution operator, since $(\overline{\lambda}\mathcal{I} - T)Q_\lambda(T)v = S_R^{-1}(\lambda, T)v$, for $v \in X$. $\qquad\square$

# 5.2 The class of operators $\mathbf{A}(T)$ and some properties

We now introduce a class of closed operators which will be useful in the sequel.

**Definition 5.2.1** (The class $\mathbf{A}(T)$). Let $X$ be a two-sided quaternionic Banach space and let $\mathcal{U}_T(t)$ be the strongly continuous quaternionic semigroup generated by $T$ where $T : \mathcal{D}(T) \subset X \to X$ is a right linear closed quaternionic operator. We denote by $\mathbf{A}(T)$ the class of closed right linear quaternionic operators $A$ that satisfy the conditions

(1) $\mathcal{D}(A) \supseteq \mathcal{D}(T)$.

(2) For every $t > 0$ there exists a positive constant $C(t)$ such that

$$\|Ae^{-\lambda t}\mathcal{U}_T(t)v\| \leq C(t)\|v\|$$

for $v \in \mathcal{D}(T)$ and for $\mathrm{Re}(\lambda) > \omega_0$, where $\omega_0$ is defined in (5.4).

(3) The constant $C(t)$ can be chosen such that $\int_0^1 C(t)dt$ exists and is finite.

**Lemma 5.2.2.** *Let* $X$ *be a two-sided quaternionic Banach space and let* $\mathcal{U}_T(t)$ *be the strongly continuous quaternionic semigroup generated by* $T$ *where* $T : \mathcal{D}(T) \subset X \to X$ *is a right linear closed quaternionic operator. Assume*

(1) $\overline{\lambda} \in \rho_L(T + P)$, *and* $\mathrm{Re}(\lambda) > \omega_0$, *where* $\omega_0$ *is defined in* (5.4),

(2) $B_\lambda : \mathcal{D}(T^2) \to X$, *and*

(3) $B_\lambda(\overline{\lambda}\mathcal{I} - T - P)^{-1} \in \mathbf{A}(T)$,

*where $B_\lambda$ is defined in (5.1). Then we have*

(a) $\mathcal{D}(B_\lambda(\overline{\lambda}\mathcal{I} - T - P)^{-1}e^{-t\lambda}) \supseteq \bigcup_{t>0} \mathcal{U}_T(t)X.$

(b) *The map $v \mapsto B_\lambda(\overline{\lambda}\mathcal{I} - T - P)^{-1}e^{-t\lambda}\mathcal{U}_T(t)v$ for $v \in \mathcal{D}(T)$ has a unique extension to a bounded quaternionic operator defined on all $X$. (We will denote the extension with the same symbol.)*

(c) $B_\lambda(\overline{\lambda}\mathcal{I} - T - P)^{-1}e^{-t\lambda}\mathcal{U}_T(t)v$ *is continuous in $t$ for $t > 0$ and for every $v \in X$. Moreover, if $\omega_0$ is defined in (5.4), then*

$$\limsup_{t\to\infty} \frac{\ln\|B_\lambda(\overline{\lambda}\mathcal{I} - T - P)^{-1}e^{-t\lambda}\mathcal{U}_T(t)\|}{t} \leq \omega_0,$$

*for* $\mathrm{Re}(\lambda) > 0.$

(d) *Since $\mathrm{Re}(\lambda) > \omega_0$, then*

$$B_\lambda \mathcal{Q}_\lambda^{-1}(T)v = \int_0^\infty B_\lambda(\overline{\lambda}\mathcal{I} - T - P)^{-1}\, e^{-t\lambda}\, \mathcal{U}_T(t)\, v\, dt, \quad v \in X.$$

*Proof.* To prove (a) let $v_0 \in X$ such that $v_0 = \lim_{n\to\infty} v_n$, where $v_n \in \mathcal{D}(T)$ we can make this choice since $\mathcal{D}(T)$ is dense in $X$ thanks to Lemma 4.1.9. Then $\mathcal{U}_T(t)v_n \to \mathcal{U}_T(t)v_0$ and

$$B_\lambda(\overline{\lambda}\mathcal{I} - T - P)^{-1}e^{-t\lambda}\mathcal{U}_T(t)v_n \to B_\lambda(\overline{\lambda}\mathcal{I} - T - P)^{-1}e^{-t\lambda}\mathcal{U}_T(t)v_0.$$

Since $B_\lambda(\overline{\lambda}\mathcal{I} - T - P)^{-1}$ is closed because it belongs to $\mathbf{A}(T)$, we have

$$e^{-t\lambda}\mathcal{U}_T(t)v_0 \in \mathcal{D}(B_\lambda(\overline{\lambda}\mathcal{I} - T - P)^{-1})$$

for $t \geq 0$, $\mathrm{Re}(\lambda) > 0$ and

$$B_\lambda(\overline{\lambda}\mathcal{I} - T - P)^{-1}e^{-t\lambda}\left(\mathcal{U}_T(t)v_0\right) = \left(B_\lambda(\overline{\lambda}\mathcal{I} - T - P)^{-1}e^{-t\lambda}(\mathcal{U}_T(t))\right)v_0.$$

Point (b) follows from condition (2) in Definition 5.2.1 and the Principle of extension by continuity (see Theorem 12.0.10 below).

To prove Point (c) let $0 < \delta < t$. The continuity follows from the semigroup properties since

$$B_\lambda(\overline{\lambda}\mathcal{I} - T - P)^{-1}e^{-t\lambda}\mathcal{U}_T(t)v = B_\lambda(\overline{\lambda}\mathcal{I} - T - P)^{-1}e^{-t\lambda}\mathcal{U}_T(\delta)\mathcal{U}_T(t - \delta)v.$$

The second part follows from

$$\ln\|B_\lambda(\overline{\lambda}\mathcal{I} - T - P)^{-1}e^{-t\lambda}\mathcal{U}_T(t)\| \leq \ln\|B_\lambda(\overline{\lambda}\mathcal{I} - T - P)^{-1}e^{-t\lambda}\mathcal{U}_T(\delta)\|$$
$$+ \ln\|\mathcal{U}_T(t - \delta)\|$$

so

$$
\limsup_{t\to\infty} \frac{\ln \| B_\lambda(\overline{\lambda}\mathcal{I} - T - P)^{-1} e^{-t\lambda} \mathcal{U}_T(t) \|}{t}
$$
$$
\leq \lim_{t\to\infty} \frac{\ln \| B_\lambda(\overline{\lambda}\mathcal{I} - T - P)^{-1} e^{-t\lambda} \mathcal{U}_T(\delta) \|}{t}
$$
$$
+ \lim_{t\to\infty} \frac{\ln \| \mathcal{U}_T(t - \delta) \|}{t}
$$
$$
= \omega_0,
$$

where we have used the fact that

$$
\lim_{t\to\infty} \frac{\ln \| B_\lambda(\overline{\lambda}\mathcal{I} - T - P)^{-1} e^{-t\lambda} \mathcal{U}_T(\delta) \|}{t} = 0
$$

since $\mathrm{Re}(\lambda) > \omega_0$. Statement (d) follows from Theorem 12.0.18.                    □

**Lemma 5.2.3.** *Let $X$ be a two-sided quaternionic Banach space and let $\mathcal{U}_T(t)$ be the strongly continuous quaternionic semigroup generated by $T$ where $T : \mathcal{D}(T) \subset X \to X$ is a right linear closed quaternionic operator. Let us assume that*

(1) *$h \in C((0,\infty), X) \cap L^1((0,\infty), X)$ and*

(2) *$P : \mathcal{D}(P) \subset X \to X$ is a right linear closed quaternionic operator such that $B_\lambda(\overline{\lambda}\mathcal{I} - T - P)^{-1}$, for $\overline{\lambda} \in \rho_L(T + P)$ and $\mathrm{Re}(\lambda) > \omega_0$, belongs to the class $\mathbf{A}(T)$, where $B_\lambda$ is defined in (5.1).*

*If we define*

$$
g(t) := \int_0^t e^{-\lambda(t-s)} \mathcal{U}_T(t-s) h(s)ds, \quad t \geq 0, \ \mathrm{Re}(\lambda) > 0, \tag{5.6}
$$

*then $g \in \mathcal{D}(B_\lambda(\overline{\lambda}\mathcal{I} - T - P)^{-1})$ and we have*

$$
B_\lambda(\overline{\lambda}\mathcal{I} - T - P)^{-1} g(t) = \int_0^t B_\lambda(\overline{\lambda}\mathcal{I} - T - P)^{-1} e^{-\lambda(t-s)} \mathcal{U}_T(t-s) h(s)ds. \tag{5.7}
$$

*Moreover, $g$ and $B_\lambda(\overline{\lambda}\mathcal{I} - T - P)^{-1} g(t)$ are continuous functions of $t$ for $t > 0$.*

*Proof.* The integral that defines $g$ exists for every $t \geq 0$ since $\| \mathcal{U}_T(t) \|$ is bounded in every finite interval by Proposition 4.1.4. For all $s < t$, the function

$$
s \mapsto e^{-\lambda(t-s)} \mathcal{U}_T(t-s) h(s)
$$

belongs to $\mathcal{D}(B_\lambda(\overline{\lambda}\mathcal{I} - T - P)^{-1})$ by point (a) in Lemma 5.2.2.

Thus by Theorem 12.0.18, we will show that $\int_0^t e^{-\lambda(t-s)} \mathcal{U}_T(t-s) h(s)ds$ belongs to $\mathcal{D}(B_\lambda(\overline{\lambda}\mathcal{I} - T - P)^{-1})$ and it will also prove the formula (5.7) when we show that the function

$$
s \mapsto B_\lambda(\overline{\lambda}\mathcal{I} - T - P)^{-1} e^{-\lambda(t-s)} \mathcal{U}_T(t-s) h(s)
$$

is integrable over the interval $[0, t]$. Moreover, observe that by definition for $\overline{\lambda} \in \rho_L(T + P)$ the operator $(\overline{\lambda}\mathcal{I} - T - P)^{-1}$ is continuous. From the Principle of Uniform Boundedness, see Theorem 12.0.9, and from Lemma 5.2.2 (b) it follows that $\|B_\lambda(\overline{\lambda}\mathcal{I} - T - P)^{-1}e^{-\lambda t}\mathcal{U}_T(t)\|$ is bounded on every interval that does not contain the origin. Let $0 < t_1 < t$ so that the function

$$s \mapsto \|B_\lambda(\overline{\lambda}\mathcal{I} - T - P)^{-1}e^{-\lambda(t-s)}\mathcal{U}_T(t - s)\|$$

is bounded and $\|h(\cdot)\|$ is integrable on the interval $0 \leq s \leq t_1$ while $\|h(\cdot)\|$ is bounded and

$$s \mapsto \|B_\lambda(\overline{\lambda}\mathcal{I} - T - P)^{-1}e^{-\lambda(t-s)}\mathcal{U}_T(t - s)\|$$

is integrable on the interval $t_1 \leq s \leq t$ by Proposition 4.1.4 and Definition 5.2.1 (3).

To see that $B_\lambda(\overline{\lambda}\mathcal{I} - T - P)^{-1}g(t)$ is continuous for $t > 0$, assume $0 < 2\delta < t_0$ and set

$$M_1 = \sup_{t_0 - 2\delta \leq s \leq t_0 + \delta} \|B_\lambda(\overline{\lambda}\mathcal{I} - T - P)^{-1}e^{-\lambda s}\mathcal{U}_T(s)\|.$$

Then

$$\|B_\lambda(\overline{\lambda}\mathcal{I} - T - P)^{-1}e^{-\lambda(t-s)}\mathcal{U}_T(t - s)h(s)\| \leq M_1\|h(s)\|,$$

if $|t - t_0| \leq \delta$. Consequently from Lebesgue dominated convergence theorem (see, for example, [110]),

$$\lim_{t \to t_0} \int_0^\delta B_\lambda(\overline{\lambda}\mathcal{I} - T - P)^{-1}e^{-\lambda(t-s)}\mathcal{U}_T(t - s)h(s)ds$$

$$= \int_0^\delta B_\lambda(\overline{\lambda}\mathcal{I} - T - P)^{-1}e^{-\lambda(t_0-s)}\mathcal{U}_T(t_0 - s)h(s)ds.$$

We can write

$$\int_\delta^t B_\lambda(\overline{\lambda}\mathcal{I} - T - P)^{-1}e^{-\lambda(t-s)}\mathcal{U}_T(t - s)h(s)ds$$

$$= \int_0^{t_0} B_\lambda(\overline{\lambda}\mathcal{I} - T - P)^{-1}e^{-\lambda s}\mathcal{U}_T(s)h(t - s)\chi_{[0,t-\delta]}(s)ds$$

where $\chi_{[0,t-\delta]}$ is the characteristic function of the interval $[0, t - \delta]$ and if we set

$$M_2 := \sup_{\delta \leq s \leq t_0 + \delta} \|h(s)\|,$$

we obtain that the norm of the integral on the right satisfies the estimate

$$\left\| \int_0^{t_0} B_\lambda(\overline{\lambda}\mathcal{I} - T - P)^{-1}e^{-\lambda s}\mathcal{U}_T(s)h(t - s)\chi_{[0,t-\delta]}(s)ds \right\|$$

$$\leq M_2\|B_\lambda(\overline{\lambda}\mathcal{I} - T - P)^{-1}e^{-\lambda s}\mathcal{U}_T(s)\|.$$

Thus,

$$\lim_{t \to t_0} \int_\delta^t B_\lambda(\bar{\lambda}\mathcal{I} - T - P)^{-1}\mathcal{U}_T(t - s)e^{-\lambda(t-s)}h(s)ds$$

$$= \int_\delta^{t_0} B_\lambda(\bar{\lambda}\mathcal{I} - T - P)^{-1}\mathcal{U}_T(t_0 - s)e^{-\lambda(t_0-s)}h(s)ds.$$

Combining this result with the limit above, we see that $B_\lambda(\bar{\lambda}\mathcal{I} - T - P)^{-1}g(t)$ is continuous at the arbitrary point $t_0 > 0$. The result just proved, if applied to the case when $B_\lambda(\bar{\lambda}\mathcal{I} - T - P)^{-1}$ is replaced by the identity operator $\mathcal{I}$, shows that $g$ is continuous. $\square$

## 5.3 Perturbation of the generator

We define some operators that will be useful in the sequel.

**Definition 5.3.1.** For $\bar{\lambda} \in \rho_L(T + P)$ let us define the operator

$$W_0(t) := (\bar{\lambda}\mathcal{I} - T - P)^{-1}e^{-\lambda t}\mathcal{U}_T(t)$$

and the convolution

$$(W_0 * B_\lambda W_0)(t) := \int_0^t W_0(t - s)B_\lambda W_0(s)\, ds,$$

where $B_\lambda$ is defined in (5.1).

**Theorem 5.3.2.** *Let $X$ be a two-sided quaternionic Banach space and let $T : \mathcal{D}(T) \subset X \to X$ be the generator of the strongly continuous semigroup $\{\mathcal{U}_T(t)\}_{t \geq 0}$. Let $P : \mathcal{D}(P) \subset X \to X$ be a quaternionic closed operator and let $B_\lambda$ be the operator defined in (5.1). We assume that*

(1) *$\bar{\lambda} \in \rho_L(T + P)$, and $\mathrm{Re}(\lambda) > \omega_0$, where $\omega_0$ is defined in (5.4),*

(2) *$\mathcal{D}(P) \supseteq \mathcal{D}(T)$,*

(3) *$B_\lambda : \mathcal{D}(T^2) \to X$,*

(4) *$B_\lambda(\bar{\lambda}\mathcal{I} - T - P)^{-1} \in \mathbf{A}(T)$,*

(5) *there exists a positive function $\mathcal{K}_\lambda$ such that $\|(\bar{\lambda}\mathcal{I} - T - P)^{-1}\| \leq \mathcal{K}_\lambda$, for $\mathrm{Re}(\lambda) > \omega_0$, and*

(6) *$\|B_\lambda \mathcal{Q}_\lambda^{-1}(T)\| < 1$, for some $\lambda \in \rho_S(T)$, where $\mathcal{Q}_\lambda^{-1}(T)$ is the pseudo-resolvent operator.*

*Then $T + P$, defined on $\mathcal{D}(T)$, is closed and it is the infinitesimal generator of the semigroup $\mathcal{U}_{T+P}(t)$. Moreover, we have the following representation*

$$\mathcal{U}_{T+P}(t)v = e^{\lambda t}(\bar{\lambda}\mathcal{I} - T - P)W(t)v, \quad v \in X, \tag{5.8}$$

*where*

$$W(t)v = \sum_{m=0}^{\infty} W_m(t)v, \quad v \in X, \tag{5.9}$$

$$W_0(t)v = (\bar{\lambda}\mathcal{I} - T - P)^{-1}e^{-\lambda t}\mathcal{U}_T(t)v, \quad v \in X, \tag{5.10}$$

$$W_m(t)v := W_0 * B_\lambda W_{m-1}(t)v, \quad m \in \mathbb{N}, \ v \in X. \tag{5.11}$$

*Proof.* We break the proof into several steps.

Consider the inductive construction (5.10)–(5.11) and write (5.11) explicitly as

$$W_m(t)v = \int_0^t W_0(t-\tau)B_\lambda W_{m-1}(\tau)v d\tau, \quad v \in X. \tag{5.12}$$

We define the functions

$$\chi(t) := \|(\bar{\lambda}\mathcal{I} - T - P)^{-1}e^{-\lambda t}\mathcal{U}_T(t)\| \tag{5.13}$$

and

$$\psi(t) := \|B_\lambda(\bar{\lambda}\mathcal{I} - T - P)^{-1}e^{-\lambda t}\mathcal{U}_T(t)\|. \tag{5.14}$$

By Proposition 4.1.4 the function $\chi(t)$ is measurable and $\psi(t)$ is measurable thanks to conditions (2) and (3) in Definition 5.2.1 and by Proposition 4.1.4. Thanks to Proposition 4.1.6, if $\omega > \omega_0$, where $\omega_0$ is defined in (5.4), there exists an $M_\omega < \infty$ such that

$$\|\mathcal{U}_T(t)\| \le M_\omega e^{\omega t}.$$

By assumption (5), we get

$$\chi(t) = \|(\bar{\lambda}\mathcal{I} - T - P)^{-1}e^{-\lambda t}\mathcal{U}_T(t)\| \le \|(\bar{\lambda}\mathcal{I} - T - P)^{-1}\| \, |e^{-\lambda t}| \, \|\mathcal{U}_T(t)\|$$
$$\le K_\lambda e^{-Re(\lambda t)}M_\omega e^{\omega t}.$$

By Proposition 4.1.4 and Definition 5.2.1 (3) we have that

$$\int_0^\beta \psi(t)dt < \infty$$

for every $\beta > 0$.

Now put inductively

$$\psi^{(1)}(t) = \psi(t), \quad \psi^{(n)}(t) = (\psi^{(n-1)} * \psi)(t).$$

By Lemma 12.0.17 (c) we see inductively that all the functions $\psi^{(n)}(t)$ are Lebesgue integrable over every finite interval of the real positive axis. Set

$$\chi^{(0)}(t) = \chi(t), \quad \chi^{(n)}(t) = (\chi * \psi^{(n)})(t).$$

By Lemma 12.0.17 (c) the functions $\chi^{(n)}(t)$ are Lebesgue integrable over every finite interval contained in the real positive axis.

Step 1. We show that the inductive construction (5.10)-(5.11) is well defined in terms of function spaces. Indeed, for every $m \in \mathbb{N}$ and $v \in X$, we show that

(I) $W_m(t)v \in \mathcal{D}(B_\lambda)$,

(II) $W_m(t)v$ is continuous in $t$ for $t > 0$,

(III) $\|W_m(t)\| \le \chi^{(m)}(t)$,

(IV) $B_\lambda W_m(t)v$ is continuous in $t$ for $t > 0$, and

(V) $\|B_\lambda W_m(t)\| \le \psi^{(m+1)}(t)$.

For $m = 0$ conditions (I)–(V) follow from Lemmas 5.2.2 and 5.2.3. We now suppose that they hold for $m = k$. Then from (I), (IV) and (V), we have that the integral in (5.12) exists for all $t \in (0, \infty)$ and can be used to define $W_{k+1}(t)$. The properties (I), (II) and (IV), when $m = k+1$, follow from Lemma 5.2.3. We observe that we have the estimates

$$\|W_{k+1}(t)v\| = \| \int_0^t W_0(t - \tau)B_\lambda W_k(\tau)v d\tau\|$$

$$\le \|v\| \int_0^t \chi(t - \tau)\psi^{(k+1)}(\tau)d\tau$$

$$= \|v\|\chi^{(k+1)}(t)$$

and by Lemma 5.2.3 we also have

$$\|B_\lambda W_{k+1}(t)v\| = \| \int_0^t B_\lambda W_0(t - \tau)B_\lambda W_k(\tau)v d\tau\|$$

$$\le \|v\| \int_0^t \psi(t - \tau)\psi^{(k+1)}(\tau)d\tau$$

$$= \|v\|\psi^{(k+2)}(t)$$

which proves (III) and (V) for the case $m = k+1$. Consequently, the five conditions (I)–(V) are proved inductively for all $m$.

Step 2. We estimate the series $\sum_{m=0}^{\infty} \|W_m(t)\|$.
Because of (III) we have

$$\sum_{m=0}^{\infty} \|W_m(t)\| \le \sum_{m=0}^{\infty} \chi^{(m)}(t).$$

By Lemma 5.2.2 (c) for every $\omega > \omega_0$, where $\omega_0$ is defined in (5.4), there exists a constant $M_\omega < \infty$ such that

$$\psi(t) = \|B_\lambda(\bar{\lambda}\mathcal{I} - T - P)^{-1}e^{-\lambda t}\mathcal{U}_T(t)\| < M_\omega e^{\omega t}$$

for $t$ sufficiently large. On the other hand, the function $\psi(t)$ is integrable over every finite interval of $\mathbb{R}^+$ (see Proposition 4.1.4). So if we choose $\omega_1$ sufficiently large we have

$$\int_0^{\infty} e^{-\omega_1 t}\psi(t)dt < \infty.$$

From the Lebesgue Dominated Convergence Theorem we get, for $p \in \mathbb{R}^+$

$$\lim_{p\to\infty} \int_0^\infty e^{-pt}\psi(t)dt = \lim_{p\to\infty} \int_0^\infty e^{-(p-\omega_1)t}e^{-\omega_1 t}\psi(t)dt = 0$$

so that if $\omega > \omega_1$ is chosen sufficiently large, it is

$$\int_0^\infty e^{-\omega t}\psi(t)dt = \gamma < 1.$$

Since, using the notation $\mathrm{Re}(\lambda) = \lambda_0$,

$$\chi(t) = \chi^{(0)}(t) \leq K_\lambda M_\omega e^{-(\lambda_0-\omega)t},$$

we will now show by induction that

$$\chi^{(m)}(t) \leq M_\omega e^{(\omega-\lambda_0)t}\gamma_1^m, \quad \text{for some } \gamma_1 < 1.$$

For $m = 0$ and $\omega$ sufficiently large so that $\omega - \lambda_0 > \omega_1$, it is

$$\chi^{(1)}(t) = \int_0^t \chi^{(0)}(t-\tau)\psi(\tau)v d\tau$$

$$\leq K_\lambda M_\omega e^{-(\lambda_0-\omega)t} \int_0^t e^{(\lambda_0-\omega)\tau}\psi(\tau)d\tau$$

$$\leq K_\lambda M_\omega e^{(\omega-\lambda_0)t}\gamma_1$$

for some $\gamma_1 < 1$. Assume that it holds for a given $m$, then, by Lemma 12.0.17 we have

$$\chi^{(m+1)}(t) = \int_0^t \chi^{(m)}(t-\tau)\psi(\tau)v d\tau$$

$$\leq K_\lambda M_\omega e^{-(\lambda_0-\omega)t}\gamma_1^m \int_0^t e^{-(\omega-\lambda_0)\tau}\psi(\tau)d\tau$$

$$\leq K_\lambda M_\omega e^{(\omega-\lambda_0)t}\gamma_1^{m+1}$$

for $t > 0$. Since

$$\chi^{(m)}(t) = \int_0^t \chi(t-\tau)\psi^{(m)}(\tau)d\tau$$

$$\leq \int_0^t K_\lambda M_\omega e^{-(\lambda_0-\omega)(t-\tau)}\psi^{(m)}(\tau)d\tau$$

$$\leq K_\lambda M_\omega e^{-(\lambda_0-\omega)t} \int_0^t e^{(\lambda_0-\omega)\tau}\psi^{(m)}(\tau)d\tau,$$

it is clear that $\chi^{(m)}(t) \to 0$ as $t \to 0$ for $m \geq 1$. Thus, since $\|W_m(t)\| \leq \chi^{(m)}(t)$, it is also clear that $W_m(t) \to 0$ for $m \geq 1$. Recall that in the strong operator topology, we have $\lim_{t\to 0} \mathcal{U}_T(t) = \mathcal{I}$. Hence, if we put

$$W_0(0) = (\bar{\lambda}\mathcal{I} - T - P)^{-1}$$

and

$$W_m(0) = 0 \quad \text{for } m \geq 1,$$

then $W_m(t)v$ will be continuous in $t$ for $t \geq 0$ and for every $v \in X$. Moreover, we will clearly have

$$\|W_m(t)\| \leq K_\lambda M_\omega e^{-(\lambda_0 - \omega)t} \gamma_1^m \quad \text{for } t \geq 0 \text{ and } m \geq 0.$$

So it follows that the series $\sum_{m=0}^{\infty} \|W_m(t)\|$ converges absolutely and uniformly in each finite interval $[a, b]$, and

$$\sum_{m=0}^{\infty} \|W_m(t)\| \leq (1 - \gamma_1)^{-1} K_\lambda M_\omega e^{(\omega - \lambda_0)t}.$$

Since each of the terms of the series (5.9),

$$W(t)v = \sum_{m=0}^{\infty} W_m(t)v, \quad \text{for } t \geq 0,$$

is strongly continuous for $t \geq 0$, $W(t)$ is also strongly continuous and, furthermore, we have the important estimate

$$\|W(t)\| \leq (1 - \gamma_1)^{-1} K_\lambda M_\omega e^{(\omega - \lambda_0)t}.$$

Step 3. To conclude the proof, we show that

$$Q_\lambda^{-1}(T + P) = \int_0^\infty W(t)dt$$

because this implies that

$$S_R^{-1}(\lambda, T + P)v = (\bar\lambda \mathcal{I} - T - P) \int_0^\infty W(t)dtv, \quad v \in X. \tag{5.15}$$

Thanks to Proposition 5.1.3 and the fact that $(\bar\lambda \mathcal{I} - T - P)^{-1}$ is continuous for $\bar\lambda \notin \sigma_L(T)$, it is

$$Q_\lambda^{-1}(T)v = \int_0^\infty (\bar\lambda \mathcal{I} - T - P)^{-1} e^{-t\lambda} \, \mathcal{U}_T(t) \, v \, dt, \quad v \in X. \tag{5.16}$$

Using the expansion of $Q_\lambda^{-1}(T + P)$ in Proposition 5.1.2 we get

$$Q_\lambda^{-1}(T + P) = \sum_{m=0}^{\infty} ((Q_\lambda^{-1}(T)B_\lambda)^m Q_\lambda^{-1}(T). \tag{5.17}$$

Let us reason on the second term in the expansion (5.17). Using Theorem 5.2.3, we can take $B_\lambda$, under the integral so

$$\mathcal{Q}_\lambda^{-1}(T)B_\lambda \mathcal{Q}_\lambda^{-1}(T) = \int_0^\infty (\overline{\lambda}\mathcal{I} - T - P)^{-1}e^{-\lambda t}\mathcal{U}_T(t)\,dt$$
$$\times \int_0^\infty B_\lambda(\overline{\lambda}\mathcal{I} - T - P)^{-1}e^{-\lambda s}\mathcal{U}_T(s)\,ds$$

and also, for the Fubini theorem, we obtain

$$\mathcal{Q}_\lambda^{-1}(T)B_\lambda \mathcal{Q}_\lambda^{-1}(T) = \int_0^\infty dt \int_0^\infty ds(\overline{\lambda}\mathcal{I} - T - P)^{-1}e^{-\lambda t}\mathcal{U}_T(t)$$
$$\times B_\lambda(\overline{\lambda}\mathcal{I} - T - P)^{-1}e^{-\lambda s}\mathcal{U}_T(s),$$

so with a change of variable $t \to (t - s)$ we get

$$\mathcal{Q}_\lambda^{-1}(T)B_\lambda \mathcal{Q}_\lambda^{-1}(T) = \int_0^\infty dt \int_0^t (\overline{\lambda}\mathcal{I} - T - P)^{-1}e^{-\lambda(t-s)}(t - s)$$
$$\times B_\lambda(\overline{\lambda}\mathcal{I} - T - P)^{-1}e^{-\lambda s}\mathcal{U}_T(s)\,ds.$$

Using the functions introduced in Definition 5.3.1 we have

$$\mathcal{Q}_\lambda^{-1}(T) = \int_0^\infty W_0(t)\,dt$$

and

$$\mathcal{Q}_\lambda^{-1}(T)B_\lambda \mathcal{Q}_\lambda^{-1}(T) = \int_0^\infty W_0 * B_\lambda W_0(t)\,dt.$$

With these notations we get the series

$$\mathcal{Q}_\lambda^{-1}(T + P)v = \int_0^\infty W_0(t)\,dtv$$
$$+ \int_0^\infty W_0 * B_\lambda W_0(t)\,dt$$
$$+ \int_0^\infty W_0 * (B_\lambda W_0) * B_\lambda W_0)(t)\,dt + \dots.$$

We observe that $W_m(t)$, introduced in Step 1, is given by

$$W_m(t) = (W_0 * (B_\lambda W_0)^{*m})(t), \quad m = 1, 2, 3, \dots,$$

where the symbol $*m$ stands for $m$ times the convolution of $B_\lambda W_0$ with itself. With the position

$$W(t) = \sum_{m=0}^\infty W_m(t),$$

we have

$$Q_\lambda^{-1}(T+P) = \int_0^\infty W(t)dt = \sum_{m=0}^\infty \int_0^\infty W_m(t)dt, \qquad (5.18)$$

where we have used Step 2 for the uniform convergence of the series.

Now we prove (5.15) from (5.18). Take $v \in X$ and consider $(\overline{s}\mathcal{I} - T - P) \int_0^\infty W(t)dtv$. Since $\mathcal{D}(T) \subset \mathcal{D}(P)$, we have $\mathcal{D}(T+P) = \mathcal{D}(T)$ and $S_R^{-1}(\lambda, T)X = \mathcal{D}(T)$. The relation

$$((T+P)^2 - 2\lambda_0(T+P) + |\lambda|^2\mathcal{I})(\overline{\lambda}\mathcal{I} - T - P)^{-1}(\overline{\lambda}\mathcal{I} - T - P)\int_0^\infty W(t)dtv = v$$

for $\lambda \in \rho_S(T)$ and for $v \in X$, holds true because

$$((T+P)^2 - 2\lambda_0(T+P) + |\lambda|^2\mathcal{I})\int_0^\infty W(t)dtv = v, \quad v \in X$$

is a consequence of (5.18). Now consider $v \in \mathcal{D}(T)$ and

$$(\overline{\lambda}\mathcal{I} - T - P)\int_0^\infty W(t)dt \, ((T+P)^2 - 2\lambda_0(T+P) + |\lambda|^2\mathcal{I})(\overline{\lambda}\mathcal{I} - T - P)^{-1}v = v,$$

for $v \in \mathcal{D}(T)$. Since

$$\int_0^\infty W(t)dt \, ((T+P)^2 - 2\lambda_0(T+P) + |\lambda|^2\mathcal{I}) = \mathcal{I} : \mathcal{D}(T^2) \to \mathcal{D}(T),$$

because we have assumed that $\overline{\lambda} \in \rho_L(T+P)$, we have that

$$(\overline{\lambda}\mathcal{I} - T - P)(\overline{\lambda}\mathcal{I} - T - P)^{-1}v = v, \quad v \in \mathcal{D}(T).$$

So we have (5.15). $\qquad \square$

## 5.4 Comparison with the complex setting

We recall the complex version of the generation result in order to compare it with the one in the quaternionic setting. Let $X$ be a complex Banach space and let $A$ be the (complex) infinitesimal generator of a strongly continuous semigroup $U_A(t)$.

**Definition 5.4.1.** We denote by $\mathbb{P}(A)$ the class of closed operators $P$ that satisfies the conditions

(1) $\mathcal{D}(P) \supseteq \mathcal{D}(A)$.

(2) For every $t > 0$ there exists a positive constant $C(t)$ such that

$$\|PU_A(t)x\| \leq C(t)\|x\|, \quad \text{for } x \in \mathcal{D}(A).$$

(3) The constant $C(t)$ can be chosen such that $\int_0^1 C(t)dt$ exists and is finite.

Theorem 5.3.2 extends the following classical result to the quaternionic setting, see [110, p. 630]:

**Theorem 5.4.2.** *Let $A$ be the infinitesimal generator of a strongly continuous semigroup $U_A(t)$ on $X$. If $P \in \mathbb{P}(A)$ then $A + P$ defined on $\mathcal{D}(A)$ is closed and is the infinitesimal generator of the semigroup $U_{A+P}(t)$. Moreover, an explicit construction of the semigroup $U_{A+P}(t)$ is given by*

$$U_{A+P}(t) = \sum_{n \geq 0} \mathcal{R}_n(t), \quad t \geq 0 \tag{5.19}$$

*where*

$$\mathcal{R}_0(t)x = U_A(t)x, \quad \mathcal{R}_n(t)\,x = (U_A * P\mathcal{R}_{n-1})(t)\,x, \quad x \in X, \ n = 1, 2, 3, \ldots,$$

*and*

$$(U_A * P\mathcal{R}_{n-1})(t)\,x := \int_0^t U_A(t - s)P\mathcal{R}_{n-1}(s)\,x\,ds.$$

*The series (5.19) converges uniformly for $t \in [0, \tau]$ where $\tau$ is a positive fixed real number. The function $t \to \mathcal{R}_n(t)\,x$, for fixed $n \in \mathbb{N}$ and $x \in X$, is continuous for $t \geq 0$.*

For the ensuing comments, it is useful to write the first terms in the expansion of the semigroups in both the complex and the quaternionic case, which are

$$U_{A+P}(t) = U_A(t) + (U_A * PU_A)(t) + \ldots$$

and

$$\mathcal{U}_{T+P}(t) = \mathcal{U}_T(t) + \mathcal{U}_T(t) * B_\lambda(\overline{\lambda}\mathcal{I} - T - P)^{-1}e^{-\lambda t}\mathcal{U}_T(t) + \ldots,$$

respectively.

**Remark 5.4.1.** Note that, in the complex case, the expansion of $U_{A+P}(t)$ involves just the semigroup $U_A(t)$ and the perturbation operator $P$. This expansion is based on the fact that the classical resolvent operator

$$R(\lambda, A + P) := (\lambda I - A - P)^{-1}$$

for $A + P$, for $\|PR(\lambda, A)\| < 1$, is given by

$$R(\lambda, A + P) = R(\lambda, A) \sum_{n=0}^{\infty} (PR(\lambda, A))^n \tag{5.20}$$

and the main point of the matter is that the resolvent operator $R(\lambda, A)$ is the Laplace transform of $U_A(t)$.

**Remark 5.4.2.** In the quaternionic case, the expansion of (5.20) has to be replaced by the expansion of the pseudo-resolvent operator (5.2), namely

$$Q_\lambda^{-1}(T+P) = \sum_{m=0}^\infty Q_\lambda^{-1}(T)(B_\lambda Q_\lambda^{-1}(T))^m,$$

where $B_\lambda := 2\lambda_0 P - P^2 - TP - PT$ and $\|B_\lambda Q_\lambda^{-1}(T)\| < 1$. Thus, the S-resolvent operator $S_R^{-1}(\lambda, T+P)$ can be written as (see (5.3))

$$S_R^{-1}(\lambda, T+P)v = (\overline{\lambda}\mathcal{I} - T - P)\sum_{m=0}^\infty Q_\lambda^{-1}(T)(B_\lambda Q_\lambda^{-1}(T))^m v, \quad v \in X.$$

Note that the relation between $S_R^{-1}(\lambda, T+P)$ and the Laplace transform of the quaternionic evolution operator $\mathcal{U}_T(t)$, see Remark 5.1.3, involves also the left resolvent operator, in fact

$$Q_\lambda^{-1}(T)v = (\overline{\lambda}\mathcal{I} - T)^{-1}\int_0^\infty e^{-t\lambda}\,\mathcal{U}_T(t)\,v\,dt, \quad v \in X.$$

Thus, in the quaternionic setting, two spectral problems are involved.

**Remark 5.4.3.** We point out that one can also use the consistency of quaternionic spectral theory with complex spectral theory in order to develop a different approach to the perturbation theory of generators of strongly continuous semi groups. If $T$ is the quaternionic infinitesimal generator of the quaternionic semigroup $\mathcal{U}_T(T)$, then we can choose $j \in \mathbb{S}$ and consider $T$ as a $\mathbb{C}_j$-linear operator. Then $T$ is the infinitesimal generator of the complex semigroup obtained from considering $\mathcal{U}_T(t)$ as a $\mathbb{C}_j$-linear operator for each $t \geq 0$.

Now let $P$ be a quaternionic linear operator that satisfies conditions analogue to those required in the complex case in Definition 5.4.1, that is,

(i) $\mathcal{D}(T) \subset \mathcal{D}(P)$.

(ii) For each $t > 0$, there exists $C(t)$ such that $\|P\mathcal{U}_T(t)y\| \leq C(t)\|y\|$ for all $y \in \mathcal{D}(T)$.

(iii) The constants $C(t)$ can be chosen such that $\int_0^1 C(t)\,dt$ exists and is finite.

If we consider $P$ also as a $\mathbb{C}_j$-linear operator, then Theorem 5.4.2 implies that $T+P$ is the generator of a strongly continuous semigroup $\mathcal{U}_{T+P}(t)$ of $\mathbb{C}_j$-complex linear operators. However, since $T$ and $P$ are quaternionic linear, the operator $T+P$ is quaternionic linear and $\mathcal{U}_{T+P}(t)$ consists of quaternionic linear operators. (To show this, we can use quaternionic linearity since it survives the Yosida approxiation procedure.) Therefore $T+P$ generates a strongly continuous quaternionic semigroup under the above assumptions, which is furthermore given by the series (5.19). This approach does not, however, allow to obtain the central result of this chapter, namely the series expansion (5.9) that contains the quaternionic parameter $\lambda$.

**Remark 5.4.4.** We finally observe that when $\lambda$ is a real number the expansion (5.8) becomes

$$\mathcal{U}_{T+P}(t) = \mathcal{U}_T(t) + \mathcal{U}_T(t) * B_\lambda(\lambda\mathcal{I} - T - P)^{-1}\mathcal{U}_T(t) + \dots$$

where $B_\lambda$ is defined in (5.1).

## 5.5   An application

As an application, we study a quaternionic differential equation in the space of quaternionic-valued continuous functions. Consider $Y : \mathbb{R}^+ \times \mathbb{R} \to \mathbb{H}$ and the Cauchy problem

$$\frac{\partial}{\partial t}Y(t,x) = \frac{\partial^2}{\partial x^2}Y(t,x) + h(x)\frac{\partial}{\partial x}Y(t,x),$$

$$\lim_{t \to 0^+} Y(t,x) = y_0(x), \quad \text{uniformly in } x \in \mathbb{R},$$

where

$$y_0(x) = y_0(x) + y_1(x)e_1 + y_2(x)e_2 + y_3(x)e_3 : \mathbb{R} \to \mathbb{H}$$

and

$$h = h_0(x) + h_1(x)e_1 + h_2(x)e_2 + h_3(x)e_3 : \mathbb{R} \to \mathbb{H}$$

are given functions. We now need some general facts on the group of translations that can be obtained in the quaternionic setting by adapting the arguments in [110, p. 629], with obvious modifications. The group of translations defined by

$$U_A(t)Z(\tau) = Z(t+\tau)$$

is a strongly continuous group on $X = C(\overline{\mathbb{R}}, \mathbb{H})$ where $\overline{\mathbb{R}} = [-\infty, +\infty]$ and its infinitesimal generator is $A = \frac{d}{d\tau}$ with domain

$$\mathcal{D}(A) = \{Z \in C(\overline{\mathbb{R}}, \mathbb{H}) : \quad Z' \in C(\overline{\mathbb{R}}, \mathbb{H})\}.$$

To determine the S-resolvent set of $A$ we observe that, for $Z \in C(\overline{\mathbb{R}}, \mathbb{H})$, the quaternionic differential equation

$$\lambda Z(\tau) - Z'(\tau) = X(\tau)$$

must have a unique solution in $Z \in C(\overline{\mathbb{R}}, \mathbb{H})$. But $\lambda \in \mathbb{H}$ and $A$ commute with $\lambda$, since $A$ does not contain any imaginary units, so the $S$-resolvent operator reduces to $(\lambda\mathcal{I} - A)^{-1}$. The linear quaternionic differential equation for $Z$ reduces to a linear system of differential equations for the components of $Z$, so it follows that the S-spectrum is given by

$$\sigma_S(A) = \{uj : \quad u \in \mathbb{R}, \ j \in \mathbb{S}\}.$$

Consider now the operator $A^2 = \frac{\partial^2}{\partial \tau^2}$ with domain

$$\mathcal{D}(A^2) = \{Z \in C(\overline{\mathbb{R}}, \mathbb{H}) : \quad Z', Z'' \in C(\overline{\mathbb{R}}, \mathbb{H})\}.$$

With similar considerations as in Theorem XII9.7 in [110] we have that operator $A^2$ is closed and $\mathcal{D}(A^2)$ is dense in $C(\overline{\mathbb{R}}, \mathbb{H})$. Since

$$\sigma_S(A) = \{uj : \quad u \in \mathbb{R}, \ for \ j \in \mathbb{S}\},$$

by the spectral mapping theorem it follows that

$$\sigma_S(A^2) = \{u \in \mathbb{R} : \quad u < 0\}.$$

Since $A^2$ commutes with the quaternion $\lambda$, it is

$$S_R^{-1}(\lambda, A^2)Z(\tau) = (\lambda I - A^2)^{-1}Z(\tau)$$

$$= \int_{\mathbb{R}} \frac{e^{-|\theta|\sqrt{\lambda}}}{2\sqrt{\lambda}} Z(\theta + \tau)d\theta, \quad \tau \in \mathbb{R}.$$

With computations similar to those in [110, p. 640], we have an explicit formula for the evolution operator:

$$U_{A^2}(t)Z(\tau) = \frac{1}{2\sqrt{\pi t}} \int_{\mathbb{R}} e^{-\theta^2/4t} Z(\theta + \tau)d\theta, \quad t > 0. \tag{5.21}$$

Let us consider $h \in C(\overline{\mathbb{R}}, \mathbb{H})$ and the operator $P$ whose domain is

$$\mathcal{D}(P) = \Big\{ y \in C(\overline{\mathbb{R}}, \mathbb{H}) \text{ such that } y' \text{ is continuous in a neighborhood of}$$

$$\text{each point } \tau_0 \text{ for which } h(\tau_0) \neq 0 \text{ and such that } hy' \in C(\overline{\mathbb{R}}, \mathbb{H}) \Big\}$$

and it is defined by

$$(Py)(\tau) = h(\tau)y'(\tau), \quad y \in \mathcal{D}(P).$$

The operator $P$ is closed and with some computations we have the estimate

$$\|PU_{A^2}(t)Z(\tau)\| \leq \|h\| \| \frac{\partial}{\partial \tau} \frac{1}{2\sqrt{\pi t}} \int_{\mathbb{R}} e^{-\theta^2/4t} Z(\theta + \tau)d\theta \|$$

$$\leq \frac{\|h\|\|Z\|}{\sqrt{\pi t}}.$$

The above estimate shows that the conditions (2) and (3) in Definition 5.2.1 are fulfilled since they are the conditions (2) and (3) in Definition 5.4.1. This is due to the fact that, in this case, $A^2$ commute with the quaternions. In view of theorem of generation by perturbation, the solution of the Cauchy problem is

$$Y(t, x) = U_{A^2 + P}(t)y_0(x).$$

We conclude this chapter pointing out that the above result, which has been obtained by adapting the scalar case in [110], shows that this generation theorem is useful also in quaternionic quantum mechanics, see [4, p. 38], since the quaternionic version of Schrödinger equation is of the form

$$\frac{\partial}{\partial t}\psi(t,x) = -H(x)\psi(t,x)$$

where the Hamiltonian is given by

$$H(x) = H_0(x) + e_1 H_1(x) + e_2 H_2(x) + e_3 H_3(x)$$

and $\psi$ is the quaternionic wave function. Even if it is nontrivial in concrete applications of physical interest, one may consider the operator $e_1 H_1(x) + e_2 H_2(x) + e_3 H_3(x)$ as a perturbation of $H_0(x)$.

# Chapter 6

# The Phillips functional calculus

In Chapter 3, we have introduced the direct approach to the $S$-functional calculus for unbounded operators, which only requires the operator $T$ to be closed and have a nonempty $S$-resolvent set. As in the complex case, the price for these weak requirements on the operator are the relatively strong assumptions that one has to make on the class of admissible functions, namely that they are slice hyperholomorphic on the $S$-spectrum of $T$ and at infinity. However, similar to the classical case, additional knowledge about the operator allows us to extend the class of admissible functions.

In this chapter we shall assume that $T$ is the infinitesimal generator of a strongly continuous group $\{\mathcal{U}_T(t)\}_{t \in \mathbb{R}}$ and we let $\omega \geq 0$ and $M > 0$ be the constants from Theorem 4.3.1 such that

$$\sigma_S(T) \subset \{s \in \mathbb{H} : \quad -\omega \leq \operatorname{Re}(s) \leq \omega\} \quad \text{and} \quad \|\mathcal{U}_T(t)\| \leq Me^{\omega|t|}, \ t \in \mathbb{R}.$$

If $f$ is the quaternionic Laplace–Stieltjes transform of a quaternion-valued measure $\mu$ on $\mathbb{R}$, that is,

$$f(s) = \int_{\mathbb{R}} d\mu(t)\, e^{-st}, \quad -(\omega + \varepsilon) < \operatorname{Re}(s) < \omega + \varepsilon,$$

then we can formally replace the exponential in the above integral by the group $\mathcal{U}_T(t)$, which formally corresponds to $e^{tT}$, and define

$$f(T) := \int_{\mathbb{R}} d\mu(t)\mathcal{U}_T(-t).$$

The function $f$ is now slice hyperholomorphic on $\sigma_S(T)$, but not necessarily at infinity.

In the complex setting the above procedure yields the Phillips functional calculus, which was introduced in [35, 185]. In this chapter we introduce its quaternionic counterpart and study its properties and its relation with the $S$-functional calculus following the treatise in [110]. The presented results were published in [12].

© Springer Nature Switzerland AG 2019
F. Colombo, J. Gantner, *Quaternionic Closed Operators, Fractional Powers and Fractional Diffusion Processes*, Operator Theory: Advances and Applications 274,
https://doi.org/10.1007/978-3-030-16409-6_6

# 6.1   Preliminaries on quaternionic measure theory

Before we are able to define the Phillips functional calculus for quaternionic linear operators, we have to recall some facts about quaternion-valued measures and investigate their product measures. These results will be essential when we study the properties of the quaternionic Laplace–Stieltjes transform.

In [5, Section 3], the authors showed that quaternion-valued measures have properties similar to the properties of complex-valued measures. In particular, it is possible to define their variation, which has analogous properties as the variation of a complex measure, and find that the Radon–Nikodým theorem also holds true in this setting. We recall the results that we will need in the sequel taken from [5]. Since some of these results follow by adapting the classical case, we just recall them. We will add the proofs of the results related to the product of quaternionic measures that are taken from [12].

**Definition 6.1.1.** Let $(\Omega, \mathcal{A})$ be a measurable space. A quaternionic measure is a function $\mu \colon \mathcal{A} \to \mathbb{H}$ that satisfies

$$\mu\left(\bigcup_{n\in\mathbb{N}} A_n\right) = \sum_{n\in\mathbb{N}} \mu(A_n),$$

for any sequence of pairwise disjoint sets $(A_n)_{n\in\mathbb{N}} \subset \mathcal{A}$. We denote the set of all quaternionic measures on $\mathcal{A}$ by $\mathcal{M}(\Omega, \mathcal{A}, \mathbb{H})$ or simply by $\mathcal{M}(\Omega, \mathbb{H})$ or $\mathcal{M}(\Omega)$ if there is no possibility of confusion.

**Corollary 6.1.2.** *Let $(\Omega, \mathcal{A})$ be a measurable space. The set $\mathcal{M}(\Omega, \mathcal{A}, \mathbb{H})$ is a two-sided quaternionic vector space with the operations*

$$(\mu + \nu)(A) := \mu(A) + \nu(A), \quad (a\mu)(A) := a\mu(A), \quad (\mu a)(A) := \mu(A)a,$$

*for $\mu, \nu \in \mathcal{M}(\Omega, \mathcal{A}, \mathbb{H})$, $a \in \mathbb{H}$ and $A \in \mathcal{A}$.*

**Remark 6.1.1.** Let $j, i \in \mathbb{S}$ with $j \perp i$. Since $\mathbb{H} = \mathbb{C}_j + i\mathbb{C}_j$, it is immediate that a mapping $\mu : \mathcal{A} \to \mathbb{H}$ is a quaternionic measure if and only if there exist two $\mathbb{C}_j$-valued complex measures $\mu_1, \mu_2$ such that $\mu(A) = \mu_1(A) + i\mu_2(A)$ for any $A \in \mathcal{A}$. Moreover, since $\mathbb{H} = \mathbb{C}_j + \mathbb{C}_j i$, there exist $\mathbb{C}_j$-valued measures $\tilde{\mu}_1, \tilde{\mu}_2$ such that $\mu(A) = \tilde{\mu}_1(A) + \tilde{\mu}_2(A)i$ for any $A \in \mathcal{A}$.

**Definition 6.1.3.** Let $\mu \in \mathcal{M}(\Omega, \mathcal{A}, \mathbb{H})$. For all $A \in \mathcal{A}$, we denote by $\Pi(A)$ the set of all countable partitions $\pi$ of $A$ into pairwise disjoint, measurable sets $A_\ell, \ell \in \mathbb{N}$. The total variation of $\mu$ is the set function

$$|\mu|(A) := \sup\left\{ \sum_{A_\ell \in \pi} |\mu(A_\ell)| : \quad \pi \in \Pi(A) \right\} \qquad \text{for all } A \in \mathcal{A}.$$

From the definition, we easily obtain the following lemma.

**Lemma 6.1.4.** *The total variation $|\mu|$ of a measure $\mu \in \mathcal{M}(\Omega, \mathcal{A}, \mathbb{H})$ is a finite positive measure on $\Omega$. Moreover, $|a\mu| = |\mu a| = |\mu||a|$ and $|\mu + \nu| \le |\mu| + |\nu|$ for any $\mu, \nu \in \mathcal{M}(\Omega, \mathcal{A}, \mathbb{H})$ and $a \in \mathbb{H}$.*

Recall that a measure $\mu$ is called absolutely continuous with respect to a positive measure $\nu$ if $\mu(A) = 0$ for any $A \in \mathcal{A}$ with $\nu(A) = 0$. In this case, we write $\mu \ll \nu$. We denote by $L^1(\Omega, \mathcal{A}, \nu, \mathbb{H})$ the Banach space of all $\mathbb{H}$-valued functions on $\Omega$ that are integrable with respect to the positive measure $\nu$.

**Theorem 6.1.5** (Radon–Nikodým theorem for quaternionic measures). *Let $\nu$ be a $\sigma$-finite positive measure on $(\Omega, \mathcal{A})$. A quaternionic measure $\mu \in \mathcal{M}(\Omega, \mathcal{A}, \mathbb{H})$ is absolutely continuous with respect to $\nu$ if and only if there exists a function $f \in L^1(\Omega, \mathcal{A}, \nu, \mathbb{H})$ such that*

$$\mu(A) = \int_A f(x)\, d\nu(x), \quad \text{for all } A \in \mathcal{A}.$$

*Moreover, $f$ is unique and we have*

$$|\mu|(A) = \int_A |f(x)|\, d\nu(x), \quad \text{for all } A \in \mathcal{A}. \tag{6.1}$$

The identity (6.1) follows as in the classical case, cf. [192, Theorem 6.13].

**Corollary 6.1.6.** *Let $\mu \in \mathcal{M}(\Omega, \mathcal{A}, \mathbb{H})$. Then there exists an $\mathcal{A}$-measurable function $h : \Omega \to \mathbb{H}$ such that $|h(x)| = 1$ for any $x \in \Omega$ and such that $\mu(A) = \int_A h(x)\, d|\mu|(x)$ for any $A \in \mathcal{A}$.*

In order to define the quaternionic Laplace–Stieltjes transform and the Phillips functional calculus for quaternionic linear operators, we define integrals with respect to quaternionic-valued measures as in [12]. Let us again consider a quaternionic two-sided Banach space $X$ and let $\nu$ be a positive measure. We recall that

(i) $X$ becomes a real Banach space if we restrict the scalar multiplication to the real numbers.

(ii) $\mathbb{H}$ itself is a quaternionic two-sided Banach space.

So, let $\nu$ be a positive measure. Recall that in Bochner's integration theory, a function $f$ with values in $X$ is called $\nu$-measurable if there exists a sequence of functions $f_n(x) = \sum_{\ell=1}^{n} a_i \chi_{A_\ell}(x)$, where $a_\ell \in X$ and $\chi_{A_\ell}$ is the characteristic function of a measurable set $A_\ell$, such that $f_n(x) \to f(x)$ as $n \to +\infty$ for $\nu$-almost all $x \in \Omega$. The next lemma follows as a simple application of the Pettis measurability theorem, see [107].

**Lemma 6.1.7.** *Let $X$ be a quaternionic two-sided Banach space and let $\nu$ be a positive measure on $(\Omega, \mathcal{A})$. If $f : \Omega \to X$ and $g : \Omega \to \mathbb{H}$ are $\nu$-measurable, then the functions $fg$ and $gf$ are $\nu$-measurable.*

Let $\nu$ be a positive measure on $(\Omega, \mathcal{A})$. Recall that a $\nu$-measurable function on $\Omega$ with values in a real Banach space is called Bochner-integrable, if

$$\int_{\Omega} \|f(x)\| \, d\mu(x) < +\infty.$$

**Definition 6.1.8.** Let $X$ be a two-sided quaternionic Banach space, let $\mu \in \mathcal{M}(\Omega, \mathcal{A}, \mathbb{H})$ and let $h : \Omega \to \mathbb{H}$ be the function with $|h| = 1$ such that $d\mu(x) = h(x) \, d|\mu|(x)$. We call two $\mu$-measurable functions $f : \Omega \to X$ and $g : \Omega \to \mathbb{H}$ a *$\mu$-integrable pair*, if

$$\int_{\Omega} \|f\| \|g\| \, d|\mu| < +\infty.$$

In this case, we define

$$\int_{\Omega} f \, d\mu \, g := \int_{\Omega} f h g \, d|\mu| \qquad\qquad (6.2)$$

and

$$\int_{\Omega} g \, d\mu \, f = \int_{\Omega} g h f \, d|\mu|, \qquad\qquad (6.3)$$

as the integrals of a function with values in a real Banach space in the sense of Bochner.

**Remark 6.1.2.** Note that in the definition of the integrals in (6.2) and (6.3), we can replace the variation of $\mu$ by any $\sigma$-finite positive measure $\nu$ with $\mu \ll \nu$. If $h_\nu$ is the density of $\mu$ with respect to $\nu$ and $\rho_{|\mu|}$ and $\rho_\nu$ are the real-valued densities of $|\mu|$ and $\nu$ with respect to $|\mu| + \nu$. Then we have

$$\mu = h|\mu| = h\rho_{|\mu|}(|\mu| + \nu) \quad \text{and} \quad \mu = h_\nu \nu = h_\nu \rho_\nu (|\mu| + \nu).$$

Theorem 6.1.5 implies $h\rho_{|\mu|} = h_\nu \rho_\nu$ in $L^1(|\mu| + \nu)$. Therefore,

$$\begin{aligned}
\int_{\Omega} f h_\nu g \, d\nu &= \int_{\Omega} \int_{\Omega} f h_\nu g \rho_\nu \, d(|\mu| + \nu) \\
&= \int_{\Omega} \int_{\Omega} f h_\nu \rho_\nu g \, d(|\mu| + \nu) \\
&= \int_{\Omega} f h \rho_{|\mu|} g \, d(|\mu| + \nu) \\
&= \int_{\Omega} f h g \rho_{|\mu|} \, d(|\mu| + \nu) \\
&= \int_{\Omega} f h g \, d|\mu| \\
&= \int_{\Omega} f \, d\mu \, g.
\end{aligned}$$

Hence, the integral is linear in the measure: if $\mu, \nu \in \mathcal{M}(\Omega, \mathcal{A}, \mathbb{H})$, then $\mu$ and $\nu$ are absolutely continuous with respect to $\tau = |\mu| + |\nu|$. If $\rho_\mu$ and $\rho_\nu$ are the densities of $\mu$ and $\nu$ with respect to $\tau$, then

$$
\begin{aligned}
\int_\Omega f \, d(\mu + \nu) \, g &= \int_\Omega f(\rho_\mu + \rho_\nu) g \, d\tau \\
&= \int_\Omega f \rho_\mu g \, d\tau + \int_\Omega f \rho_\nu g \, d\tau \\
&= \int_\Omega f \, d\mu \, g + \int_\Omega f \, d\nu \, g.
\end{aligned}
$$

Similarly, if $a \in \mathbb{H}$ and $\mu = \rho|\mu|$, then $a\mu = a\rho|\mu|$ and so

$$
\int_\Omega f \, d(a\mu) \, g = \int_\Omega f(a\rho) g \, d|\mu| = \int_\Omega (fa)\rho g \, d|\mu| = \int_\Omega (fa) \, d\mu \, g.
$$

In the same way, one can see that $\int_\Omega f \, d(\mu a) g = \int_\Omega f \, d\mu \, (ag)$.

We finally define the product measure and the convolution of two quaternionic measures as in [12]. Also these concepts will be essential when we discuss the product rule of the quaternionic Phillips functional calculus.

**Lemma 6.1.9.** *Let $\mu \in \mathcal{M}(\Omega, \mathcal{A}, \mathbb{H})$ and $\nu \in \mathcal{M}(\Upsilon, \mathcal{B}, \mathbb{H})$. Then there exists a unique measure $\mu \times \nu$ on the product measurable space $(\Omega \times \Upsilon, \mathcal{A} \otimes \mathcal{B})$ such that*

$$
(\mu \times \nu)(A \times B) = \mu(A)\nu(B), \tag{6.4}
$$

*for all $A \in \mathcal{A}, B \in \mathcal{B}$. We call $\mu \times \nu$ the* product measure *of $\mu$ and $\nu$.*

*Proof.* Let $j, i \in \mathbb{S}$ with $j \perp i$ and let $\mu = \mu_1 + i\mu_2$ with $\mu_1, \mu_2 \in \mathcal{M}(\Omega, \mathcal{A}, \mathbb{C}_j)$ and $\nu = \nu_1 + \nu_2 i$ with $\nu_1, \nu_2 \in \mathcal{M}(\Upsilon, \mathcal{B}, \mathbb{C}_j)$. Then, there exist unique complex product measures $\mu_\ell \times \nu_\kappa \in M(\Omega_1 \times \Omega_2, \mathcal{A} \otimes \mathcal{B}, \mathbb{C}_j)$ of $\mu_\ell$ and $\nu_\ell$ for $\ell, \kappa = 1, 2$. If we set

$$
\mu \times \nu = \mu_1 \times \nu_1 + i\mu_2 \times \nu_1 + \mu_1 \times \nu_2 i + J\mu_2 \times \nu_2 i,
$$

then $\mu \times \nu$ is a quaternionic measure on $(\Omega \times \Upsilon, \mathcal{A} \otimes \mathcal{B})$ and

$$
\begin{aligned}
\mu(A)\nu(B) &= \mu_1(A)\nu_1(B) + i\mu_2(A)\nu_1(B) \\
&\quad + \mu_1(A)\nu_2(B)i + i\mu_2(A)\nu_2(B)i \\
&= \mu_1 \times \nu_1(A \times B) + i\mu_2 \times \nu_1(A \times B) \\
&\quad + \mu_1 \times \nu_2(A \times B)i + i\mu_2 \times \nu_2(A \times B)i = (\mu \times \nu)(A \times B).
\end{aligned}
$$

In order to prove the uniqueness of the product measure, assume that two quaternionic measures $\rho = \rho_1 + \rho_2 i$ and $\tau = \tau_1 + \tau_2 i$ on $(\Omega \times \Upsilon, \mathcal{A} \times \mathcal{B})$ satisfy $\rho(A \times B) = \tau(A \times B)$ whenever $A \in \mathcal{A}$ and $B \in \mathcal{B}$. Then $\rho_1(A \times B) = \tau_1(A \times B)$ and $\rho_2(A \times B) = \tau_2(A \times B)$ for $A \in \mathcal{A}$ and $B \in \mathcal{B}$. Since two complex measures on the product space $(\Omega \times \Upsilon, \mathcal{A} \otimes \mathcal{B})$ are equal if and only if they coincide on sets of the form $A \times B$, we obtain $\rho_1 = \tau_1$ and $\rho_2 = \tau_2$ and, in turn, $\rho = \rho_1 + \rho_2 i = \tau_1 + \tau_2 i = \tau$. Therefore, $\mu \times \nu$ is uniquely determined by (6.4). $\square$

**Remark 6.1.3.** Note that it is also possible to define a commuted product measure $\mu \times_c \nu$ on $(\Omega \times \Upsilon, \mathcal{A} \otimes \mathcal{B})$ that satisfies

$$(\mu \times_c \nu)(A \times B) = \nu(B)\mu(A), \quad \forall A \in \mathcal{A}, B \in \mathcal{B}.$$

This measure is different from the measure $\nu \times \mu$ that is defined on $(\Upsilon \times \Omega, \mathcal{B} \otimes \mathcal{A})$ and satisfies

$$(\nu \times \mu)(B \times A) = \nu(B)\mu(A), \quad \forall B \in \mathcal{B}, A \in \mathcal{A}.$$

**Lemma 6.1.10.** *Let $(\Omega, \mathcal{A}, \mu)$ and $(\Upsilon, \mathcal{B}, \nu)$ be quaternionic measure spaces. Then*

$$|\mu \times \nu| = |\mu| \times |\nu|.$$

*Moreover, if $d\mu(x) = f(x) \, d|\mu|(x)$ and $\nu(x) = g(x) \, d|\nu|(x)$ as in Corollary 6.1.6, then, for any $C \in \mathcal{A} \times \mathcal{B}$,*

$$(\mu \times \nu)(C) = \int_C f(s)g(t) \, d|\mu \times \nu|(s,t).$$

*Proof.* Let $f \colon \Omega \to \mathbb{H}$ and $g \colon \Upsilon \to \mathbb{H}$ with $|f| = 1$ and $|g| = 1$ be functions as in Corollary 6.1.6 such that

$$\mu(A) = \int_A f(t) \, d|\mu|(t)$$

and

$$\nu(B) = \int_B g(s) \, d|\nu|(s),$$

for all $A \in \mathcal{A}$ and $B \in \mathcal{B}$. Moreover, let $r = (t,s)$ and let $h(r) = f(t)g(s)$. Then the function $C \mapsto \int_C h(r) \, d(|\mu| \times |\nu|)(r)$ defines a measure on $(\Omega \times \Upsilon, \mathcal{A} \times \mathcal{B})$ and Fubini's theorem for positive measures implies

$$\int_{A \times B} h(r) \, d|\mu| \times |\nu|(r) = \int_A \int_B f(t)g(s) \, d|\mu|(t) \, d|\nu|(s)$$

$$= \int_A f(t) \, d|\mu|(t) \int_B g(s) \, d|\nu|(s) = \mu(A)\nu(B).$$

The uniqueness of the product measure implies $\mu \times \nu(C) = \int_C h(r) \, d(|\mu| \times |\nu|)(r)$, for any $C \in \mathcal{A} \times \mathcal{B}$. Since $|h| = |f| \, |g| = 1$, we deduce from (6.1) that

$$|\mu \times \nu|(C) = \int_C |h(r)| \, d(|\mu| \times |\nu|)(r) = (|\mu| \times |\nu|)(C),$$

for all $C \in \mathcal{A} \times \mathcal{B}$.                                               $\square$

Splitting the measure $\mu$ into two complex components and applying the respective result for complex measures, we obtain the transformation rule for integrals with respect to a pushforward measure stated in the following lemma.

**Lemma 6.1.11.** *Let $(\Omega, \mathcal{A}, \mu)$ be a quaternionic measure space, let $(\Upsilon, \mathcal{B})$ be a measurable space and let $\phi : \Omega \to \Upsilon$ be a measurable function. If a function $f : \Upsilon \to X$ with values in a quaternionic Banach space $X$ is integrable with respect to the image measure $\mu^{\phi}(B) := \mu \left( \phi^{-1}(B) \right)$ and $f \circ \phi$ is integrable with respect to $\mu$, then*

$$\int_{\Upsilon} f \, d\mu^{\phi} = \int_{\Omega} f \circ \phi \, d\mu. \tag{6.5}$$

**Definition 6.1.12.** We denote the Borel sets on a topological space $X$ by $\mathrm{B}(X)$. In particular, we denote the Borel sets on $\mathbb{R}$, $\mathbb{C}$ and $\mathbb{H}$ by $\mathrm{B}(\mathbb{R})$, $\mathrm{B}(\mathbb{C})$ and $\mathrm{B}(\mathbb{H})$, respectively.

We recall that, for any Borel set $E \in \mathrm{B}(\mathbb{R})$, the set

$$P(E) := \{(u, v) \in \mathbb{R}^2 : \quad u + v \in E\}$$

is a Borel subset of $\mathbb{R}^2$.

**Definition 6.1.13.** Let $\mu, \nu$ be quaternionic measures on $\mathrm{B}(\mathbb{R})$. The convolution $\mu * \nu$ of $\mu$ and $\nu$ is the image measure of $\mu \times \nu$ under the mapping $\phi : \mathbb{R}^2 \to \mathbb{R}, (u, v) \mapsto u + v$, that is, for any $E \in \mathrm{B}(\mathbb{R})$, we set

$$(\mu * \nu)(E) := (\mu \times \nu)(P(E)).$$

The following corollary is immediate.

**Corollary 6.1.14.** *Let $\mu, \nu, \rho \in \mathcal{M}(\mathbb{R}, \mathrm{B}(\mathbb{R}), \mathbb{H})$ and let $a, b \in \mathbb{H}$. Then*

(i) *$(\mu + \nu) * \rho = \mu * \rho + \nu * \rho$ and $\mu * (\nu + \rho) = \mu * \nu + \mu * \rho$ and*

(ii) *$(a\mu) * \nu = a(\mu * \nu)$ and $\mu * (\nu a) = (\mu * \nu)a$.*

Then we prove the following results.

**Corollary 6.1.15.** *Let $\mu, \nu \in \mathcal{M}(\mathbb{R}, \mathrm{B}(\mathbb{R}), \mathbb{H})$. Then the estimate*

$$|\mu * \nu|(E) \leq (|\mu| * |\nu|)(E)$$

*holds true for all $E \in \mathrm{B}(\mathbb{R})$.*

*Proof.* Let $E \in \mathrm{B}(\mathbb{R})$ and let $\pi \in \Pi(E)$ be a countable measurable partition of $E$. Then

$$\sum_{E_\ell \in \pi} |(\mu * \nu)(E_\ell)| = \sum_{E_\ell \in \pi} |(\mu \times \nu)(P(E_\ell))|$$

$$\leq \sum_{E_\ell \in \pi} |\mu \times \nu|(P(E_\ell)) = |\mu \times \nu|(P(E)),$$

and taking the supremum over all possible partitions $\pi \in \Pi(E)$ yields

$$|\mu * \nu|(E) \leq |\mu \times \nu|(P(E)) = (|\mu| \times |\nu|)(P(E)) = (|\mu| * |\nu|)(E). \qquad \square$$

**Corollary 6.1.16.** *Let $\mu, \nu \in \mathcal{M}(\mathbb{R}, B(\mathbb{R}), \mathbb{H})$ and let $F : \mathbb{R} \to X$ be integrable with respect to $\mu * \nu$ and such that $\int_{-\infty}^{+\infty} \int_{-\infty}^{+\infty} \|F(s+t)\| \, d|\mu|(s) \, d|\nu|(t) < +\infty$. Then*

$$\int_{\mathbb{R}} F(r) \, d(\mu * \nu)(r) = \int_{\mathbb{R}} \int_{\mathbb{R}} F(s+t) \, d\mu(s) \, d\nu(t).$$

*Proof.* Our assumptions and Definition 6.1.13 allow us to apply Lemma 6.1.11 with $\phi(s,t) = s + t$. If $\mu(A) = \int_A f(t) \, d|\mu|(t)$ and $\nu(A) = \int_A g(s) \, d|\nu|(s)$, then the product measure satisfies

$$(\mu \times \nu)(B) = \int_B f(s) g(t) \, d(|\mu| \times |\nu|)(s,t)$$

by Lemma 6.1.10. Applying Fubini's theorem yields

$$
\begin{aligned}
\int_{\mathbb{R}} F(r) \, d(\mu * \nu)(r) &= \int_{\mathbb{R}} F(\phi(s,t)) \, d(\mu \times \nu)(s,t) \\
&= \int_{\mathbb{R}} F(\phi(s,t)) f(s) g(t) \, d|\mu \times \nu|(s,t) \\
&= \int_{\mathbb{R}} F(s+t) f(s) g(t) \, d|\mu|(s) \, d|\nu|(t) \\
&= \int_{\mathbb{R}} \int_{\mathbb{R}} F(s+t) \, d\mu(s) \, d\nu(t). \qquad \square
\end{aligned}
$$

## 6.2   Functions of the generator of a strongly continuous group

In the following we assume that $T \in \mathcal{K}(X)$ is the infinitesimal generator of the strongly continuous group $\{\mathcal{U}_T(t)\}_{t \in \mathbb{R}}$ of operators on a two-sided Banach space $X$. By Theorem 4.3.1, there exist positive constants $M > 0$ and $\omega \geq 0$ such that $\|\mathcal{U}_T(t)\| \leq M e^{\omega|t|}$ and such that the $S$-spectrum of the infinitesimal generator $T$ lies in the strip

$$W_\omega := \{s \in \mathbb{H} : \quad -\omega < \mathrm{Re}(s) < \omega\}.$$

Moreover, we have

$$S_R^{-1}(s, T) = \int_0^{+\infty} e^{-ts} \mathcal{U}_T(t) \, dt, \quad \mathrm{Re}(s) > \omega$$

and

$$S_R^{-1}(s, T) = -\int_{-\infty}^0 e^{-ts} \mathcal{U}_T(t) \, dt, \quad \mathrm{Re}(s) < -\omega.$$

**Definition 6.2.1.** We denote by $\mathbf{S}(T)$ the family of all quaternionic measures $\mu$ on $B(\mathbb{R})$ such that

$$\int_{\mathbb{R}} d|\mu|(t) \, e^{(\omega + \varepsilon)|t|} < +\infty$$

for some $\varepsilon = \varepsilon(\mu) > 0$. The function

$$\mathcal{L}(\mu)(s) = \int_{\mathbb{R}} d\mu(t)\, e^{-st}$$

with domain

$$W_{\omega + \varepsilon} := \{s \in \mathbb{H} : \quad -(\omega + \varepsilon) < \mathrm{Re}(s) < (\omega + \varepsilon)\}$$

is called the quaternionic two-sided (right) Laplace–Stieltjes transform of $\mu$.

**Definition 6.2.2.** We denote by $\mathbf{X}(T)$ the set of quaternionic two-sided Laplace–Stieltjes transforms of measures in $\mathbf{S}(T)$.

**Lemma 6.2.3.** *Let $\mu, \nu \in \mathbf{S}(T)$ and $a \in \mathbb{H}$.*

(i) *The measures $a\mu$ and $\mu + \nu$ belong to $\mathbf{S}(T)$ and*

$$\mathcal{L}(a\mu) = a\mathcal{L}(\mu), \quad \mathcal{L}(\mu + \nu) = \mathcal{L}(\mu) + \mathcal{L}(\nu).$$

(ii) *The measures $\mu * \nu$ belongs to $\mathbf{S}(T)$. If $\nu$ is real-valued, then*

$$\mathcal{L}(\mu * \nu) = \mathcal{L}(\mu)\mathcal{L}(\nu).$$

*Proof.* Let $\varepsilon = \min\{\varepsilon(\mu), \varepsilon(\nu)\}$. Lemma 6.1.4 implies

$$\int_{\mathbb{R}} d|a\mu|\, e^{|t|(\omega + \varepsilon)} = |a| \int_{\mathbb{R}} d|\mu|\, e^{|t|(\omega + \varepsilon)} < +\infty$$

and

$$\int_{\mathbb{R}} d|\mu + \nu|\, e^{|t|(\omega + \varepsilon)} \le \int_{\mathbb{R}} d|\mu|\, e^{|t|(\omega + \varepsilon)} + \int_{\mathbb{R}} d|\nu|\, e^{|t|(\omega + \varepsilon)} < +\infty.$$

Thus, $a\mu$ and $\mu + \nu$ belong to $\mathbf{S}(T)$. The relations $\mathcal{L}(a\mu) = a\mathcal{L}(\mu)$ and $\mathcal{L}(\mu + \nu) = \mathcal{L}(\mu) + \mathcal{L}(\nu)$ follow from the left linearity of the integral in the measure.

The variation of the convolution of $\mu$ and $\nu$ satisfies $|\mu * \nu|(E) \le (|\mu| * |\nu|)(E)$ for any Borel set $E \in B(\mathbb{R})$, cf. Corollary 6.1.15. In view of Corollary 6.1.16, we have

$$\int_{\mathbb{R}} d|\mu * \nu|(r) e^{(w + \varepsilon)|r|} \le \int_{\mathbb{R}} \int_{\mathbb{R}} d|\mu|(s)\, d|\nu|(t) e^{(w + \varepsilon)|s + t|}$$

$$\le \int_{\mathbb{R}} d|\mu|(s)\, e^{(w + \varepsilon)|s|} \int_{\mathbb{R}} d|\nu|(t)\, e^{(w + \varepsilon)|t|} < +\infty.$$

Therefore, $\mu * \nu \in \mathbf{S}(T)$. If $\nu$ is real-valued, then $\nu$ commutes with $e^{-st}$ and Fubini's theorem implies for $s \in \mathbb{H}$ with $-(w + \varepsilon) < \mathrm{Re}(s) < w + \varepsilon$

$$\mathcal{L}(\mu * \nu)(s) = \int_{\mathbb{R}} d(\mu * \nu)(r)\, e^{-sr} = \int_{\mathbb{R}} \int_{\mathbb{R}} d\mu(t)\, d\nu(u)\, e^{-s(t + u)}$$

$$= \int_{\mathbb{R}} d\mu(t)\, e^{-st} \int_{\mathbb{R}} d\nu(u)\, e^{-su} = \mathcal{L}(\mu)(s)\, \mathcal{L}(\nu)(s). \qquad \square$$

**Theorem 6.2.4.** *Let $f \in \mathbf{X}(T)$ with $f(s) = \int_{\mathbb{R}} d\mu(t)\, e^{-st}$, for any $s$ in the strip*

$$W_{\omega+\varepsilon} := \{s \in \mathbb{H}: \quad -(\omega+\varepsilon) < \mathrm{Re}(s) < \omega+\varepsilon\}.$$

(i)  *The function $f$ is right slice hyperholomorphic on the strip $W_{\omega+\varepsilon}$.*

(ii)  *For any $n \in \mathbb{N}$, the measure $\mu^n$ defined by*

$$\mu^n(E) = \int_{E} d\mu(t)\,(-t)^n, \quad \text{for } E \in \mathbf{B}(\mathbb{R})$$

*belongs to $\mathbf{S}(T)$ and, for $s$ with $-(\omega+\varepsilon) < \mathrm{Re}(s) < \omega+\varepsilon$, we have*

$$\partial_S{}^n f(s) = \int_{\mathbb{R}} d\mu^n(t)\, e^{-st} = \int_{\mathbb{R}} d\mu(t)\,(-t)^n e^{-st}, \tag{6.6}$$

*where $\partial_S f$ denotes the slice derivative of $f$.*

*Proof.* In the proof we will make use of the same kind of arguments as in [110, Lemma 2, p. 642]. For every $n \in \mathbb{N}$ and every $0 < \varepsilon_1 < \varepsilon$ there exists a constant $K$ such that

$$|t|^n e^{(\omega+\varepsilon_1)|t|} \le K e^{(\omega+\varepsilon)|t|}, \quad t \in \mathbb{R}.$$

Since $\mu \in \mathbf{S}(T)$, we have

$$\int_{\mathbb{R}} d|\mu^n|(t)\, e^{(\omega+\varepsilon_1)|t|} = \int_{\mathbb{R}} d|\mu|(t)\, |t|^n e^{(\omega+\varepsilon_1)|t|} \le K \int_{\mathbb{R}} d|\mu|(t)\, e^{(\omega+\varepsilon)|t|} < +\infty$$

and so $\mu^n \in \mathbf{S}(T)$. The function $f$ is a right slice function as, for $s = u + jv$,

$$f(s) = \int_{\mathbb{R}} d\mu(t) e^{-t(u+jv)} = \int_{\mathbb{R}} d\mu(t) e^{-tu} \cos(v) - \int_{\mathbb{R}} d\mu(t) e^{-tu} \sin(v) j$$

when we set

$$\alpha(u,v) := \int_{\mathbb{R}} d\mu(t) e^{-tu} \cos(v)$$

and

$$\beta(u,v) := -\int_{\mathbb{R}} d\mu(t) e^{-tu} \sin(v)$$

satisfy the compatibility condition (2.4). For any $s = u + jv \in W_{\omega+\varepsilon}$, we have

$$\lim_{\mathbb{C}_j \ni p \to s} (f_j(p) - f_j(s))(p - s)^{-1} = \lim_{\mathbb{C}_j \ni p \to s} \int_{\mathbb{R}} d\mu(t)\, \frac{e^{-pt} - e^{-st}}{p - s}.$$

If $p$ is sufficiently close to $s$ such that also $p \in W_{\omega+\varepsilon}$, then the simple calculation

$$|e^{-pt} - e^{-st}| = \left| \int_{0}^{1} e^{-ts - t\xi(p-s)} t(p-s)\, d\xi \right| \le |t| e^{(\omega+\varepsilon)|t|} |p - s|,$$

yields the estimate

$$\frac{|e^{-pt} - e^{-st}|}{|p - s|} \leq |t|e^{(\omega + \varepsilon)|t|},$$

which allows us to apply Lebesgue's theorem of dominated convergence in order to exchange limit and integration. We obtain

$$\lim_{\mathbb{C}_j \ni p \to s} (f_j(p) - f_j(s))(p - s)^{-1} = \int_{\mathbb{R}} d\mu(t)\,(-t)e^{-st} = \int_{\mathbb{R}} d\mu^1(t)\,e^{-st}. \quad (6.7)$$

Consequently, the restriction $f_j$ of $f$ to the complex plane $\mathbb{C}_j$ is right holomorphic and, by Lemmas 2.1.5 and 2.1.9, the function $f$ is in turn right slice hyperholo-morphic on the strip $\{s \in \mathbb{H} : -(\omega + \varepsilon) < \mathrm{Re}(s) < \omega + \varepsilon\}$. Moreover, (6.7) implies

$$\partial_S f(s) = \int_{\mathbb{R}} d\mu^1(t)\,e^{-st}$$

for $-(\omega + \varepsilon) < \mathrm{Re}(s) < \omega + \varepsilon$. By induction we get (6.6). $\qquad\square$

**Definition 6.2.5** (The Phillips functional calculus). Let $T \in \mathcal{K}(X)$ be the infinites-imal generator of the strongly continuous group $\{\mathcal{U}_T(t)\}_{t \in \mathbb{R}}$ of operators on a two-sided Banach space $X$. For $f \in \mathbf{X}(T)$ with

$$f(s) = \int_{\mathbb{R}} d\mu(t)\,e^{-st} \quad \text{for } -(\omega + \varepsilon) < \mathrm{Re}(s) < \omega + \varepsilon,$$

and $\mu \in \mathbf{S}(T)$, we define the right linear operator $f(T)$ on $X$ by

$$f(T) = \int_{\mathbb{R}} d\mu(t)\,\mathcal{U}_T(-t). \quad (6.8)$$

**Remark 6.2.1.** In particular for $p \in \mathbb{H}$ with $\mathrm{Re}(p) < -\omega$ the function $s \mapsto S_R^{-1}(p, s)$ belongs to $\mathbf{S}(T)$. Set $d\mu_p(t) = -\chi_{[0,+\infty)}(t)e^{tp}\,dt$, where $\chi_A$ denotes the character-istic function of a set $A$. If $\mathrm{Re}(p) < \mathrm{Re}(s)$, then

$$\mathcal{L}(\mu_p)(s) = \int_{\mathbb{R}} d\mu_p(t)\,e^{-ts} = -\int_0^{+\infty} e^{tp}e^{-ts}\,dt = -S_L^{-1}(s, p) = S_R^{-1}(p, s)$$

and

$$\mathcal{L}(\mu_p)(T) = \int_{\mathbb{R}} d\mu_p(t)\,\mathcal{U}(-t)$$

$$= -\int_0^{+\infty} e^{tp}\,\mathcal{U}(-t)\,dt$$

$$= -\int_{-\infty}^0 e^{-tp}\,\mathcal{U}(t)\,dt = S_R^{-1}(p, T).$$

For $p \in \mathbb{H}$ with $\omega < \mathrm{Re}(p)$ set $d\mu_p(t) = \chi_{(-\infty,0]}(t)e^{tp}\,dt$. Similar computations show that also in this case $S_R^{-1}(p, s) = \mathcal{L}(\mu_p)(s) \in \mathbf{S}(T)$ if $\mathrm{Re}(s) < \mathrm{Re}(p)$ and $\mathcal{L}(\mu_p)(T) = S_R^{-1}(p, T)$.

**Theorem 6.2.6.** *For any $f \in \mathbf{X}(T)$, the operator $f(T)$, defined in (6.8), is bounded.*

*Proof.* Let $f(s) = \int_{\mathbb{R}} d\mu(t)\, e^{-st} \in \mathbf{X}(T)$ with $\mu \in \mathbf{S}(T)$. Since $\|\mathcal{U}_T(t)\| \leq M e^{w|t|}$, we have

$$\|f(T)\| \leq \int_{\mathbb{R}} d|\mu|(t)\, \|\mathcal{U}_T(-t)\| \leq M \int_{\mathbb{R}} d|\mu|(t)\, e^{w|t|} < +\infty. \qquad \square$$

**Lemma 6.2.7.** *Let $f(T)$ be the operator defined in (6.8). Let $f = \mathcal{L}(\mu)$ and $g = \mathcal{L}(\nu)$ belong to $\mathbf{X}(T)$ and let $a \in \mathbb{H}$.*

(i) *We have $(af)(T) = af(T)$ and $(f + g)(T) = f(T) + g(T)$.*

(ii) *If $g$ is an intrinsic function, then $\nu$ is real-valued and $(fg)(T) = f(T)g(T)$.*

*Proof.* The statement (i) follows immediately from Lemma 6.2.3 and the left linearity of the integral (6.8) in the measure. In order to show (ii), we assume that $g = \mathcal{L}(\nu)$ is intrinsic. Then the measure $\nu$ is real-valued and Lemma 6.2.3 gives $fg = \mathcal{L}(\mu * \nu) \in \mathbf{X}(T)$. We find

$$
\begin{aligned}
(fg)(T)v &= \int_{\mathbb{R}} d(\mu * \nu)(r)\,\mathcal{U}_T(-r) \\
&= \int_{\mathbb{R}} \int_{\mathbb{R}} d\mu(s)\, d\nu(t)\,\mathcal{U}_T(-(s+t)) \\
&= \int_{\mathbb{R}} d\mu(s)\,\mathcal{U}_T(-s) \int_{\mathbb{R}} d\nu(t)\,\mathcal{U}_T(-t) = f(T)g(T),
\end{aligned}
$$

where we use that $\mathcal{U}_T(-s)$ and $\nu$ commute because $\nu$ is real-valued. $\qquad \square$

## 6.3 Comparison with the $S$-Functional Calculus

A natural question that arises is regarding the relation between the Phillips functional calculus introduced in Definition 6.2.5 and the $S$-functional calculus for closed operators. In this section we show that the two functional calculi coincide if the function $f$ is slice hyperholomorphic at infinity. In order to prove this, we need a specialized version of the Residue theorem that fits into our setting.

**Lemma 6.3.1.** *Let $O \subset \mathbb{H}$ be an axially symmetric open set, let $f : O \setminus [p] \to \mathbb{H}$ be right slice hyperholomorphic and let $g : O \to X$ be left slice hyperholomorphic such that $p = u + jv \in O$ is a pole of order $n_f \geq 0$ of the $\mathbb{H}$-valued right holomorphic function $f_j := f|_{O \cap \mathbb{C}_j}$. If $\varepsilon > 0$ is such that $\overline{B_\varepsilon(p)} \cap \mathbb{C}_j \subset O$, then*

$$\frac{1}{2\pi} \int_{\partial(B_\varepsilon(p) \cap \mathbb{C}_j)} f(s)\, ds_j\, g(s) = \sum_{k=0}^{n_f - 1} \frac{1}{k!} \mathrm{Res}_p \left( f_j(s)(s-p)^k \right) \left( \partial_s^k g(p) \right).$$

*Proof.* Since $f$ is right slice hyperholomorphic, its restriction $f_j$ is a vector-valued holomorphic function on $O \cap \mathbb{C}_j$ if we consider $\mathbb{H}$ as a vector space over $\mathbb{C}_j$ by restricting the multiplication with quaternions on the right to $\mathbb{C}_j$. Similarly, since $g$ is left slice hyperholomorphic, its restriction $g_j := g|_{O \cap \mathbb{C}_j}$ is an $X$-valued holomorphic function if we consider $X$ as a complex vector space over $\mathbb{C}_j$ by restricting the left scalar multiplication to $\mathbb{C}_j$. Consequently, if we set $\rho = \mathrm{dist}(p, \partial(O \cap \mathbb{C}_j))$, then

$$f_j(s) = \sum_{k=-n_f}^{+\infty} a_k(s-p)^k \quad \text{and} \quad g_j(s) = \sum_{k=0}^{+\infty}(s-p)^k b_k \tag{6.9}$$

for $s \in (B_\rho(p) \cap \mathbb{C}_j) \setminus \{p\}$ with $a_k \in \mathbb{H}$ and $b_k \in X$. These series converge uniformly on $\partial(B_\varepsilon(p) \cap \mathbb{C}_j)$ for any $0 < \varepsilon < \rho$. Thus,

$$\frac{1}{2\pi} \int_{\partial(B_\varepsilon(p) \cap \mathbb{C}_j)} f(s)\, ds_j\, g(s)$$

$$= \frac{1}{2\pi} \int_{\partial(B_\varepsilon(p) \cap \mathbb{C}_j)} \left( \sum_{k=0}^{+\infty} a_{k-n_f}(s-p)^{k-n_f} \right) ds_j \left( \sum_{j=0}^{+\infty}(s-p)^j b_j \right)$$

$$= \sum_{k=0}^{+\infty} \sum_{j=0}^{k} a_{k-j-n_f} \left( \frac{1}{2\pi} \int_{\partial(B_\varepsilon(p) \cap \mathbb{C}_j)} (s-p)^{k-j-n_f}\, ds_j\, (s-p)^j \right) b_j$$

$$= \sum_{j=0}^{n_f-1} a_{-(j+1)} b_j,$$

since $\frac{1}{2\pi} \int_{\partial(B_\varepsilon(p) \cap \mathbb{C}_j)} (s-p)^{k-n_f}\, ds_j$ equals 1 if $k - n_f = -1$ and 0 otherwise. Finally, we observe that $a_{-k} = \mathrm{Res}_p \left( f_j(s)(s-p)^{k-1} \right)$ and $b_k = \frac{1}{k!} \partial_S^k g_j(p)$ by their definition in (6.9). $\square$

In order to compute the integral in the $S$-functional calculus, we recall the definition of the strip

$$W_c := \{ s \in \mathbb{H} : \quad -c < \mathrm{Re}(s) < c \} \quad \text{for } c > 0$$

and we introduce the set $\partial(W_c \cap \mathbb{C}_j)$ for $j \in \mathbb{S}$. It consists of the two lines $s = c + j\tau$ and $s = -c - j\tau$ with $\tau \in \mathbb{R}$.

**Proposition 6.3.2.** *Let $\alpha$ and $c$ be real numbers such that $\omega < c < |\alpha|$. For any vector $y \in \mathcal{D}(T^2)$, we have*

$$\mathcal{U}_T(t)y = \frac{1}{2\pi} \int_{\partial(W_c \cap \mathbb{C}_j)} e^{ts}(\alpha - s)^{-2}\, ds_j\, S_R^{-1}(s, T)(\alpha \mathcal{I} - T)^2 y. \tag{6.10}$$

*Proof.* We recall that

$$S_R^{-1}(s, T) = \int_0^\infty e^{-ts} \mathcal{U}_T(t)\, dt, \quad \mathrm{Re}(s) > \omega.$$

Since $\|\mathcal{U}_T(t)\| \leq Me^{\omega|t|}$, we get a bound for the $S$-resolvent operator by

$$\|S_R^{-1}(s,T)\| = M \int_0^\infty e^{(\omega-\mathrm{Re}(s))t}\, dt, \quad \mathrm{Re}(s) > \omega \qquad (6.11)$$

which assures that $\|S_R^{-1}(s,T)\|$ is uniformly bounded on $\{s \in \mathbb{H} : \mathrm{Re}(s) > \omega + \varepsilon\}$ for any $\varepsilon > 0$. A similar consideration gives a uniform bound of $\|S_R^{-1}(s,T)\|$ on $\{s \in \mathbb{H} : \mathrm{Re}(s) < -(\omega+\varepsilon)\}$. Thanks to such bound, the integral in (6.10) is well defined since the $(\alpha-s)^{-2}$ goes to zero with order $1/|s|^2$ as $s \to \infty$. We set

$$F(t)y = \frac{1}{2\pi} \int_{\partial(W_c \cap \mathbb{C}_j)} e^{ts}(\alpha-s)^{-2}\, ds_j\, S_R^{-1}(s,T)(\alpha\mathcal{I} - T)^2 y$$

for $y \in \mathcal{D}(T^2)$. We show that $F(t)y = \mathcal{U}_T(t)y$ using the Laplace transform and we first assume $t > 0$. If $\mathrm{Re}(p) > c$, then

$$\int_0^\infty e^{-pt} F(t)y\, dt$$

$$= \frac{1}{2\pi} \int_0^\infty e^{-pt} \int_{\partial(W_c \cap \mathbb{C}_j)} e^{ts}(\alpha-s)^{-2}\, ds_j\, S_R^{-1}(s,T)(\alpha\mathcal{I} - T)^2 y\, dt$$

$$= \frac{1}{2\pi} \int_{\partial(W_c \cap \mathbb{C}_j)} \left( \int_0^{+\infty} e^{-pt} e^{ts} dt \right) (\alpha-s)^{-2}\, ds_j\, S_R^{-1}(s,T)(\alpha\mathcal{I} - T)^2 y.$$

Now observe that

$$\int_0^\infty e^{-pt} e^{ts}\, dt = S_R^{-1}(p,s),$$

so we have

$$\int_0^\infty e^{-pt} F(t)y\, dt$$

$$= \frac{1}{2\pi} \int_{\partial(W_c \cap \mathbb{C}_j)} S_R^{-1}(p,s)(\alpha-s)^{-2}\, ds_j\, S_R^{-1}(s,T)(\alpha\mathcal{I} - T)^2 y.$$

We point out that the function $s \mapsto S_R^{-1}(p,s)(\alpha-s)^{-2}$ is right slice hyperholomorphic for $s \notin [p] \cup \{\alpha\}$ and that the function $s \mapsto S_R^{-1}(s,T)(\alpha\mathcal{I} - T)^2 y$ is left slice hyperholomorphic on $\rho_S(T)$. Observe that the integrand is such that $(\alpha-s)^{-2}$ goes to zero with order $1/|s|^2$ as $s \to \infty$. By applying Cauchy's integral theorem, we can replace the path of integration by small negatively oriented circles of radius $\delta > 0$ around the singularities of the integrand in the plane $\mathbb{C}_j$. These singularities are $\alpha$, $p_j = p_0 + jp_1$ and $\overline{p}_j$ if $j \neq \pm j_p$. We obtain

$$\int_0^\infty e^{-pt} F(t) y\, dt$$

$$= -\frac{1}{2\pi} \int_{\partial(U_\delta(\alpha) \cap \mathbb{C}_j)} S_R^{-1}(p, s)(\alpha - s)^{-2}\, ds_j\, S_R^{-1}(s, T)(\alpha\mathcal{I} - T)^2 y$$

$$- \frac{1}{2\pi} \int_{\partial(U_\delta(p_j) \cap \mathbb{C}_j)} S_R^{-1}(p, s)(\alpha - s)^{-2}\, ds_j\, S_R^{-1}(s, T)(\alpha\mathcal{I} - T)^2 y$$

$$- \frac{1}{2\pi} \int_{\partial(U_\delta(\overline{p_j}) \cap \mathbb{C}_j)} S_R^{-1}(p, s)(\alpha - s)^{-2}\, ds_j\, S_R^{-1}(s, T)(\alpha\mathcal{I} - T)^2 y.$$

Observe that the integrand has a pole of order 2 at $\alpha$ and poles of order 1 at $p_j$ and $\overline{p}_j$ (except if $j = \pm j_p$). Applying Lemma 6.3.1 with $f(s) = S_R^{-1}(p, s)(\alpha - s)^{-2}$ and $g(s) = S_R^{-1}(s, T)(\alpha\mathcal{I} - T)^2 y$ yields therefore

$$\int_0^\infty e^{-pt} F(t) y\, dt = -\mathrm{Res}_\alpha \left( S_R^{-1}(p, s)(\alpha - s)^{-2} \right) S_R^{-1}(\alpha, T)(\alpha\mathcal{I} - T)^2 y$$

$$- \mathrm{Res}_\alpha \left( S_R^{-1}(p, s)(s - \alpha)^{-1} \right) \left( \partial_s S_R^{-1}(\alpha, T)(\alpha\mathcal{I} - T)^2 y \right)$$

$$- \mathrm{Res}_{p_j} \left( S_R^{-1}(p, s)(\alpha - s)^{-2} \right) S_R^{-1}(p_j, T)(\alpha\mathcal{I} - T)^2 y$$

$$- \mathrm{Res}_{\overline{p}_j} \left( S_R^{-1}(p, s)(\alpha - s)^{-2} \right) S_R^{-1}(\overline{p}_j, T)(\alpha\mathcal{I} - T)^2 y.$$

We calculate the residues of the function $f(s) = S_R^{-1}(p, s)(\alpha - s)^{-2}$. Since it has a pole of order two at $\alpha$, we have

$$\mathrm{Res}_\alpha(f_j) = \lim_{\mathbb{C}_j \ni s \to \alpha} \frac{\partial}{\partial s} f_j(s)(s - \alpha)^2 = \lim_{\mathbb{C}_j \ni s \to \alpha} \frac{\partial}{\partial s} S_R^{-1}(p, s) = (p - \alpha)^{-2},$$

where the last identity holds because $\alpha$ is real, and

$$\mathrm{Res}_\alpha(f_j(s)(s - \alpha)) = \lim_{\mathbb{C}_j \ni s \to \alpha} f_j(s)(s - \alpha)^2 = S_R^{-1}(p, \alpha).$$

The point $p_j = p_0 + jp_1$ is a pole of order 1. Thus, setting $s_{j_p} = s_0 + j_p s_1 \in \mathbb{C}_{j_p}$ for $s = s_0 + js_1 \in \mathbb{C}_j$, we deduce from the representation formula that

$$\mathrm{Res}_{p_j}(f_j) = \lim_{\mathbb{C}_j \ni s \to p_j} f_j(s)(s - p_j) = \lim_{\mathbb{C}_j \ni s \to p_j} S_R^{-1}(p, s)(\alpha - s)^{-2}(s - p_j)$$

$$= \lim_{\mathbb{C}_j \ni s \to p_j} \left[ S_R^{-1}(p, s_{j_p})(1 - j_p j)\frac{1}{2} + S_R^{-1}(p, \overline{s_{j_p}})(1 + j_p j)\frac{1}{2} \right] (s - p_j)(\alpha - s)^{-2}$$

$$= \lim_{\mathbb{C}_j \ni s \to p_j} (p - s_{j_p})^{-1}(1 - j_p j)(s - p_j)\frac{1}{2}(\alpha - p_j)^{-2}$$

$$+ \lim_{\mathbb{C}_j \ni s \to p_j} (p - \overline{s_{j_p}})^{-1}(1 + j_p j)(s - p_j)\frac{1}{2}(\alpha - p_j)^{-2}$$

$$= \left[ \lim_{\mathbb{C}_j \ni s \to p_j} (p - s_{j_p})^{-1}(1 - j_p j)(s - p_j) \right] \frac{1}{2}(\alpha - p_j)^{-2}.$$

We compute

$$\lim_{\mathbb{C}_j \ni s \to p_j} (p - s_{j_p})^{-1}(1 - j_p j)(s - p_j)$$

$$= \lim_{\mathbb{C}_j \ni s \to p_j} (p - s_{j_p})^{-1}(1 - j_p j)(s_0 - p_0) + (p - s_{j_p})^{-1}(1 - j_p j)j(s_1 - p_1)$$

$$= \lim_{\mathbb{C}_j \ni s \to p_j} (p - s_{j_p})^{-1}(s_0 - p_0)(1 - j_p j) + (p - s_{j_p})^{-1}(s_1 - p_1)(j + j_p)$$

$$= \lim_{\mathbb{C}_j \ni s \to p_j} (p - s_{j_p})^{-1}(s_0 - p_0)(1 - j_p j) + (p - s_{j_p})^{-1}(s_1 - p_1)j_p(-j_p j + 1)$$

$$= \lim_{\mathbb{C}_j \ni s \to p_j} (p - s_{j_p})^{-1}(s_0 - p_0 + j_p(s_1 - p_1))(1 - j_p j)$$

$$= \lim_{\mathbb{C}_j \ni s \to p_j} (p - s_{j_p})^{-1}(s_{j_p} - p)(1 - j_p j) = -(1 - j_p j)$$

and finally obtain

$$\operatorname{Res}_{p_j}(f_j) = -\frac{1}{2}(1 - j_p j)(\alpha - p_j)^{-2}.$$

Replacing $j$ by $-j$ in this formula yields

$$\operatorname{Res}_{\overline{p_j}}(f_j) = -\frac{1}{2}(1 + j_p j)(\alpha - \overline{p_j})^{-2}.$$

Note that these formulas also hold true if $j = \pm j_p$. In this case either $\operatorname{Res}_{p_j}(f_j) = -(\alpha - p_j)^{-2}$ and $\operatorname{Res}_{\overline{p_j}}(f_j) = 0$ because $\overline{p_j}$ is a removable singularity of $f_j$, or vice versa. Moreover,

$$S_R^{-1}(\alpha, T)(\alpha\mathcal{I} - T)^2 y = (\alpha\mathcal{I} - T)^{-1}(\alpha\mathcal{I} - T)^2 y = (\alpha\mathcal{I} - T)y$$

and

$$\partial_S S_R^{-1}(\alpha, T)(\alpha\mathcal{I} - T)^2 y = -(\alpha\mathcal{I} - T)^{-2}(\alpha\mathcal{I} - T)^2 y = -y$$

because $\alpha$ is real and so $S_R^{-1}(\alpha, T) = (\alpha\mathcal{I} - T)^{-1}$. Putting these pieces together, we get

$$\int_0^\infty e^{-pt} F(t)y\, dt = -(p - \alpha)^{-2} S_R^{-1}(\alpha, T)(\alpha\mathcal{I} - T)^2 y$$

$$+ S_R^{-1}(p, \alpha) S_R^{-2}(\alpha, T)(\alpha\mathcal{I} - T)^2 y$$

$$+ \frac{1}{2}(1 - j_p j)(\alpha - p_j)^{-2} S_R^{-1}(p_j, T)(\alpha\mathcal{I} - T)^2 y$$

$$+ \frac{1}{2}(1 + j_p j)(\alpha - \overline{p_j})^{-2} S_R^{-1}(\overline{p_j}, T)(\alpha\mathcal{I} - T)^2 y$$

$$= -(p - \alpha)^{-2}(\alpha\mathcal{I} - T)y + (p - \alpha)^{-1} y$$

$$+ (p - \alpha)^{-2} S_R^{-1}(p, T)(\alpha\mathcal{I} - T)^2 y,$$

where that last identity follows from representation formula because the mapping

$$p \mapsto (\alpha - p)^{-2} S_R^{-1}(p, T)(\alpha\mathcal{I} - T)^2 y$$

is left slice hyperholomorphic. We factor out $(p - \alpha)^{-2}$ on the left and obtain

$$\int_0^\infty e^{-pt} F(t)y\, dt = (p - \alpha)^{-2} \left( -(\alpha\mathcal{I} - T)y + (p - \alpha)y + S_R^{-1}(p,T)(\alpha\mathcal{I} - T)^2 y \right)$$

$$= (p - \alpha)^{-2} \left( py - 2\alpha y + Ty + S_R^{-1}(p,T)(\alpha\mathcal{I} - T)^2 y \right).$$

Recall that we assumed that $y \in \mathcal{D}(T^2)$. Hence, $Ty \in \mathcal{D}(T)$ and so we can apply the right $S$-resolvent equation twice to obtain

$$S_R^{-1}(p,T)(\alpha\mathcal{I} - T)^2 y$$
$$= S_R^{-1}(p,T)(T^2 y - 2\alpha Ty + \alpha^2 y)$$
$$= pS_R^{-1}(p,T)Ty - Ty - 2\alpha p S_R^{-1}(p,T)y + 2\alpha y + \alpha^2 S_R^{-1}(p,T)y$$
$$= p^2 S_R^{-1}(p,T)y - py - Ty - 2\alpha p S_R^{-1}(p,T)y + 2\alpha y + \alpha^2 S_R^{-1}(p,T)y$$
$$= (p - \alpha)^2 S_R^{-1}(p,T)y - py + 2\alpha y - Ty.$$

So finally

$$\int_0^\infty e^{-pt} F(t)y\, dt = (p - \alpha)^{-2}(p - \alpha)^2 S_R^{-1}(p,T)y = S_R^{-1}(p,T)y.$$

Hence,

$$\int_0^\infty e^{-pt} F(t)y\, dt = S_R^{-1}(p,T)y = \int_0^\infty e^{-pt} \mathcal{U}_T(t)y\, dt,$$

for $\mathrm{Re}(p) > c$, which implies $F(t)y = \mathcal{U}_T(t)y$ for $y \in D(T^2)$ and $t \geq 0$ as a consequence of the quaternionic version of the Hahn–Banach theorem (see Corollary 12.0.7).

Applying the same reasoning to the semigroup $\{\mathcal{U}(-t)\}_{t \geq 0}$, with infinitesimal generator $-T$, we see that

$$\mathcal{U}(-t)y = \frac{1}{2\pi} \int_{\partial(W_c \cap \mathbb{C}_j)} e^{ts}(\alpha - s)^{-2}\, ds_j\, S_R^{-1}(s, -T)(\alpha\mathcal{I} + T)^2 y$$

$$= \frac{1}{2\pi} \int_{\partial(W_c \cap \mathbb{C}_j)} e^{-ts}(\alpha + s)^{-2}\, ds_j\, S_R^{-1}(s, T)(\alpha\mathcal{I} + T)^2 y,$$

where the second equality follows by substitution of $s$ by $-s$ because

$$S_R^{-1}(-s, -T) = -S_R^{-1}(s, T).$$

Replacing $\alpha$ by $-\alpha$ and $-t$ by $t$, we finally find

$$\mathcal{U}(t)y = \frac{1}{2\pi} \int_{\partial(W_c \cap \mathbb{C}_j)} e^{ts}(\alpha - s)^{-2}\, ds_j\, S_R^{-1}(s, T)(\alpha\mathcal{I} - T)^2 y$$

also for $t < 0$. $\qquad\qquad\qquad\qquad\qquad\qquad\qquad\qquad\qquad\qquad\qquad$ $\square$

**Proposition 6.3.3.** *Let $\alpha$ and $c$ be real numbers such that $\omega < c < |\alpha|$. If $f \in \mathbf{X}(T)$ is right slice hyperholomorphic on $\overline{W_c}$, then, for any $y \in D(T^2)$, we have*

$$f(T)y = \frac{1}{2\pi} \int_{\partial(W_c \cap \mathbb{C}_j)} f(s)(\alpha - s)^{-2} \, ds_j \, S_R^{-1}(s,T)(\alpha \mathcal{I} - T)^2 y. \qquad (6.12)$$

*Proof.* We recall that $f$ can be represented as

$$f(s) = \int_{\mathbb{R}} d\mu(t) \, e^{-st}$$

with $\mu \in \mathbf{S}(T)$. Using Proposition 6.3.2, we obtain

$$\frac{1}{2\pi} \int_{\partial(W_c \cap \mathbb{C}_j)} f(s)(\alpha - s)^{-2} \, ds_j \, S_R^{-1}(s,T)(\alpha \mathcal{I} - T)^2 y$$

$$= \frac{1}{2\pi} \int_{\partial(W_c \cap \mathbb{C}_j)} \int_{\mathbb{R}} d\mu(t) \, e^{-st}(\alpha - s)^{-2} \, ds_j \, S_R^{-1}(s,T)(\alpha \mathcal{I} - T)^2 y$$

$$= \int_{\mathbb{R}} d\mu(t) \left( \frac{1}{2\pi} \int_{\partial(W_c \cap \mathbb{C}_j)} e^{-st}(\alpha - s)^{-2} \, ds_j \, S_R^{-1}(s,T)(\alpha \mathcal{I} - T)^2 y \right)$$

$$= \int_{\mathbb{R}} d\mu(t) \, \mathcal{U}_T(-t)y = f(T)y.$$

Note that Fubini's theorem allows us to exchange the order of integration as the $S$-resolvent $S_R^{-1}(s,T)$ is uniformly bounded on $\partial(W_c \cap \mathbb{C}_j)$ because of (6.11). So there exists a constant $K > 0$ such that

$$\frac{1}{2\pi} \int_{\partial(W_c \cap \mathbb{C}_j)} \int_{\mathbb{R}} \|d\mu(t) \, e^{-st}(\alpha - s)^{-2} \, ds_j \, S_R^{-1}(s,T)(\alpha \mathcal{I} - T)^2 y\|$$

$$\leq \frac{1}{2\pi} \int_{\partial(W_c \cap \mathbb{C}_j)} \int_{\mathbb{R}} d|\mu|(t) \, e^{-\mathrm{Re}(s)t} \frac{1}{|\alpha - s|^{-2}} \|S_R^{-1}(s,T)\| \|(\alpha \mathcal{I} - T)^2 y\| ds$$

$$\leq K \int_{\partial(W_c \cap \mathbb{C}_j)} \int_{\mathbb{R}} d|\mu|(t) \, e^{c|t|} \frac{1}{(1 + |s|)^2} \, ds.$$

This integral is finite because, as $\mu \in \mathbf{S}(T)$, we have

$$\int_{\mathbb{R}} d|\mu|(t) \, e^{c|t|} < +\infty. \qquad \square$$

**Theorem 6.3.4.** *Let $f \in \mathbf{X}(T)$ and suppose that $f$ is right slice hyperholomorphic at infinity. Then the operator $f(T)$ defined using the Laplace transform equals the operator $f[T]$ obtained from the $S$-functional calculus.*

*Proof.* Consider $\alpha \in \mathbb{R}$ with $c < |\alpha|$ and observe that the function

$$g(s) := f(s)(\alpha - s)^{-2}$$

is right slice hyperholomorphic and, since $f$ is right slice hyperholomorphic at infinity, tends to zero with order $1/|s|^2$ as $s$ tends to infinity. The $S$-functional calculus for unbounded operators thus satisfies

$$g[T] = g(\infty)\mathcal{I} + \int_{\partial(W_c \cap \mathbb{C}_j)} g(s) \, ds_j \, S_R^{-1}(s, T).$$

By Theorem 3.5.1, we have for $y \in X$ that

$$f[T](\alpha\mathcal{I} - T)^{-2}y = \frac{1}{2\pi} \int_{\partial(W_c \cap \mathbb{C}_j)} f(s)(\alpha - s)^2 \, ds_j \, S_R^{-1}(s, T)y.$$

But by Proposition 6.3.3, it is

$$f(T)x = \frac{1}{2\pi} \int_{\partial(W_c \cap \mathbb{C}_j)} f(s)(\alpha - s)^2 \, ds_j \, S_R^{-1}(s, T)(\alpha\mathcal{I} - T)^2 x,$$

for $x \in D(T^2)$. Setting $y = (\alpha\mathcal{I} - T)^2 x$, we conclude

$$f[T]y = f(T)y, \quad \text{for } y \in D(T^2).$$

Since $D(T^2)$ is dense in $X$ and since the operators $f[T]$ and $f(T)$ are bounded we get $f[T] = f(T)$. $\qquad\square$

## 6.4 The Inversion of the Operator $f(T)$

We study the inversion of the operator $f(T)$ defined by the Phillips functional calculus via approximation with polynomials $P_n$ such that $\lim_{n \to \infty} P_n(s)f(s) = 1$. In general, the pointwise product $P_n(s)f(s)$ is not slice hyperholomorphic and therefore we must limit ourselves to intrinsic functions. The main goal of this section is to deduce sufficient conditions such that

$$\lim_{n \to +\infty} P_n(T)f(T)y = y, \quad \text{for every } y \in X.$$

**Lemma 6.4.1.** *Let* $T \in \mathcal{K}(X)$ *such that* $\rho_S(T) \cap \mathbb{R} \neq \emptyset$. *If* $D(T)$ *is closed, then* $D(T^n)$ *is dense in* $X$ *for every* $n \in \mathbb{N}$.

*Proof.* If $\alpha \in \rho_S(T) \cap \mathbb{R}$, then

$$D(T^n) = D((\alpha\mathcal{I} - T)^n) = (\alpha\mathcal{I} - T)^{-n} X.$$

Therefore, a continuous right linear functional $y^* \in X^*$ on $X$ vanishes on $D(T^n)$ if and only if the functional $y^*(\alpha\mathcal{I} - T)^{-n}$, which is defined as

$$\langle y^*(\alpha\mathcal{I} - T)^{-n}, y \rangle := \langle y^*, (\alpha\mathcal{I} - T)^{-n} y \rangle$$

for $y \in X$, vanishes on the entire space $X$. We prove the statement by induction. It is obviously true for $n = 0$, so let us choose $n \in \mathbb{N}$ and let us assume that it holds for $n - 1$. By the above arguments, a functional $y^* \in X^*$ vanishes on $\mathcal{D}(T^n)$ if and only if

$$y^*(\alpha \mathcal{I} - T)^{-n} = y^*(\alpha \mathcal{I} - T)^{-(n-1)}(\alpha \mathcal{I} - T)^{-1}$$

vanishes on $X$, which is in turn equivalent to $y^*(\alpha \mathcal{I} - T)^{-(n-1)}$ vanishing on $\mathcal{D}(T)$. Now observe that by assumption $\mathcal{D}(T)$ is dense in $X$. Hence, Corollary 12.0.8 implies that a functional $x^* \in X^*$ vanishes on $\mathcal{D}(T)$ if and only if it vanishes on all of $X$. We conclude that $y^* \in X^*$ vanishes on $\mathcal{D}(T^n)$ if and only if $y^*(\alpha \mathcal{I} - T)^{-(n-1)}$ vanishes on all of $X^*$, which is in turn equivalent to $y^*$ vanishing on $\mathcal{D}(T^{n-1})$. Since $\mathcal{D}(T^{n-1})$ is dense in $X$ by the induction hypothesis, Corollary 12.0.8 implies again that a functional $x^* \in X^*$ vanishes on $\mathcal{D}(T^{n-1})$ if and only if it vanishes on all of $X$. Therefore, we finally find that $y^*$ vanishes on $\mathcal{D}(T^n)$ if and only if it vanishes on all of $X^*$ and a final application of Corollary 12.0.8 yields that $\mathcal{D}(T^n)$ is dense in $X$. $\qquad\square$

**Lemma 6.4.2.** *Let $P$ be an intrinsic polynomial of degree $m$ and let $f$ and $P_n f$ both belong to $\mathbf{X}(T)$. Then $f(T)X \subseteq \mathcal{D}(T^m)$ and*

$$P(T)f(T)y = (Pf)(T)y, \quad \text{for all } y \in X.$$

*Proof.* We first consider the case $y \in \mathcal{D}\left(T^{m+2}\right)$. Let $\alpha, c \in \mathbb{R}$ with $w < c < |\alpha|$ and let $j \in \mathbb{S}$. The function $Pf$ is the product of two intrinsic functions and therefore intrinsic itself. By Proposition 6.3.3, Lemma 6.2.7 and Remark 6.2.1, we have

$$(\alpha \mathcal{I} - T)^{-m}(Pf)(T)y$$
$$= \frac{1}{2\pi} \int_{\partial(W_c \cap \mathbb{C}_j)} (\alpha - s)^{-m} P(s) f(s)(\alpha - s)^{-2}\, ds_j\, S_R^{-1}(s, T)(\alpha \mathcal{I} - T)^2 y.$$

We write the polynomial $P$ in the form $P(s) = \sum_{k=0}^{m} a_k(\alpha - s)^k$ with $a_k \in \mathbb{R}$. In view of Proposition 6.3.3, Lemma 6.2.7 and Remark 6.2.1 we obtain again

$$(\alpha \mathcal{I} - T)^{-m}(Pf)(T)y$$
$$= \sum_{k=0}^{m} a_k \frac{1}{2\pi} \int_{\partial(W_c \cap \mathbb{C}_j)} (\alpha - s)^{-m+k} f(s)(\alpha - s)^{-2}\, ds_j\, S_R^{-1}(s, T)(\alpha \mathcal{I} - T)^2 y$$
$$= \sum_{k=0}^{m} a_k (\alpha \mathcal{I} - T)^{-m+k} f(T)u = (\alpha \mathcal{I} - T)^{-m} \sum_{k=0}^{m} a_k(\alpha \mathcal{I} - T)^k f(T)y$$
$$= (\alpha \mathcal{I} - T)^{-m} P(T) f(T)y.$$

Consequently, $(Pf)(T)y = P(T)f(T)y$ for $y \in \mathcal{D}\left(T^{m+2}\right)$.

Now let $y \in X$ be arbitrary. Since $\mathcal{D}(T^{m+2})$ is dense in $X$ by Lemma 6.4.1, there exists a sequence $y_n \in \mathcal{D}(T^{m+2})$ with $\lim_{n \to +\infty} y_n = y$. Then $f(T)y_n \to$

$f(T)y$ and $P(T)f(T)y_n = (Pf)(T)y_n \to (Pf)(T)y$ as $n \to +\infty$. Since $P(T)$ is closed with domain $\mathcal{D}(T^m)$, it follows that $f(T)y \in \mathcal{D}(T^m)$ and $P(T)f(T)y = (Pf)(T)y$. $\qquad\square$

**Definition 6.4.3.** A sequence of intrinsic polynomials $\{P_n\}_{n \in \mathbb{N}}$ is called an *inverting sequence* for an intrinsic function $f \in \mathbf{X}(T)$ if

(i) $P_n f \in \mathbf{X}(T)$,

(ii) $|P_n(s)f(s)| \leq M$, $n \in \mathbb{N}$ for some constant $M > 0$ and

$$\lim_{n \to +\infty} P_n(s)f(s) = 1$$

in a strip $W_{\omega+\varepsilon} = \{s \in \mathbb{H} : -(\omega + \varepsilon) < \mathrm{Re}(s) \leq \omega + \varepsilon\}$,

(iii) $\|(P_n f)(T)\| \leq M$, $n \in \mathbb{N}$ for some constant $M > 0$.

**Theorem 6.4.4.** *If $\{P_n\}_{n \in \mathbb{N}}$ is an inverting sequence for an intrinsic function $f \in \mathbf{X}(T)$, then*

$$\lim_{n \to +\infty} P_n(T)f(T)y = y, \quad \forall y \in X.$$

*Proof.* First consider $y \in \mathcal{D}(T^2)$ and choose $\alpha \in \mathbb{R}$ with $\omega < |\alpha|$. Then Proposition 6.3.3 and Lemma 6.4.2 imply

$$P_n(T)f(T)y = (P_n f)(T)y$$
$$= \frac{1}{2\pi} \int_{\partial(W_{c_n} \cap \mathbb{C}_j)} P_n(s)f(s)(\alpha - s)^{-2} \, ds_j \, S_R^{-1}(s, T)(\alpha \mathcal{I} - T)^2 y,$$

for arbitrary $j \in \mathbb{S}$ and $c_n \in \mathbb{R}$ with $w < c_n < |\alpha|$ such that $P_n f$ is right slice hyperholomorphic on $\overline{W_{c_n}}$. However, we have assumed that there exists a constant $M$ such that $|P_n(s)f(s)| \leq M$ for any $n \in \mathbb{N}$ on a strip $-(\omega + \varepsilon) \leq \mathrm{Re}(s) \leq \omega + \varepsilon$. Moreover, because of (6.11), the right $S$-resolvent is uniformly bounded on any set $\{s \in \mathbb{C}_j : |\mathrm{Re}(s)| > \omega + \varepsilon'\}$ with $\varepsilon' > 0$. Applying Cauchy's integral theorem we can therefore replace $\partial(W_{c_n} \cap \mathbb{C}_j)$ for any $n \in \mathbb{N}$ by $\partial(W_c \cap \mathbb{C}_j)$ where $c$ is a real number with $\omega < c < \min\{|\alpha|, \omega + \varepsilon\}$. In particular, we can choose $c$ independent of $n$. Lebesgue's dominated convergence theorem allows us to exchange limit and integration and we obtain

$$P_n(T)f(T)y = \frac{1}{2\pi} \int_{\partial(W_c \cap \mathbb{C}_j)} (\alpha - s)^{-2} \, ds_I \, S_R^{-1}(s, T)(\alpha \mathcal{I} - T)^2 y = y.$$

If $y \in X$ does not belong to $\mathcal{D}(T^2)$, then we can choose for any $\varepsilon > 0$ a vector $y_\varepsilon \in \mathcal{D}(T^2)$ with $\|y - y_\varepsilon\| < \varepsilon$. Since the mappings $(P_n f)(T)$ are uniformly bounded by a constant $M > 0$, we get

$$\|(P_n f)(T)y - y\|$$
$$\leq \|(P_n f)(T)y - (P_n f)(T)y_\varepsilon\| + \|(P_n f)(T)y_\varepsilon - y_\varepsilon\| + \|y_\varepsilon - y\|$$
$$\leq M\|y - y_\varepsilon\| + \|(P_n f)(T)y_\varepsilon - y_\varepsilon\| + \|y_\varepsilon - y\|$$
$$\xrightarrow{n \to +\infty} M\|y - y_\varepsilon\| + \|y_\varepsilon - y\| \leq (M + 1)\varepsilon.$$

Since $\varepsilon > 0$ was arbitrary, we deduce $\lim_{n \to +\infty} \|(P_n f)(T)y - y\| = 0$ even for arbitrary $y \in X$.                   $\square$

**Corollary 6.4.5.** *Let $X$ be reflexive and let $P_n$ be an inverting sequence for an intrinsic function $f \in \mathbf{S}(T)$. A vector $y$ belongs to the range of $f(T)$ if and only if it is in $\mathcal{D}(P_n(T))$ for all $n \in \mathbb{N}$ and the sequence $\{P_n(T)y\}_{n \in \mathbb{N}}$ is bounded.*

*Proof.* If $y \in \operatorname{ran} f(T)$ with $y = f(T)x$ then Lemma 6.4.2 implies $y \in \mathcal{D}(P_n(T))$ for all $n \in \mathbb{N}$. Theorem 6.4.4 states $\lim_{n \to +\infty} P_n(T)y = x$, which implies that the sequence $(P_n(T)y)_{n \in \mathbb{N}}$ is bounded.

To prove the converse statement consider $y \in X$ such that $\{P_n(T)y\}_{n \in \mathbb{N}}$ is bounded. Since $X$ is reflexive the set $\{P_n(T)y\}_{n \in \mathbb{N}}$ is weakly sequentially compact. (The proof that a set $E$ in a reflexive quaternionic Banach space $X$ is weakly sequentially compact if and only if $E$ is bounded can be completed similarly to the classical case when $X$ is a complex Banach space, see [110, Theorem II.28].) Hence, there exists a subsequence $\{P_{n_k}(T)y\}_{k \in \mathbb{N}}$ and a vector $x \in X$ such that

$$\langle y^*, P_{n_k}(T)y \rangle \to \langle y^*, x \rangle$$

as $k \to +\infty$ for any $y^* \in X^*$. We show $y = f(T)x$. For any functional $y^* \in X^*$, the mapping $y^* f(T)$, which is defined by

$$\langle y^* f(T), w \rangle = \langle y^*, f(T)w \rangle,$$

also belongs to $X^*$. Hence,

$$\langle y^*, f(T)P_{n_k}(T)y \rangle = \langle y^* f(T), P_{n_k}(T)y \rangle \to \langle y^* f(T), x \rangle = \langle y^*, f(T)x \rangle.$$

Recall that the measure $\mu$ with $f = \mathcal{L}(\mu)$ is real-valued since $f$ is intrinsic. Therefore it commutes with the operator $P_{n_k}(T)$. Recall also that if $w \in \mathcal{D}(T^n)$ for some $n \in \mathbb{N}$, then $\mathcal{U}_T(t)w \in \mathcal{D}(T^n)$ for any $t \in \mathbb{R}$ and $\mathcal{U}_T(t)T^n w = T^n \mathcal{U}_T(t)w$. Thus,

$$P_{n_k}(T)\mathcal{U}_T(t)y = \mathcal{U}_T(t)P_{n_k}(T)y$$

because $P_{n_k}$ has real coefficients. Moreover, we can therefore exchange the integral with the unbounded operator $P_{n_k}(T)$ in the following computation

$$f(T)P_{n_k}(T)y = \int_{\mathbb{R}} d\mu(t)\,\mathcal{U}_T(-t)P_{n_k}(T)y$$

$$= P_{n_k}(T)\int_{\mathbb{R}} d\mu(t)\,\mathcal{U}_T(-t)y = P_{n_k}(T)f(T)y.$$

Theorem 6.4.4 implies for any $y^* \in X^*$

$$\langle y^*, y \rangle = \lim_{k \to \infty} \langle y^*, P_{n_k}(T)f(T)y \rangle = \lim_{k \to \infty} \langle y^*, f(T)P_{n_k}(T)y \rangle = \langle x^*, f(T)x \rangle$$

and so $y = f(T)x$ follows from Corollary 12.0.8.          $\square$

# Chapter 7

# The $H^\infty$-Functional Calculus

The $H^\infty$-functional calculus was originally introduced in [170] by Alan McIntosh. His approach was generalized to quaternionic sectorial operators that are injective and have dense range in [30]. Moreover, under the above assumptions, in [30], it is also treated the case of $n$-tuples of noncommuting operators. The $H^\infty$-functional calculus stands out among all holomorphic (resp. slice hyperholomorphic) functional calculi because it allows to define functions $f$ of an operator $T$ such that $f(T)$ is unbounded.

This chapter is based on our paper [54], where we defined the $H^\infty$-functional calculus for arbitrary sectorial operators following the strategy of [165]. This provides also the techniques to introduce fractional powers of quaternionic linear operators. The approach in [54] requires neither the injectivity of $T$ nor that $T$ has dense range. Several proofs do not need much additional work and the strategies of the complex setting can be applied in a quite straightforward way. We shall therefore, in particular, focus on the proof of the chain rule and of the the spectral mapping theorem, since more severe technical difficulties arise in these proofs.

## 7.1 The $S$-functional calculus for sectorial operators

In order to define the notion of a sectorial operator, we introduce the sector $\Sigma_\varphi$ for $\varphi \in (0, \pi]$ as

$$\Sigma_\varphi := \{s \in \mathbb{H} : \quad \arg(s) < \varphi\}.$$

**Definition 7.1.1** (Sectorial operator). Let $\omega \in [0, \pi)$. An operator $T \in \mathcal{K}(X)$ is called sectorial of angle $\omega$ if

(i) we have $\sigma_S(T) \subset \overline{\Sigma_\omega}$ and

(ii) for every $\varphi \in (\omega, \pi)$ there exists a constant $C > 0$ such that for $s \notin \overline{\Sigma_\varphi}$

$$\left\| S_L^{-1}(s, T) \right\| \leq \frac{C}{|s|} \quad \text{and} \quad \left\| S_R^{-1}(s, T) \right\| \leq \frac{C}{|s|}. \tag{7.1}$$

© Springer Nature Switzerland AG 2019
F. Colombo, J. Gantner, *Quaternionic Closed Operators, Fractional Powers and Fractional Diffusion Processes*, Operator Theory: Advances and Applications 274,
https://doi.org/10.1007/978-3-030-16409-6_7

We denote the infimum of all these constants by $C_\varphi$ and additionally by $C_{\varphi,T}$ if we also want to stress its dependence on $T$.

Next we introduce the following notations:

(a) We denote the set of all operators in $\mathcal{K}(X)$ that are sectorial of angle $\omega$ by $\mathrm{Sect}(\omega)$. Furthermore, if $T$ is a sectorial operator, we call

$$\omega_T = \min\{\omega : \quad T \in \mathrm{Sect}(\omega)\}$$

the spectral angle of $T$.

(b) A family of operators $\{T_\ell\}_{\ell \in \Lambda}$ is called uniformly sectorial of angle $\omega$ if $T_\ell \in \mathrm{Sect}(\omega)$ for all $\ell \in \Lambda$ and $\sup_{\ell \in \Lambda} C_{\varphi,T_\ell} < +\infty$ for all $\varphi \in (\omega, \pi)$.

The class of slice hyperholomorphic functions that will be considered in order to define the $H^\infty$-functional calculus is specified in the next definitions.

**Definition 7.1.2.** Let $f$ be a slice hyperholomorphic function.

(i) We say that $f$ has polynomial limit $c \in \mathbb{H}$ in $\Sigma_\varphi$ at 0 if there exists $\alpha > 0$ such that $f(p) - c = O(|p|^\alpha)$ as $p \to 0$ in $\Sigma_\varphi$ and that it has polynomial limit $\infty$ in $\Sigma_\varphi$ at 0 if $f^{-*_L}$ (resp. $f^{-*_R}$) has polynomial limit 0 at 0 in $\Sigma_\varphi$. (By (2.26) this is equivalent to $1/|f(p)| \in O(|p|^\alpha)$ for some $\alpha > 0$ as $p \to 0$ in $\Sigma_\varphi$.)

(ii) Similarly, we say that $f$ has polynomial limit $c \in \mathbb{H}_\infty$ at $\infty$ in $\Sigma_\varphi$ if $p \mapsto f(p^{-1})$ has polynomial limit $c$ at 0.

(iii) If a function has polynomial limit 0 at 0 or $\infty$, we say that it decays regularly at 0 (resp. $\infty$).

Observe that the mapping $p \mapsto p^{-1}$ leaves $\Sigma_\varphi$ invariant such that the above relation between polynomial limits at 0 and $\infty$ makes sense.

**Definition 7.1.3.** Let $\varphi \in (0, \pi]$.

(i) We define $\mathcal{SH}^\infty_{L,0}(\Sigma_\varphi)$ as the set of all bounded functions in $\mathcal{SH}_L(\Sigma_\varphi)$ that decay regularly at 0 and $\infty$.

(ii) Similarly, we define $\mathcal{SH}^\infty_{R,0}(\Sigma_\varphi)$ and $\mathcal{SH}^\infty_0(\Sigma_\varphi)$ as the set of all bounded functions in $\mathcal{SH}_R(\Sigma_\varphi)$ (resp. $\mathcal{N}(\Sigma_\varphi)$) that decay regularly at 0 and $\infty$.

The following Lemma is an immediate consequence of Theorem 2.1.3.

**Lemma 7.1.4.** Let $\varphi \in (0, \pi]$.

(i) If $f, g \in \mathcal{SH}^\infty_{L,0}(\Sigma_\varphi)$ and $a \in \mathbb{H}$, then $fa + g \in \mathcal{SH}^\infty_{L,0}(\Sigma_\varphi)$. If in addition $f \in \mathcal{SH}^\infty_0(\Sigma_\varphi)$, then $fg \in \mathcal{SH}^\infty_0(\Sigma_\varphi)$.

(ii) If $f, g \in \mathcal{SH}^\infty_{R,0}(\Sigma_\varphi)$ and $a \in \mathbb{H}$, then $af + g \in \mathcal{SH}^\infty_{R,0}(\Sigma_\varphi)$. If in addition $g \in \mathcal{SH}^\infty_0(\Sigma_\varphi)$, then $fg \in \mathcal{SH}^\infty_0(\Sigma_\varphi)$.

(iii) *The space $\mathcal{SH}_0^\infty(\Sigma_\varphi)$ is a real algebra.*

**Definition 7.1.5** (S-functional calculus for sectorial operators). Let $T \in \text{Sect}(\omega)$. For $f \in \mathcal{SH}_{L,0}^\infty(\Sigma_\varphi)$ with $\omega < \varphi < \pi$, we choose $\varphi'$ with $\omega < \varphi' < \varphi$ and $j \in \mathbb{S}$ and define

$$f(T) := \frac{1}{2\pi} \int_{\partial(\Sigma_{\varphi'} \cap \mathbb{C}_j)} S_L^{-1}(s,T) \, ds_j \, f(s). \tag{7.2}$$

Similarly, for $f \in \mathcal{SH}_{R,0}^\infty(\Sigma_\varphi)$ with $\omega < \varphi < \pi$, we choose $\varphi'$ with $\omega < \varphi' < \varphi$ and $j \in \mathbb{S}$ and define

$$f(T) := \frac{1}{2\pi} \int_{\partial(\Sigma_{\varphi'} \cap \mathbb{C}_j)} f(s) \, ds_j \, S_R^{-1}(s,T). \tag{7.3}$$

**Remark 7.1.1.** Since $T$ is sectorial of angle $\omega$, the estimates in (7.1) assure the convergence of the above integrals. A standard argument using the slice hyper-holomorphic version of Cauchy's integral theorem shows that the integrals are independent of the choice of the angle $\varphi'$, and standard slice hyperholomorphic techniques, based on the representation formula, show that they are independent of the choice of the imaginary unit $j \in \mathbb{S}$. Finally, computations as in the proof of Theorem 3.4.6 show that (7.2) and (7.3) yield the same operator if $f$ is intrinsic.

If $T \in \text{Sect}(\omega)$, then $f(T)$ in Definition 7.1.5 can be defined for any function that belongs to $\mathcal{SH}_{L,0}^\infty(\Sigma_\varphi)$ for some $\varphi \in (\omega, \pi]$. We thus introduce a notation for the space of all such functions.

**Definition 7.1.6.** Let $\omega \in (0,\pi)$. We define

$$\mathcal{SH}_{L,0}^\infty[\Sigma_\omega] = \bigcup_{\omega < \varphi \le \pi} \mathcal{SH}_{L,0}^\infty(\Sigma_\varphi),$$

$$\mathcal{SH}_{R,0}^\infty[\Sigma_\omega] = \bigcup_{\omega < \varphi \le \pi} \mathcal{SH}_{R,0}^\infty(\Sigma_\varphi),$$

$$\mathcal{SH}_0^\infty[\Sigma_\omega] = \bigcup_{\omega < \varphi \le \pi} \mathcal{SH}_0^\infty(\Sigma_\varphi).$$

The following properties of the $S$-functional calculus for sectorial operators can be proved by standard slice-hyperholomorphic techniques, see Theorem 3.5.1 or see also [30, Theorem 4.12].

**Lemma 7.1.7.** *If $T \in \text{Sect}(\omega)$, then the following statements hold true.*

(i) *If $f \in \mathcal{SH}_{L,0}^\infty[\Sigma_\omega]$ or $f \in \mathcal{SH}_{R,0}^\infty[\Sigma_\omega]$, then the operator $f(T)$ is bounded.*

(ii) *If $f,g \in \mathcal{SH}_{L,0}^\infty[\Sigma_\omega]$ and $a \in \mathbb{H}$, then $(fa+g)(T) = f(T)a + g(T)$. Similarly, if $f,g \in \mathcal{SH}_{R,0}^\infty[\Sigma_\omega]$ and $a \in \mathbb{H}$, then $(af+g)(T) = af(T) + g(T)$.*

(iii) *If $f \in \mathcal{SH}_0^\infty[\Sigma_\omega]$ and $g \in \mathcal{SH}_{L,0}^\infty[\Sigma_\omega]$, then $(fg)(T) = f(T)g(T)$. Similarly, if $f \in \mathcal{SH}_{R,0}^\infty[\Sigma_\omega]$ and $g \in \mathcal{SH}_0^\infty[\Sigma_\omega]$, then also $(fg)(T) = f(T)g(T)$.*

We recall that a closed operator $A \in \mathcal{K}(X)$ is said to commute with $B \in \mathcal{B}(X)$, if $BA \subset AB$.

**Lemma 7.1.8.** Let $T \in \text{Sect}(\omega)$ and $A \in \mathcal{K}(X)$ commute with $\mathcal{Q}_s(T)^{-1}$ and $T\mathcal{Q}_s(T)^{-1}$ for any $s \in \rho_S(T)$. Then $A$ commutes with $f(T)$ for any $f \in \mathcal{SH}_0^\infty[\Sigma_\omega]$. In particular, $f(T)$ commutes with $T$ for any $f \in \mathcal{SH}_0^\infty[\Sigma_\omega]$.

*Proof.* If $f \in \mathcal{SH}_0^\infty[\Sigma_\omega]$, then for suitable $\varphi \in (\omega, \pi)$ and $j \in \mathbb{S}$, we have

$$
\begin{aligned}
f(T) &= \frac{1}{2\pi} \int_{\partial(\Sigma_\varphi \cap \mathbb{C}_j)} f(s) \, ds_j S_R^{-1}(s, T) \\
&= \frac{1}{2\pi} \int_{-\infty}^0 f\left(-te^{j\varphi}\right)\left(-e^{j\varphi}\right)(-j)\left(-te^{-j\varphi} - T\right) \mathcal{Q}_{-te^{j\varphi}}(T)^{-1} \, dt \\
&\quad + \frac{1}{2\pi} \int_0^{+\infty} f\left(te^{-j\varphi}\right)\left(e^{-j\varphi}\right)(-j)\left(te^{j\varphi} - T\right) \mathcal{Q}_{te^{-j\varphi}}(T)^{-1} \, dt.
\end{aligned}
$$

After changing $t \mapsto -t$ in the first integral, we find

$$
\begin{aligned}
f(T) &= \frac{1}{2\pi} \int_0^{+\infty} f\left(te^{j\varphi}\right)\left(e^{j\varphi}j\right)\left(te^{-j\varphi} - T\right) \mathcal{Q}_{te^{j\varphi}}(T)^{-1} \, dt \\
&\quad + \frac{1}{2\pi} \int_0^{+\infty} f\left(te^{-j\varphi}\right)\left(-e^{-j\varphi}j\right)\left(te^{j\varphi} - T\right) \mathcal{Q}_{te^{-j\varphi}}(T)^{-1} \, dt \\
&= \frac{1}{2\pi} \int_0^{+\infty} 2\text{Re}\left[f\left(te^{j\varphi}\right) jt\right] \mathcal{Q}_{te^{j\varphi}}(T)^{-1} \, dt \\
&\quad - \frac{1}{2\pi} \int_0^{+\infty} 2\text{Re}\left[f\left(te^{j\varphi}\right) je^{j\varphi}\right] T\mathcal{Q}_{te^{j\varphi}}(T)^{-1} \, dt,
\end{aligned}
$$

where the last identity holds because $f(\bar{s}) = \overline{f(s)}$ as $f$ is intrinsic and

$$
\mathcal{Q}_{te^{j\varphi}}(T)^{-1} = \mathcal{Q}_{te^{-j\varphi}}(T)^{-1}.
$$

If now $y \in \mathcal{D}(A)$, then the fact that $A$ commutes with $\mathcal{Q}_s(T)^{-1}$ and $T\mathcal{Q}_s(T)^{-1}$ and any real scalar implies

$$
\begin{aligned}
f(T)Ay &= \frac{1}{2\pi} \int_0^{+\infty} 2\text{Re}\left[f\left(te^{j\varphi}\right) jt\right] \mathcal{Q}_{te^{j\varphi}}(T)^{-1} Ay \, dt \\
&\quad - \frac{1}{2\pi} \int_0^{+\infty} 2\text{Re}\left[f\left(te^{j\varphi}\right) je^{j\varphi}\right] T\mathcal{Q}_{te^{j\varphi}}(T)^{-1} Ay \, dt \\
&= A\frac{1}{2\pi} \int_0^{+\infty} 2\text{Re}\left[f\left(te^{j\varphi}\right) jt\right] \mathcal{Q}_{te^{j\varphi}}(T)^{-1} y \, dt \\
&\quad - A\frac{1}{2\pi} \int_0^{+\infty} 2\text{Re}\left[f\left(te^{j\varphi}\right) je^{j\varphi}\right] T\mathcal{Q}_{te^{j\varphi}}(T)^{-1} y \, dt = Af(T)y.
\end{aligned}
$$

We thus find $y \in \mathcal{D}(Af(T))$ with $f(T)Ay = Af(T)y$ and in turn $f(T)A \subset Af(T)$. $\qquad\square$

**Lemma 7.1.9.** *Let* $T \in \text{Sect}(\omega)$. *If* $\lambda \in (-\infty, 0)$ *and* $f \in \mathcal{SH}^\infty_{L,0}[\Sigma_\omega]$, *then*

$$s \mapsto (\lambda - s)^{-1} f(s) \in \mathcal{SH}^\infty_{L,0}[\Sigma_\omega]$$

*and*

$$\left((\lambda - s)^{-1} f(s)\right)(T) = (\lambda - T)^{-1} f(T) = S_L^{-1}(\lambda, T) f(T).$$

*Similarly, if* $\lambda \in (-\infty, 0)$ *and* $f \in \mathcal{SH}^\infty_{R,0}[\Sigma_\omega]$, *then*

$$s \mapsto f(s)(\lambda - s)^{-1} \in \mathcal{SH}^\infty_{R,0}[\Sigma_\omega]$$

*and*

$$\left(f(s)(\lambda - s)^{-1}\right)(T) = f(T)(\lambda - T)^{-1} = f(T) S_R^{-1}(\lambda, T).$$

*Proof.* Let $\lambda \in (-\infty, 0)$ and observe that, since $\lambda$ is real, the $S$-resolvent equation turns into

$$(\lambda - T)^{-1} S_L^{-1}(s, T) = S_R^{-1}(\lambda, T) S_L^{-1}(s, T)$$
$$= \left(S_R^{-1}(\lambda, T) - S_L^{-1}(s, T)\right)(s - \lambda)^{-1}.$$

If now $f \in \mathcal{SH}^\infty_{L,0}[\Sigma_\omega]$, then for suitable $\varphi \in (\omega, \pi)$ and $j \in \mathbb{S}$, we have

$$(\lambda \mathcal{I} - T)^{-1} f(T) = \frac{1}{2\pi} \int_{\partial(\Sigma_\varphi \cap \mathbb{C}_j)} (\lambda \mathcal{I} - T)^{-1} S_L^{-1}(s, T) \, ds_j \, f(s)$$

$$= \frac{1}{2\pi} \int_{\partial(\Sigma_\varphi \cap \mathbb{C}_j)} \left(S_R^{-1}(\lambda, T) - S_L^{-1}(s, T)\right)(s - \lambda)^{-1} \, ds_j \, f(s)$$

$$= S_R^{-1}(\lambda, T) \frac{1}{2\pi} \int_{\partial(\Sigma_\varphi \cap \mathbb{C}_j)} ds_j \, (s - \lambda)^{-1} f(s)$$

$$+ \frac{1}{2\pi} \int_{\partial(\Sigma_\varphi \cap \mathbb{C}_j)} S_L^{-1}(s, T) \, ds_j \, (\lambda - s)^{-1} f(s)$$

$$= \left((\lambda - s)^{-1} f(s)\right)(T),$$

where the last equality holds because $\frac{1}{2\pi} \int_{\partial(\Sigma_\varphi \cap \mathbb{C}_j)} ds_j \, (s - \lambda)^{-1} f(s) = 0$ by Cauchy's integral theorem. $\qquad \square$

Similar to [165], we can extend the class of functions that are admissible for this functional calculus to the analogue of the extended Riesz class.

**Definition 7.1.10.** For $0 < \varphi < \pi$, we define

$$\mathcal{E}_L(\Sigma_\varphi) = \left\{ f(p) = \tilde{f}(p) + (1+p)^{-1} a + b : \quad \tilde{f} \in \mathcal{SH}^\infty_{L,0}(\Sigma_\varphi), a, b \in \mathbb{H} \right\}$$

and similarly

$$\mathcal{E}_R(\Sigma_\varphi) = \left\{ f(p) = \tilde{f}(p) + a(1+p)^{-1} + b : \quad \tilde{f} \in \mathcal{SH}^\infty_{R,0}(\Sigma_\varphi), a, b \in \mathbb{H} \right\}.$$

Finally, we define $\mathcal{E}(\Sigma_\varphi)$ as the set of all intrinsic functions in $\mathcal{E}_L(\Sigma_\varphi)$, i.e.,

$$\mathcal{E}(\Sigma_\varphi) = \left\{ f(p) = \tilde{f}(p) + (1+p)^{-1} a + b : \quad \tilde{f} \in \mathcal{SH}^\infty_0(\Sigma_\varphi), a, b \in \mathbb{R} \right\}.$$

Keeping in mind the product rule of slice-hyperholomorphic functions, simple calculations as in the classical case show the following two corollaries, cf. [165, Lemma 2.2.3].

**Corollary 7.1.11.** *Let $0 < \varphi < \pi$.*

(i) *The set $\mathcal{E}_L(\Sigma_\varphi)$ is a quaternionic right vector space and it is closed under multiplication with functions in $\mathcal{E}(\Sigma_\varphi)$ from the left.*

(ii) *The set $\mathcal{E}_R(\Sigma_\varphi)$ is a quaternionic left vector space and it is closed under multiplication with functions in $\mathcal{E}(\Sigma_\varphi)$ from the right.*

(iii) *The set $\mathcal{E}(\Sigma_\varphi)$ is a real algebra.*

**Corollary 7.1.12.** *Let $0 < \varphi < \pi$. A function $f \in \mathcal{SH}_L(\Sigma_\varphi)$ (or $f \in \mathcal{SH}_R(\Sigma_\varphi)$ or $f \in \mathcal{N}(\Sigma_\varphi)$) belongs to $\mathcal{E}_L(\Sigma_\varphi)$ (resp. $\mathcal{E}_R(\Sigma_\varphi)$ or $\mathcal{E}(\Sigma_\varphi)$) if and only if it is bounded and has finite polynomial limits at 0 and infinity.*

**Definition 7.1.13.** For $\omega \in (0, \pi)$, we denote

$$\mathcal{E}_L[\Sigma_\omega] = \bigcup_{\omega < \varphi < \pi} \mathcal{E}_L(\Sigma_\varphi),$$

$$\mathcal{E}_R[\Sigma_\omega] = \bigcup_{\omega < \varphi < \pi} \mathcal{E}_R(\Sigma_\varphi),$$

$$\mathcal{E}[\Sigma_\omega] = \bigcup_{\omega < \varphi < \pi} \mathcal{E}(\Sigma_\varphi).$$

**Definition 7.1.14.** Let $T \in \mathrm{Sect}(\omega)$. We define for any function $f \in \mathcal{E}_L[\Sigma_\omega]$ with $f(s) = \tilde{f}(s) + (1+s)^{-1}a + b$ the bounded operator

$$f(T) := \tilde{f}(T) + (1+T)^{-1}a + \mathcal{I}b$$

and for any function $f \in \mathcal{E}_R[\Sigma_\omega]$ with $f(s) = \tilde{f}(s) + a(1+s)^{-1} + b$ the bounded operator

$$f(T) := \tilde{f}(T) + a(1+T)^{-1} + b\mathcal{I},$$

where $\tilde{f}(T)$ is intended in the sense of Definition 7.1.5.

**Lemma 7.1.15.** *Let $T \in \mathrm{Sect}(\omega)$ and let $f \in \mathcal{E}_L[\Sigma_\omega]$. If $f$ is left slice hyperholomorphic at 0 and decays regularly at infinity, then*

$$f(T) = \frac{1}{2\pi} \int_{\partial(U(r) \cap \mathbb{C}_j)} S_L^{-1}(s, T) \, ds_j \, f(s), \qquad (7.4)$$

*with $j \in \mathbb{S}$ arbitrary and $U(r) = \Sigma_\varphi \cup B_r(0)$, where $\varphi \in (\omega, \pi)$ is such that $f \in \mathcal{E}_L(\Sigma_\varphi)$ and $r > 0$ is such that $f$ is left slice hyperholomorphic on $\overline{B_r(0)}$. Moreover, if $f$ is left slice hyperholomorphic both at 0 and at infinity, then*

$$f(T) = f(\infty)\mathcal{I} + \frac{1}{2\pi} \int_{\partial(U(r,R) \cap \mathbb{C}_j)} S_L^{-1}(s, T) \, ds_j \, f(s), \qquad (7.5)$$

with $j \in \mathbb{S}$ arbitrary and $U(r, R) = U(r) \cup (\mathbb{H} \setminus B_R(0))$, where $\varphi \in (\omega, \pi)$ is such that $f \in \mathcal{E}_L(\Sigma_\varphi)$, $r > 0$ is such that $f$ is left slice hyperholomorphic on $\overline{B_r(0)}$ and $R > r$ is such that $f$ is left slice-hyperholmorphic on $\mathbb{H} \setminus B_R(0)$.

Similarly, if $f \in \mathcal{E}_R[\Sigma_\omega]$, is right slice hyperholomorphic at $0$ and decays regularly at infinity, then

$$f(T) = \frac{1}{2\pi} \int_{\partial(U(r) \cap \mathbb{C}_j)} f(s) \, ds_j \, S_R^{-1}(s, T),$$

with $j \in \mathbb{S}$ arbitrary and $U(r)$ chosen as above. Moreover, if $f \in \mathcal{E}_R[\Sigma_\omega]$ is right slice hyperholomorphic both at $0$ and at infinity, then

$$f(T) = f(\infty)\mathcal{I} + \frac{1}{2\pi} \int_{\partial(U(r,R) \cap \mathbb{C}_j)} f(s) \, ds_j \, S_R^{-1}(s, T),$$

with $j \in \mathbb{S}$ arbitrary and $U(r, R)$ is chosen as above.

*Proof.* Let us first assume that $f \in \mathcal{E}_L[\Sigma_\omega]$ is left slice hyperholomorphic at $0$ and regularly decaying at infinity. Then $f(s) = \tilde{f}(s) + (1+s)^{-1}a$, where $\tilde{f} \in \mathcal{SH}_{L,0}^\infty(\Sigma_{\varphi'})$ with $\omega < \varphi < \varphi'$, and the function $\tilde{f}$ is, moreover, left slice hyperholomorphic at $0$. For $j \in \mathbb{S}$ and $\omega < \varphi < \varphi'$, we therefore have

$$\frac{1}{2\pi} \int_{\partial(U(r) \cap \mathbb{C}_j)} S_L^{-1}(s, T) \, ds_j \, f(s)$$

$$= \frac{1}{2\pi} \int_{\partial(U(r) \cap \mathbb{C}_j)} S_L^{-1}(s, T) \, ds_j \, \tilde{f}(s) + \frac{1}{2\pi} \int_{\partial(U(r) \cap \mathbb{C}_j)} S_L^{-1}(s, T) \, ds_j \, (1+s)^{-1}a.$$

If $r' > r > 0$ is sufficiently small such that $\tilde{f}$ is left slice hyperholomorphic at $\overline{B_{r'}(0)}$, then Cauchy's integral theorem implies that the value of the first integral remains constant as $r$ varies. Letting $r$ tend to $0$, we find that this integral equals $\tilde{f}(T)$ in the sense of Definition 7.1.5. For the second integral we find that

$$\frac{1}{2\pi} \int_{\partial(U(r) \cap \mathbb{C}_j)} S_L^{-1}(s, T) \, ds_j \, (1+s)^{-1}a$$

$$= \lim_{R \to +\infty} \frac{1}{2\pi} \int_{\partial(U(r,R) \cap \mathbb{C}_j)} S_L^{-1}(s, T) \, ds_j \, (1+s)^{-1}a = (1+T)^{-1}a,$$

where the last identity can be deduced either from the compatibility of the $S$-functional calculus for closed operators with intrinsic polynomials in Lemma 3.5.3 and Theorem 3.5.1 or as in the complex case in [165, Lemma 2.3.2] from the residue theorem. In either way, we obtain (7.4).

If $f \in \mathcal{E}_L[\omega]$ is left slice hyperholomorphic both at $0$ and at infinity, then $f(s) = \tilde{f}(s) + (1+s)^{-1}a + b$ where $\tilde{f} \in \mathcal{SH}_{L,0}^\infty(\Sigma_{\varphi'})$ with $\omega < \varphi' < \pi$ is left slice hyperholomorphic both at 0 and infinity and $a, b \in \mathbb{H}$. We therefore have

$$f(\infty)\mathcal{I} + \frac{1}{2\pi} \int_{\partial(U(r,R)\cap\mathbb{C}_j)} S_L^{-1}(s,T)\,ds_j\, f(s)$$

$$= \frac{1}{2\pi} \int_{\partial(U(r,R)\cap\mathbb{C}_j)} S_L^{-1}(s,T)\,ds_j\, \tilde{f}(s)$$

$$+ f(\infty)\mathcal{I} + \frac{1}{2\pi} \int_{\partial(U(r,R)\cap\mathbb{C}_j)} S_L^{-1}(s,T)\,ds_j\, \left((1+s)^{-1}a + b\right).$$

As before, because of the left slice hyperholomorphicity of $\tilde{f}$ at 0 and infinity, Cauchy's integral theorem allows us to vary the values of $r$ and $R$ for sufficiently small $r$ and sufficiently large $R$ without changing the value of the first integral. Letting $r$ tend to 0 and $R$ tend to $\infty$, we find that this integral equals $\tilde{f}(T)$ in the sense of Definition 7.1.5. Since $f(\infty) = b$, the remaining terms, however, equal $(1+T)^{-1}a + \mathcal{I}b$, which can again either be deduced by a standard application of the the the residue theorem and Cauchy's integral theorem as in [165, Corollary 2.3.5] or from the properties of the $S$-functional calculus for closed operators since the function $s \mapsto (1+s)^{-1}a + b$ is left slice hyperholomorphic on the spectrum of $T$ and at infinity. Altogether, we find that (7.5) holds true.

The right slice hyperholomorphic case follows by analogous arguments.  □

**Corollary 7.1.16.** *The $S$-functional calculus for closed operators and the $S$-functional calculus for sectorial operators are compatible.*

*Proof.* Let $T \in \mathrm{Sect}(\omega)$. If $f \in \mathcal{E}_L[\Sigma_\omega]$ is admissible for the $S$-functional calculus for closed operators, then it is left slice hyperholomorphic at infinity such that (7.5) holds true. The set $U(r, R)$ in this representation is however a slice Cauchy domain and therefore admissible as a domain of integration in the $S$-functional calculus for closed operators. Hence, both approaches yield the same operator.  □

Definition 7.1.14 is compatible with the algebraic structures of the underlying function classes.

**Lemma 7.1.17.** *If $T \in \mathrm{Sect}(\omega)$, then the following statements hold true.*

(i) *If $f, g \in \mathcal{E}_L[\Sigma_\omega]$ and $a \in \mathbb{H}$, then $(fa+g)(T) = f(T)a + g(T)$. If $f, g \in \mathcal{E}_R[\Sigma_\omega]$ and $a \in \mathbb{H}$, then $(af + g)(T) = af(T) + g(T)$.*

(ii) *If $f \in \mathcal{E}[\Sigma_\omega]$ and $g \in \mathcal{E}_L[\Sigma_\omega]$, then $(fg)(T) = f(T)g(T)$. If $f \in \mathcal{E}_R[\Sigma_\omega]$ and $g \in \mathcal{E}[\Sigma_\omega]$, then also $(fg)(T) = f(T)g(T)$.*

*Proof.* The compatibility with the respective vector space structure is trivial. In order to show the product rule, consider $f \in \mathcal{E}[\Sigma_\omega]$ and $g \in \mathcal{E}_L[\Sigma_\omega]$ with $f(s) = \tilde{f}(s) + (1+s)^{-1}a + b$ with $\tilde{f} \in \mathcal{SH}_0^\infty[\Sigma_\omega]$ and $a, b \in \mathbb{R}$ and $g(s) = \tilde{g}(s) + (1+s)^{-1}c + d$ with $\tilde{g} \in \mathcal{SH}_{L,0}^\infty[\Sigma_\omega]$ and $c, d \in \mathbb{H}$. By Lemma 7.1.7, Lemma 7.1.9 and the identity

$$(\mathcal{I} + T)^{-2} = (\mathcal{I} + T)^{-1} - T(\mathcal{I} + T)^{-2},$$

we then have

$$
\begin{aligned}
f(T)g(T) &= \tilde{f}(T)\tilde{g}(T) + \tilde{f}(T)(\mathcal{I}+T)^{-1}c + \tilde{f}(T)d + (\mathcal{I}+T)^{-1}\tilde{g}(T)a \\
&\quad + (\mathcal{I}+T)^{-2}ac + (\mathcal{I}+T)^{-1}ad + \tilde{g}(T)b + (\mathcal{I}+T)^{-1}bc + bd\mathcal{I} \\
&= \left( \tilde{f}\tilde{g} + \tilde{f}(1+s)^{-1}c + \tilde{f}d + (1+s)^{-1}\tilde{g}a + \tilde{g}b \right)(T) \\
&\quad - T(\mathcal{I}+T)^{-2}ac + (\mathcal{I}+T)^{-1}(ad + ac + bc) + bd\mathcal{I}.
\end{aligned}
$$

Since $-s(1+s)^{-2} \in \mathcal{E}_L[\Sigma_\omega]$ is left slice hyperholomorphic at zero and infinity, Corollary 7.1.16 and the properties of the $S$-functional calculus imply

$$
(-s(1+s)^2)(T) = -T(\mathcal{I}+T)^{-2}
$$

such that

$$
\begin{aligned}
f(T)g(T) &= \left[ \tilde{f}\tilde{g} + \tilde{f}(1+s)^{-1}c + \tilde{f}d + (1+s)^{-1}\tilde{g}a + \tilde{g}b - s(1+s)^{-2}ac \right](T) \\
&\quad + (\mathcal{I}+T)^{-1}(ad + ac + bc) + bd\mathcal{I} = (fg)(T)
\end{aligned}
$$

since

$$
\begin{aligned}
(fg)(s) &= \tilde{f}(s)\tilde{g}(s) + \tilde{f}(s)(1+s)^{-1}c + \tilde{f}(s)d + (1+s)^{-1}\tilde{g}(s)a \\
&\quad + \tilde{g}(s)b - s(1+s)^{-2}ac + (1+s)^{-1}(ad + ac + bc) + bd.
\end{aligned}
$$

The product rule in the right slice hyperholomorphic case can be shown with analogous arguments. □

**Lemma 7.1.18.** *If $T \in \mathrm{Sect}(\omega)$, then the following statements hold true.*

(i) *We have $(s(1+s)^{-1})(T) = T(\mathcal{I}+T)^{-1}$.*

(ii) *If $A$ is closed and commutes with $\mathcal{Q}_s(T)^{-1}$ and $T\mathcal{Q}_s(T)^{-1}$ for all $s \in \rho_S(T)$, then $A$ commutes with $f(T)$ for any $f \in \mathcal{E}[\Sigma_\omega]$. In particular, $T$ commutes with $f(T)$ for any $f \in \mathcal{E}[\Sigma_\omega]$.*

(iii) *If $y \in \ker(T)$ and $f \in \mathcal{E}_R[\Sigma_\omega]$, then $f(A)y = f(0)y$. In particular, this holds true if $f \in \mathcal{E}[\Sigma_\omega]$.*

*Proof.* The first statement holds as

$$
\left( s(1+s)^{-1} \right)(T) = (1 - (1+s)^{-1})(T) = \mathcal{I} - (\mathcal{I}+T)^{-1} = T(\mathcal{I}+T)^{-1}
$$

and the second one follows from Lemma 7.1.8. Finally, if $y \in \ker(T)$, then

$$
\mathcal{Q}_s(T)y = \left( T^2 - 2s_0 T + |s|^2 \mathcal{I} \right) y = |s|^2 y
$$

and hence

$$
S_R^{-1}(s, T)y = (\bar{s}\mathcal{I} - T)\mathcal{Q}_s(T)^{-1}y = \bar{s}|s|^{-2}y = s^{-1}y.
$$

For $\tilde{f} \in \mathcal{SH}_{R,0}^\infty[\Sigma_\varphi]$, we hence have

$$\tilde{f}(T)y = \frac{1}{2\pi} \int_{\partial(\Sigma_\varphi \cap \mathbb{C}_j)} \tilde{f}(s) \, ds_j \, S_R^{-1}(s,T)y$$

$$= \frac{1}{2\pi} \int_{\partial(\Sigma_\varphi \cap \mathbb{C}_j)} \tilde{f}(s) \, ds_j \, s^{-1}y = 0$$

by Cauchy's integral theorem such that for

$$f(s) = \tilde{f}(s) + a(1+s)^{-1} + b$$

and $y \in \ker(T)$

$$f(T)y = \tilde{f}(T)y + a(\mathcal{I}+T)^{-1}y + b\mathcal{I}y = ay + by = f(0)y. \qquad \square$$

**Remark 7.1.2.** If $f \in \mathcal{E}_L(\Sigma_\omega)$, then we cannot expect (iii) in Lemma 7.1.18 to hold true. In this case

$$\tilde{f}(T)y = \frac{1}{2\pi} \int_{\partial(\Sigma_\varphi \cap \mathbb{C}_j)} S_L^{-1}(s,T) \, ds_j \, f(s)y,$$

but $y$ and $ds_j \, f(s)$ do not commute. So we cannot exploit the fact that $y \in \ker(T)$ to simplify $S_L^{-1}(s,T)y = s^{-1}y$. Indeed, this identity does not necessarily hold true as

$$S_L^{-1}(s,T) = \mathcal{Q}_s(T)^{-1}(\bar{s} - T)y = \mathcal{Q}_s(T)^{-1}\bar{s}y$$

for $y \in \ker(T)$. But the kernel of $T$ is in general not a left linear subspace of $T$ and hence we cannot assume $\bar{s}y \in \ker(T)$. The simplification $\mathcal{Q}_s(T)^{-1}\bar{s}y = |s|^2\bar{s}y = s^{-1}y$ is not possible.

## 7.2   The $H^\infty$-functional calculus

The $H^\infty$-functional calculus for complex linear sectorial operators in [165] applies to meromorphic functions that are regularizable. Properly defining the orders of zeros poles of slice-hyperholomorphic functions is not our goal and goes beyond the scope of this book. Hence we use the following simple definition, which is sufficient for our purposes.

**Definition 7.2.1.** Let $s \in \mathbb{H}$ and let $f$ be left slice hyperholomorphic on an axially symmetric neighborhood $[B_r(s)] \setminus [s]$ of $[s]$ with

$$[B_r(s)] = \{q \in \mathbb{H}: \quad \mathrm{dist}([s],q) < r\}$$

and assume that $f$ does not have a left slice hyperholomorphic continuation to all of $[B_r(s)]$. We say that $f$ has a pole at the sphere $[s]$ if there exists $n \in \mathbb{N}$ such that

$$q \mapsto \mathcal{Q}_s(q)^n f(q)$$

has a left slice hyperholomorphic continuation to $[B_r(s)]$ if $s \notin \mathbb{R}$ (resp. if there exists $n \in \mathbb{N}$ such that $q \mapsto (q - s)^{-n} f(q)$ has a left slice hyperholomorphic continuation to $[B_r(s)]$ if $s \in \mathbb{R}$).

**Remark 7.2.1.** If $[s]$ is a pole of $f$ and $q_n$ is a sequence with $\lim_{n\to+\infty} \mathrm{dist}(q_n, [s]) = 0$, then not necessarily $\lim_{n\to+\infty} |f(q_n)| = +\infty$. One can see this easily if $f$ is restricted to one of the complex planes $\mathbb{C}_j$. If $j, i \in \mathbb{S}$ with $i \perp j$, then the function $f_j := f|_{[B_r(s)] \cap \mathbb{C}_j}$ a meromorphic function with values in the complex (left) vector space $\mathbb{H} \cong \mathbb{C}_j + \mathbb{C}_j i$ over $\mathbb{C}_j$. It must have a pole at $s_j = s_0 + j s_1$ or $\overline{s_j} = s_0 - j s_1$. Otherwise, we could extend $f_j$ to a holomorphic function on $B_r(s) \cap \mathbb{C}_j$. The representation formula would allow us then to define a slice hyperholomorphic extension of $f$ to $B_r(s)$. However, $s_j$ and $\overline{s_j}$ are not necessarily both poles of $f_j$. Consider for instance the function

$$f(q) = S_L^{-1}(s, q) = (q^2 - 2s_0 q + |s|^2)^{-1}(\overline{s} - q),$$

which is defined on $U = \mathbb{H} \setminus [s]$. If we choose $j = j_s$, then $f|_{U \cap \mathbb{C}_j} = (s - q)^{-1}$, which obviously does not have a pole at $\overline{s}$. Hence, if $q_n \in \mathbb{C}_j$ tends to $\overline{s}$, then $|f(q_n)|$ remains bounded.

However, the representation formula implies that there exists at most one complex plane $\mathbb{C}_j$ such that only one of the points $\overline{s_j}$ and $s_j$ is a pole of $f_j$. Otherwise we could use it again to find a slice hyperholomorphic extension of $f$ to $B_r(0)$. For intrinsic functions both points $s_j$ and $\overline{s_j}$ always need to be poles of $f_j$ as in this case $f_j(\overline{q}) = \overline{f_j(q)}$. In general we therefore do not have

$$\lim_{\mathrm{dist}(q,[s]) \to 0} |f(q)| = +\infty,$$

but at least for the limit superior, the equality

$$\limsup_{\mathrm{dist}(q,[s]) \to 0} |f(q)| = +\infty$$

holds. If $f$ is intrinsic, then even $\lim_{\mathrm{dist}(q,[s]) \to 0} |f(q)| = +\infty$ holds true.

**Definition 7.2.2.** Let $U \subset \mathbb{H}$ be axially symmetric. A function $f$ is said to be left meromorphic on $U$ if there exist isolated spheres $[q_n] \subset U$ for $n \in \Theta$, where $\Theta$ is a subset of $\mathbb{N}$, such that $f|_{\tilde{U}} \in \mathcal{SH}_L(\tilde{U})$ with

$$\tilde{U} = U \setminus \bigcup_{n \in \Theta} [q_n]$$

and such that each sphere $[q_n]$ is a pole of $f$. We denote the set of all such functions by $\mathcal{M}_L(U)$ and the set of all such functions that are intrinsic by $\mathcal{M}(U)$.

For $U = \Sigma_\omega$ with $0 < \omega < \pi$, we furthermore denote

$$\mathcal{M}_L[\Sigma_\omega] = \bigcup_{\omega < \varphi < \pi} \mathcal{M}_L(\Sigma_\varphi) \quad \text{and} \quad \mathcal{M}[\Sigma_\omega] = \bigcup_{\omega < \varphi < \pi} \mathcal{M}(\Sigma_\varphi).$$

**Definition 7.2.3.** Let $T \in \text{Sect}(\omega)$.

(i) A left slice hyperholomorphic function $f$ is said to be regularisable if $f \in \mathcal{M}_L(\Sigma_\varphi)$ for some $\omega < \varphi < \pi$ and there exists $e \in \mathcal{E}(\Sigma_\varphi)$ such that $e(T)$ defined in the sense of Definition 7.1.14 is injective and $ef \in \mathcal{E}_L(\Sigma_\varphi)$. In this case we call $e$ a regulariser for $f$.

(ii) We denote the set of all regularisable functions by $\mathcal{M}_L[\Sigma_\omega]_T$. Furthermore, we denote the subset of intrinsic functions in $\mathcal{M}_L[\Sigma_\omega]_T$ by $\mathcal{M}[\Sigma_\omega]_T$.

**Lemma 7.2.4.** Let $T \in \text{Sect}(\omega)$.

(i) If $f, g \in \mathcal{M}_L[\Sigma_\omega]_T$ and $a \in \mathbb{H}$, then $fa + g \in \mathcal{M}_L[\Sigma_\omega]_T$. If furthermore $f \in \mathcal{M}[\Sigma_\omega]_T$, then also $fg \in \mathcal{M}_L[\Sigma_\omega]_T$.

(ii) The space $\mathcal{M}[\Sigma_\omega]_T$ is a real algebra.

*Proof.* If $e_1$ is a regulariser for $f$ and $e_2$ is a regulariser for $g$, then $e = e_1 e_2$ is a regulariser for $fa + g$ and also for $fg$ if $f$ is intrinsic. Hence the statement follows.                                                                              □

**Definition 7.2.5** ($H^\infty$-functional calculus). Let $T \in \text{Sect}(\omega)$. For $f \in \mathcal{M}_L[\Sigma_\omega]_T$, we define

$$f(T) := e(T)^{-1}(ef)(T),$$

where $e(T)^{-1}$ is the closed inverse of $e(T)$ and $(ef)(T)$ is intended in the sense of Definition 7.1.14.

**Theorem 7.2.6.** *The operator* $f(T) := e(T)^{-1}(ef)(T)$ *is independent of the regulariser* $e$ *and hence well-defined.*

*Proof.* If $\tilde{e}$ is a different regulariser, then $e$ and $\tilde{e}$ commute because they both belong to $\mathcal{E}[\Sigma_\omega]$. Hence,

$$\tilde{e}(T)e(T) = (\tilde{e}e)(T) = (e\tilde{e})(T) = e(T)\tilde{e}(T)$$

by Lemma 7.1.17. Inverting this equality yields $e(T)^{-1}\tilde{e}(T)^{-1} = \tilde{e}(T)^{-1}e(T)$ so

$$\begin{aligned}
e(T)^{-1}(ef)(T) &= e(T)^{-1}\tilde{e}(T)^{-1}\tilde{e}(T)(ef)(T) \\
&= e(T)^{-1}\tilde{e}(T)^{-1}(\tilde{e}ef)(T) \\
&= \tilde{e}(T)^{-1}e(T)^{-1}(e\tilde{e}f)(T) \\
&= \tilde{e}(T)^{-1}e(T)^{-1}e(T)(\tilde{e}f)(T) = \tilde{e}(T)^{-1}(\tilde{e}f)(T).
\end{aligned}$$

If $f \in \mathcal{E}_L[\Sigma_\omega]$, then we can use the constant function 1 with $1(T) = \mathcal{I}$ as a regulariser in order to see that Definition 7.2.5 is consistent with Definition 7.1.14.                                                                              □

**Remark 7.2.2.** Since we are considering right linear operators, Definition 7.2.5 is not possible for right slice hyperholomorphic functions. Right slice hyperholomorphic functions maintain slice hyperholomorphicity under multiplication with intrinsic functions from the right. A regulariser of a function $f$ would hence be a function $e$ such that $e(T)$ is injective and $fe \in \mathcal{E}_R(\Sigma_\varphi)$. The operator $f(T)$ would then be defined as $(fe)(T)e(T)^{-1}$, but this operator is only defined on ran $e(T)$ and can hence not be independent of the choice of $e$. If we consider left linear operators, the situation is of course vice versa, which is a common phenomenon in quaternionic operator theory, cf. Remark 7.1.2.

The next lemma shows that the function $f$ needs to have a proper limit behaviour at 0 if $T$ is not injective.

**Lemma 7.2.7.** *Let* $T \in \mathrm{Sect}(\omega)$ *and* $f \in \mathcal{M}_L[\Sigma_\omega]_T$. *If* $T$ *is not injective, then* $f$ *has finite polynomial limit* $f(0) \in \mathbb{H}$ *in* $\Sigma_\omega$ *at* 0. *If furthermore* $f$ *is intrinsic, then* $f(T)y = f(0)y$ *for any* $y \in \ker(T)$.

*Proof.* Assume that $T$ is not injective and let $e$ be a regulariser for $f$. Since $e(T)y = e(0)y$ for all $y \in \ker(T)$ because of (iii) in Lemma 7.1.18, we have $e(0) \neq 0$ as $e(T)$ is injective. The limit

$$e(0)f(0) := \lim_{p \to 0} e(p)f(p)$$

of $e(p)f(p)$ as $p$ tends to 0 in $\Sigma_\omega$ exists and is finite because $ef \in \mathcal{E}_L(\Sigma_\omega)$. Hence, the respective limit of $f(p) = e(p)^{-1}(e(p)f(p))$ exists too and is finite. Indeed, it is

$$f(0) = \lim_{p \to 0} f(p) = e(0)^{-1}(e(0)f(0)).$$

We find that

$$f(p) - f(0) = e(p)^{-1}\left[(e(p)f(p) - e(0)f(0)) - (e(p) - e(0))f(0)\right] = O(|p|^\alpha)$$

as $p$ tends to 0 in $\Sigma_\omega$ because both $ef$ and $e$ have polynomial limit at 0. Hence, $f$ has polynomial limit $f(0)$ at 0 in $\Sigma_\omega$.

If $f$ is intrinsic, then $ef$ is intrinsic too and $e(0)$, $(ef)(0)$ and $f(0)$ are all real. Hence, for any $y \in \ker(T)$, we have $(ef)(0)y = y(ef)(0) \in \ker(T)$. As $\ker(T)$ is a right linear subspace of $X$, we conclude that also $(ef)(0)y \in \ker(T)$ and so (iii) in Lemma 7.1.18 yields

$$f(T)y = e(T)^{-1}(ef)(T)y = e(T)^{-1}(ef)(0)y = e(0)^{-1}(ef)(0)y = f(0)y. \qquad \square$$

**Remark 7.2.3.** If $T$ is injective, then $f$ does not need to have finite polynomial limit at 0 in $\Sigma_\omega$. Indeed, the function $p \mapsto p(1+p)^{-2}$ or the function $p \mapsto p\left(1+p^2\right)^{-1}$ and their powers can then serve as regularisers that may compensate a singularity at 0. Choosing the latter as a specific regulariser yields exactly the approach chosen in [30], where the $H^\infty$-functional calculus was first introduced for quaternionic linear operators.

The proof of the following lemma is analogous to the complex proofs of Proposition 1.2.2 and Corollary 1.2.4 in [165], and does not employ any specific quaternionic techniques. For the convenience of the reader, we nevertheless give the detailed proof as this result turns out to be crucial for what follows.

**Lemma 7.2.8.** *Let $T \in \mathrm{Sect}(\omega)$.*

(i) *If $A \in \mathcal{B}(X)$ commutes with $T$, then $A$ commutes with $f(T)$ for any function $f \in \mathcal{M}[\Sigma_\omega]_T$. Moreover, if $f \in \mathcal{M}[\Sigma_\omega]_T$ and $f(T) \in \mathcal{B}(X)$, then $f(T)$ commutes with $T$.*

(ii) *If $f, g \in \mathcal{M}_L[\Sigma_\omega]_T$, then*

$$f(T) + g(T) \subset (f + g)(T).$$

*If furthermore $f \in \mathcal{M}[\Sigma_\omega]_T$, then*

$$f(T)g(T) \subset (fg)(T)$$

*with $\mathcal{D}(f(T)g(T)) = \mathcal{D}((fg)(T)) \cap \mathcal{D}(g(T))$. In particular, the above inclusion turns into an equality if $g(T) \in \mathcal{B}(X)$.*

(iii) *Let $f \in \mathcal{M}[\Sigma_\omega]_T$ and $g \in \mathcal{M}[\Sigma_\omega]$ be such that $fg \equiv 1$. Then $g \in \mathcal{M}[\Sigma_\omega]_T$ if and only if $f(T)$ is injective. In this case $f(T) = g(T)^{-1}$.*

*Proof.* If $A \in \mathcal{B}(X)$ commutes with $T$, then it commutes with $\mathcal{Q}_s(T)^{-1}$ and $T\mathcal{Q}_s(T)^{-1}$ for any $s \in \rho_S(T)$. Hence, it also commutes with $e(T)$ for any $e \in \mathcal{E}[\Sigma_\omega]$ by Lemma 7.1.18. If $f \in \mathcal{M}[\Sigma_\varphi]_T$ and $e$ is a regulariser for $f$, we thus have

$$\begin{aligned} Af(T) &= Ae(T)^{-1}(ef)(T) \\ &\subset e(T)^{-1}A(ef)(T) \\ &= e(T)^{-1}(ef)(T)A = f(T)A \end{aligned}$$

such that the first assertion in (i) holds true. Because of (i) in Lemma 7.1.18, the function $(1 + p)^{-1}$ regularizes the identity function $p \mapsto p$ and we have $p(T) = T$. Once we have shown (ii), we can hence obtain the second assertion in (i) from

$$f(T)T \subset (f(p)p)(T) = (pf(p))(T) = Tf(T).$$

In order to show (ii) assume that $f, g \in \mathcal{M}_L[\Sigma_\omega]_T$ and let $e_1$ be a regulariser for $f$ and $e_2$ be a regulariser for $g$. Then $e = e_1 e_2$ regularises both $f$ and $g$ and hence also $f + g$ such that

$$\begin{aligned} f(T) + g(T) &= e(T)^{-1}(ef)(T) + e(T)^{-1}(eg)(T) \\ &\subset e(T)^{-1}[(ef)(T) + (eg)(T)] \\ &= e(T)^{-1}(e(f + g))(T) = (f + g)(T). \end{aligned}$$

Applying this relation to the functions $f + g$ and $-g$, we find that

$$(f + g)(T) - g(T) \subset f(T)$$

and so

$$(f + g)(T) = f(T) + g(T)$$

if $g(T)$ is bounded. If even $f \in \mathcal{E}[\Sigma_\omega]_T$, then $f$ and $e_2$ are both intrinsic and hence commute. Thus

$$e(fg) = (e_1 f)(e_2 g) \in \mathcal{E}_L[\Sigma_\omega]_T$$

by Corollary 7.1.11 and so $e$ regularises $fg$. Because of (ii) in Lemma 7.1.18, the operator $(e_1 f)(T)$ commutes with $e_2(T)$ and hence also with the inverse $e_2(T)^{-1}$. Because of Lemma 7.1.17, we thus find that

$$
\begin{aligned}
f(T)g(T) &= e_1(T)^{-1}(e_1 f)(T)e_2(T)^{-1}(e_2 g)(T) \\
&\subset e_1(T)^{-1}e_2(T)^{-1}(e_1 f)(T)(e_2 g)(T) \\
&= [e_2(T)e_1(T)]^{-1}(e_1 f e_2 g)(T) \\
&= e(T)^{-1}(efg)(T) = (fg)(T).
\end{aligned}
$$

In order to prove the statement about the domains, we consider

$$y \in \mathcal{D}((fg)(T)) \cap \mathcal{D}(g(T)).$$

Then $w := (e_2 g)(T)y \in \mathcal{D}\left(e_2(T)^{-1}\right)$. Since $(e_1 f)(T)$ commutes with $e_2(T)^{-1}$, we conclude that also $(e_1 f)(T)w \in \mathcal{D}\left(e_2(T)^{-1}\right)$. Since $y \in \mathcal{D}((fg)(T))$ and

$$(fg)(T)y = e(T)^{-1}(efg)(T)y,$$

we further have $(efg)(T)y \in \mathcal{D}(e(T)^{-1})$. As

$$e(T)^{-1} = e_1(T)^{-1}e_2(T)^{-1}$$

this implies

$$e_2(T)^{-1}(efg)(T)y \in \mathcal{D}(e_1(T)^{-1}).$$

From the identity

$$
\begin{aligned}
(e_1 f)(T)g(T)y &= (e_1 f)(T)e_2(T)^{-1}w \\
&= e_2(T)^{-1}(e_1 f)(T)w = e_2(T)^{-1}(efg)(T)y
\end{aligned}
$$

we conclude that $(e_1 f)(T)g(T)y \in \mathcal{D}\left(e_1(T)^{-1}\right)$. Thus, $g(T)y \in \mathcal{D}(f(T))$ and in turn $y \in \mathcal{D}(f(T)g(T))$. Therefore,

$$\mathcal{D}(f(T)g(T)) \supset \mathcal{D}((fg)(T)) \cap \mathcal{D}(g(T)).$$

The other inclusion is trivial such that altogether we find equality. If $g(T)$ is bounded, then $\mathcal{D}(g(T)) = X$ and we find

$$\mathcal{D}(f(T)g(T)) = \mathcal{D}((fg)(T))$$

such that both operators agree.

We show now the statement (iii) and assume that $f, g \in M[\Sigma_\omega]$ with $fg \equiv 1$ and that $f$ is regularisable. If $g$ is regularisable too, then (iii) implies $g(T)f(T) \subset (gf)(T) = 1(T) = \mathcal{I}$ with

$$\mathcal{D}(g(T)f(T)) = \mathcal{D}(\mathcal{I}) \cap \mathcal{D}(f(T)) = \mathcal{D}(f(T)).$$

Hence, $f(T)$ is injective, and interchanging the role of $f$ and $g$ shows that $f(T)g(T) = \mathcal{I}$ on $\mathcal{D}(g(T))$ such that actually $f(T) = g(T)^{-1}$. Conversely, if $f(T)$ is injective and $e$ is a regulariser for $f$, then

$$(fe)g = e(fg) = e \in \mathcal{E}[\Sigma_\omega]_T.$$

Moreover, $(fe)(T)$ is injective as $f(T)$ and $e(T)$ are both injective and $(fe)(T) = f(T)e(T)$ by (ii). Thus, $fe$ is a regulariser for $g$, i.e., $g \in M[\Sigma_\omega]_T$. $\qquad\square$

Intrinsic polynomials of an operator $T$ are defined as $P[T] = \sum_{k=0}^{n} T^k a_k$ with $\mathcal{D}(P[T]) = \mathcal{D}(T^n)$ for any polynomial $P(q) = \sum_{k=0}^{n} q^k a_k$. We use the squared brackets to indicate that the operator $P[T]$ is defined via this functional calculus and not via the $H^\infty$-functional calculus. However, as the next lemma shows, both approaches are consistent.

**Lemma 7.2.9.** *The $H^\infty$-functional calculus is compatible with intrinsic rational functions. More precisely, if $r(p) = P(p)Q(p)^{-1}$ is an intrinsic rational function with intrinsic polynomials $P$ and $Q$ such that the zeros of $Q$ lie in $\rho_S(T)$, then $r \in M[\Sigma_\omega]_T$ and the operator $r(T)$ is given by $r(T) = P[T]Q[T]^{-1}$.*

*Proof.* We first prove compatibility with intrinsic polynomials. For intrinsic polynomials of degree 1 this follows from the linearity of the $H^\infty$-functional calculus and from (i) in Lemma 7.1.18, which shows that $(1 + p)^{-1}$ regularises the identity function $p \mapsto p$ and that

$$p(T) = \left((1 + p)^{-1}(T)\right)^{-1} (p(1 + p)^{-1})(T) = (\mathcal{I} + T)T(\mathcal{I} + T)^{-1} = T.$$

Let us now generalize the statement by induction and let us assume that it holds for intrinsic polynomials of degree $n$. If $P$ is a polynomial of degree $n + 1$, let us write $P(q) = Q(q)q + a$ with $a \in \mathbb{R}$ and an intrinsic polynomial $Q$ of degree $n$. The induction hypothesis implies that $Q \in M[\Sigma_\omega]_T$, that $Q(T) = Q[T]$, and that $\mathcal{D}(Q(T)) = \mathcal{D}(T^n)$. Since $M[\Sigma_\omega]_T$ is a real algebra, we find that $P$ also belongs to $M[\Sigma_\omega]_T$ and we deduce from (iii) in Lemma 7.2.8 that

$$P(T) \supset Q(T)T + a\mathcal{I} = Q[T]T + a\mathcal{I} = P[T]$$

with
$$\mathcal{D}(P[T]) = \mathcal{D}\left(T^{n+1}\right) = \mathcal{D}(Q(T)T) = \mathcal{D}(P(T)) \cap \mathcal{D}(T).$$
Hence, if we show that $\mathcal{D}(T) \supset \mathcal{D}(P(T))$, the induction is complete. In order to do this, we consider $y \in \mathcal{D}(P(T))$. Then $(\mathcal{I}+T)^{-1}y$ also belongs to $\mathcal{D}(P(T))$ because
$$(\mathcal{I}+T)^{-1}P(T) \subset P(T)(\mathcal{I}+T)^{-1}$$
by (i) in Lemma 7.2.8. But obviously also $(\mathcal{I}+T)^{-1}y \in \mathcal{D}(T)$ and hence
$$(\mathcal{I}+T)^{-1}y \in \mathcal{D}(P(T)) \cap \mathcal{D}(T) = \mathcal{D}\left(T^{n+1}\right),$$
which implies $y \in \mathcal{D}(T^n) \subset \mathcal{D}(T)$. We conclude $\mathcal{D}(T) \supset \mathcal{D}(P(T))$.

Let us now turn to arbitrary intrinsic rational functions. If $s \in \rho_S(T)$ is not real, then $\mathcal{Q}_s(T)$ is injective because $\mathcal{Q}_s(T)^{-1} \in \mathcal{B}(X)$ and hence $\mathcal{Q}_s(p)^{-1} \in \mathcal{M}[\Sigma_\omega]_T$ by (iii) in Lemma 7.2.8. Similarly, if $s \in \rho_S(T)$ is real, then
$$q \mapsto (s-q)^{-1} \in \mathcal{M}[\Sigma_\omega]_T$$
because $(s-q)(T) = (s\mathcal{I} - T)$ is injective as $(s\mathcal{I} - T)^{-1} = S_L^{-1}(s,T) \in \mathcal{B}(X)$. If now $r(q) = P(q)Q(q)^{-1}$ is an intrinsic rational function with poles in $\rho_S(T)$, then we can write $Q(q)$ as the product of such factors, namely
$$Q(q) = \prod_{\ell=1}^{N}(\lambda_\ell - q)^{n_\ell} \prod_{\kappa=1}^{M} \mathcal{Q}_{s_\kappa}(q)^{m_\kappa},$$
where $\lambda_1, \ldots, \lambda_N \in \rho_S(T)$ are the real zeros of $Q$ and $[s_1], \ldots, [s_M] \subset \rho_S(T)$ are the spherical zeros of $Q$ and $n_\ell$ and $m_\kappa$ are the orders of $\lambda_\ell$ (resp $[s_\kappa]$). Since $\mathcal{M}[\Sigma_\omega]_T$ is a real algebra, we conclude that $Q \in \mathcal{M}[\Sigma_\omega]_T$ and because of (iii) we find
$$Q^{-1}(T) = Q(T)^{-1} = Q[T]^{-1}.$$
Moreover, (ii) in Lemma 7.2.8 implies
$$Q^{-1}(T) = \prod_{\ell=1}^{N}(\lambda_\ell \mathcal{I} - T)^{-n_\ell} \prod_{\kappa=1}^{M} \mathcal{Q}_{s_\kappa}(T)^{-m_\kappa} \in \mathcal{B}(X)$$
because each of the factors in this product is bounded. Finally, we deduce from the boundedness of $Q^{-1}(T)$ and (ii) in Lemma 7.2.8 that
$$r(T) = \left(PQ^{-1}\right)(T) = P(T)Q^{-1}(T) = P[T]Q[T]^{-1} = r[T]. \qquad \square$$

## 7.3 The composition rule

Let us now turn our attention to the composition rule, which will occur at several occasions when we consider fractional powers of sectorial operators. As always in the quaternionic setting, we can only expect such a rule to hold true if the inner function is intrinsic since the composition of two slice hyperholomorphic functions is slice hyperholomorphic only if the inner function is intrinsic.

**Theorem 7.3.1** (The Composition Rule). *Let $T \in \mathrm{Sect}(\omega)$ and $g \in \mathcal{M}[\Sigma_\omega]_T$ be such that $g(T) \in \mathrm{Sect}(\omega')$. Furthermore, assume that for any $\varphi' \in (\omega', \pi)$, there exists some $\varphi \in (\omega, \pi)$ such that $g \in \mathcal{M}(\Sigma_\varphi)$ and $g(\Sigma_\varphi) \subset \overline{\Sigma_{\varphi'}}$. Then $f \circ g \in \mathcal{M}[\Sigma_\omega]_T$ for any $f \in \mathcal{M}_L[\Sigma_{\omega'}]_{g(T)}$ and*

$$(f \circ g)(T) = f(g(T)).$$

*Proof.* Let us first assume that $g \equiv c$ is constant. In this case $g(T) = c\mathcal{I}$. Since $g$ is intrinsic, we have $\bar{c} = \overline{g(s)} = g(\bar{s}) = c$ and so $c \in \mathbb{R}$. Since $g$ maps $\Sigma_\varphi$ into $\overline{\Sigma_{\varphi'}}$ for suitable $\varphi \in (\omega, \pi)$ and $\varphi' \in (\omega', \pi)$, we further find

$$c \in \overline{\Sigma_{\varphi'}} \cap \mathbb{R} = [0, +\infty).$$

If $c \neq 0$, then $(f \circ g)(p) \equiv f(c)$ and we deduce easily, for instance from Corollary 7.1.16, that

$$(f \circ g)(T) = f(c)\mathcal{I} = f(g(T)).$$

If on the other hand $c = 0$, then Lemma 7.2.7 implies that $f(0) := \lim_{p \to 0} f(p)$ as $p$ tends to $0$ in $\Sigma_\omega$ exists. Hence $f \circ g$ is well defined. It is the constant function $f \circ g \equiv f(0)$ and so $(f \circ g)(T) = f(0)\mathcal{I}$. If $f$ is intrinsic, then Lemma 7.2.7 implies

$$f(g(T)) = f(0)\mathcal{I} = (f \circ g)(T).$$

If $f$ is not intrinsic, then $f = f_0 + \sum_{\ell=1}^3 f_\ell e_\ell$ with intrinsic components $f_\ell$. Since $\ker g(T) = \ker(0\mathcal{I}) = X$, for any vector $y$, also the vectors $e_\ell y$, $\ell = 1, 2, 3$, belong to $\ker g(T)$, then we conclude, again from Lemma 7.2.7, that

$$f(g(T))y = f_0(g(T))y + \sum_{\ell=1}^3 f_\ell(g(T))e_\ell y$$

$$= f_0(0)y + \sum_{\ell=1}^3 f_\ell(0)e_\ell y$$

$$= \left( f_0(0) + \sum_{\ell=1}^3 f_\ell(0)e_\ell \right) y$$

$$= f(0)y = (f \circ g)(T)y.$$

In the following, we shall thus assume that $g$ is not constant.

Let $\varphi'$ and $\varphi$ be a couple of angles as in the assumptions of the theorem. Since $g$ is intrinsic, $g|_{\mathbb{C}_j \cap \Sigma_\varphi}$ is a non-constant holomorphic function on $\mathbb{C}_j \cap \Sigma_\varphi$. Hence, it maps the open set $g(\Sigma_\varphi \cap \mathbb{C}_j)$ to an open set. The set

$$g(\Sigma_\varphi) = [g(\Sigma_\varphi \cap \mathbb{C}_j)]$$

is therefore also open and so actually contained in $\Sigma_{\varphi'}$, not only in $\overline{\Sigma_{\varphi'}}$. In particular, we find that $f \circ g$ is defined and slice hyperholomorphic on $\Sigma_\varphi$.

We assume for the moment that $f \in \mathcal{E}_L(\Sigma_{\varphi'})$ with $\varphi' \in (\omega', \pi)$ and we choose $\varphi \in (\omega, \pi)$ such that $g \in \mathcal{M}(\Sigma_\varphi)$ and $g(\Sigma_\varphi) \subset \Sigma_{\varphi'}$. Since $f$ is bounded on $\Sigma_{\varphi'}$, the function $f \circ g$ is a bounded function in $\mathcal{SH}_L(\Sigma_\varphi)$. If $T$ is injective, then

$$e(q) = q(1+q)^{-2} \in \mathcal{E}(\Sigma_\varphi)$$

such that $e(T)T(\mathcal{I}+T)^{-2}$ is injective. Moreover, the function $q \mapsto e(q)(f \circ g)(q)$ decays regularly at 0 and infinity in $\Sigma_\varphi$ and hence belongs to $\mathcal{E}_L(\Sigma_\varphi)$. In other words, $e$ is a regulariser for $f \circ g$ and hence

$$f \circ g \in \mathcal{M}_L[\Sigma_\omega]_T.$$

If $T$ is not injective, then $g$ has polynomial limit $g(0)$ at 0 by Lemma 7.2.7. Since $g$ is intrinsic, it only takes real values on the real line and so $g(0) \in \mathbb{R}$. It furthermore maps $\Sigma_\varphi$ to $\Sigma_{\varphi'}$ and so

$$g(0) \in \overline{\Sigma_{\varphi'}} \cap \mathbb{R} = [0, +\infty).$$

Therefore $f$ has polynomial limit at $g(0)$: if $g(0) = 0$ this follows from Corollary 7.1.12, otherwise it follows from the Taylor expansion of $f$ at $g(0) \in (0, \infty)$, cf. Theorem 2.1.12. As a consequence, $f \circ g$ has polynomial limit at 0. Therefore the function

$$q \mapsto (1+q)^{-1}(f \circ g)(q)$$

belongs to $\mathcal{E}_L(\Sigma_\varphi)$. Since $(\mathcal{I}+T)^{-1}$ is injective because $-1 \in \rho_S(T)$, we find that $(1+q)^{-1}$ is a regularizer for $f \circ g$ and hence $f \circ g \in \mathcal{M}_L[\Sigma_\omega]_T$.

We have

$$f(q) = \tilde{f}(q) + (1+q)^{-1}a + b$$

with $\tilde{f} \in \mathcal{SH}^\infty_{L,0}(\Sigma_{\varphi'})$ and $a, b \in \mathbb{H}$. Because of the additivity of the functional calculus, we can treat each of these pieces separately. The case that $f \equiv b$ has already been considered above. For $f(q) = (1+q)^{-1}a$, the identity

$$(f \circ g)(T) = (\mathcal{I} + g(T))^{-1}$$

follows from (iii) in Lemma 7.2.8 because $p \mapsto 1 + g(p)$ and

$$p \mapsto (f \circ g)(p) = (1 + g(p))^{-1}$$

do both belong to $\mathcal{M}_L[\Sigma_\omega]_T$. Hence, let us assume that

$$f = \tilde{f} \in \mathcal{SH}^\infty_{L,0}(\Sigma_{\varphi'})$$

with $\varphi' \in (\omega', \pi)$.

We choose $\theta' \in (\omega', \varphi')$ and $j \in \mathbb{S}$ and set

$$\Gamma_p = \partial(\Sigma_{\theta'} \cap \mathbb{C}_j).$$

We furthermore choose $\rho' \in (\omega', \theta')$ and by our assumptions on $g$, we can find $\varphi \in (\omega, \pi)$ such that $g(\Sigma_\varphi) \subset \Sigma_{\rho'} \subsetneq \Sigma_{\theta'}$. We choose $\theta \in (\omega, \varphi)$ and set $\Gamma_s = \partial(\Sigma_\theta \cap \mathbb{C}_j)$. The subscripts $s$ and $p$ in $\Gamma_s$ and $\Gamma_p$ refer to the corresponding variable of integration in the following computations. For any $p \in \Gamma_p$, the functions

$$s \mapsto \mathcal{Q}_p(g(s))^{-1} = (g(s)^2 - 2p_0 g(s) + |p|^2)^{-1}$$

and

$$s \mapsto S_L^{-1}(p, g(s))$$

do then belong to $\mathcal{E}_L(\Sigma_\varphi)$ and we have

$$\left[\mathcal{Q}_p(g(\cdot))^{-1}\right](T) = \mathcal{Q}_p(g(T))^{-1}$$

and

$$\left[S_L^{-1}(p, g(\cdot))\right](T) = S_L^{-1}(p, g(T)).$$

Indeed, by (ii) in Lemma 7.2.8, we have

$$
\begin{aligned}
\left[\mathcal{Q}_p(g(\cdot))\right](T) &= (g^2 - 2p_0 g + |p|^2)(T) \\
&\supset g(T)^2 - 2p_0 g(T) + |p|^2 \mathcal{I} = \mathcal{Q}_p(g(T)).
\end{aligned}
\tag{7.6}
$$

Taking the closed inverses of these operators, we deduce from (iii) in Lemma 7.2.8 that

$$\left[\mathcal{Q}_p(g(\cdot))^{-1}\right](T) = \left[\mathcal{Q}_p(g(\cdot))\right](T)^{-1} \supset \mathcal{Q}_p(g(T))^{-1}. \tag{7.7}$$

Since $p \in \rho_S(T)$, the operator $\mathcal{Q}_p(g(T))^{-1}$ is bounded and so already defined on all of $X$. Hence, the inclusion $\supset$ in (7.7) and (7.6) is actually an equality and we find

$$\left[\mathcal{Q}_p(g(\cdot))^{-1}\right](T) = \mathcal{Q}_p(g(T))^{-1}.$$

From (ii) we further conclude that also

$$
\begin{aligned}
\left[S_L^{-1}(p, g(\cdot))\right](T) &= \left[\mathcal{Q}_p(g(\cdot))^{-1}\overline{p} - g(\cdot)\mathcal{Q}_p(g(\cdot))^{-1}\right](T) \\
&= \mathcal{Q}_p(g(T))^{-1}\overline{p} - g(T)\mathcal{Q}_p(g(T))^{-1} = S_L^{-1}(p, g(T)).
\end{aligned}
$$

We hence have

$$f(g(T)) = \frac{1}{2\pi}\int_{\Gamma_p} S_L^{-1}(p, g(T))\, dp_j\, f(p) = \frac{1}{2\pi}\int_{\Gamma_p} \left[S_L^{-1}(p, g(\cdot))\right](T)\, dp_j\, f(p).$$

Let us first assume that $T$ is injective. Since $f$ and in turn also $f \circ g$ are bounded, we can use $e(q) = q(\mathcal{I} + q)^{-2}$ as a regulariser for $f \circ g$. As $e$ decays regularly at 0 and infinity, also the functions $s \mapsto e(s)S_L^{-1}(p, g(s))$ decays regularly at 0 and infinity for any $p \in \Gamma_p$. Hence it belongs to $\mathcal{SH}_{L,0}^\infty(\Sigma_\varphi)$ and so

$$
\begin{aligned}
f(g(T)) &= e(T)^{-1} e(T) f(g(T)) \\
&= e(T)^{-1} \frac{1}{2\pi} \int_{\Gamma_p} e(T) S_L^{-1}(p, g(T)) \, dp_j \, f(p) \\
&= e(T)^{-1} \frac{1}{2\pi} \int_{\Gamma_p} \left[ e(\cdot) S_L^{-1}(p, g(\cdot)) \right](T) \, dp_j f(p) \\
&= e(T)^{-1} \frac{1}{(2\pi)^2} \int_{\Gamma_p} \left( \int_{\Gamma_s} S_L^{-1}(s, T) \, ds_j \, s(1+s)^{-2} S_L^{-1}(p, g(s)) \right) dp_j f(p).
\end{aligned}
\tag{7.8}
$$

We can now apply Fubini's theorem in order to exchange the order of integration: estimating the resolvent using (7.1), we find that the integrand in the above integral is bounded by the function

$$
F(s,p) := C_\theta \left| p S_L^{-1}(p, g(s)) \right| \frac{1}{|1+s|^2} \frac{|f(p)|}{|p|}.
\tag{7.9}
$$

Since $p$, $s$ and $g(s)$ belong to the same complex plane as $g$ is intrinsic, we have due to (2.26) that

$$
\left| p S_L^{-1}(p, g(s)) \right| \leq \max_{\tilde{s} \in [s]} \frac{|p|}{|p - g(\tilde{s})|} = \max \left\{ \frac{1}{|1 - p^{-1} g(s)|}, \frac{1}{|1 - p^{-1} g(\bar{s})|} \right\}.
\tag{7.10}
$$

Since $g(\Gamma_s) \subset \Sigma_{\rho'} \cap \mathbb{C}_j \subsetneq \Sigma_{\theta'} \cap \mathbb{C}_j$ and $\Gamma_p = \partial(\Sigma_{\theta'} \cap \mathbb{C}_j)$, these expressions are bounded by a constant depending on $\theta'$ and $\rho'$ but neither on $p$ nor on $s$. Hence, $\left| p S_L^{-1}(p, g(s)) \right|$ is uniformly bounded on $\Gamma_s \times \Gamma_p$ and $F(s,p)$ is in turn integrable on $\Gamma_p \times \Gamma_s$ because $f$ has polynomial limit 0 both at 0 and infinity.

After exchanging the order of integration in (7.8), we deduce from Cauchy's integral formula that

$$
\begin{aligned}
f(g(T)) &= e(T)^{-1} \frac{1}{(2\pi)^2} \int_{\Gamma_s} S_L^{-1}(s, T) \, ds_j \, s(1+s)^{-2} \left( \int_{\Gamma_p} S_L^{-1}(p, g(s)) \, dp_j f(p) \right) \\
&= e(T)^{-1} \frac{1}{2\pi} \int_{\Gamma_s} S_L^{-1}(s, T) \, ds_j \, e(s) f(g(s)) \\
&= e(T)^{-1} e(T) (f \circ g)(T) = (f \circ g)(T).
\end{aligned}
$$

Let us now consider the case that $T$ is not injective. By Lemma 7.2.7, the function $g$ has then finite polynomial limit $g(0) \in \mathbb{R}$ in $\Sigma_\varphi$ and hence the function $\tilde{g}(p) = g(p) - g(0) \in \mathcal{M}(\Sigma_\varphi)_T$ has finite polynomial limit 0 in at 0. Let us choose a regulariser $e$ for $\tilde{g}$ with polynomial limit 0 at infinity. (This is always possible: if $\tilde{e}$ is an arbitrary regulariser for $\tilde{g}$, we can choose for instance $e(s) = (1+s)^{-1} \tilde{e}(s)$.) We have then $e \tilde{g} \in \mathcal{SH}_{L,0}^\infty(\Sigma_\varphi)$. Since $g(0)$ is real, we have $S_L^{-1}(p, g(0)) = (p - g(0))^{-1}$.

Moreover $g(s)$ and $\mathcal{Q}_p(g(s))^{-1}$ commute for any $s \in \Gamma_s$. For $p \notin \overline{\Sigma_{\rho'}}$ we find thus

$$
\begin{aligned}
e(s)&S_L^{-1}(p, g(s)) - e(s)S_L^{-1}(p, g(0)) \\
&= e(s)\mathcal{Q}_p(g(s))^{-1}\left[(\bar{p} - g(s))(p - g(0)) - \mathcal{Q}_p(g(s))\right](p - g(0))^{-1} \\
&= e(s)\mathcal{Q}_p(g(s))^{-1}\Big[(\bar{p} - g(s))p - g(0)(\bar{p} - g(s)) \\
&\qquad\qquad + g(s)(\bar{p} - g(s)) - (\bar{p} - g(s))p\Big](p - g(0))^{-1} \\
&= e(s)(g(s) - g(0))S_L^{-1}(p, g(s))(p - g(0))^{-1} \\
&= e(s)\tilde{g}(s)S_L^{-1}(p, g(s))S_L^{-1}(p, g(0)).
\end{aligned}
\tag{7.11}
$$

Hence, $e$ regularises also the function $s \mapsto S_L^{-1}(p, g(s)) - S_L^{-1}(p, g(0))$ and the function $e(\cdot)\left(S_L^{-1}(p, g(\cdot)) - S_L^{-1}(p, g(0))\right)$ does even belong to $\mathcal{SH}^\infty_{L,0}(\Sigma_\varphi)$. We thus have

$$
\begin{aligned}
f(g(T)) &= e(T)^{-1}e(T)f(g(T)) \\
&= e(T)^{-1}\frac{1}{2\pi}\int_{\Gamma_p} e(T)S_L^{-1}(p, g(T))\,dp_j\,f(p) \\
&= e(T)^{-1}\frac{1}{2\pi}\int_{\Gamma_p} \left[e(\cdot)S_L^{-1}(p, g(\cdot))\right](T)\,dp_j\,f(p) \\
&= e(T)^{-1}\frac{1}{2\pi}\int_{\Gamma_p} \left[e(\cdot)\tilde{g}(\cdot)S_L^{-1}(p, g(\cdot))S_L^{-1}(p, g(0))\right](T)\,dp_j f(p) \\
&\quad + e(T)^{-1}\frac{1}{2\pi}\int_{\Gamma_p} e(T)S_L^{-1}(p, g(0))\,dp_j f(p).
\end{aligned}
$$

For the second integral, Cauchy's integral formula yields

$$
\begin{aligned}
e(T)^{-1}&\frac{1}{2\pi}\int_{\Gamma_p} e(T)S_L^{-1}(p, g(0))\,dp_j f(p) \\
&= e(T)^{-1}e(T)f(g(0)) = f(g(0))\mathcal{I},
\end{aligned}
$$

as $f$ decays regularly at infinity in $\Sigma_\theta$. For the first integral, we have

$$
\begin{aligned}
e(T)^{-1}&\frac{1}{2\pi}\int_{\Gamma_p} \left[e(\cdot)\tilde{g}(\cdot)S_L^{-1}(p, g(\cdot))S_L^{-1}(p, g(0))\right](T)\,dp_j f(p) \\
&= e(T)^{-1}\frac{1}{(2\pi)^2}\int_{\Gamma_p}\left(\int_{\Gamma_s} S_L^{-1}(s, T)\,ds_j\,e(s)\tilde{g}(s)\right. \\
&\qquad\qquad\qquad\qquad \left. \cdot S_L^{-1}(p, g(s))S_L^{-1}(p, g(0))\right)dp_j f(p)
\end{aligned}
$$

$$\stackrel{(A)}{=} e(T)^{-1}\frac{1}{(2\pi)^2}\int_{\Gamma_s} S_L^{-1}(s,T)\,ds_j$$

$$\cdot\left(\int_{\Gamma_p} e(s)\tilde{g}(s)S_L^{-1}(p,g(s))S_L^{-1}(p,g(0))\,dp_j f(p)\right)$$

$$\stackrel{(B)}{=} e(T)^{-1}\frac{1}{(2\pi)^2}\int_{\Gamma_s} S_L^{-1}(s,T)\,ds_j\,e(s)$$

$$\cdot\left(\int_{\Gamma_p} S_L^{-1}(p,g(s)) - S_L^{-1}(p,g(0))\,dp_j f(p)\right)$$

$$\stackrel{(C)}{=} e(T)^{-1}\frac{1}{2\pi}\int_{\Gamma_s} S_L^{-1}(s,T)\,ds_j\,(e(s)f(g(s)) - f(g(0)))$$

$$= e(T)^{-1}(e(T)f\circ g(T) - e(T)f(g(0))\mathcal{I}) = f\circ g(T) - f(g(0))\mathcal{I}, \qquad (7.12)$$

where the identity $(A)$ follows from Fubini's theorem, the identity $(B)$ follows from (7.11) and the identity $(C)$ finally follows from Cauchy's integral formula. Altogether, we have

$$f(g(T)) = f\circ g(T) - f(g(0))\mathcal{I} + f(g(0))\mathcal{I} = f\circ g(T).$$

In order to justify the application of Fubini's theorem in $(A)$, we observe that the integrand is bounded by the function

$$F(s,p) = C_\theta\left|pS_L^{-1}(p,g(s))\right|\frac{|e(s)\tilde{g}(s)|}{|s|}\frac{|f(p)|}{|p|}\frac{1}{|p-g(0)|},$$

where we used (7.1) in order to estimate the $S$-resolvent $S_L^{-1}(s,T)$.

If $g(0) \neq 0$, then $|p - g(0)|^{-1}$ is uniformly bounded in $p$. Just as before, $\left|pS_L^{-1}(p,g(s))\right|$ is uniformly bounded on $\Gamma_s \times \Gamma_p$. Since $\tilde{g}$ decays regularly at 0, since $e$ decays regularly at infinity and since $f$ decays regularly both at 0 and infinity, the function $F$ is hence integrable on $\Gamma_s \times \Gamma_p$ and we can apply Fubini's theorem.

If on the other hand $g(0) = 0$, then $g = \tilde{g}$ and we can write

$$F(s,p) = C_\theta\left|S_L^{-1}(p,g(s))\right|\frac{|e(s)\tilde{g}(s)|}{|s|}\frac{|f(p)|}{|p|}$$

$$= C_\theta\left|p^\alpha S_L^{-1}(p,g(s))g(s)^{1-\alpha}\right|\frac{|e(s)g(s)^\alpha|}{|s|}\frac{|f(p)|}{|p|^{1+\alpha}}, \qquad (7.13)$$

with $\alpha \in (0,1)$ such that $|f(p)|/|p|^{1+\alpha}$ is integrable. This is possible because $f$ decays regularly at 0. Just as in (7.10), we can estimate the first factor in (7.13) by

$$\left|p^\alpha S_L^{-1}(p,g(s))g(s)^{1-\alpha}\right|$$

$$\leq \max\left\{\frac{|g(s)|^{1-\alpha}}{|p|^{1-\alpha}}\frac{1}{|1-p^{-1}g(s)|}, \frac{|g(\bar{s})|^{1-\alpha}}{|p|^{1-\alpha}}\frac{1}{|1-p^{-1}g(\bar{s})|}\right\},$$

where we applied $|g(s)| = \left|\overline{g(\bar{s})}\right| = |g(\bar{s})|$ because $g$ is intrinsic. As before, this expression is uniformly bounded on $\Gamma_s \times \Gamma_p$ because $g(\Gamma_s) \subset \Sigma_{\rho'} \cap \mathbb{C}_j$. Hence, $F$ is again integrable and it is actually possible to apply Fubini's theorem.

Altogether, we have shown that $f(g(T)) = (f \circ g)(T)$ for any $f \in \mathcal{E}_L[\Sigma_{\omega'}]$. Finally, we consider a general function $f \in \mathcal{M}_L[\Sigma_{\omega'}]_{g(T)}$ that does not necessarily belong to $\mathcal{E}_L[\Sigma_{\omega'}]$. If $e$ is a regulariser for $f$, then $e$ and $ef$ both belong to $\mathcal{E}_L[\omega']$. By what we have just shown, we have $e_g := e \circ g \in \mathcal{M}[\Sigma_\omega]_T$ and $(ef)_g := (ef) \circ g \in \mathcal{M}_L[\Sigma_\omega]_T$ with $e_g(T) = e(g(T))$ and $(ef)_g(T) = (ef)(g(T))$.

Let $\tau_1$ and $\tau_2$ be regularizers for $e_g$ and $(ef)_g$. Then $\tau = \tau_1 \tau_2$ regularizes both of them and hence

$$e_g(T) = \tau^{-1}(T)(\tau e_g)(T).$$

Since $e_g(T) = (e \circ g)(T) = e(g(T))$ is injective because $e$ is a regulariser for $f$, the operator $(\tau e_g)(T)$ is injective too. Moreover, for $f_g := f \circ g$, we find $(\tau e_g)f_g = \tau(e_g f_g) = \tau(ef)_g \in \mathcal{E}_L[\omega]$ because $\tau$ was chosen to regularize both $e_g$ and $(ef)_g$. Therefore, $\tau e_g$ is a regulariser for $f_g$ and hence $f_g \in \mathcal{M}_L[\Sigma_\omega]_T$. Finally, we deduce from Lemma 7.2.8 that

$$\begin{aligned}
f(g(T)) &= e(g(T))^{-1}(ef)(g(T)) = (e_g)(T)^{-1}((ef)_g)(T) \\
&= (e_g)(T)^{-1}\tau(T)^{-1}\tau(T)((ef)_g)(T) \\
&= (\tau e_g)(T)^{-1}((\tau e)_g f_g)(T) = f_g(T) = (f \circ g)(T). \qquad \square
\end{aligned}$$

**Corollary 7.3.2.** *Let $T \in \mathrm{Sect}(\omega)$ be injective and let $f \in \mathcal{M}_L[\Sigma_\omega]$. Then we have $f \in \mathcal{M}_L[\Sigma_\omega]_T$ if and only if $p \mapsto f(p^{-1}) \in \mathcal{M}_L[\Sigma_\omega]_{T^{-1}}$ and in this case*

$$f(T) = f(p^{-1})(T^{-1}).$$

*Proof.* Since $T$ is injective, the function $p^{-1}$ belongs to $\mathcal{M}[\Sigma_\omega]_T$ and the statement follows from Theorem 7.3.1. $\qquad \square$

## 7.4   Extensions according to spectral conditions

As in the complex case, cf. [165, Section 2.5], one can extend the $H^\infty$-functional calculus for sectorial operators to a larger class of functions if the operator satisfies additional spectral conditions. We shall mention the following three cases, which are relevant in the proof of the spectral mapping theorem in Section 7.5. In order to explain them, we introduce the notation

$$\Sigma_{\varphi,r,R} = (\Sigma_\varphi \cap B_R(0)) \setminus B_r(0)$$

for $0 \le r < R \le \infty$. (We set $B_\infty(0) = \mathbb{H}$ for $R = \infty$.)

(i) If the operator $T \in \text{Sect}(\omega)$ has a bounded inverse, then $B_\varepsilon(0) \subset \rho_S(T)$ for sufficiently small $\varepsilon > 0$. We can thus define the class

$$\mathcal{E}_L^\infty(\Sigma_\varphi) = \{f = \tilde{f} + a \in \mathcal{SH}_L(\Sigma_\varphi) : \quad a \in \mathbb{H}, \ \tilde{f} \in \mathcal{SH}_L(\Sigma_\varphi) \ \text{dec. reg. at } \infty\},$$

and $\mathcal{E}^\infty(\Sigma_\varphi)$ as the set of all intrinsic functions in $\mathcal{E}_L^\infty(\Sigma_\varphi)$, where dec. reg. is short for decays regularly. For any function $f \in \mathcal{E}_L^\infty(\Sigma_\varphi)$ with $\varphi > 0$, we can define $f(T)$ as

$$f(T) = \frac{1}{2\pi} \int_{\partial(\Sigma_{\varphi,r,\infty} \cap \mathbb{C}_j)} S_L^{-1}(s,T) \, ds_j \, f(s) + a\mathcal{I},$$

with $0 < r < \varepsilon$ arbitrary. It follows as in Lemma 7.1.15 from Cauchy's integral theorem that this approach is consistent with the usual one if $f \in \mathcal{E}_L(\Sigma_\varphi)$, but the class of admissible functions $\mathcal{E}_L^\infty(\Sigma_\varphi)$ is now larger. We can further extend this functional calculus by calling a function $e \in \mathcal{E}_L^\infty(\Sigma_\varphi)$ a regulariser for a function $f \in \mathcal{M}_L(\Sigma_\varphi)$, if $e(T)$ is injective and $ef \in \mathcal{E}_L^\infty(\Sigma_\varphi)$. In this case, we define

$$f(T) = e(T)^{-1}(ef)(T).$$

Clearly, all the results shown so far still hold for this extended functional calculus since the respective proofs can be carried out in this setting with marginal and obvious modifications. Only in the case of the composition rule we have to consider several conditions, just as in the complex case, namely the combinations

  a) $T$ is sectorial and $g(T)$ is invertible and sectorial,

  b) $T$ is invertible and sectorial and $g(T)$ is sectorial,

  c) $T$ and $g(T)$ are both invertible and sectorial.

In a) and c), one needs the additional assumption $0 \notin \overline{g(\Sigma_\omega)}$ on the function $g$.

(ii) If the operator $T \in \text{Sect}(\omega)$ is bounded, then $\mathbb{H} \backslash B_\rho(0) \subset \rho_S(T)$ for sufficiently large $\rho > 0$. We can thus define the class

$$\mathcal{E}_L^0(\Sigma_\varphi) = \{f = \tilde{f} + a \in \mathcal{SH}_L(\Sigma_\varphi) : \quad a \in \mathbb{H}, \tilde{f} \in \mathcal{SH}_L(\Sigma_\varphi) \ \text{dec. reg. at } 0\}$$

and $\mathcal{E}^0(\Sigma_\varphi)$ as the set of all intrinsic functions in $\mathcal{E}_L^0(\Sigma_\varphi)$. For any function $f \in \mathcal{E}_L^0(\Sigma_\varphi)$ with $\varphi > 0$, we can define $f(T)$ as

$$f(T) = \frac{1}{2\pi} \int_{\partial(\Sigma_{\varphi,0,R} \cap \mathbb{C}_j)} S_L^{-1}(s,T) \, ds_j \, f(s) + a\mathcal{I},$$

with $0 < \rho < R$ arbitrary. As before, this approach is consistent with the usual one if $f \in \mathcal{E}_L(\Sigma_\varphi)$, but the class of admissible functions $\mathcal{E}_L^0(\Sigma_\varphi)$ is again larger than $\mathcal{E}_L(\Sigma_\varphi)$. We can further extend this functional calculus by calling $e \in \mathcal{E}_L{}^0(\Sigma_\varphi)$ a regulariser for $f \in \mathcal{M}_L(\Sigma_\varphi)$, if $e(T)$ is injective and $ef \in \mathcal{E}_L{}^0(\Sigma_\varphi)$ and define again $f(T) = e(T)^{-1}(ef)(T)$ for such $f$.

As before, all results shown so far hold for this extended functional calculus because the respective proofs can be carried out in this setting with marginal and obvious modifications. For the composition rule, we have to consider again several cases and distinguish the following situations:

a) $T$ is sectorial and $g(T)$ is bounded and sectorial,

b) $T$ is invertible and sectorial and $g(T)$ is bounded and sectorial,

c) $T$ and $g(T)$ are both bounded and sectorial,

d) $T$ is bounded and sectorial and $g(T)$ is sectorial,

e) $T$ is bounded and sectorial and $g(T)$ is invertible and sectorial.

In the cases a), b) and c), one needs the additional assumption $\infty \notin \overline{g(\Sigma_\omega)}^{\mathbb{H}_\infty}$ and in the case e) one needs the additional assumption $0 \notin \overline{g(\Sigma_\omega)}$ on the function $g$.

(iii) If finally $T \in \text{Sect}(\omega)$ is bounded and has a bounded inverse, then we can set $\mathcal{E}_L{}^{0,\infty}(\Sigma_\varphi) = \mathcal{SH}_L(\Sigma_\varphi)$ and $\mathcal{E}^{0,\infty}(\Sigma_\varphi)$ and define for such functions

$$f(T) = \frac{1}{2\pi} \int_{\partial(\Sigma_{\varphi,r,R} \cap \mathbb{C}_j)} S_L^{-1}(s,T) \, ds_j \, f(s)$$

for sufficiently small $r$ and sufficiently large $R$. Choosing regularizers in $\mathcal{E}^{0,\infty}(\Sigma_\varphi)$ gives again an extension of the $H^\infty$-functional calculus and of the two extended functional calculi presented in (i) and (ii). All the results presented so far still hold for this extended functional calculus, where the composition rule can be shown again under suitable conditions on the function $g$.

## 7.5   The spectral mapping theorem

Let us now show the spectral mapping theorem for the $H^\infty$-functional calculus. We point out that a substantial technical difficulty appears here that does not occur in the classical situation: the proof of the spectral mapping theorem in the complex setting makes use of the fact that $f(T|_{X_\sigma}) = f(T)|_{X_\sigma}$ if $\sigma$ is a spectral set and $X_\sigma$ is the invariant subspace associated with $\sigma$. However, subspaces that are invariant under right linear operators are in general only right linear subspaces, but not necessarily left linear subspaces. Hence, they are not two-sided Banach spaces and we cannot define $f(T|_{X_\sigma})$ with the techniques presented in this book because the $S$-functional calculus as defined in Chapter 3 requires the operator to act on a two-sided Banach space. The $S$-resolvents can otherwise not be defined.

Instead of using the properties of the $S$-functional calculus for $T|_{X_\sigma}$, we thus have to find a workaround and prove several steps directly, which is essentially done in Lemma 7.5.5.

We start with two technical lemmas that are necessary in order to show the spectral inclusion theorem.

**Lemma 7.5.1.** *Let $T \in \mathrm{Sect}(\omega)$ and let $s \in \mathbb{H}$. If $\mathcal{Q}_s(T)$ is injective and there exist $e \in \mathcal{M}[\Sigma_\omega]_T$ and $c \in \mathbb{H}$, $c \neq 0$ such that*

$$f(q) := \mathcal{Q}_c(e(q))\,\mathcal{Q}_s(q)^{-1} \in \mathcal{M}[\Sigma_\omega]_T$$

*and such that $e(T)$ and $f(T)$ are bounded, then $e(T)\mathcal{Q}_s(T)^{-1} = \mathcal{Q}_s(T)^{-1}e(T)$.*

*Proof.* By assumption, the operator $\mathcal{Q}_s(T)$ is injective and hence (iii) in Lemma 7.2.8 implies that $\mathcal{Q}_s^{-1} \in \mathcal{M}[\omega]_T$. Since $e(T)$ is bounded, it commutes with $T$ and so also with $\mathcal{Q}_s(T)^{-1}$. We thus have

$$e(T)\mathcal{Q}_s(T)^{-1} \subset \mathcal{Q}_s(T)^{-1}e(T).$$

In order to show that this relation is actually an equality, it is sufficient to show that $y \in \mathcal{D}\left(\mathcal{Q}_s(T)^{-1}\right)$ for any $y \in X$ with $e(T)y \in \mathcal{D}(\mathcal{Q}_s(T)^{-1})$. This is indeed the case: if $e(T)y$ belongs to $\mathcal{D}(\mathcal{Q}_s(T)^{-1})$, then there exists $x \in \mathcal{D}(\mathcal{Q}_s(T))$ with $e(T)y = \mathcal{Q}_s(T)x$. Hence,

$$\begin{aligned}
\mathcal{Q}_c(e(T))y &= e(T)^2 y - 2c_0 e(T)y + |c|^2 y \\
&= e(T)\mathcal{Q}_s(T)x - 2c_0\mathcal{Q}_s(T)x + |c|^2 y \qquad (7.14) \\
&= \mathcal{Q}_s(T)(e(T)x - 2c_0 x) + |c|^2 y,
\end{aligned}$$

where the last identity follows again from (i) in Lemma 7.2.8 because $e(T)$ is bounded and commutes with $T$ and in turn also with $\mathcal{Q}_s(T)$. Since $f(T) \in \mathcal{B}(X)$, we conclude on the other hand from (ii) of Lemma 7.2.8 that

$$\mathcal{Q}_c(e(T)) = \mathcal{Q}_s(T)\left[\mathcal{Q}_c(e(\cdot))\mathcal{Q}_s(\cdot)^{-1}\right](T) = \mathcal{Q}_s(T)f(T).$$

Due to (7.14), we then find

$$\begin{aligned}
y &= \frac{1}{|c|^2}\left(\mathcal{Q}_c(e(T))y - \mathcal{Q}_s(T)(e(T)x - 2c_0 x)\right) \\
&= \mathcal{Q}_s(T)\frac{1}{|c|^2}\left(f(T)y - e(T)x + 2c_0 x\right).
\end{aligned}$$

Hence, $y$ belongs to $\mathcal{D}(\mathcal{Q}_s(T)^{-1})$ and the statement follows. $\qquad\square$

**Lemma 7.5.2.** *Let $T \in \mathrm{Sect}(\omega)$ and let $f \in \mathcal{M}_L[\Sigma_\omega]_T$. For any $s \in \overline{\Sigma_\omega} \setminus \{0\}$ there exists a regulariser $e$ for $f$ with $e(s) \neq 0$.*

*Proof.* Let $\tilde{e}$ be an arbitrary regulariser of $f$ such that $\tilde{e} \in \mathcal{E}[\Sigma_\omega]$, $\tilde{e}f \in \mathcal{E}_L[\Sigma_\omega]$ and $\tilde{e}(T)$ is injective. If $\tilde{e}(s) \neq 0$, then we can set $e = \tilde{e}$ and we are done. Otherwise, recall that $[s]$ is a spherical zero of $\tilde{e}$ and that its order is a finite number $n \in \mathbb{N}$ since $e \neq 0$ as $e(T)$ is injective. We define now $e(q) := \mathcal{Q}_s^{-n}(q)e(q)$ with $\mathcal{Q}_s(q) = q^2 - 2s_0 q + |s|^2$. Then $e \in \mathcal{E}[\Sigma_\omega]$ with $e(s) \neq 0$ and $ef = \mathcal{Q}_s^{-n}\tilde{e}f \in \mathcal{E}_L[\Sigma_\omega]$. Furthermore, by (ii) in Lemma 7.2.8, we have $\tilde{e}(T) = \mathcal{Q}_s(T)e(T)$. Since $\tilde{e}(T)$ is injective, we deduce that also $e(T)$ is injective. Hence, $e$ is a regulariser for $f$ with $e(s) \neq 0$.                                                                                      □

**Lemma 7.5.3.** *Let $T \in \mathrm{Sect}(\omega)$ and let $s \in \overline{\Sigma_\omega}$ with $s \neq 0$. If $f(T)$ has a bounded inverse for some $f \in \mathcal{M}[\Sigma_\omega]_T$ with $f(s) = 0$, then $s \in \rho_S(T)$.*

*Proof.* Let $f$ be as above and let us first show that $\mathcal{Q}_s(T) = T^2 - 2s_0 T + |s|^2 I$ is injective and hence invertible as a closed operator. By Lemma 7.5.2, there exists a regulariser $e$ for $f$ with $c := e(s) \neq 0$. We have $ef \in \mathcal{E}[\Sigma_\omega]$ with $(ef)(s) = 0$. Since all zeros of intrinsic functions are spherical zeros, we find that also $h = ef\mathcal{Q}_s^{-1} = \mathcal{Q}_s^{-1}ef \in \mathcal{E}[\Sigma_\omega]$. The product rule (ii) in Lemma 7.2.8 implies therefore

$$h(T)\mathcal{Q}_s(T) \subset (h\mathcal{Q}_s)(T) = (ef)(T) = (fe)(T) = f(T)e(T),$$

where $ef = fe$ because both functions are intrinsic. Since $e(T)$ and $f(T)$ are both injective, we find that $\mathcal{Q}_s(T)$ is injective. Moreover, $e$ is also a regulariser for $\mathcal{Q}_s^{-1}f$. Now observe that the function

$$g(q) := \mathcal{Q}_c(e(q))\mathcal{Q}_s(q)^{-1} = (e(q)^2 - 2c_0 e(q) + |c|^2)(q^2 - 2s_0 q + |s|^2)^{-1}$$

belongs to $\mathcal{E}[\Sigma_\omega]$. Indeed, by Corollary 7.1.11, the space $\mathcal{E}[\Sigma_\omega]$ is a real algebra such $\mathcal{Q}_c(e(q)) = e(q)^2 - 2c_0 e(q) + |c|^2$ belongs to it as $e$ does. The function $\mathcal{Q}_c(e(q))$ however has a spherical zero at $s$ because $e(s) = c$ such that $g(q) = \mathcal{Q}_c(e(q))\mathcal{Q}_s^{-1}(q)$ is bounded and hence belongs to $\mathcal{E}[\Sigma_\omega]$ by Corollary 7.1.12. In particular, this implies that $g(T)$ is bounded.

We deduce from Lemma 7.5.1 that $e(T)\mathcal{Q}_s(T)^{-1} = \mathcal{Q}_s(T)^{-1}e(T)$ and inverting both sides of this equation yields $\mathcal{Q}_s(T)e(T)^{-1} = e(T)^{-1}\mathcal{Q}_s(T)$. The product rule in (ii) of Lemma 7.2.8, the boundedness of $h(T) = (e\mathcal{Q}_s^{-1}f)(T)$ and the fact that $\mathcal{Q}_s^{-1}$ and $e$ commute because both are intrinsic functions imply

$$\begin{aligned}
f(T) &= e(T)^{-1}(ef)(T) \\
&= e(T)^{-1}\left(\mathcal{Q}_s e\mathcal{Q}_s^{-1}f\right)(T) \\
&= e(T)^{-1}\mathcal{Q}_s(T)\left(e\mathcal{Q}_s^{-1}f\right)(T) \\
&= \mathcal{Q}_s(T)e(T)^{-1}(e\mathcal{Q}_s^{-1}f)(T) \\
&= \mathcal{Q}_s(T)(\mathcal{Q}_s^{-1}f)(T).
\end{aligned}$$

Since $f(T)$ is surjective, we find that $\mathcal{Q}_s(T)$ is surjective too. Hence, $\mathcal{Q}_s(T)^{-1}$ is an everywhere defined closed operator and thus bounded by the closed graph theorem. Consequently $s \in \rho_S(T)$.                                                            □

**Proposition 7.5.4.** *If $T \in \mathrm{Sect}(\omega)$ and $f \in \mathcal{M}[\Sigma_\omega]_T$, then*

$$f(\sigma_S(T) \setminus \{0\}) \subset \sigma_{SX}(f(T)).$$

*Proof.* Let $s \in \sigma_S(T) \setminus \{0\}$ and set $c := f(s)$. If $c \neq \infty$, then Lemma 7.5.3 implies that

$$\mathcal{Q}_c(f(T))^2 = f(T)^2 - 2c_0 f(T) + |c|^2 \mathcal{I}$$

does not have a bounded inverse because $g = f^2 - 2c_0 f + |c|^2$ belongs to $\mathcal{M}[\Sigma_\omega]_T$ and satisfies $g(c) = 0$. Hence, $c = f(s)$ belongs to $\sigma_S(f(T))$ for $s \in \sigma_S(T) \setminus \{0\}$ with $f(s) \neq \infty$.

If on the other hand $c = \infty$, then suppose that $c \notin \sigma_{SX}(f(T))$, i.e., that $f(T)$ is bounded. In this case there exists $p \in \mathbb{H}$ such that $\mathcal{Q}_p(f(T))$ has a bounded inverse. By (iii) in Lemma 7.2.8, this implies $g(q) = \mathcal{Q}_p(f(q))^{-1} \in \mathcal{M}[\Sigma_\omega]_T$. The operator $g(T)$ is invertible as $g(T)^{-1} = \mathcal{Q}_p(f(T))$ belongs to $\mathcal{B}(X)$ because $f(T)$ is bounded. Moreover, since $g(s) = 0$ as $f(s) = \infty$, another application of Lemma 7.5.3 yields $s \in \rho_S(T)$. But this contradicts our assumption $s \in \sigma_S(T) \setminus \{0\}$. Hence, we must have $c \in \sigma_{SX}(f(T))$. $\qquad\square$

We have so far shown the spectral inclusion theorem for spectral values not equal to 0 nor $\infty$. These two values need a special treatment. They also need additional assumptions on the function $f$ for a spectral inclusion theorem to hold as we shall see in the following. (The assumptions presented here might, however, not be the most general ones that are possible, cf. [165].)

First, we have to show a technical lemma. We recall that if $\sigma \subset \sigma_{SX}(T)$ is a spectral set, then $E_\sigma := \chi_\sigma(T)$ is by Theorem 3.7.8 a projection that commutes with $T$, i.e., it is a projection onto a right-linear subspace of $X$ that is invariant under $T$. If $\infty \notin \sigma$, then we can choose a bounded slice Cauchy domain $U_\sigma \subset \mathbb{H}$ such that $\sigma \subset U_\sigma$ and such that $(\sigma_S(T) \setminus \sigma) \cap U_\sigma = \emptyset$. The projection $E_\sigma$ is then given by

$$E_\sigma = \frac{1}{2\pi} \int_{\partial(U_\sigma \cap \mathbb{C}_j)} ds_j \, S_R^{-1}(s, T) = \frac{1}{2\pi} \int_{\partial(U_\sigma \cap \mathbb{C}_j)} S_L^{-1}(p, T) \, dp_j. \qquad (7.15)$$

If on the other hand $\infty \in \sigma$, then we can choose an unbounded slice Cauchy domain $U_\sigma \subset \mathbb{H}$ such that $\sigma \subset U_\sigma$ and such that $(\sigma_S(T) \setminus \sigma) \cap U_\sigma = \emptyset$. The projection $E_\sigma$ is then given by

$$E_\sigma = \mathcal{I} + \frac{1}{2\pi} \int_{\partial(U_\sigma \cap \mathbb{C}_j)} ds_j \, S_R^{-1}(s, T) = \mathcal{I} + \frac{1}{2\pi} \int_{\partial(U_\sigma \cap \mathbb{C}_j)} S_L^{-1}(p, T) \, dp_j.$$

**Lemma 7.5.5.** *Let $T \in \mathrm{Sect}(\omega)$ be unbounded and assume that $\sigma_S(T)$ is bounded. Furthermore, let $E_\infty$ be the spectral projection onto the invariant subspace associated with $\infty$. If $f \in \mathcal{M}[\Sigma_\omega]_T$ has polynomial limit 0 at infinity, then*

$$\{f(T)\}_\infty := f(T) E_\infty$$

*is a bounded operator that is given by the slice hyperholomorphic Cauchy integral*

$$\{f(T)\}_\infty = \int_{\partial(\Sigma_\varphi \backslash B_r(0)) \cap \mathbb{C}_j} f(s)\, ds_j\, S_R^{-1}(s,T), \qquad (7.16)$$

*where $B_r(0)$ is the ball centered at 0 with $r > 0$ sufficiently large such that it contains $\sigma_S(T)$ and any singularity of $f$. Moreover, for two such functions, we have*

$$\{f(T)\}_\infty \{g(T)\}_\infty = \{(fg)(T)\}_\infty. \qquad (7.17)$$

*Proof.* Let us first assume that $f \in \mathcal{E}[\Sigma_\omega]$, i.e., $f \in \mathcal{E}(\Sigma_\varphi)$ with $\omega < \varphi < \pi$. Since $f$ decays regularly at infinity, it is of the form $f(s) = \tilde{f}(s) + a(1+s)^{-1}$ with $a \in \mathbb{R}$ and $f \in \mathcal{SH}_0^\infty(\Sigma_\varphi)$. The operator $\tilde{f}(T)$ is given by the slice hyperholomorphic Cauchy integral

$$\tilde{f}(T) = \frac{1}{2\pi} \int_{\partial(\Sigma_{\varphi'} \cap \mathbb{C}_j)} \tilde{f}(s)\, ds_j\, S_R^{-1}(s,T) \qquad (7.18)$$

with $j \in \mathbb{S}$ and $\varphi' \in (\omega, \varphi)$. Let now $r_1 < r_2$ be such that $\sigma_S(T) \subset B_r(0)$. Cauchy's integral theorem allows us to replace the path of integration in (7.18) by the union of $\Gamma_{s,1} = \partial(\Sigma_{\varphi'} \cap B_{r_1}(0)) \cap \mathbb{C}_j$ and $\Gamma_{s,2} = \partial(\Sigma_{\varphi'} \backslash B_{r_2}(0)) \cap \mathbb{C}_j$ such that

$$\tilde{f}(T) = \frac{1}{2\pi} \int_{\Gamma_{s,1}} \tilde{f}(s)\, ds_j\, S_R^{-1}(s,T) + \frac{1}{2\pi} \int_{\Gamma_{s,2}} \tilde{f}(s)\, ds_j\, S_R^{-1}(s,T). \qquad (7.19)$$

Let us choose $R \in (r_1, r_2)$. Since $\sigma_{SX}(T) = \sigma_S(T) \cup \{\infty\}$, we have $E_\infty = \mathcal{I} - E_{\sigma_S(T)}$ and the spectral projection $E_{\sigma_S(T)}$ is given by the slice hyperholomorphic Cauchy integral (7.15) along $\Gamma_p = \partial(B_R(0) \cap \mathbb{C}_j)$. The subscripts $s$ and $p$ in $\Gamma_{s,1}$, $\Gamma_{s,2}$ and $\Gamma_p$ are chosen in order to indicate the corresponding variable of integration in the following computation.

If we write the operators $\tilde{f}(T)$ and $E_{\sigma_S(T)}$ in terms of the slice hyperholomorphic Cauchy integrals defined above, we find that

$$\begin{aligned}
\tilde{f}(T) E_{\sigma_s(T)} &= \frac{1}{2\pi} \int_{\Gamma_{s,1}} \tilde{f}(s)\, ds_j\, S_R^{-1}(s,T) \frac{1}{2\pi} \int_{\Gamma_p} S_L^{-1}(p,T)\, dp_j \\
&\quad + \frac{1}{2\pi} \int_{\Gamma_{s,2}} \tilde{f}(s)\, ds_j\, S_R^{-1}(s,T) \frac{1}{2\pi} \int_{\Gamma_p} S_L^{-1}(p,T)\, dp_j.
\end{aligned} \qquad (7.20)$$

If we apply the $S$-resolvent equation in the first integral, which we denote by $\Psi_1$

for neatness, we find

$$
\begin{aligned}
\Psi_1 = {} & \frac{1}{(2\pi)^2} \int_{\Gamma_{s,1}} \tilde{f}(s)\, ds_j\, S_R^{-1}(s,T) \int_{\Gamma_p} p\left(p^2 - 2s_0 p + |s|^2\right)^{-1} dp_j \\
& - \frac{1}{(2\pi)^2} \int_{\Gamma_{s,1}} \tilde{f}(s)\, ds_j\, \bar{s} S_R^{-1}(s,T) \int_{\Gamma_p} \left(p^2 - 2s_0 p + |s|^2\right)^{-1} dp_j \\
& - \frac{1}{(2\pi)^2} \int_{\Gamma_p} \left( \int_{\Gamma_{s,1}} \tilde{f}(s)\, ds_j \left(S_L^{-1}(p,T) p \right.\right. \\
& \qquad\qquad \left.\left. - \bar{s} S_L^{-1}(p,T)\right) \left(p^2 - 2s_0 p + |s|^2\right)^{-1}\right) dp_j.
\end{aligned}
\tag{7.21}
$$

For $s \in \Gamma_s$, the functions

$$
p \mapsto \left(p^2 - 2s_0 p + |s|^2\right)^{-1} \quad \text{and} \quad p \mapsto p\left(p^2 - 2s_0 p + |s|^2\right)^{-1}
$$

are rational functions on $\mathbb{C}_j$ that have two singularities, namely $s = s_0 + j s_1$ and $\bar{s} = s_0 - j s_1$. Since we chose $r_1 < R$, these singularities lie inside of $B_R(0)$ for any $s \in \Gamma_s$. As $\Gamma_p = \partial(B_R(0) \cap \mathbb{C}_j)$, the residue theorem yields

$$
\begin{aligned}
& \frac{1}{2\pi} \int_{\Gamma_p} p\left(p^2 - 2s_0 p + |s|^2\right)^{-1} dp_j \\
& = \lim_{\mathbb{C}_j \ni p \to s} p(p - \bar{s})^{-1} + \lim_{\mathbb{C}_j \ni p \to \bar{s}} p(p - s)^{-1} = 1
\end{aligned}
$$

and

$$
\begin{aligned}
& \frac{1}{2\pi} \int_{\Gamma_p} \left(p^2 - 2s_0 p + |s|^2\right)^{-1} dp_j \\
& = \lim_{\mathbb{C}_j \ni p \to s} (p - \bar{s})^{-1} + \lim_{\mathbb{C}_j \ni p \to \bar{s}} (p - \bar{s})^{-1} = 0,
\end{aligned}
$$

where $\lim_{\mathbb{C}_j \ni p \to s} \tilde{f}(p)$ denotes the limit of $\tilde{f}(p)$ as $p$ tends to $s$ in $\mathbb{C}_j$. If we apply the identity (2.49) with $B = S_L^{-1}(p,T)$ in the third integral in (7.21), it turns into

$$
\begin{aligned}
& \frac{1}{(2\pi)^2} \int_{\Gamma_p} \left( \int_{\Gamma_{s,1}} \tilde{f}(s)\, ds_j \left(s^2 - 2p_0 s + |p|^2\right)^{-1} s S_L^{-1}(p,T)\right) dp_j \\
& - \frac{1}{(2\pi)^2} \int_{\Gamma_p} \left( \int_{\Gamma_{s,1}} \tilde{f}(s)\, ds_j \left(s^2 - 2p_0 s + |p|^2\right)^{-1} S_L^{-1}(p,T) \bar{p}\right) dp_j = 0.
\end{aligned}
$$

The last identity follows from Cauchy's integral theorem because $\tilde{f}(s)$ is right slice hyperholomorphic and the functions $s \mapsto (s^2 - 2p_0 s + |p|^2)^{-1} S_L^{-1}(p,T)$ and

$s \mapsto s(s^2 - 2p_0 s + |p|^2)^{-1} S_L^{-1}(p, T)$ are left slice hyperholomorphic on $\Sigma_{\varphi'} \cap B_{r_1}(0)$ for any $p \in \Gamma_p$ as we chose $R > r_1$. Hence, we find

$$\Psi_1 = \frac{1}{2\pi} \int_{\Gamma_{s,1}} \tilde{f}(s) \, ds_j \, S_R^{-1}(p, T).$$

The second integral in (7.20), which we denote by $\Psi_2$ for neatness, turns after an application of the $S$-resolvent equation into

$$
\begin{aligned}
\Psi_2 = {} & \frac{1}{(2\pi)^2} \int_{\Gamma_{s,2}} \tilde{f}(s) \, ds_j \, S_R^{-1}(s, T) \int_{\Gamma_p} p \left( p^2 - 2s_0 p + |s|^2 \right)^{-1} dp_j \\
& - \frac{1}{(2\pi)^2} \int_{\Gamma_{s,2}} \tilde{f}(s) \, ds_j \, \bar{s} S_R^{-1}(s, T) \int_{\Gamma_p} \left( p^2 - 2s_0 p + |s|^2 \right)^{-1} dp_j \\
& - \frac{1}{(2\pi)^2} \int_{\Gamma_{s,2}} \left( \int_{\Gamma_p} \tilde{f}(s) \, ds_j \, (S_L^{-1}(p, T)p - \right. \\
& \qquad \left. - \bar{s} S_L^{-1}(p, T)) \left( p^2 - 2s_0 p + |s|^2 \right)^{-1} \right) dp_j.
\end{aligned}
\tag{7.22}
$$

Since we chose $R < r_2$, the singularities of $p \mapsto \left( p^2 - 2s_0 p + |s|^2 \right)^{-1}$ and $p \mapsto p \left( p^2 - 2s_0 p + |s|^2 \right)^{-1}$ lie outside of $\overline{B_R(0)}$ for any $s \in \Gamma_{s,2}$. Hence, these functions are right slice hyperholomorphic on $\overline{B_R(0)}$ and so Cauchy's integral theorem implies that the first two integrals in (7.22) vanish. Since $\tilde{f}$ decays regularly at infinity, since (7.1) holds true and since $\Gamma_p$ is a path of finite length, we can apply Fubini's theorem and exchange the order of integration in the third integral of (7.22). After applying the identity (2.49), we find

$$
\begin{aligned}
\Psi_2 = {} & \frac{1}{(2\pi)^2} \int_{\Gamma_p} \left( \int_{\Gamma_{s,2}} \tilde{f}(s) ds_j \left( s^2 - 2p_0 s + |p|^2 \right)^{-1} \right. \\
& \qquad \left. \cdot \left( s S_L^{-1}(p, T) - S_L^{-1}(p, T) \bar{p} \right) \right) dp_j.
\end{aligned}
$$

However, this integral also vanishes: as $f$ decays regularly at infinity, the integrand decays sufficiently fast so that we can use Cauchy's integral theorem to transform the path of integration and write

$$
\begin{aligned}
& \int_{\Gamma_{s,2}} \tilde{f}(s) ds_j \left( s^2 - 2p_0 s + |p|^2 \right)^{-1} \left( s S_L^{-1}(p, T) - S_L^{-1}(p, T) \bar{p} \right) \\
& = \lim_{\rho \to +\infty} \int_{\partial(U_\rho \cap \mathbb{C}_j)} \tilde{f}(s) \, ds_j \left( s^2 - 2p_0 s + |p|^2 \right)^{-1} \left( s S_L^{-1}(p, T) - S_L^{-1}(p, T) \bar{p} \right) = 0
\end{aligned}
$$

where $U_\rho = (\Sigma_\varphi \setminus B_{r_2(0)}) \cap B_\rho(0)$ for $\rho > r_2$. The last identity follows again from Cauchy's integral theorem because the singularities $p$ and $\bar{p}$ of the functions

$s \mapsto (s^2 - 2p_0 s + |p|^2)^{-1}$ and $s \mapsto (s^2 - 2p_0 s + |p|^2)^{-1} s$ lie outside of $\overline{U_\rho}$ because we chose $R < r_2$.

Putting these pieces together, we find that

$$\tilde{f}(T) E_{\sigma_S(T)} = \frac{1}{2\pi} \int_{\Gamma_{s,1}} \tilde{f}(p) \, dp_j \, S_R^{-1}(p, T). \tag{7.23}$$

We therefore deduce from (7.19) and $E_\infty = \mathcal{I} - E_{\sigma_S(T)}$ that

$$\tilde{f}(T) E_\infty = \tilde{f}(T) - \tilde{f}(T) E_{\sigma_S(T)} = \frac{1}{2\pi} \int_{\Gamma_{s,2}} \tilde{f}(p) \, dp_j \, S_R^{-1}(p, T). \tag{7.24}$$

Let us now consider the operator $a(\mathcal{I} + T)^{-1}$. Since it is slice hyperholomorphic on $\sigma_S(T)$ and at infinity, it is admissible for the $S$-fuctional calculus. If we set $\chi_{\{\infty\}}(s) := \chi_{\mathbb{H} \setminus U_R(0)}$—that is $\chi_{\{\infty\}}(s) = 1$ if $s \notin U_R(0)$ and $\chi_{\{\infty\}}(s) = 0$ if $s \in U_R(0)$—then $\chi_{\{\infty\}}(T) = E_\infty$ via the $S$-functional calculus. The product rule of the $S$-functional calculus yields $a(\mathcal{I} + T)^{-1} E_\infty = g(T)$ with $g(s) = a(1 + s) \chi_{\{\infty\}}(s)$. If we set

$$U_{\rho,1} := (\Sigma_\varphi \setminus B_{r_2}(0)) \cup (\mathbb{H} \setminus B_\rho(0)) \quad \text{and} \quad U_{\rho,2} = (\Sigma_\varphi \cap B_{r_1}(0)) \cup B_\varepsilon(0)$$

with $0 < \varepsilon < 1$ sufficiently small, then $U_\rho = U_{\rho,1} \cup U_{\rho,2}$ is an unbounded slice Cauchy domain that contains $\sigma_S(T)$ and such that $g$ is slice hyperholomorphic on $\overline{U_\rho}$. Hence,

$$a(\mathcal{I} + T)^{-1} E_\infty = g(\infty) \mathcal{I} + \frac{1}{2\pi} \int_{\partial(U_\rho \cap \mathbb{C}_j)} dp_j \, S_R^{-1}(p, T)$$

$$= \frac{1}{2\pi} \int_{\partial(U_{\rho,1} \cap \mathbb{C}_j)} a(1 + s) \, dp_j \, S_R^{-1}(p, T)$$

and letting $\rho$ tend to infinity, we finally find

$$a(\mathcal{I} + T)^{-1} E_\infty = \frac{1}{2\pi} \int_{\Gamma_{s,2}} a(1 + s) \, dp_j \, S_R^{-1}(p, T). \tag{7.25}$$

Adding (7.24) and (7.25), we find that (7.16) holds true for $f \in \mathcal{E}[\Sigma_\omega]$.

Now let $f$ be an arbitrary function in $\mathcal{M}[\Sigma_\omega]_T$ that decays regularly at infinity and let $e$ be a regulariser for $f$. We can assume that $e$ decays regularly at infinity; otherwise, we can replace $e$ by $s \mapsto (1 + s)^{-1} e(s)$, which is a regulariser for $f$ with this property. We expect that

$$\begin{aligned} f(T) E_\infty &= e^{-1}(T)(ef)(T) E_\infty \\ &= e^{-1}(T) \{(ef)(T)\}_\infty \\ &\overset{(*)}{=} e^{-1}(T) \{e(T)\}_\infty \{f(T)\}_\infty \\ &= e^{-1}(T) e(T) E_\infty \{f(T)\}_\infty \\ &= E_\infty \{f(T)\}_\infty \overset{(**)}{=} \{f(T)\}_\infty \end{aligned} \tag{7.26}$$

such that (7.16) holds true. Then, the boundedness of $f(T)E_\infty$ also follows from the boundedness of the integral $\{f(T)\}_\infty$. The second and the fourth of the above equalities follow from the above arguments since $ef$ and $e$ both belong to $\mathcal{E}[\Sigma_\varphi]$ and decay regularly at infinity. The equalities marked with $(*)$ and $(**)$ however remain to be shown.

Let $\omega < \varphi_2 < \varphi_1 < \varphi$ be such that $e, f \in \mathcal{E}(\Sigma_\varphi)$ and let $r_1 < r_2$ be such that $B_{r_1}(0)$ contains $\sigma_S(T)$ and any singularity of $f$. We set $U_s = \Sigma_{\varphi_1} \backslash B_{r_1}(0)$ and $U_p = \Sigma_{\varphi_2} \backslash B_{r_2}(0)$, where the subscripts $s$ and $p$ indicate again the respective variable of integration in the following computation. An application of the $S$-resolvent equation shows then that, using the notation $\mathcal{Q}_s(p)^{-1} = (p^2 - 2s_0 p + |s|^2)^{-1}$,

$$
\begin{aligned}
&\{e(T)\}_\infty \{f(T)\}_\infty \\
&= \frac{1}{2\pi} \int_{\partial(U_s \cap \mathbb{C}_j)} e(s)\, ds_j\, S_R^{-1}(s,T) \frac{1}{2\pi} \int_{\partial(U_p \cap \mathbb{C}_j)} S_L^{-1}(p,T)\, dp_j\, f(p) \\
&= \frac{1}{(2\pi)^2} \int_{\partial(U_s \cap \mathbb{C}_j)} e(s)\, ds_j S_R^{-1}(s,T) \int_{\partial(U_p \cap \mathbb{C}_j)} p\mathcal{Q}_s(p)^{-1}\, dp_j f(p) \\
&\quad + \frac{1}{(2\pi)^2} \int_{\partial(U_s \cap \mathbb{C}_j)} e(s)\, ds_j S_R^{-1}(s,T) \int_{\partial(U_p \cap \mathbb{C}_j)} \mathcal{Q}_s(p)^{-1}\, dp_j f(p) \\
&\quad + \frac{1}{(2\pi)^2} \int_{\partial(U_s \cap \mathbb{C}_j)} e(s)\, ds_j\, \left(\bar{s} S_L^{-1}(p,T) - S_L^{-1}(p,T)p\right) \\
&\qquad\qquad \cdot \int_{\partial(U_p \cap \mathbb{C}_j)} \mathcal{Q}_s(p)^{-1}\, dp_j f(p).
\end{aligned}
$$

Because of our choice of $U_s$ and $U_p$, the singularities of $p \mapsto (p^2 - s_0 p + |s|^2)^{-1}$ lie outside $U_p$ for any $s \in \partial(U_s \cap \mathbb{C}_j)$ such that $p \mapsto (p^2 - 2s_0 p + |s|)^{-1}$ and $p \mapsto p(p^2 - 2s_0 p + |s|)^{-1}$ are right slice hyperholomorphic on $\overline{U_p}$ for any such $s$. Since $f$ also decays regularly in $U_p$ at infinity, Cauchy's integral theorem implies that the first two of the above integrals equal zero. The fact that $e$ and $f$ decay polynomially at infinity allows us to exchange the order of integration in the third integral, such that

$$
\begin{aligned}
\{e(T)\}_\infty \{f(T)\}_\infty = \frac{1}{(2\pi)^2} \int_{\partial(U_p \cap \mathbb{C}_j)} &\left[ \int_{\partial(U_s \cap \mathbb{C}_j)} e(s)\, ds_j \right. \\
&\left. \cdot \left(\bar{s} S_L^{-1}(p,T) - S_L^{-1}(p,T)p\right) \mathcal{Q}_s(p)^{-1} \right] dp_j f(p).
\end{aligned}
$$

If $p \in \partial(U_p \cap \mathbb{C}_j)$, then $p$ lies for sufficiently large $\rho$ in the bounded axially symmetric Cauchy domain $U_{s,\rho} = U_s \cap B_\rho(0)$. Since $f$ is an intrinsic function on $\overline{U_{s,\rho}}$,

Lemma 2.2.24 implies

$$\frac{1}{2\pi}\int_{\partial(U_s\cap\mathbb{C}_j)} e(s)\,ds_j\left(\bar{s}S_L^{-1}(p,T)-S_L^{-1}(p,T)p\right)Q_s(p)^{-1}$$

$$=\lim_{\rho\to\infty}\frac{1}{2\pi}\int_{\partial(U_s\cap B_\rho(0)\cap\mathbb{C}_j)} e(s)\,ds_j\left(\bar{s}S_L^{-1}(p,T)-S_L^{-1}(p,T)p\right)Q_s(p)^{-1}$$

$$=S_L^{-1}(p,T)e(p).$$

Recalling the equivalence of right and left slice hyperholomorphic Cauchy integrals for intrinsic functions, cf. Remark 3.4.2, we finally find that

$$\{e(T)\}_\infty\{f(T)\}_\infty=\frac{1}{2\pi}\int_{\partial(U_p\cap\mathbb{C}_j)} S_L^{-1}(p,T)\,dp_j e(p)f(p)=\{(ef)(T)\}_\infty.$$

Hence, the identity $(*)$ in (7.26) is true.

Similar arguments show that also $(**)$ holds true. We choose $0<R<r$ such that $B_R(0)$ contains $\sigma_S(T)$ and all singularities of $f(T)$ and we choose $\omega<\varphi'<\varphi$ such that $f\in\mathcal{E}(\Sigma_{\varphi'})$ and set $U_p:=\Sigma_{\varphi'}\setminus B_r(0)$. An application of the $S$-resolvent equation shows that

$$E_{\sigma_s(T)}\{f(T)\}_\infty$$

$$=\frac{1}{2\pi}\int_{\partial(B_R(0)\cap\mathbb{C}_j)} ds_j\,S_R^{-1}(s,T)\frac{1}{2\pi}\int_{\partial(U_p\cap\mathbb{C}_j)} S_L^{-1}(p,T)\,dp_j\,f(p)$$

$$=\frac{1}{(2\pi)^2}\int_{\partial(B_R(0)\cap\mathbb{C}_j)} ds_j\,S_R^{-1}(s,T)\int_{\partial(U_p\cap\mathbb{C}_j)} pQ_s(p)^{-1}\,dp_j\,f(p)$$

$$-\frac{1}{(2\pi)^2}\int_{\partial(B_R(0)\cap\mathbb{C}_j)} ds_j\,\bar{s}S_R^{-1}(s,T)\int_{\partial(U_p\cap\mathbb{C}_j)} Q_s(p)^{-1}\,dp_j\,f(p)$$

$$+\frac{1}{(2\pi)^2}\int_{\partial(B_R(0)\cap\mathbb{C}_j)}\left[\int_{\partial(U_p\cap\mathbb{C}_j)} ds_j\right.$$

$$\left.\cdot\left(\bar{s}S_L^{-1}(p,T)-S_L^{-1}(p,T)p\right)Q_s(p)^{-1}\right]dp_j\,f(p).$$

Again, the first two integrals vanish as a consequence of Cauchy's integral theorem because the poles of the function $p\mapsto(p^2-2s_0p+|s|^2)^{-1}$ lie outside of $\overline{U}_p$ for any $s\in\partial(B_R(0)\cap\mathbb{C}_j)$ and $f$ decays regularly at infinity. Because of (7.1) and the regular decay of $f$ at infinity, we can, however, apply Fubini's theorem to exchange the order of integration in the third integral and find

$$E_{\sigma_S(T)}\{f(T)\}_\infty=\frac{1}{(2\pi)^2}\int_{\partial(U_p\cap\mathbb{C}_j)}\left[\int_{\partial(B_R(0)\cap\mathbb{C}_j)} ds_j\right.$$

$$\left.Q_s(p)^{-1}\left(sS_L^{-1}(p,T)-S_L^{-1}(p,T)\bar{p}\right)\right]dp_j\,f(p).$$

As the functions $s \mapsto (s^2 - 2p_0 s + |p|^2)^{-1}$ and $s \mapsto (s^2 - 2p_0 s + |p|^2)^{-1} s$ are right slice hyperholomorphic on $\overline{B_R(0)}$ for any $p \in \partial(U_p \cap \mathbb{C}_j)$, this integral also vanishes due to Cauchy's integral theorem. Consequently, the identity $(**)$ in (7.26) holds also true as

$$E_\infty \{f(T)\}_\infty = \{f(T)\}_\infty - E_\sigma \{f(T)\}_\infty = \{f(T)\}_\infty.$$

Finally, we point out that the above computations, which proved that

$$\{(ef)(T)\}_\infty = \{e(T)\}_\infty \{f(T)\}_\infty,$$

did not require that $e \in \mathcal{E}[\Sigma_\omega]$. They also work if $e$ belongs to $\mathcal{M}[\Sigma_\omega]_T$ and decays regularly at infinity. Hence the same calculations show that (7.17) holds true.   $\square$

**Theorem 7.5.6.** *Let* $T \in \mathrm{Sect}(\omega)$ *and* $s \in \{0, \infty\}$. *If* $f \in \mathcal{M}[\Sigma_\omega]_T$ *has polynomial limit* $c$ *at* $s$ *and* $s \in \sigma_{SX}(T)$, *then* $c \in \sigma_{SX}(f(T))$.

*Proof.* If $c \neq \infty$, then $c \in \mathbb{R}$ because, as an intrinsic function, $f$ takes only real values on the real line. We can hence consider the function $f - c$ instead of $f$ because

$$\sigma_{SX}(f(T)) = \sigma_{SX}(f(T) - c\mathcal{I}) + c$$

so that it is sufficient to consider the cases $c = 0$ or $c = \infty$.

Let us start with the case $c = 0$ and $s = \infty$. If $\infty \in \overline{\sigma_S(T) \setminus \{0\}}^{\mathbb{H}_\infty}$, then

$$0 \in \overline{f(\sigma_S(T) \setminus \{0\})}^{\mathbb{H}_\infty} \subset \sigma_{SX}(f(T))$$

because $f(\sigma_S(T) \setminus \{0\}) \subset \sigma_{SX}(f(T))$ by Proposition 7.5.4 and the latter is a closed subset of $\mathbb{H}_\infty$. In case $\infty \notin \overline{\sigma_S(T)}^{\mathbb{H}_\infty}$, we show that $0 \notin \sigma_{SX}(f(T))$ implies that $T$ is bounded so that even $\infty \notin \sigma_{SX}(T)$. Let us hence assume that $\infty \notin \overline{\sigma_S(T)}^{\mathbb{H}_\infty}$ and that $0 \notin \sigma_{SX}(f(T))$. In this case, there exists $R > 0$ such that $\sigma_S(T)$ is contained in the open ball $B_R(0)$ of radius $R$ centered at zero. The integral

$$E_{\sigma_S(T)} := \frac{1}{2\pi} \int_{\partial(B_R(0) \cap \mathbb{C}_j)} ds_j \, S_R^{-1}(s, T)$$

defines then a bounded projection that commutes with $T$, namely the spectral projection associated with the spectral set $\sigma_S(T) \subset \sigma_{SX}(T)$ that is obtained from the $S$-functional calculus. The compatibility of the $S$-functional calculus with polynomials in $T$ moreover implies

$$T E_{\sigma_S(T)} = (s \chi_{\sigma_S(T)})(T) = \frac{1}{2\pi} \int_{\partial(B_R(0) \cap \mathbb{C}_j)} s \, ds_j \, S_R^{-1}(s, T) \in \mathcal{B}(X),$$

where $\chi_{\sigma_S(T)}(s)$ denotes the characteristic function of an arbitrary axially symmetric bounded set that contains $\overline{B_R(0)}$.

Set $E_\infty := \mathcal{I} - E_{\sigma_S(T)}$ and let $X_\infty := E_\infty X$ be the range of $E_\infty$. Since $T$ commutes with $E_\infty$, the operator $T_\infty := T|_{X_\infty}$ is a closed operator on $X_\infty$ with domain $\mathcal{D}(T_\infty) = \mathcal{D}(T) \cap X_\infty$. Moreover, we conclude from the properties of the projections that

$$\sigma_{SX}(T_\infty) = \sigma_{SX}(T) \setminus \sigma_S(T) \subset \{\infty\}$$

and so in particular

$$\sigma_S(T_\infty) = \sigma_{SX}(T_\infty) \setminus \{\infty\} = \emptyset. \tag{7.27}$$

Now observe that $f(T)$ commutes with $E_\infty$ because of (i) in Lemma 7.2.8. Hence, $f(T)$ leaves $X_\infty$ invariant and $f(T)_\infty := f(T)|_{X_\infty}$ defines a closed operator on $X_\infty$ with domain $\mathcal{D}(f(T)_\infty) = \mathcal{D}(f(T)) \cap X_\infty$. (Note that $f(T)_\infty$ intuitively corresponds to $f(T_\infty)$. The $S$-functional calculus as introduced in this book is, however, only defined on two-sided Banach spaces. As $X_\infty$ is only a right-linear subspace of $X$ and hence not a two-sided Banach space, we can not define the operator $f(T_\infty)$, cf. the remark at the beginning of Section 7.5.) Since $f(T)$ is invertible because we assumed $0 \notin \sigma_{SX} f(T)$, the operator $f(T)_\infty$ is invertible too and its inverse is $f(T)^{-1}|_{X_\infty} \in \mathcal{B}(X_\infty)$.

Our goal is now to show that $T_\infty$ is bounded. Since any bounded operator on a nontrivial Banach space has non-empty $S$-spectrum, we can conclude from (7.27) that $X_\infty = \{0\}$. Since $f$ decays regularly at infinity, there exists $n \in \mathbb{N}$ such that $sf^n(s) \in \mathcal{M}[\omega]_T$ decays regularly at infinity too. Because of Lemma 7.2.8, the operators $Tf^n(T)$ and $(sf^n)(T)$ both commute with $E_\infty$. Hence, they leave $X_\infty$ invariant and we find, again because of Lemma 7.2.8, that

$$Tf^n(T)|_{X_\infty} \subset (sf^n)(T)|_{X_\infty} \in \mathcal{B}(X_\infty)$$

with

$$\begin{aligned}
\mathcal{D}\left(Tf^n(T)|_{X_\infty}\right) &= \mathcal{D}\left(Tf^n(T)\right) \cap X_\infty \\
&= \mathcal{D}\left((sf^n)(T)\right) \cap \mathcal{D}\left(f^n(T)\right) \cap X_\infty \\
&= \mathcal{D}\left((sf^n)(T)|_{X_\infty}\right) \cap \mathcal{D}\left(f^n(T)|_{X_\infty}\right).
\end{aligned}$$

But since $sf^n$ and $f^n$ both decay regularly at infinity in $\Sigma_\varphi$, Lemma 7.5.5 implies that $f^n(T)|_{X_\infty}$ and $(sf^n)(T)|_{X_\infty}$ are both bounded linear operators on $X_\infty$. Hence, their domain is the entire space $X_\infty$ and we find that

$$Tf^n(T)|_{X_\infty} = (sf^n)(T)|_{X_\infty} \in \mathcal{B}(X_\infty).$$

Finally, observe that Lemma 7.5.5 also implies that $f^n(T)|_{X_\infty} = (f(T)|_{X_\infty})^n$. As $f(T)|_{X_\infty}$ has a bounded inverse on $X_\infty$, namely $f(T)^{-1}|_{X_\infty}$, we find that $T_\infty \in \mathcal{B}(X_\infty)$ too. As pointed out above, this implies $X_\infty = \{0\}$.

Altogether we find that $X = X_{\sigma_S(T)} := E_{\sigma_S(T)} X$ such that $T = T|_{X_{\sigma_S(T)}}$ belongs to $\mathcal{B}(X_{\sigma_S(T)}) = \mathcal{B}(X)$ and in turn

$$\infty \notin \sigma_{SX}(T) \quad \text{if } 0 = f(\infty) \notin \sigma_{SX}(f(T)).$$

Now let us consider the case that $s = 0$ and $c = 0$, that is $f(0) = 0$. If $0$ does not belong to $\sigma_{SX}(f(T))$, then $f(T)$ has a bounded inverse. Let $e$ be a regulariser for $f$ such that $ef \in \mathcal{E}[\Sigma_\omega]$. Since $f(T) = e(T)^{-1}(ef)(T)$ is injective, the operator $(ef)(T)$ must be injective too. As the function $ef$ has polynomial limit $0$ at $0$, we conclude from Lemma 7.2.7 that even $T$ is injective. If we define $\tilde{f}(q) := f(q^{-1})$, then $\tilde{f}$ has polynomial limit $0$ at $\infty$ and $\tilde{f}(T^{-1})$ is invertible as $\tilde{f}(T^{-1}) = f(T)$ by Corollary 7.3.2. Hence, $0 = \tilde{f}(\infty) \notin \sigma_{SX}(\tilde{f}(T^{-1}))$ and arguments as the ones above show that $\infty \notin \sigma_{SX}(T^{-1})$ such that $T^{-1} \in \mathcal{B}(X)$. Thus, $T$ has a bounded inverse and in turn $0 \notin \sigma_{SX}(T)$ if $0 = f(0) \notin \sigma_{SX}(f(T))$.

Finally, let us consider the case $c = f(s) = \infty$ with $s = 0$ or $s = \infty$ and let us assume that $\infty \notin \sigma_{SX}(f(T))$, that is that $f(T)$ is bounded. If we choose $a \in \mathbb{R}$ with $|a| > \|f(T)\|$, then $a \in \rho_S(f(T))$ and hence $a\mathcal{I} - f(T)$ has a bounded inverse. By (iii) in Lemma 7.2.8, the function $g(q) := (a - f(q))^{-1}$ belongs to $\mathcal{M}[\Sigma_\omega]_T$. Moreover, $g(T)$ is invertible and $g(T)$ has polynomial limit $0$ at $s$. As we have shown above, this implies $s \notin \sigma_{SX}(T)$, which concludes the proof.  $\square$

Combining Proposition 7.5.4 and Theorem 7.5.6, we arrive at the following theorem.

**Theorem 7.5.7.** *Let $T \in \mathrm{Sect}(\omega)$. If $f \in \mathcal{M}[\Sigma_\omega]_T$ and $f$ has polynomial limits at $\sigma_{SX}(T) \cap \{0, \infty\}$, then*

$$f(\sigma_{SX}(T)) \subset \sigma_{SX}(f(T)).$$

Let us now consider the inverse inclusion. We start with the following auxiliary lemma.

**Lemma 7.5.8.** *Let $T \in \mathrm{Sect}(\omega)$ and let $f \in \mathcal{M}[\Sigma_\omega]_T$ have finite polynomial limits at $\{0, \infty\} \cap \sigma_{SX}(T)$ in $\Sigma_\varphi$ for some $\varphi \in (\omega, \pi)$. Furthermore, assume that all poles of $f$ are contained in $\rho_S(T)$.*

(i) *If $\{0, \infty\} \subset \sigma_{SX}(T)$, then $f(T)$ is defined by the $H^\infty$-functional calculus for sectorial operators.*

(ii) *If $0 \in \sigma_{SX}(T)$ but $\infty \notin \sigma_{SX}(T)$, then $f(T)$ is defined by the extended $H^\infty$-functional calculus for bounded sectorial operators.*

(iii) *If $\infty \in \sigma_{SX}(T)$ but $0 \notin \sigma_{SX}(T)$, then $f(T)$ is defined by the extended $H^\infty$-functional calculus for invertible sectorial operators.*

(iv) *If $0, \infty \notin \sigma_{SX}(T)$, then $f(T)$ is defined by the $H^\infty$-functional calculus for bounded and invertible sectorial operators.*

*In all of these cases $f(T) \in \mathcal{B}(X)$.*

*Proof.* Let us first consider the case (i), i.e., we assume that $\{0, \infty\} \subset \sigma_{SX}(T)$. Since $f$ has polynomial limits at $0$ and $\infty$ in $\Sigma_\omega$, the function $f$ has only finitely many poles $[s_1], \ldots, [s_n]$ in $\overline{\Sigma_\omega}$. Moreover, for suitably large $m_1 \in \mathbb{N}$, the function $f_1(q) = (1+q)^{-2m_1} \mathcal{Q}_{s_1}(q)^{m_1} f(q)$ has also polynomial limits at $0$ and $\infty$ and poles at $[s_2], \ldots, [s_n]$ but it does not have a pole at $[s_1]$. Moreover, if we set $r_1(q) =$

$(1+q)^{-2m_1}\mathcal{Q}_{s_1}(q)^{m_1}$, then $r_1(T)$ is bounded and injective because $[s_1] \subset [\rho_S(T))]$. We can now repeat this argument and find inductively $m_2, \ldots, m_n$ such that, after setting $r_\ell(q) = (1+q)^{-2m_\ell}\mathcal{Q}_{s_\ell}(q)^{m_\ell}$ for $\ell = 2, \ldots, n$ and $r := r_n \cdots r_1$, the function $\tilde{f} = rf$ belongs to $\mathcal{M}[\Sigma_\omega]_T$, has polynomial limits at $0$ and $\infty$ and does not have any poles in $\overline{\Sigma_\omega}$. Hence, it belongs to $\mathcal{E}[\Sigma_\omega]$. Moreover, $r$ belongs to $\mathcal{E}[\Sigma_\omega]$ too and since $r(T) = r_n(T) \cdots r_1(T)$ is the product of invertible operators, it is invertible itself. Hence, $r$ regularises $f$ such that $f(T)$ is defined in terms of the $H^\infty$-functional calculus. Moreover, $f(T) = r(T)^{-1}\tilde{f}(T)$ is bounded as it is the product of two bounded operators.

Similar arguments show the other cases: in (ii), for example, the function $f$ has polynomial limit at $0$ but not at $\infty$, such that the poles of $f$ may accumulate at $\infty$. However, we integrate along the boundary of $\Sigma_{\omega,0,R} = \Sigma_\omega \cap B_R(0)$ in $\mathbb{C}_j$ for sufficiently large $R$ when we define the $H^\infty$-functional calculus for bounded sectorial operators. Hence, only finitely many poles are contained in $\Sigma_{\omega,0,R}$ and therefore relevant. Thus, we can apply the above strategy again in order to show that $f$ is regularised by a rational intrinsic function and that $f(T)$ is defined and a bounded operator. Similar, we can argue for (iii) and (iv), where the poles of $f$ may accumulate at $0$ (resp. at $0$ and $\infty$), but only finitely many of them are relevant. □

**Proposition 7.5.9.** *Let $T \in \mathrm{Sect}(\omega)$. If $f \in \mathcal{M}[\Sigma_\omega]_T$ has polynomial limits at any point in $\sigma_S(T) \cap \{0, \infty\}$, then*

$$f(\sigma_{SX}(T)) \supset \sigma_{SX}(f(T)).$$

*Proof.* Let $s \in \mathbb{H}$ with $s \notin f(\sigma_{SX}(T))$. The function $q \mapsto \mathcal{Q}_s(f(q))^{-1}$ belongs then to $\mathcal{M}[\Sigma_\omega]_T$ and has finite polynomial limits at $\sigma_{SX}(T) \cap \{0, \infty\}$. Moreover, the set of poles of $\mathcal{Q}_s(f(\cdot))$ as an element of $\mathcal{M}[\Sigma_\omega]$ consists of those spheres $[q]$ in $\overline{\Sigma_\omega} \setminus \{0\}$ for which $f([q]) = [f(q)] = [s]$ and it is contained in the $S$-resolvent set of $T$ as we chose $s \notin f(\sigma_{SX}(T))$. From Lemma 7.5.8 we therefore deduce that $\mathcal{Q}_s(f(T))^{-1}$ is defined and belongs to $\mathcal{B}(X)$. Hence, $\mathcal{Q}_s(f(T))$ has a bounded inverse and so $s \in \sigma_{SX}(f(T))$.

If finally $s = \infty \notin f(\sigma_{SX}(T))$, then the poles of $f$ are contained in the $S$-resolvent set of $T$. Hence, Lemma 7.5.8 implies that $f(T)$ is a bounded operator and in turn $s = \infty \notin \sigma_{SX}(f(T))$. □

Combining Theorem 7.5.7 and Proposition 7.5.9, we obtain the following spectral mapping theorem

**Theorem 7.5.10** (Spectral Mapping Theorem). *Let $T \in \mathrm{Sect}(\omega)$ and let $f \in \mathcal{M}[\Sigma_\omega]_T$ have polynomial limits at $\{0, \infty\} \cap \sigma_{SX}(T)$. Then*

$$f(\sigma_{SX}(T)) = \sigma_{SX}(f(T)).$$

# Chapter 8

# Fractional powers of quaternionic linear operators

Fractional powers of operators aroused the interest of mathematicians in the 1960s in [36, 167–169, 204, 205] and have been extensively studied since then. They have applications in the theory of semi-groups [164], they allow to define interpolation spaces [113] and they provide the theoretical background for defining fractional evolution equations [48]. In this chapter, we generalize three classical approaches for defining fractional powers of operators to the quaternionic setting. Fractional powers of vector operators are a natural tool to define fractional diffusion and evolution problems, as we well see in the next chapters. The results of this chapter are taken from our papers [52–54].

We recall that the slice hyperholomorphic logarithm on $\mathbb{H}$ was in Example 2.1.14 defined as

$$\log q := \ln|q| + j\arg(q) \quad \text{for } q = u + jv \in \mathbb{H} \setminus (-\infty, 0], \tag{8.1}$$

where $\arg(q) = \arccos(u/|q|) \in [0, \pi]$ is the unique angle such that $q = |q|e^{j\arg(q)}$ for $x \in \mathbb{H} \setminus (-\infty, 0]$. For $\alpha \in \mathbb{R}$, we furthermore defined the fractional power $q^\alpha$ of $q = u + jv \in \mathbb{H} \setminus (-\infty, 0]$ in Example 2.1.15 as

$$q^\alpha := e^{\alpha \log q} = e^{\alpha(\ln|q| + j\arg(q))}. \tag{8.2}$$

The logarithm and fractional powers with real exponents are intrinsic slice hyperholomorphic functions on their domain of definition.

Since $q^\alpha$ is a slice hyperholomorphic function for $\alpha \in \mathbb{R}$, our setting provides suitable techniques for defining fractional powers of quaternionic linear operators. In the complex setting, several approaches for defining fractional powers of sectorial operators are known and they make different assumptions on the respective operator. In this chapter, we generalize three of these approaches to the quaternionic setting.

© Springer Nature Switzerland AG 2019

F. Colombo, J. Gantner, *Quaternionic Closed Operators, Fractional Powers and Fractional Diffusion Processes*, Operator Theory: Advances and Applications 274, https://doi.org/10.1007/978-3-030-16409-6_8

1) In Section 8.1 we follow the approach of [113] and define fractional powers of invertible sectorial operators directly via a slice hyperholomorphic Cauchy integral. This approach allows to define interpolation spaces, but the theory of interpolation spaces in the quaternionic setting does not show any significant difference to the complex one. The results presented in this section can be found in [52].

2) In Section 8.2, we follow [165] and develop the most general approach to fractional powers in the quaternionic setting. We use the $H^\infty$-functional calculus introduced in Chapter 7 in order to introduce fractional powers of arbitrary sectorial operators and show several of their properties. The results in this section are part of [53, 54].

3) In Section 8.3, we finally follow [167] and introduce fractional powers of exponent $\alpha \in (0, 1)$ indirectly. We first find integral representations for operators that should correspond to the $S$-resolvents of $T^\alpha$. Then we show that there exists actually an operator such that these integrals coincide with their resolvents. In particular, this implies that the famous Kato formula for the resolvent of the fractional power of an operator has an analogue in the quaternionic setting. The results in this section can again be found in [52].

## 8.1　A direct approach to fractional powers of invertible sectorial operators with negative exponent

In the following we assume that $T$ is a densely defined closed quaternionic right linear operator such that $(-\infty, 0] \subset \rho_S(T)$ and such that there exists a positive constant $M > 0$ such that

$$\|S_R^{-1}(s, T)\| \le \frac{M}{1 + |s|}, \quad \text{for } s \in (-\infty, 0]. \tag{8.3}$$

Obviously, this is the case if $T$ is a sectorial operator in the sense of Definition 7.1.1 and has a bounded inverse. In order to show that the converse is also true, we recall the notation

$$\Sigma_\varphi = \{s \in \mathbb{H} : \quad \arg(s) < \varphi\}$$

for the sector of angle $\varphi \in (0, \pi)$ around the positive real axis.

**Lemma 8.1.1.** *The estimate* (8.3) *implies the existence of constants* $a > 0$, $\varphi \in (0, \pi)$ *and* $M_n > 0$ *for* $n \in \mathbb{N}$ *such that* $\sigma_S(T)$ *is contained in the translated sector*

$$\Sigma_\varphi + a := \{s \in \mathbb{H} : \quad \arg(s - a) < \varphi\}$$

*and such that, for any* $s \notin \Sigma_\varphi + a$ *and any* $n \in \mathbb{N}$, *we have*

$$\|S_R^{-n}(s, T)\| \le \frac{M_n}{(1 + |s|)^n} \quad \text{and} \quad \|S_L^{-n}(s, T)\| \le \frac{M_n}{(1 + |s|)^n}. \tag{8.4}$$

*Proof.* By Proposition 3.5.7, we have

$$\partial_S{}^k S_R^{-1}(s,T) = (-1)^k k!\, S_R^{-(k+1)}(s,T), \quad \text{for } k \in \mathbb{N}. \tag{8.5}$$

Since the slice derivative of a left slice hyperholomorphic function is again left slice hyperholomorphic, the function $s \mapsto S_R^{-n}(s,T)$ is a left slice hyperholomorphic function on $\rho_S(T)$ with values in $\mathcal{B}(X)$ for any $n \in \mathbb{N}$. From the identity (8.5), we deduce

$$\partial_S{}^k S_R^{-n}(s,T) = \partial_S{}^{k+n-1} \frac{(-1)^{n-1}}{(n-1)!} S_R^{-1}(s,T)$$

$$= \frac{(-1)^k (k+n-1)!}{(n-1)!} S_R^{-(k+n)}(s,T).$$

When we apply Theorem 2.1.12 in order to expand $S_R^{-n}(s,T)$ into a Taylor series at a real point $\alpha \in \rho_S(T)$, we therefore get

$$S_R^{-n}(s,T) = \sum_{k=0}^{+\infty} \frac{1}{k!} (s-\alpha)^k \partial_S{}^k S_R^{-n}(\alpha,T)$$

$$= \sum_{k=0}^{+\infty} \frac{1}{k!} (s-\alpha)^k \frac{(-1)^k (k+n-1)!}{(n-1)!} S_R^{-(k+n)}(\alpha,T) \tag{8.6}$$

$$= \sum_{k=0}^{+\infty} (-1)^k \binom{n+k-1}{k} (s-\alpha)^k S_R^{-(n+k)}(\alpha,T)$$

on any ball $B_r(\alpha)$ contained in $\rho_S(T)$. Since $\alpha$ is real, $S_R^{-n}(\alpha,T) = \left(S_R^{-1}(\alpha,T)\right)^n$ and thus $\|S_R^{-n}(\alpha,T)\| \leq \|S_R^{-1}(\alpha,T)\|^n$. The ratio test and the estimate (8.3) therefore imply that this series converges on the ball with radius $(1+|\alpha|)/M$ centered at $\alpha$ for $\alpha \in (-\infty,0]$. In particular, considering the case $n = 1$, we deduce from Theorem 3.2.11 that any such ball is contained in $\rho_S(T)$. Otherwise, the above series would give a nontrivial slice hyperholomorphic continuation of $S_R^{-1}(s,T)$, which cannot exist by Theorem 3.2.11.

Set $a = \min\left\{\frac{1}{4M}, 1\right\}$. Then the closed ball $\overline{B_a(0)}$ is contained in $\rho_S(T)$ and for any $s \in B_a(0)$, we have the estimate

$$\|S_R^{-n}(s,T)\| \leq \sum_{k=0}^{+\infty} \binom{n+k-1}{k} |s|^k \|S_R^{-(n+k)}(0,T)\|$$

$$\leq \sum_{k=0}^{+\infty} \binom{n+k-1}{k} \frac{1}{(4M)^k} M^{n+k} \frac{(1+|s|)^{n+k}}{(1+|s|)^{n+k}}$$

$$= \frac{2^n M^n}{(1+|s|)^n} \sum_{k=0}^{+\infty} \binom{n+k-1}{k} \frac{1}{2^k} = \frac{4^n M^n}{(1+|s|)^n},$$

where the last equation follows from the Taylor series expansion

$$(1-q)^{-n} = \sum_{k=0}^{+\infty} \binom{n+k-1}{k} q^k \quad \text{for } |q| < 1.$$

Now set $\varphi_0 = \pi - \arctan(\frac{1}{2M})$ and consider the sector $\Sigma_{\varphi_0}$. If $s = s_0 + j_s s_1 \notin \Sigma_{\varphi_0}$, we have $0 \le s_1 \le |s_0|/(2M)$ and from the power series expansion (8.6) of $S_R^{-n}(s, T)$ at $s_0$, we conclude

$$\left\| S_R^{-n}(s, T) \right\| \le \sum_{k=0}^{+\infty} \binom{n+k-1}{k} |s_1|^k \left\| S_R^{-1}(s_0, T) \right\|^{n+k}$$

$$\le \sum_{k=0}^{+\infty} \binom{n+k-1}{k} \left( \frac{|s_0|}{2M} \right)^k \left( \frac{M}{1+|s_0|} \right)^{n+k}$$

$$\le \left( \frac{M}{1+|s_0|} \right)^n \sum_{k=0}^{+\infty} \binom{n+k-1}{k} \frac{1}{2^k} = \frac{2^n M^n}{(1+|s_0|)^n}.$$

Since $|s| \le |s_0| + |s_1| \le (1 + \frac{1}{2M})|s_0|$, we get

$$\left\| S_R^{-n}(s, T) \right\| \le \frac{2^n M^n}{\left( 1 + (1 + \frac{1}{2M})^{-1} |s| \right)^n} \le \frac{\left( 1 + \frac{1}{2M} \right)^n 2^n M^n}{(1+|s|)^n}.$$

Hence, the estimate

$$\left\| S_R^{-n}(s, T) \right\| \le \frac{M_n}{(1+|s|)^n} \tag{8.7}$$

with

$$M_n := \left( 1 + \frac{1}{2M} \right)^n 4^n M^n$$

holds true for any $s \notin \Omega := \Sigma_{\varphi_0} \setminus B_a(0)$. Now observe that the sector $\Sigma_\varphi + a$ with

$$\varphi := \arctan \left( \frac{a \sin \varphi_0}{a(-1 + \cos \varphi_0)} \right)$$

contains the set $\Omega$. Hence, if $s \notin \Sigma_\varphi + a$, then $s \notin \Omega$ and so $s \in \rho_S(T)$ and (8.7) holds true.

Since $S_L^{-1}(s, T) = S_R^{-1}(s, T)$ for $s \in (-\infty, 0]$, the estimate (8.3) applies also to the left $S$-resolvent. Thus, we can use analogous arguments to prove that the the estimate for the left $S$-resolvent in (8.4) also holds true with these constants.   $\square$

**Definition 8.1.2.** Let $j \in \mathbb{S}$ and let $\Sigma_\varphi + a$ be the sector obtained from Lemma 8.1.1. Let $\theta \in (\varphi, \pi)$ and choose a piecewise smooth path $\Gamma$ in $\mathbb{H} \setminus (\Sigma_\varphi + a)$ that goes in $\mathbb{C}_j$ from $\infty e^{j\theta}$ to $\infty e^{-j\theta}$ and avoids the negative real axis $(-\infty, 0]$. For $\alpha > 0$, we define

$$T^{-\alpha} := \frac{1}{2\pi} \int_\Gamma s^{-\alpha} \, ds_j \, S_R^{-1}(s, T). \tag{8.8}$$

**Theorem 8.1.3.** *For any $\alpha > 0$, the operator $T^{-\alpha}$ is bounded and independent of the choice of $j \in \mathbb{S}$, of $\theta \in (\theta_0, \pi)$ and of the concrete path $\Gamma$ in $\mathbb{C}_j$ and therefore well-defined.*

*Proof.* The estimate (8.4) assures that the integral in (8.8) exists and defines a bounded right-linear operator. Since $s \mapsto s^{-\alpha}$ is right slice hyperholomorphic and $s \mapsto S_R^{-1}(s, T)$ is left slice hyperholomorphic, the independence of the choice of $\theta$ and the independence of the choice of the path $\Gamma$ in the complex plane $\mathbb{C}_j$ follow from Cauchy's integral theorem.

In order to show that $T^{-\alpha}$ is independent of the choice of the imaginary unit $j \in \mathbb{S}$, we consider an arbitrary imaginary unit $i \in \mathbb{S}$ with $i \neq j$. If $\Sigma_\varphi + a$ is the sector obtained from Lemma 8.1.1, then let $\varphi < \theta_s < \theta_p < \pi$ and set $U_s := \Sigma_{\theta_s} \setminus B_{a/2}(0)$ and $U_p := \Sigma_{\theta_p} \setminus B_{a/3}(0)$. (The indices $s$ and $p$ are chosen in order to indicate the variable of integration over the boundary of the respective set in the following calculation.) Then $U_p$ and $U_s$ are slice domains that contain $\sigma_S(T)$, and both $\partial(U_s \cap \mathbb{C}_j)$ and $\partial(U_p \cap \mathbb{C}_i)$ are paths that are admissible in Definition 8.1.2.

Observe that $s \mapsto s^{-\alpha}$ is right slice hyperholomorphic on $\overline{U_p}$ and that, by our choices of $U_p$ and $U_s$, we have $s \in U_p$ for any $s \in \partial(U_s \cap \mathbb{C}_j)$. If we choose $r > 0$ large enough, then $s \in U_p \cap B_r(0)$ and we obtain from Cauchy's integral formula that

$$s^\alpha = \lim_{r \to +\infty} \frac{1}{2\pi} \int_{\partial(U_p \cap B_r(0)) \cap \mathbb{C}_j} p^{-\alpha} \, dp_i \, S_R^{-1}(p, s)$$

$$= \frac{1}{2\pi} \int_{\partial(U_p \cap \mathbb{C}_i)} p^{-\alpha} \, dp_i \, S_R^{-1}(p, s),$$

where the second equation holds since $p^{-\alpha} \to 0$ uniformly as $p \to \infty$ in $U_p$. For the operator $T^{-\alpha}$, we thus obtain

$$T^{-\alpha} = \frac{1}{2\pi} \int_{\partial(U_s \cap \mathbb{C}_j)} s^{-\alpha} \, ds_j \, S_R^{-1}(s, T)$$

$$= \frac{1}{(2\pi)^2} \int_{\partial(U_s \cap \mathbb{C}_j)} \left( \int_{\partial(U_p \cap \mathbb{C}_i)} p^{-\alpha} \, dp_i \, S_R^{-1}(p, s) \right) ds_j \, S_R^{-1}(s, T). \quad (8.9)$$

We now apply Fubini's theorem, but we postpone the estimate that justifies this to the end of the proof as it is very technical and requires long computations. By interchanging the order of integration, we get

$$T^{-\alpha} = \frac{1}{2\pi} \int_{\partial(U_p \cap \mathbb{C}_i)} p^{-\alpha} \, dp_i \left( \frac{1}{2\pi} \int_{\partial(U_s \cap \mathbb{C}_j)} S_R^{-1}(p, s) \, ds_j \, S_R^{-1}(s, T) \right)$$

$$= \frac{1}{2\pi} \int_{\partial(U_p \cap \mathbb{C}_i)} p^{-\alpha} \, dp_i \, S_R^{-1}(p, T),$$

where the last equation follows as an application of the $S$-functional calculus since $S_R^{-1}(p,\infty) = \lim_{s\to\infty} S_R^{-1}(p,s) = 0$. Hence, the operator $T^{-\alpha}$ is also independent of the choice of the imaginary unit $j \in \mathbb{S}$ provided that it is actually possible to apply Fubini's theorem in (8.9). In order to show that the integrand in (8.9) is absolutely integrable, we consider the parametrizations $\Gamma_s$ and $\Gamma_p$ of $\partial(U_s \cap \mathbb{C}_j)$ and $\partial(U_p \cap \mathbb{C}_i)$ that are given by

$$\Gamma_s(r) = \begin{cases} \Gamma_s^+(r) := re^{-\theta_s j}, & r \in [a/2, +\infty), \\ \Gamma_s^0(r) := \frac{a}{2}e^{-\frac{2\theta_s}{a}jr} & r \in (-a/2, a/2), \\ \Gamma_s^-(r) := re^{\theta_s j}, & r \in (-\infty, -a/2], \end{cases}$$

and

$$\Gamma_p(t) = \begin{cases} \Gamma_p^+(t) := te^{-\theta_p i}, & t \in [a/3, +\infty), \\ \Gamma_p^0(t) := \frac{a}{3}e^{-\frac{3\theta_p}{a}it} & t \in (-a/3, a/3), \\ \Gamma_p^-(t) := te^{\theta_p i}, & t \in (-\infty, -a/3]. \end{cases}$$

Then,

$$\frac{1}{(2\pi)^2} \int_{\Gamma_s} \int_{\Gamma_p} \left\| p^{-\alpha} \, dp_i \, S_R^{-1}(p,s) \, ds_j \, S_R^{-1}(s,T) \right\|$$

$$= \sum_{\tau,\nu\in\{-,0,+\}} \int_{\Gamma_s^\tau} \int_{\Gamma_p^\nu} \left\| p^{-\alpha} \, dp_i \, S_R^{-1}(p,s) \, ds_j \, S_R^{-1}(s,T) \right\|$$

and it is sufficient to estimate each of the terms in the sum separately. Applying the representation formula allows us to estimate

$$|S_R^{-1}(p,s)| \leq \frac{1}{2}|1-ji|\frac{1}{|p_j - s|} + \frac{1}{2}|1+ji|\frac{1}{|\overline{p_j} - s|} \leq \frac{2}{|p_j - s_j|}, \tag{8.10}$$

where $p_j = u + jv$ for $p = u + j_p v$. By applying (8.4), we find for any $\tau, \nu \in \{+, -\}$ that

$$\int_{\Gamma_s^\tau} \int_{\Gamma_p^\nu} \left\| p^{-\alpha} \, dp_i \, S_R^{-1}(p,s) \, ds_j \, S_R^{-1}(s,T) \right\|$$

$$= \int_{\frac{a}{2}}^{+\infty} \int_{\frac{a}{3}}^{+\infty} t^{-\alpha} \frac{2}{|te^{\theta_p j} - re^{\theta_s j}|} \frac{M_1}{1+r} \, dt \, dr$$

$$= \int_{\frac{a}{2}}^{+\infty} \int_{\frac{a}{3}}^{+\infty} \frac{t^{-\alpha}}{|t - re^{(\theta_s - \theta_p)j}|} \frac{2M_1}{1+r} \, dt \, dr$$

$$= \int_{\frac{a}{2}}^{+\infty} \int_{\frac{a}{3}}^{+\infty} \frac{\left(\frac{t}{r}\right)^{-\alpha}}{\left|\frac{t}{r} - e^{(\theta_s - \theta_p)j}\right|} \, dt \, \frac{1}{r^{1+\alpha}} \frac{2M_1}{1+r} \, dr$$

$$= \int_{\frac{a}{2}}^{+\infty} \int_{\frac{a}{3r}}^{+\infty} \frac{\mu^{-\alpha}}{\left|\mu - e^{(\theta_s - \theta_p)j}\right|} \, d\mu \, \frac{1}{r^\alpha} \frac{2M_1}{1+r} \, dr.$$

The modulus of $\mu - e^{(\theta_s - \theta_p)j}$ can be estimated from below by the absolute value of its real part or by the absolute value of its imaginary part, and therefore,

$$
\int_{\Gamma_s^\tau} \int_{\Gamma_p^\nu} \left\| p^{-\alpha} \, dp_i \, S_R^{-1}(p, s) \, ds_j \, S_R^{-1}(s, T) \right\|
$$

$$
\leq \int_{\frac{a}{2}}^{+\infty} \int_2^{+\infty} \frac{\mu^{-\alpha}}{\mu - \cos(\theta_s - \theta_p)} \, d\mu \, \frac{1}{r^\alpha} \frac{2M_1}{1 + r} \, dr
$$

$$
+ \int_{\frac{a}{2}}^{+\infty} \int_{\frac{a}{3r}}^2 \frac{\mu^{-\alpha}}{\sin(\theta_p - \theta_s)} \, d\mu \, \frac{1}{r^\alpha} \frac{2M_1}{1 + r} \, dr
$$

$$
= \underbrace{\int_2^{+\infty} \frac{\mu^{-\alpha}}{\mu - \cos(\theta_s - \theta_p)} \, d\mu}_{=:C_1 < +\infty} \underbrace{\int_{\frac{a}{2}}^{+\infty} \frac{1}{r^\alpha} \frac{2M_1}{1 + r} \, dr}_{=:C_2 < +\infty}
$$

$$
+ \int_{\frac{a}{2}}^{+\infty} \frac{1}{\sin(\theta_p - \theta_s)} \left( \frac{2^{1-\alpha}}{1 - \alpha} - \frac{a^{1-\alpha}}{(1 - \alpha)3^{1-\alpha}} \frac{1}{r^{1-\alpha}} \right) \frac{1}{r^\alpha} \frac{2M_1}{1 + r} \, dr
$$

$$
= C_1 C_2 + \frac{2M_1}{\sin(\theta_p - \theta_s)} \frac{2^{1-\alpha}}{1 - \alpha} \int_{\frac{a}{2}}^{+\infty} \frac{r^{-\alpha}}{1 + r} \, dr
$$

$$
+ \frac{2M_1}{\sin(\theta_p - \theta_s)} \frac{a^{1-\alpha}}{(1 - \alpha)3^{1-\alpha}} \int_{\frac{a}{2}}^{+\infty} \frac{1}{r(1 + r)} \, d\mu \, dr
$$

where each of these integrals is finite.

For $\tau = 0$ and $\nu = +$, we can again use (8.10) to estimate

$$
\int_{\Gamma_s^0} \int_{\Gamma_p^+} \left\| p^{-\alpha} \, dp_i \, S_R^{-1}(p, s) \, ds_j \, S_R^{-1}(s, T) \right\|
$$

$$
\leq \int_{-\frac{a}{2}}^{\frac{a}{2}} \int_{\frac{a}{3}}^{+\infty} t^{-\alpha} \left| S_R^{-1} \left( te^{i\theta_p}, \frac{a}{2} e^{-\frac{2\theta_s}{a} rj} \right) \right| \frac{\theta_s M_1}{1 + \frac{a}{2}} \, dt \, dr
$$

$$
\leq \int_{-\frac{a}{2}}^{\frac{a}{2}} \int_{\frac{a}{3}}^{+\infty} t^{-\alpha} \frac{2}{\left| te^{j\theta_p} - \frac{a}{2} e^{\frac{2\theta_s}{a} rj} \right|} \frac{\theta_s M_1}{1 + \frac{a}{2}} \, dt \, dr
$$

$$
= \frac{2\theta_s M_1}{1 + \frac{a}{2}} \int_{-\frac{a}{2}}^{\frac{a}{2}} \int_{\frac{a}{3}}^{+\infty} \frac{t^{-\alpha}}{\left| t - \frac{a}{2} e^{\left( \frac{2\theta_s}{a} r - \theta_p \right) j} \right|} \, dt \, dr.
$$

Since $0 < \theta_s < \theta_p < \pi$, the distance $\delta$ of the set

$$
\left\{ \frac{a}{2} e^{\left( \frac{2\theta_s}{a} r - \theta_p \right) j} : \quad -\frac{a}{2} < r < \frac{a}{2} \right\}
$$

to the positive real axis is greater than zero, and hence,

$$\int_{\Gamma_s^0}\int_{\Gamma_p^+}\left\|p^{-\alpha}\,dp_i\,S_R^{-1}(p,s)\,ds_j\,S_R^{-1}(s,T)\right\|$$

$$\leq \frac{2\theta_s M_1}{1+\frac{a}{2}}\int_{-\frac{a}{2}}^{\frac{a}{2}}\int_{\frac{a}{3}}^{a}\frac{t^{-\alpha}}{\delta}\,dt\,dr + \frac{2\theta_s M_1}{1+\frac{a}{2}}\int_{-\frac{a}{2}}^{\frac{a}{2}}\int_{a}^{+\infty}t^{-\alpha}\frac{1}{t-\frac{a}{2}}\,dt\,dr$$

$$= \frac{2\theta_s a M_1}{\delta\left(1+\frac{a}{2}\right)}\int_{\frac{a}{3}}^{a}t^{-\alpha}\,dt + \frac{2\theta_s a M_1}{1+\frac{a}{2}}\int_{a}^{+\infty}\frac{t^{-\alpha}}{t-\frac{a}{2}}\,dt,$$

where again these integrals are finite. A similar computation can be done for the case $\tau = 0$ and $\nu = -$.

For $\tau = +$ and $\nu = 0$, we apply once more (8.10) and obtain

$$\int_{\Gamma_s^+}\int_{\Gamma_p^0}\left\|p^{-\alpha}\,dp_i\,S_R^{-1}(p,s)\,ds_j\,S_R^{-1}(s,T)\right\|$$

$$\leq \int_{\frac{a}{2}}^{+\infty}\int_{-\frac{a}{3}}^{\frac{a}{3}}\left(\frac{a}{3}\right)^{-\alpha}\theta_p\left|S_R^{-1}\left(\frac{a}{3}e^{\frac{-3\theta_p}{a}it},re^{-\theta_s j}\right)\right|\frac{M_1}{1+r}\,dt\,dr$$

$$\leq 2\left(\frac{a}{3}\right)^{-\alpha}\theta_p M_1\int_{\frac{a}{2}}^{+\infty}\int_{-\frac{a}{3}}^{\frac{a}{3}}\frac{1}{\left|\frac{a}{3}e^{\frac{3\theta_p}{a}jt}-re^{\theta_s j}\right|}\frac{1}{1+r}\,dt\,dr$$

$$= 2\left(\frac{a}{3}\right)^{-\alpha}\theta_p M_1\int_{\frac{a}{2}}^{+\infty}\int_{-\frac{a}{3}}^{\frac{a}{3}}\frac{1}{\left|r-\frac{a}{3}e^{\left(\frac{3\theta_p}{a}t-\theta_s\right)j}\right|}\frac{1}{1+r}\,dt\,dr.$$

Estimating the modulus of the denominator from below with the modulus of its real part, we obtain

$$\int_{\Gamma_s^+}\int_{\Gamma_p^0}\left\|p^{-\alpha}\,dp_i\,S_R^{-1}(p,s)\,ds_j\,S_R^{-1}(s,T)\right\|$$

$$\leq 2\left(\frac{a}{3}\right)^{-\alpha}\theta_p M_1\int_{\frac{a}{2}}^{+\infty}\int_{-\frac{a}{3}}^{\frac{a}{3}}\frac{1}{\left|r-\frac{a}{3}\cos\left(\frac{3\theta_p}{a}t-\theta_s\right)\right|}\frac{1}{1+r}\,dt\,dr$$

$$\leq 2\left(\frac{a}{3}\right)^{-\alpha}\theta_p M_1\int_{\frac{a}{2}}^{+\infty}\int_{-\frac{a}{3}}^{\frac{a}{3}}\frac{1}{r-\frac{a}{3}}\frac{1}{1+r}\,dt\,dr$$

$$= \frac{4a}{3}\left(\frac{a}{3}\right)^{-\alpha}\theta_p M_1\int_{\frac{a}{2}}^{+\infty}\frac{1}{r-\frac{a}{3}}\frac{1}{1+r}\,dr,$$

and this last integral is finite. The estimate of the case $\tau = -$ and $\nu = 0$ can be done in a similar way.

Finally, the summand for $\tau = 0$ and $\nu = 0$ consists of the integral of a continuous function over a bounded domain and is therefore finite.

Putting these pieces together, we obtain that the integrand in (8.9) is absolutely integrable, which allows us to apply Fubini's theorem in order to exchange the order of integration. □

If $\alpha \in \mathbb{N}$, then $s^{-\alpha}$ is right slice hyperholomorphic at infinity. The following Corollary immediately follows as an application of the $S$-functional calculus.

**Corollary 8.1.4.** *If $\alpha \in \mathbb{N}$, then the operator $T^{-\alpha}$ defined in (8.8) coincides with the $\alpha$-th inverse power of $T$.*

If we follow the arguments of the proof of Theorem 5.27 in [113, Chapter II], we obtain an integral representation of $T^{-\alpha}$ that is almost identical to the one derived for the complex case: the only difference is the constant in front of the integral. This is due to the different choice of the branch of the logarithm that is used in [113] in order to define the fractional powers. As pointed out in Remark 2.1.3, it is not possible to define different branches of the logarithm in the quaternionic slice hyperholomorphic setting. In Corollary 8.1.7 we, however, obtain an integral representation that is clearly different from any integral representation known from the classical complex setting.

**Theorem 8.1.5.** *Let $n \in \mathbb{N}$. For $\alpha \in (0, n+1)$ with $\alpha \notin \mathbb{N}$, the operator $T^{-\alpha}$ defined in (8.8) has the representation*

$$T^{-\alpha} = (-1)^{n+1}\frac{\sin(\alpha\pi)}{\pi}\frac{n!}{(n-\alpha)\cdots(1-\alpha)}\int_0^{+\infty} t^{n-\alpha} S_R^{-(n+1)}(-t, T)\, dt. \quad (8.11)$$

*Proof.* Let $a$ and $\varphi$ be the constants obtained from Lemma 8.1.1. For $b \in (0, a)$ and $\theta \in (\varphi, \pi)$, we can choose $U = \Sigma_\theta + b$ and integrate over the boundary $\partial(U \cap \mathbb{C}_j)$ of $U$ in $\mathbb{C}_j$ for some $j \in \mathbb{S}$ in the integral representation of $T^{-\alpha}$. The boundary consists of the path

$$\gamma(t) = \begin{cases} b - te^{j\theta}, & t \in (-\infty, 0], \\ b + te^{-j\theta}, & t \in (0, \infty), \end{cases}$$

and hence

$$\begin{aligned} T^{-\alpha} &= \frac{1}{2\pi}\int_{-\infty}^0 (b - te^{j\theta})^{-\alpha}(-j)(-e^{j\theta})S_R^{-1}(b - te^{j\theta}, T)\, dt \\ &\quad + \frac{1}{2\pi}\int_0^{+\infty}(b + te^{-j\theta})^{-\alpha}(-j)e^{-j\theta}S_R^{-1}(b + te^{-j\theta}, T)\, dt \\ &= \frac{j}{2\pi}\int_0^{+\infty}(b + te^{j\theta})^{-\alpha}e^{j\theta}S_R^{-1}(b + te^{j\theta}, T)\, dt \\ &\quad - \frac{j}{2\pi}\int_0^{+\infty}(b + te^{-j\theta})^{-\alpha}e^{-j\theta}S_R^{-1}(b + te^{-j\theta}, T)\, dt. \end{aligned}$$

Integrating $n$ times by parts yields

$$
T^{-\alpha} = \frac{n!}{(n-\alpha)\cdots(1-\alpha)}\frac{j}{2\pi}\int_0^{+\infty}(b+te^{j\theta})^{n-\alpha}e^{j\theta}S_R^{-(n+1)}(b+te^{j\theta},T)\,dt
$$
$$
-\frac{n!}{(n-\alpha)\cdots(1-\alpha)}\frac{j}{2\pi}\int_0^{+\infty}(b+te^{-j\theta})^{n-\alpha}e^{-j\theta}S_R^{-(n+1)}(b+te^{-j\theta},T)\,dt.
$$

Because of the estimate (8.4), we can apply Lebesgue's dominated convergence theorem with dominating function

$$
f(t) = \begin{cases} C(1+t^{n-\alpha}) & \text{if } t \le 1, \\ Ct^{-\alpha-1} & \text{if } t > 1, \end{cases}
$$

where $C > 0$ is a sufficiently large constant. Taking the limit $b \to 0$, we obtain

$$
T^{-\alpha} = \frac{n!}{(n-\alpha)\cdots(1-\alpha)}\frac{j}{2\pi}\int_0^{+\infty}t^{n-\alpha}e^{j\theta(n-\alpha)}e^{j\theta}S_R^{-(n+1)}(te^{j\theta},T)\,dt
$$
$$
-\frac{n!}{(n-\alpha)\cdots(1-\alpha)}\frac{j}{2\pi}\int_0^{+\infty}t^{n-\alpha}e^{-j\theta(n-\alpha)}e^{-j\theta}S_R^{-(n+1)}(te^{-j\theta},T)\,dt
$$
$$\tag{8.12}$$

and then, taking the limit $\theta \to \pi$, we get

$$
T^{-\alpha} = -\frac{n!}{(n-\alpha)\cdots(1-\alpha)}\frac{j}{2\pi}\int_0^{+\infty}t^{n-\alpha}e^{j\pi(n-\alpha)}S_R^{-(n+1)}(-t,T)\,dt
$$
$$
+\frac{n!}{(n-\alpha)\cdots(1-\alpha)}\frac{j}{2\pi}\int_0^{+\infty}t^{n-\alpha}e^{-j\pi(n-\alpha)}S_R^{-(n+1)}(-t,T)\,dt
$$
$$
= (-1)^{n+1}\frac{\sin(\alpha\pi)}{\pi}\frac{n!}{(n-\alpha)\cdots(1-\alpha)}\int_0^{+\infty}t^{n-\alpha}S_R^{-(n+1)}(-t,T)\,dt,
$$

where the last equation follows from the identity

$$
-je^{j\pi(n-\alpha)} + je^{-j\pi(n-\alpha)} = \sin((n-\alpha)\pi) = (-1)^{n+1}\sin(\alpha\pi). \qquad \square
$$

**Corollary 8.1.6.** *We have $\mathcal{I}^{-\alpha} = \mathcal{I}$ for $\alpha \ge 0$, where $\mathcal{I}$ denotes the identity operator on $X$.*

*Proof.* If $\alpha \in \mathbb{N}$, the equality follows immediately from Corollary 8.1.4. For $\alpha \notin \mathbb{N}$, we consider $n \in \mathbb{N}$ with $\alpha \in (0, n+1)$. Since $S_R^{-1}(s,\mathcal{I}) = (s-1)^{-1}\mathcal{I}$, we then have

$$
\mathcal{I}^{-\alpha} = (-1)^{n+1}\frac{\sin(\alpha\pi)}{\pi}\frac{n!}{(n-\alpha)\cdots(1-\alpha)}\int_0^{+\infty}\frac{t^{n-\alpha}}{(-t-1)^{n+1}}\,dt\,\mathcal{I}
$$
$$
= \frac{\sin(\alpha\pi)}{\pi}\frac{n!}{(n-\alpha)\cdots(1-\alpha)}\int_0^{+\infty}\frac{t^{n-\alpha}}{(t+1)^{n+1}}\,dt\,\mathcal{I}.
$$

By formula number 3.194, in the table of integrals from [161], we have

$$\int_0^{+\infty} \frac{t^{n-\alpha}}{(t+1)^{n+1}}\, dt = B(n-\alpha+1,\alpha) = \frac{(n-\alpha)\cdots(1-\alpha)}{n!}\,\frac{\pi}{\sin(\pi\alpha)}, \qquad (8.13)$$

where $B(x,y)$ denotes the Beta function, and hence $\mathcal{I}^{-\alpha} = \mathcal{I}$. $\qquad\square$

**Corollary 8.1.7.** *Let* $\alpha \in (0,1)$. *Then*

$$T^{-\alpha} = -\frac{\sin(\alpha\pi)}{\pi} \int_0^{+\infty} t^{-\alpha} S_R^{-1}(-t,T)\, dt. \qquad (8.14)$$

**Corollary 8.1.8.** *For* $\alpha \in (0, n+1)$, *the operators* $T^{-\alpha}$ *are uniformly bounded by the constant* $M_{n+1}$ *obtained from Lemma 8.1.1.*

*Proof.* From (8.11), Lemma 8.1.1 and (8.13), we obtain the estimate

$$\|T^{-\alpha}\| \le \frac{\sin(\alpha\pi)}{\pi}\,\frac{n!}{(n-\alpha)\cdots(1-\alpha)} \int_0^{+\infty} t^{n-\alpha}\,\frac{M_{n+1}}{(1+t)^{n+1}}\, dt = M_{n+1}. \qquad \square$$

**Corollary 8.1.9.** *Assume that* $\sigma_S(T) \subset \{s \in \mathbb{H} : \mathrm{Re}(s) > 0\}$ *and that* $\varphi$ *in Lemma 8.1.1 can be chosen lower or equal to* $\pi/2$. *For* $\alpha \in (0,1)$, *we then have*

$$T^{-\alpha} = \frac{1}{\pi} \int_0^{+\infty} \tau^{-\alpha} \left( \cos\left(\frac{\alpha\pi}{2}\right) T + \sin\left(\frac{\alpha\pi}{2}\right) \tau\mathcal{I} \right) (T^2 + \tau^2)^{-1}\, d\tau.$$

*Proof.* By our assumptions, we can choose $n = 0$ and $\theta = \pi/2$ in (8.12). Since $e^{j\frac{\pi}{2}} = j$ and $e^{-j\frac{\pi}{2}} = -j$, we then have

$$T^{-\alpha} = \frac{j}{2\pi} \int_0^{+\infty} t^{-\alpha} e^{-j\frac{\alpha-1}{2}\pi} S_R^{-1}(jt,T)\, dt - \frac{j}{2\pi} \int_0^{+\infty} t^{-\alpha} e^{j\frac{\alpha-1}{2}\pi} S_R^{-1}(-jt,T)\, dt.$$

We observe that

$$S_R^{-1}(\pm tj,T) = -(T \pm tj\mathcal{I})(T^2 + t^2)^{-1},$$

and so

$$T^{-\alpha} = \frac{j}{2\pi} \int_0^{+\infty} t^{-\alpha} \left( -e^{-j\frac{\alpha-1}{2}\pi}(T+tj\mathcal{I}) + e^{j\frac{\alpha-1}{2}\pi}(T-tj\mathcal{I}) \right) (T^2 + t^2)^{-1}\, dt.$$

Some easy simplifications show

$$-e^{-j\frac{\alpha-1}{2}\pi}(T+tj\mathcal{I}) + e^{j\frac{\alpha-1}{2}\pi}(T-tj\mathcal{I}) = -2j \left[ \cos\left(\frac{\alpha\pi}{2}\right) T + 2\sin\left(\frac{\alpha\pi}{2}\right) t\mathcal{I} \right],$$

and in turn

$$T^{-\alpha} = \frac{1}{\pi} \int_0^{+\infty} t^{-\alpha} \left( \cos\left(\frac{\alpha\pi}{2}\right) T + \sin\left(\frac{\alpha\pi}{2}\right) t\mathcal{I} \right) (T^2 + t^2)^{-1}\, dt. \qquad \square$$

Observe that $s \mapsto s^{-\alpha}$ is intrinsic slice hyperholomorphic. Hence, we could also use the left $S$-resolvent operator to define fractional powers of $T$. Indeed, this yields exactly the same operator.

**Proposition 8.1.10.** *Let* $\alpha > 0$ *and let* $\Gamma$ *be an admissible path as in Definition 8.1.2. The operator* $T^{-\alpha}$ *satisfies*

$$T^{-\alpha} = \frac{1}{2\pi} \int_{\Gamma} S_L^{-1}(s, T) \, ds_j \, s^{-\alpha}. \tag{8.15}$$

*Proof.* We can prove this statement in two different ways. We can either perform computations as (3.4.6) in order to show that the slice hyperholomorphic Cauchy integral in (8.15) equals the one in (8.8).

As an alternative approach, we can perform computations analogue to those in the proof of Theorem 8.1.5 to show that, for $n \in \mathbb{N}$ and $\alpha \in (0, n+1)$ with $\alpha \notin \mathbb{N}$, one has

$$\frac{1}{2\pi} \int_{\Gamma} S_L^{-1}(s, T) \, ds_j \, s^{-\alpha}$$

$$= (-1)^{n+1} \frac{\sin(\alpha\pi)}{\pi} \frac{n!}{(n-\alpha)\cdots(1-\alpha)} \int_0^{+\infty} S_L^{-(n+1)}(-t, T) t^{n-\alpha} \, dt.$$

But for real $t$ one has $S_L^{-1}(-t, T) = (-t - T)^{-1} = S_R^{-1}(-t, T)$, and in turn this integral equals

$$(-1)^{n+1} \frac{\sin(\alpha\pi)}{\pi} \frac{n!}{(n-\alpha)\cdots(1-\alpha)} \int_0^{+\infty} t^{n-\alpha} S_R^{-(n+1)}(-t, T) \, dt = T^{-\alpha},$$

where the last equation follows from Theorem 8.1.5. If $\alpha \in \mathbb{N}$, the statement follows immediately from the $S$-functional calculus and Corollary 8.1.4 because $s^{-\alpha}$ is intrinsic slice hyperholomorphic at infinity. $\square$

**Theorem 8.1.11.** *The family* $\{T^{-\alpha}\}_{\alpha>0}$ *has the semigroup property* $T^{-\alpha}T^{-\beta} = T^{-(\alpha+\beta)}$.

*Proof.* Let $\varphi$ and $a$ be the constants obtained from Lemma 8.1.1. We choose $\theta_p$ and $\theta_s$ such that $\max\{\varphi, \pi/2\} < \theta_p < \theta_s < \pi$ and we choose $a_p$ and $a_s$ with $0 < a_s < a_p < a$ and $a_p$ sufficiently small such that $\overline{B_{a_p}(0)} \cap \overline{\Sigma_\varphi} + a = \emptyset$. Then the sets

$$G_p = \Sigma_{\theta_p} \setminus \overline{B_{a_p}(0)} \quad \text{and} \quad G_s = \Sigma_{\theta_s} \setminus \overline{B_{a_s}(0)}$$

satisfy $\sigma_S(T) \subset G_p$ and $\overline{G_p} \subset G_s$ and for $j \in \mathbb{S}$ their boundaries $\partial(G_p \cap \mathbb{C}_j)$ and $\partial(G_s \cap \mathbb{C}_j)$ are admissible paths as in Definition 8.1.2. The subscripts $p$ and $s$ refer again to the respective variables of integration in the following calculation.

The $S$-resolvent equation and Proposition 8.1.10 imply

$$T^{-\alpha}T^{-\beta} = \frac{1}{(2\pi)^2} \int_{\partial(G_s \cap \mathbb{C}_j)} s^{-\alpha} \, ds_j \, S_R^{-1}(s, T) \int_{\partial(G_p \cap \mathbb{C}_j)} S_L^{-1}(p, T) \, dp_j \, p^{-\beta}$$

$$= \frac{1}{(2\pi)^2} \int_{\partial(G_s \cap \mathbb{C}_j)} s^{-\alpha} \, ds_j \int_{\partial(G_p \cap \mathbb{C}_j)} S_R^{-1}(s,T) p \mathcal{Q}_s(p)^{-1} dp_j \ p^{-\beta}$$

$$- \frac{1}{(2\pi)^2} \int_{\partial(G_s \cap \mathbb{C}_j)} s^{-\alpha} \, ds_j \int_{\partial(G_p \cap \mathbb{C}_j)} S_L^{-1}(p,T) p \mathcal{Q}_s(p)^{-1} dp_j \ p^{-\beta}$$

$$- \frac{1}{(2\pi)^2} \int_{\partial(G_s \cap \mathbb{C}_j)} s^{-\alpha} \, ds_j \int_{\partial(G_p \cap \mathbb{C}_j)} \bar{s} S_R^{-1}(s,T) \mathcal{Q}_s(p)^{-1} dp_j \ p^{-\beta}$$

$$+ \frac{1}{(2\pi)^2} \int_{\partial(G_s \cap \mathbb{C}_j)} s^{-\alpha} \, ds_j \int_{\partial(G_p \cap \mathbb{C}_j)} \bar{s} S_L^{-1}(p,T) \mathcal{Q}_s(p)^{-1} dp_j \ p^{-\beta},$$

where we use the notation $\mathcal{Q}_s(p)^{-1} = (p^2 - 2\text{Re}(s)p + |s|^2)^{-1}$. But since the functions $p \mapsto p\mathcal{Q}_s(p)^{-1} p^{-\beta}$ and $p \mapsto \mathcal{Q}_s(p)^{-1} p^{-\beta}$ are holomorphic on an open set that contains $\overline{G_p \cap \mathbb{C}_j}$ and since they tend uniformly to zeros as $p \to \infty$ in $G_p$, Cauchy's integral theorem implies

$$\frac{1}{(2\pi)^2} \int_{\partial(G_s \cap \mathbb{C}_j)} s^{-\alpha} \, ds_j \int_{\partial(G_p \cap \mathbb{C}_j)} S_R^{-1}(s,T) p \mathcal{Q}_s(p)^{-1} dp_j \ p^{-\beta} = 0$$

and

$$- \frac{1}{(2\pi)^2} \int_{\partial(G_s \cap \mathbb{C}_j)} s^{-\alpha} \, ds_j \int_{\partial(G_p \cap \mathbb{C}_j)} \bar{s} S_R^{-1}(s,T) \mathcal{Q}_s(p)^{-1} \, dp_j \ p^{-\beta} = 0.$$

It follows that

$$T^{-\alpha} T^{-\beta}$$

$$= -\frac{1}{(2\pi)^2} \int_{\partial(G_s \cap \mathbb{C}_j)} s^{-\alpha} \, ds_j \int_{\partial(G_p \cap \mathbb{C}_j)} S_L^{-1}(p,T) p \mathcal{Q}_s(p)^{-1} dp_j \ p^{-\beta} \tag{8.16}$$

$$+ \frac{1}{(2\pi)^2} \int_{\partial(G_s \cap \mathbb{C}_j)} s^{-\alpha} \, ds_j \int_{\partial(G_p \cap \mathbb{C}_j)} \bar{s} S_L^{-1}(p,T) \mathcal{Q}_s(p)^{-1} dp_j \ p^{-\beta}.$$

Let us assume that we can apply Fubini's theorem in order to exchange the order of integration. We then find that

$$T^{-\alpha} T^{-\beta} = \frac{1}{(2\pi)^2} \int_{\partial(G_s \cap \mathbb{C}_j)} s^{-\alpha} \, ds_j$$

$$\cdot \int_{\partial(G_p \cap \mathbb{C}_j)} [\bar{s} S_L^{-1}(p,T) - S_L^{-1}(p,T) p] \mathcal{Q}_s(p)^{-1} dp_j \ p^{-\beta}.$$

Applying Lemma 2.2.24 with $B = S_L^{-1}(p,T)$, we obtain

$$T^{-\alpha} \, T^{-\beta} = \frac{1}{2\pi} \int_{\partial(G_p \cap \mathbb{C}_j)} S_L^{-1}(p,T) dp_j \ p^{-\alpha} \ p^{-\beta}$$

$$= \frac{1}{2\pi} \int_{\partial(G_p \cap \mathbb{C}_j)} S_L^{-1}(p,T) dp_j \ p^{-(\alpha+\beta)} = T^{-(\alpha+\beta)}.$$

What remains to show is that we can actually apply Fubini's theorem in (8.16). For $\tau \in \{s, p\}$ we therefore decompose $\partial(G_\tau \cap \mathbb{C}_j) = \Gamma_\tau^- \cup \Gamma_\tau^- \cup \Gamma_\tau^+$ with

$$
\begin{aligned}
\Gamma_\tau^- &= \{-re^{j\theta_\tau}, r \in (-\infty, -a_\tau]\}, \\
\Gamma_\tau^0 &= \{a_\tau e^{-j\theta}, \theta \in (-\theta_\tau, \theta_\tau)\}, \\
\Gamma_\tau^+ &= \{re^{-j\theta_\tau}, r \in [a_\tau, +\infty)\},
\end{aligned}
$$

such that

$$
\begin{aligned}
T^{-\alpha} &\, T^{-\beta} \\
=& \sum_{u,v \in \{+,0,-\}} -\frac{1}{(2\pi)^2} \int_{\Gamma_s^u} s^{-\alpha} \, ds_j \int_{\Gamma_p^v} S_L^{-1}(p, T) p \mathcal{Q}_s(p)^{-1} dp_j \; p^{-\beta} \\
&+ \sum_{u,v \in \{+,0,-\}} \frac{1}{(2\pi)^2} \int_{\Gamma_s^u} s^{-\alpha} \, ds_j \int_{\Gamma_p^v} \overline{s} S_L^{-1}(p, T) \mathcal{Q}_s(p)^{-1} dp_j \; p^{-\beta}.
\end{aligned}
\tag{8.17}
$$

Since $p$ and $s$ commute, we have

$$
\mathcal{Q}_s(p)^{-1} = (p^2 - 2s_0 p + |s|^2)^{-1} = (p - s)^{-1}(p - \overline{s})^{-1}
$$

and thus for $u = +$ and $v = +$

$$
\begin{aligned}
\int_{\Gamma_s^+} & s^{-\alpha} \, ds_j \int_{\Gamma_p^+} S_L^{-1}(p, T) p \mathcal{Q}_s(p)^{-1} dp_j \; p^{-\beta} \\
=& \int_{a_s}^{+\infty} r^{-\alpha} e^{j\alpha\theta_s} e^{-j\theta_s}(-j) \int_{a_p}^{+\infty} S_L^{-1}(te^{-j\theta_p}, T) \\
& \cdot te^{-j\theta_p} \left(te^{-j\theta_p} - re^{-j\theta_s}\right)^{-1} \left(te^{-j\theta_p} - re^{j\theta_s}\right)^{-1} e^{-j\theta_p}(-j) t^{-\beta} e^{j\beta\theta_p} \, dt \, dr.
\end{aligned}
$$

Using the estimate

$$
\|S_L^{-1}(s, T)\| \leq \frac{M_1}{1 + |s|}
\tag{8.18}
$$

obtained from Lemma 8.1.1 and setting

$$
C := \sup_{t \in [0, +\infty)} \frac{M_1 t}{1 + t} < +\infty,
\tag{8.19}
$$

we find that the integral of the norm of the integrand is lower or equal to

$$
\begin{aligned}
\int_{a_s}^{+\infty} & \int_{a_p}^{+\infty} r^{-\alpha} \frac{M_1 t}{1 + t} \frac{1}{|t - re^{j(\theta_p - \theta_s)}|} \frac{1}{|t - re^{j(\theta_p + \theta_s)}|} t^{-\beta} \, dt \, dr \\
\leq C & \int_{a_s}^{+\infty} \int_{a_p}^{+\infty} \frac{r^{-(1+\alpha)}}{\left|\frac{t}{r} - e^{-j(\theta_s - \theta_p)}\right|} \frac{t^{-(1+\beta)}}{\left|\frac{r}{t} - e^{-j(\theta_s + \theta_p)}\right|} \, dt \, dr.
\end{aligned}
\tag{8.20}
$$

Since $\pi/2 < \theta_p < \theta_s < \pi$, we have $0 < \theta_s - \theta_p < \pi$ and $\pi < \theta_s + \theta_p < 2\pi$. Therefore,

$$\inf_{\xi \in \mathbb{R}} |\xi - e^{-j(\theta_s \pm \theta_p)}| < \inf_{\xi \in \mathbb{R}} \left| \mathrm{Im}\left( \xi - e^{-j(\theta_s \pm \theta_p)} \right) \right| = \inf_{\xi \in \mathbb{R}} |\sin(\theta_s \pm \theta_p)| \neq 0$$

and so

$$K_1 := \sup_{\xi \in [0, +\infty)} \frac{1}{|\xi - e^{-j(\theta_s - \theta_p)}|} < +\infty$$

$$K_2 := \sup_{\xi \in [0, +\infty)} \frac{1}{|\frac{\tau}{t} - e^{-j(\theta_s + \theta_p)}|} < +\infty.$$

$$\tag{8.21}$$

As $\alpha, \beta > 0$ and $a_s, a_p > 0$, we conclude from (8.20) that

$$\int_{a_s}^{+\infty} \int_{a_p}^{+\infty} r^{-\alpha} \frac{M_1 t}{1+t} \frac{1}{|t - re^{j(\theta_p - \theta_s)}|} \frac{1}{|t - re^{j(\theta_p + \theta_s)}|} t^{-\beta} \, dt \, dr$$

$$\leq C K_1 K_2 \int_{a_s}^{+\infty} r^{-(1+\alpha)} \, dr \int_{a_p}^{+\infty} t^{-(1+\beta)} \, dt < +\infty.$$

$$\tag{8.22}$$

The second integral in (8.17) with $u = +$ and $v = +$ is

$$\int_{\Gamma_s^+} s^{-\alpha} \, ds_j \int_{\Gamma_p^+} \bar{s} S_L^{-1}(p, T) \mathcal{Q}_s(p)^{-1} \, dp_j \, p^{-\beta}$$

$$= \int_{a_s}^{+\infty} r^{-\alpha} e^{j\alpha\theta_s} e^{-j\theta_s} (-j) \int_{a_p}^{+\infty} re^{j\theta_s} S_L^{-1}(te^{j\theta_p}, T)$$

$$\cdot (te^{-j\theta_p} - re^{-j\theta_s})^{-1} (te^{-j\theta_p} - re^{j\theta_s})^{-1} e^{-j\theta_p} (-j) t^{-\beta} e^{j\beta\theta_p} \, dt \, dr.$$

With (8.18) and (8.21), we can estimate the integral of the norm of the integrand by

$$\int_{a_s}^{+\infty} \int_{a_p}^{+\infty} r^{-\alpha+1} \frac{M_1}{1+t} \frac{1}{|te^{-j(\theta_p - \theta_s)} - r|} \frac{1}{|te^{-j(\theta_p + \theta_s)} - r|} t^{-\beta} \, dt \, dr$$

$$\leq M_1 \int_{a_s}^{+\infty} r^{-(1+\alpha)} \int_{a_p}^{+\infty} \frac{t^{-\beta}}{1+t} \frac{1}{|\frac{t}{r} - e^{-j(\theta_s - \theta_p)}|} \frac{1}{|\frac{t}{r} - e^{j(\theta_p + \theta_s)}|} \, dt \, dr$$

$$\leq M_1 K_1 K_2 \int_{a_s}^{+\infty} r^{-(1+\alpha)} \, dr \int_{a_p}^{+\infty} \frac{t^{-\beta}}{1+t} \, dt < +\infty.$$

For $u = +$ and $v = 0$, we have

$$\int_{\Gamma_s^+} s^{-\alpha} \, ds_j \int_{\Gamma_p^0} S_L^{-1}(p, T) p \mathcal{Q}_s(p)^{-1} \, dp_j \, p^{-\beta}$$

$$= \int_{a_s}^{+\infty} r^{-\alpha} e^{j\alpha\theta_s} e^{-j\theta_s} (-j) \int_{-\theta_p}^{\theta_p} S_L^{-1}(a_p e^{-j\theta}, T) a_p e^{-j\theta}$$

$$\cdot (a_p e^{-j\theta} - re^{-j\theta_s})^{-1} (a_p e^{-j\theta} - re^{j\theta_s})^{-1} a_p e^{-j\theta} (-j)^2 a_p^{-\beta} e^{j\beta\theta} \, d\theta \, dr$$

and, again using (8.18), we find that the integral of the absolute value of the integrand is lower or equal to

$$\int_{a_s}^{+\infty} r^{-\alpha} \int_{-\theta_p}^{\theta_p} \frac{M_1 a_p^{2-\beta}}{1+a_p} \frac{1}{\left|r-a_p e^{-j(\theta-\theta_s)}\right|} \frac{1}{\left|r-a_p e^{-j(\theta+\theta_s)}\right|} \, d\theta \, dr.$$

Since $\pi/2 < \theta_p < \theta_s < \pi$, the distance $\delta$ between the set

$$\left\{ a_p e^{-j(\theta+\theta_s)} : \quad \theta \in [-\theta_p, \theta_p] \right\} \cup \left\{ a_p e^{-j(\theta-\theta_s)} : \quad \theta \in [-\theta_p, \theta_p] \right\} \qquad (8.23)$$

and the positive real axis is greater than zero and hence the above integral can be estimated by

$$\int_{a_s}^{a_s+2a_p} r^{-\alpha} \int_{-\theta_p}^{\theta_p} \frac{M_1 a_p^{2-\beta}}{1+a_p} \frac{1}{\delta^2} \, dr \, d\theta$$

$$+ \int_{a_s+2a_p}^{+\infty} \frac{r^{-\alpha}}{(r-a_p)^2} \int_{-\theta_p}^{\theta_p} \frac{M_1 a_p^{2-\beta}}{1+a_p} \, d\theta \, dr < +\infty.$$

For the second integral in (8.17) with $u = +$ and $v = 0$, we have

$$\int_{\Gamma_s^+} s^{-\alpha} \, ds_j \int_{\Gamma_p^0} \bar{s} S_L^{-1}(p,T) \mathcal{Q}_s(p)^{-1} \, dp_j \, p^{-\beta}$$

$$= \int_{a_s}^{+\infty} r^{-\alpha} e^{j\alpha\theta_s} e^{-j\theta_s} (-j) \int_{-\theta_p}^{\theta_p} r e^{j\theta_s} S_L^{-1}(a_p e^{-j\theta}, T)$$

$$\cdot \left(a_p e^{-j\theta} - r e^{-j\theta_s}\right)^{-1} \left(a_p e^{-j\theta} - r e^{j\theta_s}\right)^{-1} a_p e^{-j\theta}(-j)^2 a_p^{-\beta} e^{j\beta\theta} \, d\theta \, dr$$

Using (8.18), we can estimate the integral of the absolute value of the integrand by

$$\int_{a_s}^{+\infty} r^{1-\alpha} \int_{-\theta_p}^{\theta_p} \frac{a_p^{1-\beta} M_1}{1+a_p} \frac{1}{\left|a_p e^{-j(\theta-\theta_s)} - r\right|} \frac{1}{\left|a_p e^{-j(\theta+\theta_s)} - r\right|} \, d\theta \, dr$$

$$\leq \int_{a_s}^{a_s+2a_p} r^{1-\alpha} \int_{-\theta_p}^{\theta_p} \frac{a_p^{1-\beta} M_1}{1+a_p} \frac{1}{\delta^2} \, d\theta \, dr$$

$$+ \int_{a_s+2a_p}^{+\infty} \frac{r^{1-\alpha}}{(r-a_p)^2} \int_{-\theta_p}^{\theta_p} \frac{a_p^{1-\beta} M_1}{1+a_p} \, d\theta \, dr < +\infty,$$

where $\delta > 0$ is again the distance between the set in (8.23) and the positive real axis. Similar estimates hold true if $u = -$ and $v = 0$.

If $u = 0$ and $v = +$, then

$$\int_{\Gamma_s^0} s^{-\alpha} \, ds_j \int_{\Gamma_p^+} S_L^{-1}(p, T) p Q_s(p)^{-1} dp_j \; p^{-\beta}$$

$$= \int_{-\theta_s}^{\theta_s} a_s^{-\alpha} e^{j\alpha\theta} a_s e^{-j\theta} (-j)^2 \int_{a_p}^{+\infty} S_L^{-1}(te^{-j\theta_p}, T) t e^{-j\theta_s}$$

$$\cdot \left(te^{-j\theta_p} - a_s e^{-j\theta}\right)^{-1} \left(te^{-j\theta_p} - a_s e^{j\theta}\right)^{-1} e^{-j\theta_s}(-j) t^{-\beta} e^{j\beta\theta_p} \, dt \, d\theta.$$

Once more (8.18) allows us to estimate the integral of the absolute value of the integrand by

$$\int_{-\theta_s}^{\theta_s} a_s^{1-\alpha} \int_{a_p}^{+\infty} \frac{M_1 t}{1+t} \frac{1}{\left|t - a_s e^{j(\theta_p - \theta)}\right|} \frac{1}{\left|t - a_s e^{j(\theta_p + \theta)}\right|} t^{-\beta} \, dt \, d\theta$$

$$\leq C \int_{-\theta_s}^{\theta_s} a_s^{1-\alpha} \int_{a_p}^{+\infty} \frac{t^{-\beta}}{(t - a_s)^2} \, dt \, d\theta < +\infty,$$

where $C$ is again as in (8.19), and the second inequality follows because $a_s < a_p$. For the second integral in (8.17) with $u = 0$ and $v = +$, we similarly have

$$\int_{\Gamma_s^0} s^{-\alpha} \, ds_j \int_{\Gamma_p^+} \bar{s} S_L^{-1}(p, T) Q_s(p)^{-1} dp_j \; p^{-\beta}$$

$$= \int_{-\theta_s}^{\theta_s} a_s^{-\alpha} e^{j\alpha\theta} a_s e^{-j\theta} (-j)^2 \int_{a_p}^{+\infty} a_s e^{j\theta_s} S_L^{-1}(te^{-j\theta_p}, T)$$

$$\cdot \left(te^{-j\theta_p} - a_s e^{-j\theta}\right)^{-1} \left(te^{-j\theta_p} - a_s e^{j\theta}\right)^{-1} e^{-j\theta_s}(-j) t^{-\beta} e^{j\beta\theta_p} \, dt \, d\theta.$$

As above, we can estimate the integral of the absolute value of the integrand by

$$\int_{-\theta_s}^{\theta_s} a_s^{2-\alpha} \int_{a_p}^{+\infty} \frac{M_1}{1+t} \frac{1}{\left|t - a_s e^{j(\theta_p - \theta)}\right|} \frac{1}{\left|t - a_s e^{j(\theta_p + \theta)}\right|} t^{-\beta} \, dt \, d\theta$$

$$\leq C \int_{-\theta_s}^{\theta_s} a_s^{2-\alpha} \int_{a_p}^{+\infty} \frac{t^{-(1+\beta)}}{(t - a_s)^2} \, dt \, d\theta < +\infty,$$

where the last inequality follows again because $a_s < a_p$. Similar estimates hold for the case $u = 0$ and $v = -$.

Finally, the integrals in (8.17) with $u = 0$ and $v = 0$ are absolutely convergent since, in this case, we integrate a continuous and hence bounded function over a bounded domain.

Putting these pieces together, we obtain that we can actually apply Fubini's theorem in (8.16) in order to exchange the order of integration. □

**Lemma 8.1.12.** *Since $\mathcal{D}(T)$ is dense in $X$, the semigroup $\{T^{-\alpha}\}_{\alpha \geq 0}$ is strongly continuous.*

*Proof.* We first consider $y \in \mathcal{D}(T)$. For $\alpha \in (0,1)$, we have

$$S_R^{-1}(t,T)y - S_R^{-1}(t,\mathcal{I})y = S_R^{-1}(t,T)S_R^{-1}(t,\mathcal{I})(Ty - \mathcal{I}y)$$

if $t \in \mathbb{R}$. Hence, we deduce from Corollary 8.1.7 that

$$
\begin{aligned}
& T^{-\alpha}y - \mathcal{I}^{-\alpha}y \\
&= -\frac{\sin(\alpha\pi)}{\pi}\int_0^{+\infty} t^{-\alpha}S_R^{-1}(-t,T)y\,dt + \frac{\sin(\alpha\pi)}{\pi}\int_0^{+\infty} t^{-\alpha}S_R^{-1}(-t,\mathcal{I})y\,dt \\
&= -\frac{\sin(\alpha\pi)}{\pi}\int_0^{+\infty} t^{-\alpha}S_R^{-1}(t,T)S_R^{-1}(t,\mathcal{I})(Ty - \mathcal{I}y)\,dt \xrightarrow{\alpha\to 0} 0
\end{aligned}
$$

because $\sin(\alpha\pi) \to 0$ as $\alpha \to 0$ and the integral is uniformly bounded for $\alpha \in [0,1/2]$ due to (8.3). Since $\mathcal{I}^{-\alpha} = \mathcal{I}$ by Corollary 8.1.6, we get $T^{-\alpha}y \to y$ as $\alpha \to 0$ for any $y \in \mathcal{D}(T)$.

For arbitrary $y \in X$ and $\varepsilon > 0$, there exists $y_\varepsilon \in \mathcal{D}(T)$ with $\|y - y_\varepsilon\| < \varepsilon$ because $\mathcal{D}(T)$ is dense in $X$. Corollary 8.1.8 therefore implies

$$
\begin{aligned}
\lim_{\alpha\to 0}\|Ty - y\| &\le \lim_{\alpha\to 0}\|Ty - T^{-\alpha}y_\varepsilon\| + \|T^{-\alpha}y_\varepsilon - y_\varepsilon\| + \|y_\varepsilon - y\| \\
&\le (M_1 + 1)\|y - y_\varepsilon\| \le (M_1 + 1)\varepsilon.
\end{aligned}
$$

Since $\varepsilon > 0$ was arbitrary, we deduce that $T^{-\alpha}y \to y$ as $\alpha \to 0$ even for arbitrary $y \in X$. This is equivalent to the strong continuity of the semigroup $(T^{-\alpha})_{\alpha \ge 0}$.  □

**Proposition 8.1.13.** *The operator $T^{-\alpha}$ is injective for any $\alpha > 0$.*

*Proof.* For $\alpha > 0$ choose $\beta > 0$ with $n = \alpha + \beta \in \mathbb{N}$. Then $T^{-\beta}T^{-\alpha} = T^{-n}$ and in turn $T^n T^{-\beta}T^{-\alpha} = \mathcal{I}$, which implies the injectivity of $T^{-\alpha}$.  □

The previous proposition allows us to define powers of $T$ also for $\alpha > 0$.

**Definition 8.1.14.** For $\alpha > 0$, we define the operator $T^\alpha$ as the inverse of the operator $T^{-\alpha}$, which is defined on $\mathcal{D}(T^\alpha) = \operatorname{ran}(T^{-\alpha})$.

**Corollary 8.1.15.** *Let $\alpha, \beta \in \mathbb{R}$. Then the operators $T^\alpha T^\beta$ and $T^{\alpha+\beta}$ agree on $\mathcal{D}(T^\gamma)$ with $\gamma = \max\{\alpha, \beta, \alpha + \beta\}$.*

*Proof.* If $\alpha, \beta \ge 0$ and $y \in \mathcal{D}(T^{\alpha+\beta})$ then, since $T^{-(\alpha+\beta)} = T^{-\beta}T^{-\alpha}$ by Theorem 8.1.11, we have

$$T^\alpha T^\beta y = T^\alpha T^\beta (T^{-\beta}T^{-\alpha}T^{\alpha+\beta})y = (T^\alpha T^\beta T^{-\beta}T^{-\alpha})T^{\alpha+\beta}y = T^{\alpha+\beta}y.$$

The other cases follow in a similar way.  □

With these definitions it is possible to establish a theory of interpolation spaces for strongly continuous quaternionic semigroups analogue to the one for complex operator semigroups. Since the proofs follow the lines of this classical case and the quaternionic theory does not show any significant difference to the complex one, we refer to Chapter II in [113] for an overview on these results.

## 8.2 Fractional powers via the $H^\infty$-functional calculus

The most general strategy for introducing fractional powers of complex linear operators is via the $H^\infty$-functional calculus. It applies to arbitrary sectorial operators and does not require further assumptions on the operator. Following [165], we show now that this is also possible in the quaternionic setting.

Let $T \in \mathrm{Sect}(\omega)$ and let $\alpha \in (0, +\infty)$. The function $s \mapsto s^\alpha$ then obviously belongs to $\mathcal{M}[\Sigma_\omega]_T$ and we can define $T^\alpha$ using the quaternionic $H^\infty$-functional calculus introduced in Chapter 7. Precisely, we can choose $n \in \mathbb{N}$ with $n > \alpha$ and find

$$T^\alpha := s^\alpha(T) = (\mathcal{I} + T)^n \left( s^\alpha (1+s)^{-n} \right)(T), \tag{8.24}$$

where $(s^\alpha(1+s)^{-n})(T)$ is defined via a slice hyperholomorphic Cauchy integral as in (7.2) or (7.3).

**Definition 8.2.1.** Let $T \in \mathrm{Sect}(\omega)$ and $\alpha > 0$. We call the operator defined in (8.24) the fractional power with exponent $\alpha$ of $T$.

The following properties are immediate consequences of the properties of the $H^\infty$-functional calculus.

**Lemma 8.2.2.** *Let $T \in \mathrm{Sect}(\omega)$ and let $\alpha \in (0, +\infty)$.*

(i) *If $T$ is injective, then $\left(T^{-1}\right)^\alpha = (T^\alpha)^{-1}$. Thus $0 \in \rho_S(T)$ if and only if $0 \in \rho_S(T^\alpha)$.*

(ii) *Any bounded operator that commutes with $T$ commutes also with $T^\alpha$.*

(iii) *The spectral mapping theorem holds, namely*

$$\sigma_S(T^\alpha) = \{s^\alpha : \quad s \in \sigma_S(T)\}.$$

Another important property is the analyticity in the exponent. Observe that, although in the complex case the mapping $\alpha \mapsto T^\alpha$ is holomorphic in $\alpha$, we cannot expect slice hyperholomorphicity here because the fractional powers are only defined for real exponents, cf. Example 2.1.15.

**Proposition 8.2.3.** *If $T \in \mathrm{Sect}(\omega)$, then the following statements hold true.*

(i) *If $T$ is bounded, then $T^\alpha$ is bounded too and the mapping $\Lambda : \alpha \to T^\alpha$ is analytic on $(0, +\infty)$ and has a left and a right slice hyperholomorphic extension to $\mathbb{H}^+ = \{s \in \mathbb{H} : \mathrm{Re}(s) > 0\}$. In particular, for any $\alpha_0 \in (0, +\infty)$ the Taylor series expansion of $\Lambda$ at $\alpha_0$ converges on $(0, 2\alpha_0)$.*

(ii) *If $n \in \mathbb{N}$ and $0 < \alpha < n$, then $\mathcal{D}(T^n) \subset \mathcal{D}(T^\alpha)$. For each $y \in \mathcal{D}(T^n)$, the mapping $\Lambda_y : \alpha \mapsto T^\alpha y$ is analytic on $(0, n)$ and the power series expansion of $\Lambda_y$ at $\alpha_0 \in (0, n)$ converges on $(-r_{\alpha_0} + \alpha, \alpha + r_{\alpha_0})$ with $r_{\alpha_0} = \min\{\alpha_0, n - \alpha_0\}$. Hence, $\Lambda_y$ has a left and a right slice hyperholomorphic expansion to the set $\bigcup_{\alpha_0 \in (0,n)} B_{r_{\alpha_0}}(\alpha_0)$.*

*Proof.* Let us first show (ii). If $n \in \mathbb{N}$ and $\alpha \in (0, n)$, then

$$T^\alpha = (\mathcal{I} + T)^n \left(s^\alpha (1+s)^{-n}\right)(T).$$

If $y \in \mathcal{D}(T^n)$, then the operators $T^n$ and $(s^\alpha(1+s)^{-n})(T)$ commute because of (i) in Lemma 7.2.8 such that $T^\alpha y = (s^\alpha(1+s)^{-n})(T)(\mathcal{I}+T)^n y$ and hence $y \in \mathcal{D}(T^\alpha)$.

Let now $\alpha_0 \in (0, n)$ and set $r := r_{\alpha_0} = \min\{\alpha_0, n - \alpha_0\}$. The Taylor series expansion of $\alpha \mapsto s^\alpha$ at $\alpha_0$ is $s^\alpha = \sum_{n=0}^{+\infty} \frac{(\alpha-\alpha_0)^k}{k!} s^{\alpha_0} \log(s)^k$ and converges on $(0, 2\alpha_0)$. If $\varepsilon \in (0,1)$ and $\alpha \in (0,n)$ with $|\alpha - \alpha_0| < (1-\varepsilon)r$, then we have, after choosing $\varphi \in (\omega, \pi)$, that

$$T^\alpha y = \left(s^\alpha(1+s)^{-n}\right)(T)(\mathcal{I}+T)^n y$$

$$= \frac{1}{2\pi} \int_{\partial(\Sigma_\varphi \cap \mathbb{C}_j)} s^\alpha(1+s)^{-n}\, ds_j\, S_R^{-1}(s,T)(\mathcal{I}+T)^n y$$

$$= \frac{1}{2\pi} \int_{\partial(\Sigma_\varphi \cap \mathbb{C}_j)} \sum_{k=0}^{+\infty} \frac{(\alpha-\alpha_0)^k}{k!} s^{\alpha_0} \tag{8.25}$$

$$\cdot \log(s)^k (1+s)^{-n}\, ds_j\, S_R^{-1}(s,T)(\mathcal{I}+T)^n y.$$

We want to apply the theorem of dominated convergence in order to exchange the integral and the series. Using (7.1) we find that $\tilde\Psi(s) = M\,\|(\mathcal{I}+T)^n y\|\,\Psi(s)$ with

$$\Psi(s) := \sum_{k=0}^{+\infty} \frac{|\alpha-\alpha_0|^k}{k!}|s|^{\alpha_0-1}\frac{|\log(s)|^k}{|1+s|^n}$$

is a dominating function for the integrand in (8.25). In order to show the integrability of $\Psi(s)$ along $\partial(\Sigma_\varphi \cap \mathbb{C}_j)$, we choose $C_{est} > 1$ such that $(1-\varepsilon)C_{est} < 1$ and $0 < t_0 < 1$ and $1 < t_1$ such that

$$|\ln(t) + j\theta| < C_{est}|\ln(t)|, \quad \forall t \in (0, t_0] \cup [t_1, +\infty). \tag{8.26}$$

We then have

$$\frac{1}{2}\int_{\partial(\Sigma_\varphi \cap \mathbb{C}_j)} \psi(s)\, d|s_j| = \int_0^{+\infty} \sum_{k=0}^{+\infty} \frac{|\alpha-\alpha_0|^k}{k!} t^{\alpha_0-1}\frac{|\ln(t)+j\varphi|^k}{|1+te^{j\varphi}|^n}\, dt$$

$$\leq \sum_{k=0}^{+\infty} \frac{|\alpha-\alpha_0|^k}{k!}\left(C_0 C_{est}^k \int_0^{t_0} t^{\alpha_0-1}(-\ln(t))^k\, dt\right.$$

$$\left. + C_0 C_1^k \int_{t_0}^{t_1} t^{\alpha_0-1}\, dt + C_2 C_{est}^k \int_{t_1}^{+\infty} t^{\alpha_0-(n+1)} \ln(t)^k\, dt\right),$$

with the constants

$$C_0 := \max_{t \in [0,t_1]} \frac{1}{|1+te^{j\varphi}|^n}, \quad C_1 := \max_{t \in [t_0,t_1]} |\ln(t)+j\varphi|$$

and a constant $C_2 > 0$ such that

$$\frac{1}{|1 + te^{j\varphi}|} < \frac{C_2}{t}, \quad \forall t \in [t_1, +\infty).$$

Since

$$\int_0^{t_1} t^{\alpha_0 - 1}(-\ln(t))^k \, dt \leq \int_{-\infty}^0 e^{\alpha_0 \xi}(-\xi)^k \, d\xi = \frac{k!}{\alpha_0^{k+1}}$$

and similarly

$$\int_{t_1}^{+\infty} t^{\alpha_0 - (n+1)} \ln(t)^k \, dt \leq \int_1^{+\infty} e^{-(n-\alpha_0)\xi} \xi^k \, d\xi = \frac{k!}{(n - \alpha_0)^{k+1}},$$

we can further estimate

$$\frac{1}{2} \int_{\partial(\Sigma_\varphi \cap \mathbb{C}_i)} \Psi(s) \, d|s_j|$$

$$\leq \sum_{k=0}^{+\infty} \frac{|\alpha - \alpha_0|^k}{k!} \left( C_0 C_{est}^k \frac{k!}{\alpha_0^{k+1}} + C_0 C_1^k \left( \frac{t_1^{\alpha_0} - t_0^{\alpha_0}}{\alpha_0} \right) + C_2 C_{est}^k \frac{k!}{(n - \alpha_0)^{k+1}} \right).$$

As $|\alpha - \alpha_0| < (1 - \varepsilon)r = (1 - \varepsilon)\min\{\alpha_0, n - \alpha_0\}$, we finally find

$$\frac{1}{2} \int_{\partial(\Sigma_\varphi \cap \mathbb{C}_i)} \Psi(s) \, d|s_j| \leq \frac{C_0}{\alpha_0} \sum_{k=0}^{+\infty} ((1 - \varepsilon)C_{est})^k$$

$$+ \frac{C_0(t_1^{\alpha_0} - t_0^{\alpha_0})}{\alpha_0} \sum_{k=0}^{+\infty} \frac{(C_1|\alpha - \alpha_0|)^k}{k!} + \frac{C_2}{\alpha_0} \sum_{k=0}^{+\infty} ((1 - \varepsilon)C_{est})^k.$$

Since $(1 - \varepsilon)C_1 < 1$, these series are finite and hence $\tilde{\Psi}$ is an integrable majorant of the integrand in (8.25). We can thus exchange the series and the integral in (8.25) such that

$$T^\alpha y = \sum_{k=0}^{+\infty} \frac{(\alpha - \alpha_0)^k}{k!} \frac{1}{2\pi} \int_{\partial(\Sigma_\varphi \cap \mathbb{C}_j)} s^{\alpha_0} \log(s)^k.$$

$$\cdot (1 + s)^{-n} \, ds_j \, S_R^{-1}(s, T)(\mathcal{I} + T)^n y,$$

where this series converges uniformly for $|\alpha - \alpha_0| < (1 - \varepsilon)r$. Since $\varepsilon \in (0, 1)$ was arbitrary, we obtain the statement.

Let us now consider (i). If $T$ is bounded, then (8.24) is the composition of two bounded operators and hence bounded. With arguments similar to those used above, one can show that the power series expansion of $\Lambda$ at $\alpha_0$ converges in $\mathcal{B}(X)$ on $(0, 2\alpha_0)$. If we write the scalar variable $(\alpha - \alpha_0)$ in the power series expansion on the left or on the right side of the coefficients and extend $\alpha$ to a quaternionic variable, we find that $\Lambda$ has a left (resp. a right) slice hyperholomorphic extension to $B_{\alpha_0}(\alpha_0)$. Finally, any point in $\mathbb{H}^+$ is contained in a ball of this form, and hence we find that we can extend $\Lambda$ to a left or to a right slice hyperholomorphic function on all of $\mathbb{H}^+$. $\qquad\square$

We show now that the usual computational rules that we expect to hold for fractional powers of an operator hold true with our approach.

**Proposition 8.2.4** (First Law of Exponents). *Let $T \in \mathrm{Sect}(\omega)$. For all $\alpha, \beta > 0$, the identity $T^{\alpha+\beta} = T^{\alpha}T^{\beta}$ holds. In particular $\mathcal{D}(T^{\gamma}) \subset \mathcal{D}(T^{\alpha})$ for $0 < \alpha < \gamma$.*

*Proof.* Because of (ii) in Lemma 7.2.8, we have $T^{\alpha}T^{\beta} \subset T^{\alpha+\beta}$ with

$$\mathcal{D}\left(T^{\alpha}T^{\beta}\right) = \mathcal{D}\left(T^{\alpha+\beta}\right) \cap \mathcal{D}\left(T^{\beta}\right).$$

We choose $n \in \mathbb{N}$ with $\alpha, \beta < n$ and define the bounded operators

$$\Lambda_{\alpha} := \left(s^{\alpha}(1+s)^{-n}\right)(T) \qquad \text{and} \qquad \Lambda_{\beta} := \left(s^{\beta}(1+s)^{-n}\right)(T).$$

If now $y \in \mathcal{D}\left(T^{\alpha+\beta}\right)$, then (ii) in Lemma 7.2.8 implies

$$\begin{aligned}
T^{\alpha+\beta}y &= (\mathcal{I}+T)^{2n}(\mathcal{I}+T)^{-2n}T^{\alpha+\beta}y \\
&= (\mathcal{I}+T)^{2n}T^{\alpha+\beta}(\mathcal{I}+T)^{-2n}y \\
&= (\mathcal{I}+T)^{2n}\left(s^{\alpha+\beta}(1+s)^{-2n}\right)(T)y = (\mathcal{I}+T)^{2n}\Lambda_{\alpha}\Lambda_{\beta}y
\end{aligned}$$

and hence $\Lambda_{\alpha}\Lambda_{\beta}y \in \mathcal{D}\left((\mathcal{I}+T^{2n})\right) = \mathcal{D}\left(T^{2n}\right)$. Since $\left(s^{n-\alpha}(1+s)^{-n}\right)(T)$ commutes with $T^{2n}$ because of (i) in Lemma 7.2.8, we thus find

$$T^{n}(\mathcal{I}+T)^{-2n}\Lambda_{\beta}y = \left(s^{n+\beta}(1+s)^{-3n}\right)(T)y = \left(s^{n-\alpha}(1+s)^{-n}\right)(T)\Lambda_{\alpha}\Lambda_{\beta}y$$

belongs to $\mathcal{D}\left(T^{2n}\right)$. Since $T$ and $T(\mathcal{I}+T)^{-1}$ commute, we have

$$T(\mathcal{I}+T)^{-1}Ty = T^{2}(\mathcal{I}+T)y$$

and hence $y \in \mathcal{D}(T)$ implies $T(\mathcal{I}+T)^{-1}y \in \mathcal{D}(T)$. If on the other hand $T(\mathcal{I}+T)^{-1}y \in \mathcal{D}(T)$, then the identity

$$T(\mathcal{I}+T)^{-1}y = y - (\mathcal{I}+T)^{-1}y$$

implies $y \in \mathcal{D}(T)$ and hence $y \in \mathcal{D}(T)$ if and only if $T(\mathcal{I}+T)^{-1}y \in \mathcal{D}(T)$. By induction, we find that $y \in \mathcal{D}(T^{m})$ if and only if $T^{n}(\mathcal{I}+T)^{-n}y \in \mathcal{D}(T^{m})$. We thus conclude that $(\mathcal{I}+T)^{-n}\Lambda_{\beta}y \in \mathcal{D}(T^{2n})$, which implies that $\Lambda_{\beta}y \in \mathcal{D}(T^{n})) = \mathcal{D}((\mathcal{I}+T)^{n})$. Thus, $T^{\beta}y = (\mathcal{I}+T)^{n}\Lambda_{\beta}y$ is defined, such that $y \in \mathcal{D}\left(T^{\beta}\right)$ for any $y \in \mathcal{D}(T^{\alpha+\beta})$. We conclude that

$$\mathcal{D}\left(T^{\alpha}T^{\beta}\right) = \mathcal{D}\left(T^{\alpha+\beta}\right) \cap \mathcal{D}\left(T^{\beta}\right) = \mathcal{D}\left(T^{\alpha+\beta}\right)$$

and so $T^{\alpha}T^{\beta} = T^{\alpha+\beta}$. $\qquad\square$

**Proposition 8.2.5** (Scaling Property). *Let $T \in \mathrm{Sect}(\omega)$ and let $\Lambda = [\delta_1, \delta_2] \subset (0, \pi/\omega)$ be a compact interval. Then the family $\{T^\alpha\}_{\alpha \in \Lambda}$ is uniformly sectorial of angle $\delta_2 \omega$. In particular, for every $\alpha \in (0, \pi/\omega_T)$, the operator $T^\alpha$ is sectorial with spectral angle $\omega_{T^\alpha} = \alpha \omega_T$.*

*Proof.* The second statement obviously follows from the first by choosing $\lambda = [\alpha, \alpha]$. Because of (iii) in Lemma 8.2.2, we know that

$$\sigma_S(T^\alpha) = (\sigma_S(T))^\alpha \subset \overline{\Sigma_{\alpha\omega}} \subset \overline{\Sigma_{\delta_2\omega}}$$

for $\alpha \in \Lambda$. What remains to show are the uniform estimates (7.1) for the $S$-resolvents.

We choose $\varphi \in (\delta_2\omega, \pi)$. In order to show that $\|S_L^{-1}(s, T^\alpha)s\|$ is uniformly bounded for $s \notin \Sigma_\varphi$ and $\alpha \in \Lambda$, we define for $\alpha \in \Lambda$ and $s \notin \Sigma_\varphi$ the function

$$\Psi_{s,\alpha}(p) = S_L^{-1}(s, p^\alpha) s + S_L^{-1}\left(-|s|^{\frac{1}{\alpha}}, p\right) |s|^{\frac{1}{\alpha}}$$
$$= \mathcal{Q}_s(p^\alpha)^{-1} \left(|s|^{\frac{1}{\alpha}} + p\right)^{-1} \left(p(\overline{s} - p^\alpha)s + p^\alpha(\overline{s} - p^\alpha)|s|^{\frac{1}{\alpha}}\right). \tag{8.27}$$

This function belongs to $\mathcal{SH}_{L,0}^\infty[\Sigma_\omega]$: as $s \notin \Sigma_\varphi$, it is left slice hyperholomorphic on $\Sigma_{\theta_0}$ with $\theta_0 := \min\{\alpha^{-1}\varphi, \pi\} > \omega$. The first line in (8.27) implies that $\Psi_{s,\alpha}$ has polynomial limit 0 at infinity because $S_L^{-1}(s, p^\alpha)$ and $S_L^{-1}(|s|^{1/\alpha}, p)$ have polynomial limit 0 at infinity and the second line in (8.27) implies that $\Psi_{s,\alpha}$ has polynomial limit 0 at 0 because $\mathcal{Q}_s(p^\alpha)^{-1}$ and $(|s|^{1/\alpha} - p)^{-1}$ are bounded for $p$ sufficiently close to 0. Since the function

$$S_L^{-1}\left(|s|^{1/\alpha}, p\right) = \left(|s|^{1/\alpha} - p\right)^{-1}$$

belongs to $\mathcal{E}_L[\Sigma_\omega]$, we find that also

$$S_L^{-1}(s, p^\alpha)s = \Psi_{s,\alpha}(p) + S_L^{-1}(|s|^{1/\alpha}, T)|s|^{1/\alpha}$$

belongs to $\mathcal{E}_L[\Sigma_\omega]$ and that

$$S_L^{-1}(s, T^\alpha)s = S_L^{-1}\left(|s|^{\frac{1}{\alpha}}, T\right) |s|^{\frac{1}{\alpha}} + \Psi_{s,\alpha}(T).$$

The function $\Psi_{s,\alpha}$ satisfies the scaling property $\Psi_{t^\alpha s,\alpha}(tp) = \Psi_{s,\alpha}(p)$ and so we have

$$\Psi_{\frac{s}{|s|},\alpha}\left(|s|^{-\frac{1}{\alpha}}, p\right) = \Psi_{s,\alpha}(p).$$

If we choose $\theta \in (\omega, \min\{\pi, \delta_2^{-1}\varphi\})$ and $j = j_s$, we therefore find that

$$\|S_L^{-1}(s, T^\alpha)s\| \leq C_{\theta',T} + \left\|\Psi_{s/|s|,\alpha}\left(|s|^{-\frac{1}{\alpha}}T\right)\right\|$$
$$\leq C_{\theta',T} + \frac{1}{2\pi}\left\|\int_{\partial(\Sigma_\theta \cap \mathbb{C}_j)} S_L^{-1}(p, T) \, ds_j \, \Psi_{s/|s|,\alpha}\left(|s|^{-\frac{1}{\alpha}}p\right)\right\|$$
$$\leq C_{\theta',T} + \frac{C_{\theta',T}}{2\pi}\int_{\partial(\Sigma_\theta \cap \mathbb{C}_j)} |p|^{-1}d|p| \, |\Psi_{s/|s|,\alpha}(p)|,$$

where $C_{\theta',T}$ is the respective constant in (7.1) for some $\theta' \in (\omega,\theta)$, which is independent of $s$ and $\alpha \in \Lambda$. Hence, if we are able to show that

$$\sup\left\{\int_{\partial(\Sigma_\theta \cap \mathbb{C}_{j_s})} |p|^{-1} d|p| \, |\Psi_{s,\alpha}(p)| : \quad |s| = 1, \; s \notin \Sigma_\varphi, \; \alpha \in \Lambda \right\} < +\infty, \quad (8.28)$$

then we are done. Since we integrate along a path in the complex plane $\mathbb{C}_{j_s}$, we find that $p$ and $s$ commute and $\Psi_{s,\alpha}(p)$ simplifies to

$$\Psi_{s,\alpha}(p) = (s - p^\alpha)^{-1} \left(p + |s|^{1/\alpha}\right)^{-1} \left(ps + |s|^{1/\alpha} p^\alpha\right).$$

As $|s| = 1$, we can therefore estimate

$$|\Psi_{s,\alpha}(p)| \leq \frac{|p|^{1-\varepsilon}}{|s - p^\alpha|} \frac{|p|^\varepsilon}{|1 + p|} + \frac{|p|^{\alpha-\varepsilon}}{|s - p^\alpha|} \frac{|p|^\varepsilon}{|1 + p|} \leq K \frac{|p|^\varepsilon}{|1 + p|}$$

with $\varepsilon \in (0, \delta_1)$, because $|p|^{1-\varepsilon}/|s - p^\alpha|$ and $|p|^{\alpha-\varepsilon}/|s - p^\alpha|$ are uniformly bounded by some constant $K > 0$ for our parameters $s$, $\alpha$ and $p$. Thus, we have an estimate for the integrand in (8.28) that is independent of the parameters such that (8.28) is actually true.

   With analogous arguments using the right slice hyperholomorphic version of the $S$-functional calculus for sectorial operators, we can show that $\|sS_R^{-1}(s, T^\alpha)\|$ is also uniformly bounded for $s \notin \overline{\Sigma_\varphi}$ and $\alpha \in \Lambda$. Since $\varphi \in (\delta_2\omega, \pi)$ was arbitrary, the proof is finished.                                                                          $\square$

   As immediate consequences of Proposition 8.2.5 and the composition rule Theorem 7.3.1, we obtain the following two results.

**Proposition 8.2.6.** *Let* $T \in \mathrm{Sect}(\omega)$ *for some* $\omega \in (0, \pi)$ *and let* $\alpha \in (0, \pi/\omega)$ *and* $\varphi \in (\omega, \pi/\alpha)$. *If* $f \in \mathcal{SH}_{L,0}^\infty(\Sigma_{\alpha\varphi})$ *(or* $f \in \mathcal{M}_L[\Sigma_{\alpha\omega}]_{T^\alpha}$*), then the function* $p \mapsto f(p^\alpha)$ *belongs to* $\mathcal{SH}_{L,0}^\infty(\Sigma_\varphi)$ *(resp.* $\mathcal{M}_L[\Sigma_\omega]_T$*) and*

$$f(T^\alpha) = (f(p^\alpha))(T).$$

**Corollary 8.2.7** (Second Law of Exponents). *Let* $T \in \mathrm{Sect}(\omega)$ *with* $\omega \in (0, \pi)$ *and let* $\alpha \in (0, \pi/\omega)$. *For all* $\beta > 0$, *we have*

$$(T^\alpha)^\beta = T^{\alpha\beta}.$$

**Corollary 8.2.8.** *Let* $T \in \mathrm{Sect}(\omega)$ *and* $\gamma > 0$. *For any* $y \in \mathcal{D}(T^\gamma)$, *the mapping* $\Lambda_y : \alpha \mapsto T^\alpha y$ *defined on* $(0, \gamma)$ *is analytic in* $\alpha$. *Moreover, the power series expansion of* $\Lambda_y$ *at any point* $\alpha_0 \in (0, \gamma)$ *converges on* $(-r + \alpha_0, \alpha_0 + r)$ *with* $r = \min\{\gamma - \alpha_0, \alpha_0\}$.

*Proof.* Let $n > \gamma$ and set $A := T^{\gamma/n}$. Because of Corollary 8.2.7, we have $T^\alpha = A^{\alpha n/\gamma}$. If $y \in \mathcal{D}(T^\gamma)$, then $y \in \mathcal{D}(A^n)$ and the mapping $\Upsilon(\beta) := A^\beta y$ is analytic on

$(0, n)$ by Proposition 8.2.3. The radius of convergence of its power series expansion at $\beta_0 \in (0, n)$ is greater than or equal to $r' = \min\{\beta_0, n - \beta_0\}$. Hence, $\Lambda_y(\alpha) = \Upsilon(n\alpha/\gamma)$ is also an analytic function and the radius of convergence of its power series expansion at any point $\alpha_0 \in (0, \gamma)$ is greater than or equal to $\min\{\alpha_0, \gamma - \alpha_0\}$, which is exactly what we wanted to show. $\qquad\square$

We conclude this section with the generalization of the famous Balakrishnan representation of fractional powers and some of its consequences. This formula was introduced in [36] as one of the first approaches for defining fractional powers of sectorial operators.

**Theorem 8.2.9** ((Quaternionic) Balakrishnan Representation). *Let $T \in \mathrm{Sect}(\omega)$. For $0 < \alpha < 1$, we have*

$$T^\alpha y = \frac{\sin(\alpha\pi)}{\pi} \int_0^{+\infty} t^{\alpha-1}(t\mathcal{I} + T)^{-1}Ty\,dt, \quad \forall y \in \mathcal{D}(T). \tag{8.29}$$

*More generally, for $0 < \alpha < n \leq m$, we have*

$$T^\alpha y = \frac{\Gamma(m)}{\Gamma(\alpha)\Gamma(m - \alpha)} \int_0^{+\infty} t^{\alpha-1}[T(t\mathcal{I} + T)^{-1}]^m y\,dt, \quad \forall y \in \mathcal{D}(T^n). \tag{8.30}$$

*Proof.* We first show (8.29) and hence assume that $\alpha \in (0, 1)$. For $y \in \mathcal{D}(T)$, we have, because of (ii) in Lemma 7.2.8 and with arbitrary $\varphi \in (\omega, \pi)$ and $\varepsilon > 0$, that

$$
\begin{aligned}
T^\alpha y &= \left(p^\alpha(p + \varepsilon)^{-1}\right)(T)(T + \varepsilon\mathcal{I})y \\
&= \left(p^\alpha(p + \varepsilon)^{-1}\right)(T)Ty + \varepsilon\left(p^\alpha(p + \varepsilon)^{-1}(1 + p)^{-1}\right)(T)(\mathcal{I} + T)y \\
&= \frac{1}{2\pi} \int_{\partial(\Sigma_\varphi \cap \mathbb{C}_j)} s^{\alpha-1}s(s + \varepsilon)^{-1}\,ds_j\, S_R^{-1}(s, T)Ty \\
&\quad + \frac{1}{2\pi} \int_{\partial(\Sigma_\varphi \cap \mathbb{C}_j)} s^\alpha \varepsilon(s + \varepsilon)^{-1}(1 + s)^{-1}\,ds_j\, S_R^{-1}(s, T)(\mathcal{I} + T)y.
\end{aligned}
$$

Now observe that there exists a positive constant $K < +\infty$ such that

$$\left|\varepsilon(s + \varepsilon)^{-1}\right| \leq \frac{K}{|s|}, \quad \forall \varepsilon > 0,\ s \in \partial(\Sigma_\varphi \cap \mathbb{C}_j).$$

Together with the estimate (7.1), this implies that the integrand in the second integral is bounded for all $\varepsilon > 0$ by the function

$$s \mapsto KC_{\varphi,T}|s|^{\alpha-1}(|1 + s|)^{-1}\|(\mathcal{I} + T)y\|,$$

which is integrable along $\partial(\Sigma_\varphi \cap \mathbb{C}_j)$ because of the assumption $\alpha \in (0, 1)$. Hence, we can apply Lebesgue's theorem in order to exchange the integral with the limit and find that the second integral vanishes as $\varepsilon$ tends to 0. In the first integral on the other hand, we find that

$$
\begin{aligned}
S_R^{-1}(s,T)Ty &= (\bar{s}\mathcal{I} - T)\mathcal{Q}_s(T)^{-1}Ty \\
&= \bar{s}T\mathcal{Q}_s(T)^{-1}y - T^2\mathcal{Q}_s(T)^{-1}y \\
&= \bar{s}T\mathcal{Q}_s(T)^{-1} - \mathcal{Q}_s(T)\mathcal{Q}_s(T)^{-1}y + \left(-2s_0T + |s|^{-1}\mathcal{I}\right)\mathcal{Q}_s(T)^{-1}y \qquad (8.31)\\
&= -y + \bar{s}T\mathcal{Q}_s(T)^{-1}y - \bar{s}T\mathcal{Q}_s(T)^{-1}y + s(\bar{s}\mathcal{I} - T)\mathcal{Q}_s(T)^{-1}y \\
&= -y + sS_R^{-1}(s,T)y.
\end{aligned}
$$

Hence, the function $s \mapsto S_R^{-1}(s,T)Ty$ for $s \in \partial(\Sigma_\varphi \cap \mathbb{C}_j)$ is bounded at 0 because of (7.1). Since it decays as $|s|^{-1}$ for $s \to \infty$ and since the function $s \mapsto s(s+\varepsilon)^{-1}$ is uniformly bounded in $\varepsilon$ on $\partial(\Sigma_\varphi \cap \mathbb{C}_j)$, we can apply Lebesgue's theorem also in the first integral in order to take the limit as $\varepsilon \to 0$ and obtain

$$
T^\alpha y = \frac{1}{2\pi} \int_{\partial(\Sigma_\varphi \cap \mathbb{C}_j)} s^{\alpha-1}\, ds_j\, S_R^{-1}(s,T)Ty.
$$

Choosing the standard parametrization of the path of integration, we thus find

$$
\begin{aligned}
T^\alpha y = \frac{1}{2\pi} \int_{-\infty}^{0} \left(-te^{j\varphi}\right)^{\alpha-1} e^{j\varphi} j S_R^{-1}\left(te^{j\varphi},T\right) Ty\, dt \\
+ \frac{1}{2\pi} \int_{0}^{+\infty} \left(te^{-j\varphi}\right)^{\alpha-1} e^{-j\varphi}(-j)S_R^{-1}\left(te^{-j\varphi},T\right) Ty\, dt.
\end{aligned}
$$

Once more (7.1) and the fact that $S_R^{-1}(s,T)Ty$ is bounded at 0 allow us to apply Lebesgue's theorem in order to take the limit as $\varphi$ tends to $\pi$. We finally find after a change of variables in the first integral that

$$
\begin{aligned}
T^\alpha y &= \frac{1}{2\pi} \int_{0}^{+\infty} t^{\alpha-1}\left(-e^{j\pi\alpha}\right)(-j)S_R^{-1}\left(-te^{j\pi},T\right) Ty\, dt \\
&\quad + \frac{1}{2\pi} \int_{0}^{+\infty} t^{\alpha-1}e^{-j\pi\alpha}(-j)S_R^{-1}\left(te^{-j\pi},T\right) Ty\, dt \\
&= -\frac{\sin(\alpha\pi)}{\pi} \int_{0}^{+\infty} t^{\alpha-1}S_R^{-1}(-t,T)Ty\, dt,
\end{aligned}
$$

which equals (8.29) as $S_R^{-1}(-t,T) = (-t\mathcal{I} - T)^{-1} = -(t\mathcal{I} + T)^{-1}$ for $t \in \mathbb{R}$.

Let us now prove (8.30) and let us for now assume that $n-1 < \alpha < n$ and $n = m$. For $y \in \mathcal{D}(T^n)$, we then have

$$
T^\alpha y = T^{\alpha-(n-1)}T^{n-1}y = \frac{\sin((\alpha-n+1)\pi)}{\pi} \int_{0}^{+\infty} t^{\alpha-n}(t\mathcal{I}+T)^{-1}T^n y\, dt.
$$

Integrating $n - 1$ times by parts, we find

$$T^\alpha y = \frac{(n-1)!\sin((\alpha - n + 1)\pi)}{\pi(\alpha - n + 1)\cdots(\alpha - 1)} \int_0^{+\infty} t^{\alpha-1}(tI + T)^{-n}T^n y \, dt$$

$$= \frac{\Gamma(n)}{\Gamma(\alpha)\Gamma(n - \alpha)} \int_0^{+\infty} t^{\alpha-1}(tI + T)^{-n}T^n y \, dt, \qquad (8.32)$$

where the second identity follows from the identities $\sin(z\pi)/\pi = 1/(\Gamma(z)\Gamma(1 - z))$ and $z\Gamma(z) = \Gamma(z + 1)$ for the gamma function. Hence, the identity (8.30) holds true if $n - 1 < \alpha < n = m$.

Now observe that, because of (7.1) and because

$$(tI + T)^{-n}T^n y = \left((tI + T)^{-1}T\right) y$$

is bounded near 0 due to (8.31), the integral (8.32) defines a real analytic function in $\alpha$ on the entire interval $(0, n)$. From Proposition 8.2.3 and the identity principle for real analytic functions, we conclude that (8.30) holds also if $0 < \alpha < n = m$.

Finally, let us show by induction on $m$ that (8.30) holds true for any $m \geq n$. For $m = n$ we have just shown this identity, so let us assume that it holds true for some $m \geq n$. We introduce the notation

$$c_m := \frac{\Gamma(m)}{\Gamma(m - \alpha)\Gamma(\alpha)} \quad \text{and} \quad I_m := \int_0^{+\infty} t^{\alpha-1}\left[T(tI + T)^{-1}\right]^m y \, dt$$

so that $T^\alpha y = c_m I_m$. We want to show that $T^\alpha y = c_{m+1}I_{m+1}$. By integration by parts, we deduce

$$I_m = \left(\frac{t^\alpha}{\alpha}\left[T(tI + T)^{-1}\right]^m y\right)\Big|_0^{+\infty}$$

$$+ \frac{m}{\alpha}\int_0^{+\infty} t^\alpha \left[T(tI + T)^{-1}\right]^m (tI + T)^{-1}y \, dt$$

$$= \frac{m}{\alpha}\int_0^{+\infty} t^\alpha \left[T(tI + T)^{-1}\right]^m (tI + T)^{-1}y \, dt$$

$$= \frac{m}{\alpha}\int_0^{+\infty} t^{\alpha-1}\left(\left[T(tI + T)^{-1}\right]^m y - \left[T(tI + T)^{-1}\right]^{m+1}y\right) dt$$

$$= \frac{m}{\alpha}(I_m - I_{m+1}).$$

Hence, $I_m = \frac{m}{m-\alpha}I_{m+1}$ and so

$$T^\alpha y = c_m I_m = c_m \frac{m}{m - \alpha}I_{m+1} = c_{m+1}I_{m+1}.$$

The induction is complete. $\qquad\qquad\qquad\qquad\qquad\qquad\qquad\qquad\qquad\qquad\qquad\square$

## 8.2.1   Fractional powers with negative real part

If $\alpha < 0$, the fractional power $p^\alpha$ has polynomial limit infinity at $0$ in any sector $\Sigma_\varphi$ with $\varphi > \pi$. Because of Lemma 7.2.7, it does not belong to $\mathcal{M}_L[\Sigma_\omega]_T$ if $T$ is not injective. If on the other hand $T$ is injective, then it is regularizable by some power of $p\,(1+p)^{-2}$ such that $p^\alpha \in \mathcal{M}_L[\Sigma_\omega]_T$. We can thus define $T^\alpha$ for injective sectorial operators via the $H^\infty$-functional calculus.

**Definition 8.2.10.** Let $T \in \mathrm{Sect}(\omega)$ be injective. For any $\alpha \in \mathbb{R}$, we call the operator $T^\alpha := (p^\alpha)(T)$ the fractional power with exponent $\alpha$ of $T$.

The properties of the fractional powers of $T$ in this case are again analogue to the complex case, cf. [165]. We state the most important properties for the sake of completeness, but we omit the proofs since they are either immediate consequences of the preceding results or can be shown with exactly the same arguments as in the complex case, without making use of any quaternionic techniques.

**Proposition 8.2.11.** *Let $T \in \mathrm{Sect}(\omega)$ be injective and let $\alpha, \beta \in \mathbb{R}$.*

(i) *The operator $T^\alpha$ is injective and $(T^\alpha)^{-1} = T^{-\alpha} = \left(T^{-1}\right)^\alpha$.*

(ii) *We have $T^\alpha T^\beta \subset T^{\alpha+\beta}$ with $\mathcal{D}\left(T^\alpha T^\beta\right) = \mathcal{D}\left(T^\beta\right) \cap \mathcal{D}\left(T^{\alpha+\beta}\right)$.*

(iii) *If $\overline{\mathcal{D}(T)} = X = \overline{\mathrm{ran}(T)}$, then $T^{\alpha+\beta} = \overline{T^\alpha T^\beta}$.*

(iv) *If $0 < \alpha < 1$, then*

$$T^{-\alpha} y = \frac{\sin(\alpha\pi)}{\pi} \int_0^{+\infty} t^{-\alpha}(t\mathcal{I} + T)^{-1} y\, dt, \quad \forall y \in \mathrm{ran}(T).$$

(v) *If $\alpha \in \mathbb{R}$ with $|\alpha| < \pi/\omega$, then $T^\alpha \in \mathrm{Sect}(|\alpha|\omega)$ and for all $\beta \in \mathbb{R}$*

$$(T^\alpha)^\beta = \left(T^{\alpha\beta}\right).$$

(vi) *If $0 < \alpha_1, \alpha_2$, then $\mathcal{D}(T^{\alpha_2}) \cap \mathrm{ran}(T^{\alpha_1}) \subset \mathcal{D}(T^\alpha)$ for each $\alpha \in (-\alpha_1, \alpha_2)$, the mapping $\alpha \mapsto T^\alpha y$ is analytic on $(-\alpha_1, \alpha_2)$ for any $y \in \mathcal{D}(T^{-\alpha_2}) \cap \mathrm{ran}(T^{\alpha_1})$.*

**Remark 8.2.1.** Observe that (iv) of Proposition 8.2.11 and Corollary 8.1.7 together with the semigroup property imply that the direct approach in Section 8.1 and the approach via the $H^\infty$-functional calculus are consistent.

**Proposition 8.2.12** (Komatsu Representation). *Let $T \in \mathrm{Sect}(\omega)$ be injective. For any $y \in \mathcal{D}(A) \cap \mathrm{ran}(A)$ and any $\alpha \in (-1, 1)$, one has*

$$
\begin{aligned}
T^\alpha y &= \frac{\sin(\alpha\pi)}{\pi} \left[ \frac{1}{\alpha} y - \frac{1}{1+\alpha} T^{-1} y \right. \\
&\qquad \left. + \int_0^1 t^{\alpha+1}(t\mathcal{I} + T)^{-1} T^{-1} y\, dt + \int_1^{+\infty} t^{\alpha-1}(t\mathcal{I} + T)^{-1} T y\, dt \right] \\
&= \frac{\sin(\alpha\pi)}{\pi} \left[ \frac{1}{\alpha} y + \int_0^1 t^{-\alpha}(\mathcal{I} + tT)^{-1} T y\, dt - \int_0^1 t^\alpha \left(\mathcal{I} + tT^{-1}\right)^{-1} T^{-1} y\, dt \right].
\end{aligned}
$$

## 8.3   Kato's Formula for the $S$-Resolvents

Fractional powers of quaternionic linear operators can, as in the complex setting, also be introduced indirectly following the approach of Kato in [167]. In particular, this leads to the analogue of the famous Kato formula for the $S$-resolvents.

**Definition 8.3.1.** A densely defined closed operator $T$ on $X$ is of type $(M, \omega)$ with $M > 0$ and $\omega \in (0, \pi)$ if $T \in \mathrm{Sect}(\omega)$ and $M$ is the uniform bound of $\|tS_R^{-1}(-t, T)\|$ on the negative real axis, that is,

$$\|tS_R^{-1}(t, T)\| \leq M, \quad \text{for } t \in (-\infty, 0). \tag{8.33}$$

**Proposition 8.3.2.** *Let $T$ be of type $(M, \omega)$ with $M > 0$ and $\omega \in (0, \pi)$. Let $0 < \alpha < 1$ and let $\pi > \phi_0 > \max(\alpha\pi, \omega)$. The parameter integral*

$$F_\alpha(p, T) = \frac{\sin(\alpha\pi)}{\pi} \int_0^{+\infty} t^\alpha (p^2 - 2pt^\alpha \cos(\alpha\pi) + t^{2\alpha})^{-1} S_R^{-1}(-t, T)\, dt \tag{8.34}$$

*defines a $\mathcal{B}(X)$-valued function on $\mathbb{H} \setminus \Sigma_{\phi_0}$ in $p$ that is left slice hyperholomorphic.*

*Proof.* For any compact subset $K$ of $\mathbb{H} \setminus \Sigma_{\phi_0}$, we have $\min_{p \in K} \arg(p) > \alpha\pi$ and thus there exists some $\delta_K > 0$ such that

$$\left| p^2 - 2pt^\alpha \cos(\alpha\pi) + t^{2\alpha} \right| = \left| p - t^\alpha e^{j_p \alpha\pi} \right| \left| p - t^\alpha e^{-j_p \alpha\pi} \right| \geq \delta_K \tag{8.35}$$

for $p \in K$ and $t \geq 0$. For the same reason, we can find a constant $C_K > 0$ such that

$$\sup_{\substack{t \in [0, +\infty) \\ p \in K}} \left| p^2 - 2pt^\alpha \cos(\alpha\pi) + t^{2\alpha} \right|^{-1} t^{2\alpha} = \sup_{\substack{t \in [0, +\infty) \\ p \in K}} \frac{1}{\left| \frac{p}{t^\alpha} - e^{j_p \alpha\pi} \right|} \frac{1}{\left| \frac{p}{t^\alpha} - e^{-j_p \alpha\pi} \right|} < C_K$$

and hence

$$\left| p^2 - 2pt^\alpha \cos(\alpha\pi) + t^{2\alpha} \right|^{-1} \leq C_K t^{-2\alpha}, \quad t \in [1, \infty), \ p \in K. \tag{8.36}$$

Now consider $p \in \mathbb{H} \setminus \Sigma_{\phi_0}$ and let $K$ be a compact neighborhood of $p$. The integral in (8.34) converges absolutely and hence defines a bounded operator: because of (8.33) and the above estimates, we have for $s \in K$, and thus in particular for $p$ itself, that

$$\|F_\alpha(s, T)\| \leq \frac{\sin(\alpha\pi)}{\pi} \int_0^{+\infty} t^\alpha \left| s^2 - 2st^\alpha \cos(\alpha\pi) + t^{2\alpha} \right|^{-1} \frac{M}{t}\, dt$$

$$\leq \frac{M \sin(\alpha\pi)}{\delta_K \pi} \int_0^1 t^{\alpha-1}\, dt + \frac{M \sin(\alpha\pi) C_K}{\pi} \int_1^{+\infty} t^{-\alpha-1}\, dt < +\infty.$$

Using (8.35) and (8.36), one can derive analogous estimates for the partial derivatives of the integrand $p \mapsto t^\alpha (p^2 - 2st^\alpha \cos(\alpha\pi) + t^{2\alpha})^{-1} S_R^{-1}(-t, T)$ with respect to $p_0$ and $p_1$.

Since these estimates are uniform on the neighborhood $K$ of $p$, we can exchange differentiation and integration in order to compute the partial derivatives $\frac{\partial}{\partial p_0} F_\alpha(p, T)$ and $\frac{\partial}{\partial p_1} F_\alpha(p, T)$ of $F_\alpha(\cdot, T)$ at $p$. The integrand is however left slice hyperholomorphic and therefore also $F_\alpha(p, T)$ is left slice hyperholomorphic. $\square$

**Lemma 8.3.3.** *Let $T$ be of type $(M, \omega)$ with $M > 0$ and $\omega \in (0, \pi)$, let $0 < \alpha < 1$ and assume that $0 \in \rho_S(T)$. Moreover, let $\phi_0$ and $F_\alpha(p, T)$ be defined as in Proposition 8.3.2. If $\Gamma$ is a piecewise smooth path that goes from $\infty e^{j\theta}$ to $\infty e^{-j\theta}$ in $(\mathbb{H} \setminus \Sigma_{\phi_0}) \cap \mathbb{C}_j$ and avoids the negative real axis $(-\infty, 0]$ for some $j \in \mathbb{S}$ and some $\theta \in (\phi_0, \pi]$, then*

$$F_\alpha(p, T) = \frac{1}{2\pi} \int_\Gamma S_R^{-1}(p, s^\alpha)\, ds_j\, S_R^{-1}(s, T). \tag{8.37}$$

*Proof.* First of all, observe that the function $s \mapsto S_R^{-1}(p, s^\alpha)$ is the composition of the intrinsic function $s \mapsto s^\alpha$ defined on $\mathbb{H} \setminus (-\infty, 0]$ and the right slice hyperholomorphic function $s \mapsto S_R^{-1}(p, s)$ defined on $\mathbb{H} \setminus [p]$. This composition is in particular well defined on all of $\mathbb{H} \setminus (-\infty, 0]$, because $s^\alpha$ maps $\mathbb{H} \setminus (-\infty, 0]$ to the set $\{s \in \mathbb{H} : \arg(s) < \alpha\pi\}$, which is contained in the domain of $S_R^{-1}(p, s^\alpha)$ because $\arg(p) > \phi_0 > \alpha\pi$ by assumption. By Theorem 2.1.3, the function $s \mapsto S_R^{-1}(p, s^\alpha)$ is therefore right slice hyperholomorphic on $\mathbb{H} \setminus (-\infty, 0]$.

An estimate similar to the one in the proof of Proposition 8.3.2, moreover, assures that the integral in (8.37) converges absolutely. It thus follows from Theorem 2.1.20 that the value of the integral in (8.37) is the same for any choice of $\Gamma$ and any choice of $\theta$. Let us denote the value of this integral by $\mathfrak{J}_\alpha(p, T)$.

Since $0 \in \rho_S(T)$, the open ball $B_\varepsilon(0)$ is contained in $\rho_S(T)$ if $\varepsilon > 0$ is small enough. For $\theta \in (\phi_0, \pi)$, we set $U(\varepsilon, \theta) := \Sigma_\theta \setminus B_\varepsilon(0)$. Then,

$$\mathfrak{J}_\alpha(p, T) = \frac{1}{2\pi} \int_{\partial(U(\varepsilon, \theta) \cap \mathbb{C}_j)} S_R^{-1}(p, s^\alpha)\, ds_j\, S_R^{-1}(s, T).$$

We assumed that $0 \in \rho_S(T)$, and hence the right $S$-resolvent is bounded near $0$, which allows us to take the limit $\varepsilon \to 0$. We obtain

$$\mathfrak{J}_\alpha(p, T) = \frac{1}{2\pi} \int_{\partial(\Sigma_\theta \cap \mathbb{C}_j)} S_R^{-1}(p, s^\alpha)\, ds_j\, S_R^{-1}(s, T)$$

$$= -\frac{1}{2\pi} \int_0^{+\infty} S_R^{-1}\left(p, t^\alpha e^{j\alpha\theta}\right) e^{j\theta}(-j) S_R^{-1}\left(te^{j\theta}, T\right)\, dt$$

$$+ \frac{1}{2\pi} \int_0^{+\infty} S_R^{-1}\left(p, t^\alpha e^{-j\alpha\theta}\right) e^{-j\theta}(-j) S_R^{-1}\left(te^{j\theta}, T\right)\, dt$$

$$= -\frac{1}{2\pi} \int_0^{+\infty} \left( p^2 - 2t^\alpha \cos(\alpha\theta) + t^{2\alpha} \right)^{-1}$$
$$\cdot \left( p - t^\alpha e^{-j\alpha\theta} \right) e^{j\theta} (-j) S_R^{-1} \left( t e^{j\theta}, T \right) dt$$
$$+ \frac{1}{2\pi} \int_0^{+\infty} \left( p^2 - 2t^\alpha \cos(\alpha\theta) + t^{2\alpha} \right)^{-1}$$
$$\cdot \left( p - t^\alpha e^{j\alpha\theta} \right) e^{-j\theta} (-j) S_R^{-1} \left( t e^{-j\theta}, T \right) dt.$$

Again an estimate analogue to the one in the proof of Proposition 8.3.2 allows us to take the limit as $\theta$ tends to $\pi$ and we obtain

$$\mathfrak{J}_\alpha(p, T)$$
$$= -\frac{1}{2\pi} \int_0^{+\infty} \left( p^2 - 2t^\alpha \cos(\alpha\pi) + t^{2\alpha} \right)^{-1}$$
$$\cdot \left( p - t^\alpha e^{-j\alpha\pi} \right) e^{j\pi} (-j) S_R^{-1} \left( t e^{j\pi}, T \right) dt$$
$$+ \frac{1}{2\pi} \int_0^{+\infty} \left( p^2 - 2t^\alpha \cos(\alpha\pi) + t^{2\alpha} \right)^{-1}$$
$$\cdot \left( p - t^\alpha e^{j\alpha\pi} \right) e^{-j\pi} (-j) S_R^{-1} \left( t e^{-j\pi}, T \right) dt$$
$$= \frac{\sin(\alpha\pi)}{\pi} \int_0^{+\infty} t^\alpha (p^2 - 2pt^\alpha \cos(\alpha\pi) + t^{2\alpha})^{-1} S_R^{-1}(-t, T) dt$$
$$= F_\alpha(p, T). \qquad \qquad \Box$$

**Lemma 8.3.4.** Let $T$ be of type $(M, \omega)$ with $M > 0$ and $\omega \in (0, \pi)$. Let $0 < \alpha < 1$ and let $\phi_0$ and $F_\alpha(p, T)$ be defined as in Proposition 8.3.2. We have

$$F_\alpha(\mu, T) - F_\alpha(\lambda, T) = (\lambda - \mu) F_\alpha(\mu, T) F_\alpha(\lambda, T), \quad for \ \lambda, \mu \in (-\infty, 0]. \quad (8.38)$$

*Proof.* Assume first that $0 \in \rho_S(T)$. Any real $\lambda$ commutes with $S_R^{-1}(-t, T)$ and thus we have

$$F_\alpha(\lambda, T) = \frac{\sin(\alpha\pi)}{\pi} \int_0^{+\infty} S_L^{-1}(-t, T)(\lambda^2 - 2\lambda t^\alpha \cos(\alpha\pi) + t^{2\alpha})^{-1} t^\alpha \, dt$$

because $S_R^{-1}(-t, T) = (-t\mathcal{I} - T)^{-1} = S_L^{-1}(-t, T)$ as $t$ is real. Computations analogue to those in the proof of Lemma 8.3.3 show that $F_\alpha(\lambda, T)$ can thus be represented as

$$F_\alpha(\lambda, T) = \frac{1}{2\pi} \int_\Gamma S_L^{-1}(s, T) \, ds_j \, S_L^{-1}(\lambda, s^\alpha), \qquad (8.39)$$

where $\Gamma$ is any path as in Lemma 8.3.3.

Now let $\varepsilon > 0$ such that $\overline{B_\varepsilon(0)} \subset \rho_S(T)$, choose $j \in \mathbb{S}$ and set

$$U_s := \Sigma_{\theta_s} \setminus B_{\varepsilon_s}(0) \quad \text{and} \quad U_p := \Sigma_{\theta_p} \setminus B_{\varepsilon_p}(0)$$

with $0 < \varepsilon_s < \varepsilon_p < \varepsilon$ and $\phi_0 < \theta_p < \theta_s < \pi$. Then $\overline{U_p} \subset U_s$ and $\Gamma_s = \partial(U_s \cap \mathbb{C}_j)$ and $\Gamma_p = \partial(U_p \cap \mathbb{C}_j)$ are paths as in Lemma 8.3.3. Moreover, since $T$ is of type $(M, \omega)$ with $0 \in \rho_S(T)$, we can find a constant $C$ such that

$$\|S_R^{-1}(s, T)\| < C/(1 + |s|),$$

for $s \in (-\infty, 0]$. By Lemma 8.1.1 we may choose $\varepsilon_p$, $\varepsilon_s$, $\theta_p$ and $\theta_s$ such that

$$\|S_R^{-1}(s, T)\| \leq \frac{M_1}{1 + |s|}, s \in \Gamma_s \quad \text{and} \quad \|S_L^{-1}(p, T)\| \leq \frac{M_1}{1 + |p|}, \ p \in \Gamma_p, \qquad (8.40)$$

for some constant $M_1 > 0$. Lemma 8.3.3 and (8.39) then imply

$$F_\alpha(\mu, T) F_\alpha(\lambda, T)$$
$$= \frac{1}{(2\pi)^2} \int_{\Gamma_s} \int_{\Gamma_p} S_R^{-1}(\mu, s^\alpha)\, ds_j\, S_R^{-1}(s, T) S_L^{-1}(p, T)\, dp_j\, S_L^{-1}(\lambda, p^\alpha).$$

Applying the $S$-resolvent equation yields

$$F_\alpha(\mu, T) F_\alpha(\lambda, T)$$
$$= \frac{1}{(2\pi)^2} \int_{\Gamma_s} \int_{\Gamma_p} S_R^{-1}(\mu, s^\alpha)\, ds_j\, S_R^{-1}(s, T) p \mathcal{Q}_s(p)^{-1}\, dp_j\, S_L^{-1}(\lambda, p^\alpha)$$
$$- \frac{1}{(2\pi)^2} \int_{\Gamma_s} \int_{\Gamma_p} S_R^{-1}(\mu, s^\alpha)\, ds_j\, S_L^{-1}(p, T) p \mathcal{Q}_s(p)^{-1}\, dp_j\, S_L^{-1}(\lambda, p^\alpha)$$
$$- \frac{1}{(2\pi)^2} \int_{\Gamma_s} \int_{\Gamma_p} S_R^{-1}(\mu, s^\alpha)\, ds_j\, \bar{s} S_R^{-1}(s, T) \mathcal{Q}_s(p)^{-1}\, dp_j\, S_L^{-1}(\lambda, p^\alpha)$$
$$+ \frac{1}{(2\pi)^2} \int_{\Gamma_s} \int_{\Gamma_p} S_R^{-1}(\mu, s^\alpha)\, ds_j\, \bar{s} S_L^{-1}(p, T) \mathcal{Q}_s(p)^{-1}\, dp_j\, S_L^{-1}(\lambda, p^\alpha),$$

where we use the notation $\mathcal{Q}_s(p)^{-1} = (p^2 - 2\mathrm{Re}(s)p + |s|^2)^{-1}$ for neatness. Since $p \mapsto \mathcal{Q}_s(p)^{-1} S_L^{-1}(\lambda, p^\alpha)$ and $p \mapsto p \mathcal{Q}_s(p)^{-1} S_L^{-1}(\lambda, p^\alpha)$ are holomorphic on $\overline{U_p} \cap \mathbb{C}_j$ and tend uniformly to zero as $p$ tends to infinity in $U_p$, we deduce from Cauchy's integral theorem that the first and the third of the above integrals equal zero. The estimate (8.40) allows us to apply Fubini's theorem in order to exchange the order of integration so that we are left with

$$F_\alpha(\mu, T) F_\alpha(\lambda, T) = \frac{1}{2\pi} \int_{\Gamma_p} \left[ \frac{1}{2\pi} \int_{\Gamma_s} S_R^{-1}(\mu, s^\alpha)\, ds_j \right.$$
$$\left. \cdot \left( \bar{s} S_L^{-1}(p, T) - S_L^{-1}(p, T) p \right) \mathcal{Q}_s(p)^{-1} \right] dp_j\, S_L^{-1}(\lambda, p^\alpha). \quad (8.41)$$

We want to apply Lemma 2.2.24 and thus define the set $U_{s,r} := U_s \cap B_r(0)$ for $r > 0$, which is a bounded slice Cauchy domain. Its boundary $\partial(U_{s,r} \cap \mathbb{C}_j)$ in

$\mathbb{C}_j$ consists of $\Gamma_{s,r} := \Gamma_s \cap B_r(0)$ and the set $C_r := \{re^{j\varphi} : -\theta_s \leq \varphi \leq \theta_s\}$. If $p \in \Gamma_p$, then $p \in U_{s,r}$ for sufficiently large $r$ because $\overline{U}_p \subset U_s$. Since the function $s \mapsto S_R^{-1}(\mu, s^\alpha) = (\mu - s^\alpha)^{-1}$ is intrinsic because $\mu$ is real, we can therefore apply Lemma 2.2.24 and obtain for any such $r$

$$S_L^{-1}(p, T) S_R^{-1}(\mu, p^\alpha)$$
$$= \frac{1}{2\pi} \int_{\partial(U_{s,r} \cap \mathbb{C}_j)} S_R^{-1}(\mu, s^\alpha) \, ds_j \left(\bar{s} S_L^{-1}(p, T) - S_L^{-1}(p, T) p\right) \mathcal{Q}_s(p)^{-1}$$
$$= \frac{1}{2\pi} \int_{\Gamma_{s,r}} S_R^{-1}(\mu, s^\alpha) \, ds_j \left(\bar{s} S_L^{-1}(p, T) - S_L^{-1}(p, T) p\right) \mathcal{Q}_s(p)^{-1}$$
$$+ \frac{1}{2\pi} \int_{C_r} S_R^{-1}(\mu, s^\alpha) \, ds_j \left(\bar{s} S_L^{-1}(p, T) - S_L^{-1}(p, T) p\right) \mathcal{Q}_s(p)^{-1}.$$

As $r$ tends to infinity the integral over $C_r$ vanishes and hence

$$S_L^{-1}(p, T) S_R^{-1}(\mu, p^\alpha)$$
$$= \lim_{r \to +\infty} \frac{1}{2\pi} \int_{\Gamma_{s,r}} S_R^{-1}(\mu, s^\alpha) \, ds_j \left(\bar{s} S_L^{-1}(p, T) - S_L^{-1}(p, T) p\right) \mathcal{Q}_s(p)^{-1}$$
$$= \frac{1}{2\pi} \int_{\Gamma_s} S_R^{-1}(\mu, s^\alpha) \, ds_j \left(\bar{s} S_L^{-1}(p, T) - S_L^{-1}(p, T) p\right) \mathcal{Q}_s(p)^{-1}.$$

Applying this identity in (8.41), we obtain

$$F_\alpha(\mu, T) F_\alpha(\lambda, T) = \frac{1}{2\pi} \int_{\Gamma_p} S_L^{-1}(p, T) \, dp_j \, S_R^{-1}(\mu, p^\alpha) S_L^{-1}(\lambda, p^\alpha)$$

because $S_R^{-1}(\mu, p^\alpha)$ and $dp_j$ commute as $\mu$ is real. Since also $\lambda$ is real, we have

$$S_R^{-1}(\mu, p^\alpha) S_L^{-1}(\lambda, p^\alpha) = \frac{1}{\mu - p^\alpha} \frac{1}{\lambda - p^\alpha}$$
$$= \frac{1}{\lambda - \mu} \left(\frac{1}{\mu - p^\alpha} - \frac{1}{\lambda - p^\alpha}\right) = (\lambda - \mu)^{-1} \left(S_L^{-1}(\mu, p^\alpha) - S_L^{-1}(\lambda, p^\alpha)\right)$$

and thus, recalling (8.39), we obtain

$$F_\alpha(\mu, T) F_\alpha(\lambda, T)$$
$$= (\lambda - \mu)^{-1} \left(\frac{1}{2\pi} \int_{\Gamma_p} S_L^{-1}(p, T) \, dp_j \, S_L^{-1}(\mu, p^\alpha)\right.$$
$$\left. - \frac{1}{2\pi} \int_{\Gamma_p} S_L^{-1}(p, T) \, dp_j \, S_L^{-1}(\lambda, p^\alpha)\right)$$
$$= (\lambda - \mu)^{-1} \left(F_\alpha(\mu, T) - F_\alpha(\lambda, T)\right).$$

If $0 \notin \rho_S(T)$, then we consider the operator $T + \varepsilon \mathcal{I}$ for small $\varepsilon > 0$. This operator satisfies $0 \in \rho_S(T + \varepsilon \mathcal{I}) = \rho_S(T) + \varepsilon$ and hence (8.38) applies. Moreover, for real $t$, we have

$$S_R^{-1}(-t, T + \varepsilon \mathcal{I}) = S_R^{-1}(-(t + \varepsilon), T).$$

The estimate

$$\|S_R^{-1}(-t, T + \varepsilon \mathcal{I})\| \le \frac{M}{t + \varepsilon} \le \frac{M}{t}$$

therefore allows us to apply Lebesgue's dominated convergence theorem to see that

$$
\begin{aligned}
& F_\alpha(p, T + \varepsilon \mathcal{I}) \\
&= \frac{\sin(\alpha \pi)}{\pi} \int_0^{+\infty} t^\alpha (p^2 - 2pt^\alpha \cos(\alpha \pi) + t^{2\alpha})^{-1} S_R^{-1}(-t, T + \varepsilon \mathcal{I}) \, dt \\
& \xrightarrow{\varepsilon \to 0} \frac{\sin(\alpha \pi)}{\pi} \int_0^{+\infty} t^\alpha (p^2 - 2pt^\alpha \cos(\alpha \pi) + t^{2\alpha})^{-1} S_R^{-1}(-t, T) \, dt \\
&= F_\alpha(p, T).
\end{aligned}
$$

Consequently, we have

$$
\begin{aligned}
& F_\alpha(\mu, T) - F_\alpha(\lambda, T) \\
&= \lim_{\varepsilon \to 0} F_\alpha(\mu, T + \varepsilon \mathcal{I}) - F_\alpha(\lambda, T + \varepsilon \mathcal{I}) \\
&= \lim_{\varepsilon \to 0} (\lambda - \mu) F_\alpha(\mu, T + \varepsilon \mathcal{I}) F_\alpha(\lambda, T + \varepsilon \mathcal{I}) \\
&= (\lambda - \mu) F_\alpha(\mu, T) F_\alpha(\lambda, T),
\end{aligned}
$$

for $\lambda, \mu \in (-\infty, 0]$ also in this case. $\qquad\qquad\qquad \square$

**Theorem 8.3.5.** *Let $T$ be of type $(M, \omega)$, let $\alpha \in (0, 1)$ and let $\phi_0 > \max(\alpha\pi, \omega)$. There exists a densely defined closed operator $B_\alpha$ such that*

$$S_R^{-1}(p, B_\alpha) = F_\alpha(p, T) \quad for \quad p \in \mathbb{H} \setminus \Sigma_{\phi_0},$$

*where $F_\alpha(p, T)$ is the operator-valued function defined by the integral (8.37). Moreover, $B_\alpha$ is of type $(M, \alpha\omega)$.*

*Proof.* From identity (8.38) it follows immediately that $F_\alpha(\mu, T)$ and $F_\alpha(\lambda, T)$ commute and have the same kernel. Rewriting this equation in the form

$$F_\alpha(\mu, T) = F_\alpha(\lambda, T) \left( \mathcal{I} + (\lambda - \mu) F_\alpha(\mu, T) \right) \tag{8.42}$$

shows that $\mathrm{ran}(F_\alpha(\mu, T)) \subset \mathrm{ran}(F_\alpha(\lambda, T))$ and exchanging the roles of $\mu$ and $\lambda$ yields $\mathrm{ran}(F_\alpha(\mu, T)) = \mathrm{ran}(F_\alpha(\lambda, T))$. Hence, $\mathrm{ran}(F_\alpha(\mu, T))$ does not depend on $\mu$ and so we denote it by $\mathrm{ran}(F_\alpha(\cdot, T))$. We show now that

$$\lim_{\mathbb{R} \ni \mu \to -\infty} \mu F_\alpha(\mu, T) y = y, \quad \text{for all } y \in X, \tag{8.43}$$

where $\lim_{\mathbb{R} \ni \mu \to -\infty} \mu F_\alpha(\mu, T)y$ denotes the limit as $\mu$ tends to $-\infty$ in $\mathbb{R}$. From (8.43), we easily deduce that $\mathrm{ran}(F_\alpha(\cdot, T))$ is dense in $X$ because

$$X = \overline{\bigcup_{\mu \in (-\infty, 0]} \mathrm{ran}(F_\alpha(\mu, T))} = \overline{\mathrm{ran}(F_\alpha(\cdot, T))}.$$

We consider first $y \in \mathcal{D}(T)$. Since

$$\int_0^{+\infty} \frac{t^{\alpha-1}}{\mu^2 - 2\mu t^\alpha \cos(\alpha\pi) + t^{2\alpha}} \, dt = -\frac{\pi}{\mu \sin(\alpha\pi)}, \quad \text{for } \mu \le 0, \tag{8.44}$$

it is

$$\mu F_\alpha(\mu, T)y - y$$
$$= -\frac{\sin(\alpha\pi)}{\pi} \int_0^{+\infty} \frac{-\mu t^{\alpha-1}}{\mu^2 - 2\mu t^\alpha \cos(\alpha\pi) + t^{2\alpha}} \left( tS_R^{-1}(-t, T)y + y \right) dt.$$

For $-\mu \ge 1$ and $t \in (0, +\infty)$, we can estimate

$$\frac{-\mu t^{\alpha-1}}{\mu^2 - 2\mu t^\alpha \cos(\alpha\pi) + t^{2\alpha}} = \frac{-\mu t^{\alpha-1}}{\mu^2 \sin(\alpha\pi)^2 + (\mu \cos(\alpha\pi) - t^\alpha)^2}$$
$$\le \frac{-\mu t^{\alpha-1}}{\mu^2 \sin(\alpha\pi)^2} \le \frac{t^{\alpha-1}}{\sin(\alpha\pi)^2}$$

and due to (8.33) we have $\|tS_R^{-1}(-t, T)y + y\| \le (M+1)\|y\|$. On the other hand, since $y \in \mathcal{D}(T)$, it is

$$tS_R^{-1}(-t, T)y + y = -S_R^{-1}(-t, T)Ty \tag{8.45}$$

and so, again due to (8.33), we can also estimate

$$\|tS_R^{-1}(-t, T)y + y\| \le \frac{\|Ty\|}{t}.$$

We can hence apply Lebesgue's dominated convergence theorem with dominating function

$$f(t) = \begin{cases} \dfrac{K}{\sin(\alpha\pi)^2} t^{\alpha-1}, & \text{for } t \in (0, 1), \\[2ex] \dfrac{K}{\sin(\alpha\pi)^2} t^{-\alpha-1}, & \text{for } t \in [1, +\infty), \end{cases}$$

with $K > 0$ large enough, in order to exchange the integral with the limit for $\mu \to -\infty$ in $\mathbb{R}$. In view of (8.45), we obtain

$$\lim_{\mathbb{R} \ni \mu \to -\infty} \mu F_\alpha(\mu, T)y - y$$
$$= -\frac{\sin(\alpha\pi)}{\pi} \int_0^{+\infty} \lim_{\mathbb{R} \ni \mu \to -\infty} \frac{\mu t^{\alpha-1}}{\mu^2 - 2\mu t^\alpha \cos(\alpha\pi) + t^{2\alpha}} S_R^{-1}(-t, T)Ty \, dt = 0.$$

For arbitrary $y \in X$ and $\varepsilon > 0$ consider a vector $y_\varepsilon \in \mathcal{D}(T)$ with $\|y - y_\varepsilon\| < \varepsilon$. Because of (8.33) and (8.44), we have the uniform estimate

$$\|\mu F_\alpha(\mu, T)\| \leq \frac{-\mu \sin(\alpha\pi)}{\pi} \int_0^{+\infty} \frac{t^\alpha}{\mu^2 - 2\mu t^\alpha \cos(\alpha\pi) + t^{2\alpha}} \frac{M}{t} dt = M. \quad (8.46)$$

Therefore,

$$\lim_{\mathbb{R} \ni \mu \to -\infty} \|\mu F_\alpha(\mu, T) y - y\|$$

$$\leq \lim_{\mathbb{R} \ni \mu \to -\infty} \|\mu F_\alpha(\mu, T)\| \|y - y_\varepsilon\| + \|F_\alpha(\mu, T) y_\varepsilon - y_\varepsilon\| + \|y_\varepsilon - y\|$$

$$\leq (M + 1)\varepsilon.$$

Since $\varepsilon > 0$ was arbitrary, we deduce that (8.43) also holds true for arbitrary $y \in X$.

Overall, we obtain that $\mathrm{ran}(F_\alpha(\cdot, T))$ is dense in $X$. The identity (8.43), moreover, implies that $\ker(F_\alpha(\cdot, T)) = \{0\}$ because $y = \lim_{\mathbb{R} \ni \mu \to -\infty} F_\alpha(\mu, T) y = 0$ for any $y \in \ker(F_\alpha(\cdot, T))$.

We consider now an arbitrary point $\mu_0 \in (-\infty, 0)$. By the above arguments, the mapping $F_\alpha(\mu_0, T) : X \to \mathrm{ran}(F_\alpha(\cdot, T))$ is invertible. Hence, we can define the operator $B_\alpha := \mu_0 \mathcal{I} - F_\alpha(\mu_0, T)^{-1}$ that maps $\mathcal{D}(B_\alpha) = \mathrm{ran}(F_\alpha(\mu_0, T))$ to $X$. Apparently, $B_\alpha$ has dense domain and $S_R^{-1}(\mu_0, B_\alpha) = F_\alpha(\mu_0, B_\alpha)$. For $\mu \in (-\infty, 0]$, we can apply (8.42) and (8.38) in order to obtain

$$(\mu \mathcal{I} - B_\alpha) F_\alpha(\mu, T)$$

$$= ((\mu - \mu_0)\mathcal{I} + (\mu_0 \mathcal{I} - B_\alpha)) F_\alpha(\mu_0, T)(\mathcal{I} + (\mu_0 - \mu) F_\alpha(\mu, T))$$

$$= \mathcal{I} + (\mu - \mu_0)(F_\alpha(\mu_0, T) + (\mu_0 - \mu) F_\alpha(\mu_0, T) F_\alpha(\mu, T) - F_\alpha(\mu, T)) = \mathcal{I}.$$

A similar calculation shows that $F_\alpha(\mu, T)(\mu \mathcal{I} - B_\alpha) y = y$ for all $y \in \mathcal{D}(B_\alpha)$. We conclude that $S_R^{-1}(\mu, B_\alpha) = F_\alpha(\mu, T)$ for any $\mu \in (-\infty, 0)$. Since $p \mapsto F_\alpha(p, T)$ and $p \mapsto S_R^{-1}(p, B_\alpha)$ are left slice hyperholomorphic and agree on $(-\infty, 0)$, Theorem 2.1.6 implies $S_R^{-1}(p, B_\alpha) = F_\alpha(p, T)$ for any $p \in \mathbb{H} \setminus \overline{\Sigma_{\phi_0}}$.

Finally, in order to show that $B_\alpha$ is of type $(M, \alpha\omega)$, we choose an arbitrary imaginary unit $j \in \mathbb{S}$ and consider the restriction of $S_R^{-1}(\cdot, B_\alpha)$ to the plane $\mathbb{C}_j$. This restriction is a holomorphic function with values in the left vector space $\mathcal{B}(X)$ over $\mathbb{C}_j$. We show now that this restriction has a holomorphic continuation to the sector $(\mathbb{H} \setminus \overline{\Sigma_{\alpha\omega}}) \cap \mathbb{C}_j$. Since this sector is symmetric with respect to the real axis, we can apply Lemma 2.1.9 and obtain a left slice hyperholomorphic continuation of $S_R^{-1}(p, B_\alpha)$ to the sector $\mathbb{H} \setminus \overline{\Sigma_{\alpha\omega}}$. By Theorem 3.2.11, this implies in particular that $\mathbb{H} \setminus \overline{\Sigma_{\alpha\omega}} \subset \rho_S(T)$.

The above considerations showed that we can represent $S_R^{-1}(p, B_\alpha)$ for any point $p \in (\mathbb{H} \setminus \overline{\Sigma_\omega}) \cap \mathbb{C}_j$ using Kato's formula (8.34). Rewriting this formula as a path integral over the path $\gamma_0(t) = t e^{j\pi}, t \in [0, +\infty)$, we obtain

$$S_R^{-1}(p, B_\alpha) = -\frac{\sin(\alpha\pi)}{\pi} \int_{\gamma_0} \frac{z^\alpha e^{-j\pi\alpha}}{(z^\alpha e^{-j2\alpha\pi} - p)(z^\alpha - p)} S_R^{-1}(z, T)\, dz,$$

where $z$ denotes a complex variable in $\mathbb{C}_j$ and $z \mapsto z^\alpha$ is a branch of a complex $\alpha$-th power of $z$ that is holomorphic on $\mathbb{C}_j \setminus [0, \infty)$. To be more precise, let us choose $\left(re^{j\theta}\right)^\alpha = r^\alpha e^{j\theta\alpha}$ with $\theta \in (0, 2\pi)$. (This is, however, not the restriction of the quaternionic function $s \mapsto s^\alpha$ defined in Example 2.1.15 to the plane $\mathbb{C}_j$, cf. Remark 2.1.3.)

Observe that for fixed $p$, the integrand is holomorphic on $D_0 := (\mathbb{H} \setminus \overline{\Sigma_\omega}) \cap \mathbb{C}_j$. Hence, by applying Cauchy's Integral Theorem, we can exchange the path of integration $\gamma_0$ by a suitable path $\gamma_\kappa(t) = te^{j(\pi-\kappa)}, t \in [0, +\infty)$, and obtain

$$S_R^{-1}(p, B_\alpha) = -\frac{\sin(\alpha\pi)}{\pi} \int_{\gamma_\kappa} \frac{z^\alpha e^{-j\pi\alpha}}{(z^\alpha e^{-j2\alpha\pi} - p)(z^\alpha - p)} S_R^{-1}(z, T) \, dz. \qquad (8.47)$$

On the other hand, for any $\kappa \in (-\omega, \omega)$, such integral defines a holomorphic function on the sector $D_\kappa := \{p \in \mathbb{C}_j : \alpha(\pi - \kappa) < \arg p < 2\pi - \alpha(\pi + \kappa)\}$, where the convergence of the integral is guaranteed because the operator $T$ is of type $(M, \omega)$. The above argument showed that this function coincides with $S_R^{-1}(p, B_\alpha)$ on the common domain $D_0 \cap D_\kappa$, and hence $p \mapsto S_R^{-1}(p, B_\alpha)$ has a holomorphic continuation $F_j$ to

$$D = \bigcup_{\kappa \in (-\omega, \omega)} D_\kappa = \{p \in \mathbb{C}_j : \quad \alpha(\pi - \kappa) < \arg_{\mathbb{C}_j}(p) < 2\pi - \alpha(\pi - \kappa)\}.$$

This set is symmetric with respect to the real axis and, as mentioned above, we deduce from Lemma 2.1.9 that there exists a left slice hyperholomorphic extension $F$ of $F_j$ to the axially symmetric hull $[D] = \mathbb{H} \setminus \overline{\Sigma_{\alpha\omega}}$ of $D$. Consequently, $\mathbb{H} \setminus \overline{\Sigma_{\alpha\omega}} \subset \rho_S(B_\alpha)$ and $F$ coincides with $S_R^{-1}(\cdot, B_\alpha)$ on $\mathbb{H} \setminus \overline{\Sigma_{\alpha\omega}}$.

In order to show that $\|pS_R(p, B_\alpha)\|$ is bounded on every sector $\mathbb{H} \setminus \overline{\Sigma_\theta}$ with $\theta \in (\omega\alpha, 0)$, we consider first a set

$$D_{\kappa, \delta} := \{p \in \mathbb{C}_j : \quad \delta + \alpha(\pi - \kappa) < \arg p < 2\pi - \alpha(\pi + \kappa) - \delta\}$$

with $\kappa \in (-\omega, \omega)$ and small $\delta > 0$. For $p \in D_{\kappa, \delta}$ with $\phi = \arg_{\mathbb{C}_j}(p) \in (0, 2\pi)$, we may represent $pS_R^{-1}(p, B_\alpha)$ by means of (8.47) and estimate

$$\frac{\pi}{\sin(\alpha\pi)} \|pS_R^{-1}(p, B_\alpha)\|$$

$$\leq |p| \int_0^{+\infty} \frac{r^\alpha}{|(r^\alpha e^{-j(\pi+\kappa)\alpha} - p)(r^\alpha e^{j(\pi-\kappa)\alpha} - p)|} \|S_R^{-1}(re^{j(\pi-\kappa)}, T)\| \, dr$$

$$= |p| \int_0^{+\infty} \frac{r^\alpha}{|(r^\alpha - |p|e^{j(\phi+(\pi+\kappa)\alpha)})(r^\alpha - |p|e^{j(\phi-(\pi-\kappa)\alpha)})|} \|S_R^{-1}(re^{j(\pi-\kappa)}, T)\| \, dr.$$

The operator $T$ is of type $(\omega, M)$ and hence there exists a constant $M_\kappa > 0$ such that $\|S_R^{-1}(re^{j(\pi-\kappa)}, T)\| \leq M/r$. Substituting $\tau = r^\alpha/|p|$ yields

$$\|pS_R^{-1}(p, B_\alpha)\| \leq \frac{\sin(\alpha\pi)}{\pi} \int_0^{+\infty} \frac{M_\kappa}{|(\tau - e^{j(\phi+(\pi+\kappa)\alpha)})(\tau - e^{j(\phi-(\pi-\kappa)\alpha)})|} \, d\tau.$$

This integral is uniformly bounded for

$$\phi \in (\delta + \alpha(\pi - \kappa), 2\pi - \alpha(\pi + \kappa) - \delta)$$

such that there exists a constant that depends only on $\kappa$ and $\delta$ such that

$$\|pS(p, B_\alpha)\| \le C(\kappa, \delta), \quad \text{for } p \in D_{\kappa, \delta}.$$

Now consider a sector $\mathbb{H} \setminus \overline{\Sigma_\theta}$ with $\theta \in (\omega\alpha, \pi)$. Then there exist $(\kappa_\ell, \delta_\ell)$ with $\ell = 1, \ldots, n$ such that $\mathbb{H} \setminus \overline{\Sigma_\theta} \cap \mathbb{C}_j \subset \bigcup_{\ell=1}^{n} D_{\kappa_\ell, \delta_\ell}$ and hence

$$\|pS_R^{-1}(p, B_\alpha)\| \le C := \max_{1 \le \ell \le n} C(\kappa_\ell, \delta_\ell), \quad \text{for } p \in \mathbb{H} \setminus \overline{\Sigma_\theta} \cap \mathbb{C}_j.$$

For arbitrary $p = u + iv \in \mathbb{H} \setminus \overline{\Sigma_\theta} \cap \mathbb{C}_j$, set $p_j = u + jv$. Then the representation formula implies

$$\|pS_R^{-1}(p, T)\| \le \frac{1}{2}\|(1 - ij)p_j S_R^{-1}(p_j, B_\alpha)\| + \frac{1}{2}\|(1 + ij)\overline{p_j} S_R^{-1}(\overline{p_j}, B_\alpha)\| \le 2C.$$

Finally, the estimate $\|tS_R^{-1}(-t, B_\alpha)\| \le M/t$ follows immediately from (8.46).  $\square$

**Definition 8.3.6.** Let $T \in \mathcal{K}(X)$ be of type $(\omega, M)$. For $\alpha \in (0, 1)$, we define

$$T^\alpha := B_\alpha.$$

**Corollary 8.3.7.** *Definition 8.3.6 is consistent with Definition 8.1.14 and Definition 8.2.1.*

*Proof.* Let $T \in \mathcal{K}(X)$, let $\alpha \in (0, 1)$ and let $T^\alpha$ be the operator obtained from Definition 8.3.6. If $\|S_R^{-1}(s, T)\| \le K/(1 + |s|)$ for $s \in (-\infty, 0]$, then we can apply Lebesgue's dominated convergence theorem in order to pass to the limit as $p$ tends to 0 in Kato's formula (8.34) for the right $S$-resolvent of $T^\alpha$. We obtain

$$(T^\alpha)^{-1} = -S_R^{-1}(0, T^\alpha) = -\frac{\sin(\alpha\pi)}{\pi} \int_0^{+\infty} t^{-\alpha} S_R^{-1}(-t, T), \, dt = T^{-\alpha},$$

where the last equality follows from Corollary 8.1.7 (resp. from (8.2.11)).  $\square$

# Chapter 9

# The fractional heat equation using quaternionic techniques

The development of the spectral theory of the Nabla operator opens the way to a large class of fractional diffusion problems, and some of them will be treated in the next chapter. Indeed, the main aim of this chapter is to show how our theory, for the case of the Nabla operator, reproduces known results. Since it is very general, it allows us to manipulate a very large class of new fractional diffusion processes. The results presented in this chapter were originally proved in [53, 54]. Precisely, if $v(x,t)$ is the temperature at the point $x \in \mathbb{R}^3$ and the time $t > 0$ and $\kappa$ is the thermal diffusivity of the considered material, then the heat equation

$$\partial_t v(x,t) - \kappa \Delta v(x,t) = 0, \tag{9.1}$$

where $\Delta = \sum_{\ell=1}^{3} \partial_{x_\ell}^2$ with $x = (x_1, x_2, x_3)^T$, describes the evolution of the temperature distribution in space and time. (For mathematical treatment, one usually sets $\kappa = 1$ and we will emulate this.) This model has, however, several unphysical properties, so scientists have tried to modify it. One approach has been the introduction of the fractional heat equation. In order to modify the properties of the equation, researchers replaced the negative Laplacian in (9.1) by its fractional powers of exponent $\alpha$ and considered the evolution equation

$$\frac{\partial}{\partial t} v(x,t) + (-\Delta)^\alpha v(x,t) = 0. \tag{9.2}$$

There are different approaches for defining the fractional Laplace operator, but each approach leads to a global integral operator, which, in contrast to the local differential operator $\Delta$, is able to take long distance effects into account.

We want to develop a similar approach for defining fractional evolution equations with the generalized gradient. What we show here is that, if we replace the gradient in Fourier's law of conductivity by its fractional power instead of directly

© Springer Nature Switzerland AG 2019

F. Colombo, J. Gantner, *Quaternionic Closed Operators, Fractional Powers and Fractional Diffusion Processes*, Operator Theory: Advances and Applications 274,

https://doi.org/10.1007/978-3-030-16409-6_9

replacing the negative Laplacian by its fractional power in (9.1), we get the same equation. Indeed, this would lead to the equation

$$\frac{\partial}{\partial t} v(x,t) - \operatorname{div}(\nabla^\alpha v(x,t)) = 0,$$

with suitable interpretation of the symbol $\nabla^\alpha$. Our initial task is to understand the definition of the fractional powers $\nabla^\alpha$ according to our theory developed in the previous chapters where we identified the gradient with the quaternionic Nabla operator.

Additionally in this chapter, we develop the spectral theory of the quaternionic Nabla operator on $L^2(\mathbb{R}^3, \mathbb{H})$. We find that the previously developed theory is not directly applicable because the Nabla operator does not belong to the class of sectorial operators. We therefore present a slightly modified approach and show that this allows us to reproduce the fractional heat equation (9.2) using quaternionic techniques. Finally, we give an example for a more general operator with non-constant coefficients that can be treated with our methods.

## 9.1   Spectral properties of the Nabla operator

The gradient of a function $v : \mathbb{R}^3 \to \mathbb{R}$ is the vector-valued function

$$\nabla v(x) = \begin{pmatrix} \partial_{x_1} v(x) \\ \partial_{x_2} v(x) \\ \partial_{x_3} v(x) \end{pmatrix}, \quad \text{for } x = (x_1, x_2, x_3).$$

If we identify $\mathbb{R}$ with the set of real quaternions and $\mathbb{R}^3$ with the set of purely imaginary quaternions, this corresponds to the quaternionic Nabla operator

$$\nabla = \partial_{x_1} e_1 + \partial_{x_2} e_2 + \partial_{x_3} e_3.$$

In the following, we shall often denote the standard basis of the quaternions by $\mathbf{I} := e_1$, $\mathbf{J} := e_2$ and $\mathbf{K} := e_3 = \mathbf{IJ} = -\mathbf{JI}$. This suggests a relation with the complex theory, which we shall use excessively. With this notation, we have

$$\nabla = \partial_{x_1} \mathbf{I} + \partial_{x_2} \mathbf{J} + \partial_{x_3} \mathbf{K}.$$

We study the properties of a quaternionic Nabla operator on the space $L^2(\mathbb{R}^3, \mathbb{H})$ of all square-integrable quaternion-valued functions on $\mathbb{R}^3$, which is a quaternionic right Hilbert space when endowed with the scalar product

$$\langle w, v \rangle = \int_{\mathbb{R}^3} \overline{w(x)} v(x) \, dx.$$

On this space, the Nabla operator is closed and has dense domain. This follows immediately from its representation (9.4) in the Fourier space that we derive in the proof of Theorem 9.1.1.

Let $v \in L^2(\mathbb{R}^3, \mathbb{H})$ and write $v(x) = v_1(x) + v_2(x)\mathbf{J}$ with two $\mathbb{C}_\mathsf{I}$-valued functions $v_1$ and $v_2$. As $|v(x)|^2 = |v_1(x)|^2 + |v_2(x)|^2$, we have

$$\|v\|^2_{L^2(\mathbb{R}^3, \mathbb{H})} = \|v_1\|^2_{L^2(\mathbb{R}^3, \mathbb{C}_\mathsf{I})} + \|v_2\|^2_{L_2(\mathbb{R},^3, \mathbb{C}_\mathsf{I})}, \tag{9.3}$$

where $L^2(\mathbb{R}^3, \mathbb{H})$ denotes the complex Hilbert space over $\mathbb{C}_\mathsf{I}$ of all square-integrable $\mathbb{C}_\mathsf{I}$-valued functions on $\mathbb{R}^3$. Hence, $v \in L^2(\mathbb{R}^3, \mathbb{H})$ if and only if $v_1, v_2 \in L^2(\mathbb{H}, \mathbb{C}_\mathsf{I})$.

**Theorem 9.1.1.** *The $S$-spectrum of $\nabla$ as an operator on $L^2(\mathbb{R}^3, \mathbb{H})$ is*

$$\sigma_S(\nabla) = \mathbb{R}.$$

*Proof.* Let us consider $L^2(\mathbb{R}^3, \mathbb{H})$ as a Hilbert space over $\mathbb{C}_\mathsf{I}$ by restricting the right scalar multiplication to $\mathbb{C}_\mathsf{I}$ and setting

$$\langle w, v \rangle_\mathsf{I} := \{\langle w, v \rangle_{L^2(\mathbb{R}^3, \mathbb{H})}\}_\mathsf{I}.$$

Here $\{\cdot\}_\mathsf{I}$ denotes the $\mathbb{C}_\mathsf{I}$-part of a quaternion: if $a = a_1 + a_2\mathbf{J} = a_1 + \mathbf{J}\overline{a_2}$ with $a_1, a_2 \in \mathbb{C}_\mathsf{I}$, then $\{a\}_\mathsf{I} := a_1$. If we write $v, w \in L^2(\mathbb{R}^3, \mathbb{H})$ as $v = v_1 + \mathbf{J}v_2$ and $w = w_1 + \mathbf{J}w_2$ with $v_1, v_2, w_1, w_2 \in L^2(\mathbb{R}^3, \mathbb{C}_\mathsf{I})$, then

$$
\begin{aligned}
\langle w, v \rangle_{L^2(\mathbb{R}^3, \mathbb{H})} &= \int_{\mathbb{R}^3} \overline{(w_1(x) + \mathbf{J}w_2(x))}(v_1(x) + \mathbf{J}v_2(x)) \, dx \\
&= \int_{\mathbb{R}^3} \overline{w_1(x)}v_1(x) \, dx + \int_{\mathbb{R}^3} \overline{w_2(x)}(-\mathbf{J})v_1(x) \, dx \\
&\quad + \int_{\mathbb{R}^3} \overline{w_1(x)}\mathbf{J}v_2(x) \, dx + \int_{\mathbb{R}^3} \overline{w_2(x)}(-\mathbf{J}^2)v_2(x) \, dx \\
&= \int_{\mathbb{R}^3} \overline{w_1(x)}v_1(x) \, dx + \int_{\mathbb{R}^3} \overline{w_2(x)}v_2(x) \, dx \\
&\quad + \mathbf{J}\left(-\int_{\mathbb{R}^3} w_2(x)v_1(x) \, dx + \int_{\mathbb{R}^3} w_1(x)v_2(x) \, dx\right).
\end{aligned}
$$

Therefore, we have

$$\langle w, v \rangle_\mathsf{I} := \langle w_1, v_1 \rangle_{L^2(\mathbb{R}^3, \mathbb{C}_\mathsf{I})} + \langle w_2, v_2 \rangle_{L^2(\mathbb{R}^3, \mathbb{C}_\mathsf{I})}$$

and hence $L^2(\mathbb{R}^3, \mathbb{H})$ considered as a $\mathbb{C}_\mathsf{I}$-complex Hilbert space with the scalar product $\langle \cdot, \cdot \rangle_\mathsf{I}$ equals $L^2(\mathbb{R}^3, \mathbb{C}_\mathsf{I}) \oplus L^2(\mathbb{R}^3, \mathbb{C}_\mathsf{I})$. Moreover, because of (9.3), the quaternionic scalar product $\langle \cdot, \cdot \rangle$ and the $\mathbb{C}_\mathsf{I}$-complex scalar product $\langle \cdot, \cdot \rangle_\mathsf{I}$ induce the same norm on $L^2(\mathbb{R}^3, \mathbb{H})$. Applying the Nabla operator to $v = v_1 + \mathbf{J}v_2$, we find

$$
\begin{aligned}
\nabla v(x) &= (\mathbf{I}\partial_{x_1} + \mathbf{J}\partial_{x_2} + \mathbf{K}\partial_{x_3})(v_1(x) + \mathbf{J}v_2(x)) \\
&= \mathbf{I}\partial_{x_1}v_1(x) + \mathbf{J}\partial_{x_2}v_1(x) + \mathbf{K}\partial_{x_3}v_1(x) \\
&\quad + \mathbf{I}\partial_{x_1}\mathbf{J}v_2(x) + \mathbf{J}\partial_{x_2}\mathbf{J}v_2(x) + \mathbf{K}\partial_{x_3}\mathbf{J}v_2(x) \\
&= \mathbf{I}\partial_{x_1}v_1(x) - \partial_{x_2}v_2(x) - \mathbf{I}\partial_{x_3}v_2(x) \\
&\quad + \mathbf{J}\left(-\mathbf{I}\partial_{x_1}v_2(x) + \partial_{x_2}v_1(x) - \mathbf{I}\partial_{x_3}v_1(x)\right).
\end{aligned}
$$

Writing this in terms of the components $L^2(\mathbb{R}^3, \mathbb{H}) \cong L^2(\mathbb{R}^3, \mathbb{C}_\mathsf{l}) \oplus L^2(\mathbb{R}^3, \mathbb{C}_\mathsf{l})$, we obtain

$$\nabla \begin{pmatrix} v_1(x) \\ v_2(x) \end{pmatrix} = \begin{pmatrix} \mathsf{l}\partial_{x_1} v_1(x) - \partial_{x_2} v_2(x) - \mathsf{l}\partial_{x_3} v_2(x) \\ -\mathsf{l}\partial_{x_1} v_2(x) + \partial_{x_2} v_1(x) - \mathsf{l}\partial_{x_3} v_1(x) \end{pmatrix}.$$

If we apply the Fourier transform on $L^2(\mathbb{R}^3, \mathbb{C}_\mathsf{l})$ componentwise, this turns into

$$\widehat{\nabla} \begin{pmatrix} \widehat{v}_1(x) \\ \widehat{v}_2(x) \end{pmatrix} = \begin{pmatrix} -\xi_1 & -\mathsf{l}\xi_2 + \xi_3 \\ \mathsf{l}\xi_2 + \xi_3 & \xi_1 \end{pmatrix} \begin{pmatrix} \widehat{v}_1(\xi) \\ \widehat{v}_2(\xi) \end{pmatrix}. \tag{9.4}$$

Hence, in the Fourier space, the Nabla operator corresponds to the multiplication operator $M_G : \widehat{v} \mapsto G\widehat{v}$ on $\widehat{X} := L^2(\mathbb{R}^3, \mathbb{C}_\mathsf{l}) \oplus L^2(\mathbb{R}^3, \mathbb{C}_\mathsf{l})$ that is generated by the matrix valued function

$$G(\xi) := \begin{pmatrix} -\xi_1 & -\mathsf{l}\xi_2 + \xi_3 \\ \mathsf{l}\xi_2 + \xi_3 & \xi_1 \end{pmatrix}. \tag{9.5}$$

For $s \in \mathbb{C}_\mathsf{l}$, we find

$$s\mathcal{I}_{\widehat{X}} - G(\xi) = \begin{pmatrix} s + \xi_1 & \mathsf{l}\xi_2 - \xi_3 \\ -\mathsf{l}\xi_2 - \xi_3 & s - \xi_1 \end{pmatrix}.$$

For $s \in \mathbb{C}_\mathsf{l}$, the inverse of $s\mathcal{I}_{\widehat{X}} - M_G$ is hence given by the multiplication operator $M_{(s\mathcal{I}-G)^{-1}}$ determined by the matrix-valued function

$$(s\mathcal{I}_{\widehat{X}} - G(\xi))^{-1} = \frac{1}{s^2 - \xi_1^2 - \xi_2^2 - \xi_3^2} \begin{pmatrix} s - \xi_1 & -\mathsf{l}\xi_2 + \xi_3 \\ \mathsf{l}\xi_2 + \xi_3 & s + \xi_1 \end{pmatrix}.$$

This operator is bounded if and only if the function $\xi \mapsto (s\mathcal{I} - G(\xi))^{-1}$ is bounded on $\mathbb{R}^3$, that is if and only $s \notin \mathbb{R}$. Hence, $\sigma(M_G) = \mathbb{R}$.

The componentwise Fourier transform $\Psi$ is a unitary $\mathbb{C}_\mathsf{l}$-linear operator from the space $L^2(\mathbb{R}^3, \mathbb{H}) \cong L^2(\mathbb{R}^3, \mathbb{C}_\mathsf{l}) \oplus L^2(\mathbb{R}^2, \mathbb{C}_\mathsf{l})$ to $\widehat{X}$ under which $\nabla$ corresponds to $M_G$, that is $\nabla = \Psi^{-1} M_G \Psi$. The spectrum $\sigma_{\mathbb{C}_\mathsf{l}}(\nabla)$ of $\nabla$ considered as a $\mathbb{C}_\mathsf{l}$-linear operator on $L^2(\mathbb{R}^3, \mathbb{H})$, therefore, equals $\sigma_{\mathbb{C}_\mathsf{l}}(\nabla) = \sigma(M_G) = \mathbb{R}$. By Theorem 3.1.8, we however have $\sigma_{\mathbb{C}_\mathsf{l}}(\nabla) = \sigma_S(\nabla) \cap \mathbb{C}_\mathsf{l}$ and so $\sigma_S(\nabla) = \mathbb{R}$.  $\square$

The above result shows that the gradient does not belong to the class of sectorial operators as $(-\infty, 0) \not\subset \rho_S(T)$, so the theory developed in Chapter 8 is not directly applicable. Even worse, we cannot find any other slice hyperholomorphic functional calculus that allows us to define fractional powers $\nabla^\alpha$ of $\nabla$ because the scalar function $s^\alpha$ is not slice hyperholomorphic on $(-\infty, 0]$ and hence not slice hyperholomorphic on $\sigma_S(\nabla)$.

However, we shall now show another characterization of the $S$-spectrum of the Nabla operator on the quaternionic right Hilbert space $L^2(\mathbb{R}^3, \mathbb{H})$ that makes use of the relation $\nabla^2 = -\Delta$ and will be fundamental later on.

If $j, i \in \mathbb{S}$ with $i \perp j$, then any $v \in L^2(\mathbb{R}^3, \mathbb{H})$ can be written as $v = v_1 + v_2 i$ with components $v_1, v_2$ in $L^2(\mathbb{R}^3, \mathbb{C}_j)$, i.e., $L^2(\mathbb{R}^3, \mathbb{H}) = L^2(\mathbb{R}^3, \mathbb{C}_j) \oplus L^2(\mathbb{R}^3, \mathbb{C}_j)i$.

Contrary to the decomposition $v = v_1 + iv_1$, which we used in the proof of Theorem 9.1.1 with $j = \mathsf{I}$ and $i = \mathsf{J}$, this decomposition is not compatible with
the $\mathbb{C}_j$-right vector space structure of $L^2(\mathbb{R}^3, \mathbb{H})$ as $va = v_1 a + v_2 \bar{a} i$ for any
$a \in \mathbb{C}_j$. However, this identification has a different advantage: any closed $\mathbb{C}_j$-
linear operator $A : \mathcal{D}(A) \subset L^2(\mathbb{R}^3, \mathbb{C}_{\mathsf{I}}) \to L^2(\mathbb{R}^3, \mathbb{C}_{\mathsf{I}})$ extends to a closed $\mathbb{H}$-
linear operator on $L^2(\mathbb{R}^3, \mathbb{H})$ with domain $\mathcal{D}(A) \oplus \mathcal{D}(A)i$, namely to the operator
$A(v_1 + v_2 i) := A(v_1) + A(v_2)i$. Moreover, if $A$ is bounded, then its extension to
$L^2(\mathbb{R}^3, \mathbb{H})$ has the same norm as $A$. We shall denote an operator on $L^2(\mathbb{R}^3, \mathbb{C}_j)$
and its extension to $L^2(\mathbb{R}^3, \mathbb{H}) = L^2(\mathbb{R}^3, \mathbb{C}_j) \oplus L^2(\mathbb{R}^3, \mathbb{C}_j)i$ via componentwise application by the same symbol. This will not cause any confusion as it will be clear
from the context to which we refer.

**Theorem 9.1.2.** *Let* $\Delta$ *be the Laplace operator on* $L^2(\mathbb{H}, \mathbb{C}_j)$ *and let* $R_z(-\Delta)$ *be
the resolvent of* $-\Delta$ *at* $z \in \mathbb{C}_j$. *We have*

$$\sigma_S(\nabla)^2 = \left\{ s^2 \in \mathbb{H} : \quad s \in \sigma_S(T) \right\} = \sigma(-\Delta) \tag{9.6}$$

*and*

$$\mathcal{Q}_{c,s}(\nabla)^{-1} = R_{s^2}(-\Delta), \quad \forall s \in \mathbb{C}_j \setminus \mathbb{R}. \tag{9.7}$$

*Proof.* Since the components of $\nabla$ commute and $e_\kappa e_\ell = -e_\ell e_\kappa$ for $1 \leq \kappa, \ell \leq 3$
with $\kappa \neq \ell$, we have

$$\nabla^2 = \sum_{\ell, \kappa=1}^{3} \partial_{x_\ell} \partial_{x_\kappa} e_\ell e_\kappa$$

$$= \sum_{\ell=1}^{3} -\partial_{x_\ell}^2 + \sum_{1 \leq \ell < \kappa \leq 3} (\partial_{x_\ell} \partial_{x_\kappa} - \partial_{x_\kappa} \partial_{x_\ell}) e_\ell e_\kappa$$

$$= \sum_{\ell=1}^{3} -\partial_{x_\ell}^2 = -\Delta.$$

As $\nabla_0 = 0$, we have $\overline{\nabla} = -\nabla$ and in turn

$$\mathcal{Q}_{c,s}(\nabla) = s^2 \mathcal{I} - 2s\nabla_0 + \nabla\overline{\nabla} = s^2 \mathcal{I} - \nabla^2 = s^2 \mathcal{I} - (-\Delta)$$

Hence, $\mathcal{Q}_{c,s}(\nabla)$ is invertible if and only if $s^2 \mathcal{I} - (-\Delta)$ is invertible. In this case

$$\mathcal{Q}_{c,s}(\nabla) = (s^2 \mathcal{I} - (-\Delta))^{-1} = R_{s^2}(-\Delta). \qquad \square$$

## 9.2 A relation with the fractional heat equation

As one can easily verify, the Nabla operator is self-adjoint on $L^2(\mathbb{R}^3, \mathbb{H})$. From
the spectral theorem for unbounded normal quaternionic linear operators (see the
paper [14] or the book [57]), we hence deduce the existence of a unique spectral

measure $E$ on $\sigma_S(\nabla) = \mathbb{R}$, the values of which are orthogonal quaternionic linear projections on $L^2(\mathbb{R}^3, \mathbb{H})$, such that

$$\nabla = \int_{\mathbb{R}} s \, dE(s).$$

Using the measurable functional calculus for intrinsic slice functions (see [14] or the book [57]), it is now possible to define $P_\alpha(s) = s^\alpha \chi_{[0,+\infty)}(s)$ of $T$ as

$$P_\alpha(\nabla) = \int_{\mathbb{R}} s^\alpha \chi_{[0,+\infty)}(s) \, dE(s),$$

where $\chi_{[0,+\infty)}$ denotes the characteristic function of the set $[0, +\infty)$. This corresponds to defining $\nabla^\alpha$, at least on the subspace associated with the spectral values $[0, +\infty)$, on which $s^\alpha$ is defined. (We stress that, even with the measurable functional calculus, the operator $\nabla^\alpha$ cannot be defined because $s^\alpha$ is not defined on $(-\infty, 0)$.)

We shall now give an integral representation for this operator via an approach similar to the one of the slice hyperholomorphic $H^\infty$-functional calculus. Surprisingly, this yields a possibility to obtain the fractional heat equation via quaternionic operator techniques applied to the Nabla operator. For $\alpha \in (0, 1)$, we define

$$P_\alpha(\nabla)v := \frac{1}{2\pi} \int_{-j\mathbb{R}} S_L^{-1}(s, \nabla) \, ds_j \, s^{\alpha-1} \nabla v, \quad \forall v \in \mathcal{D}(\nabla). \tag{9.8}$$

Intuitively, this corresponds to Balakrishnan's formula for $\nabla^\alpha$, where only spectral values on the positive real axis, i.e., points where $s^\alpha$ is actually defined, are taken into account, because the path of integration surrounds only the positive real axis.

**Theorem 9.2.1.** *The integral (9.8) converges for any $v \in \mathcal{D}(\nabla)$ and hence defines a quaternionic linear operator on $L^2(\mathbb{R}^3, \mathbb{H})$.*

*Proof.* If we write the integral (9.8) explicitly, we have

$$
\begin{aligned}
P_\alpha(\nabla)v &= \frac{1}{2\pi} \int_{-\infty}^{+\infty} S_L^{-1}(-jt, \nabla)\,(-j)^2\,(-jt)^{\alpha-1}\nabla v \\
&= -\frac{1}{2\pi} \int_0^{+\infty} S_L^{-1}(-jt, \nabla)(-jt)^{\alpha-1}\nabla v \, dt \\
&\quad - \frac{1}{2\pi} \int_0^{+\infty} S_L^{-1}(jt, \nabla)(jt)^{\alpha-1}\nabla v \, dt \qquad (9.9) \\
&= -\frac{1}{2\pi} \int_0^{+\infty} S_L^{-1}(-jt, \nabla)t^{\alpha-1}e^{-j\frac{(\alpha-1)\pi}{2}}\nabla v \, dt \\
&\quad - \frac{1}{2\pi} \int_0^{+\infty} S_L^{-1}(jt, \nabla)t^{\alpha-1}e^{j\frac{(\alpha-1)\pi}{2}}\nabla v \, dt,
\end{aligned}
$$

where $P_\alpha(\nabla)v$ is defined if and only if the last two integrals converge in $L^2(\mathbb{R}^3, \mathbb{H})$.

Let us consider $L^2(\mathbb{R}^3, \mathbb{H})$ as a Hilbert space over $\mathbb{C}_j$ as in the proof of Theorem 9.1.1. If we write $v \in L^2(\mathbb{R}^3, \mathbb{H})$ as $v = v_1 + iv_2$ with $v_1, v_2 \in L^2(\mathbb{R}, \mathbb{C}_j)$ and apply the Fourier transform componentwise, we obtain an isometric $\mathbb{C}_j$-linear isomorphism $\Psi : v \mapsto (\widehat{v_1}, \widehat{v_2})^T$ between $L^2(\mathbb{R}^3, \mathbb{H})$ and

$$\widehat{X} := L^2(\mathbb{R}^3, \mathbb{C}_j) \oplus L^2(\mathbb{R}^3, \mathbb{C}_j).$$

For any quaternionic linear operator $T$ on $L^2(\mathbb{R}^3, \mathbb{H})$, the composition $\Psi T \Psi^{-1}$ is a $\mathbb{C}_j$-linear operator on $\widehat{X}$ with $\mathcal{D}(\Psi T \Psi^{-1}) = \Psi \mathcal{D}(T)$.

Applying $\nabla$ to $v \in \mathcal{D}(\nabla) \subset L^2(\mathbb{R}^3, \mathbb{H})$, corresponds to applying the multiplication operator $M_G$ associated with the matrix-valued function $G(\xi)$ defined in (9.5) to $\widehat{v}(\xi) = (\widehat{v_1}(\xi), \widehat{v_2}(\xi))^T$. Hence, $\nabla = \Psi^{-1} M_G \Psi$ and

$$\begin{aligned}
\Psi \mathcal{D}(\nabla) = \mathcal{D}(M_G) &= \left\{ \widehat{v} \in \widehat{X} : G(\xi)\widehat{v}(\xi) \in \widehat{X} \right\} \\
&= \left\{ \widehat{v} \in \widehat{X} : |\xi|\widehat{v}(\xi) \in \widehat{X} \right\}.
\end{aligned} \tag{9.10}$$

The last identity holds, for $\widehat{v}(\xi) = (\widehat{v_1}(\xi), \widehat{v_2}(\xi))^T \in \widehat{X}$, as straightforward computations show that

$$\begin{aligned}
|G(\xi)\widehat{v}(\xi)|^2 &= \left| \begin{pmatrix} -\xi_1\widehat{v_1}(\xi) + (-j\xi_2 + \xi_3)\widehat{v_2}(\xi) \\ (j\xi_2 + \xi_3)\widehat{v_1}(\xi) + \xi_1\widehat{v_2}(\xi) \end{pmatrix} \right|^2 \\
&= (\xi_1^2 + \xi_2^2 + \xi_3^2)(|\widehat{v_1}(\xi)|^2 + |\widehat{v_2}(\xi)|^2) = |\xi|^2|\widehat{v}(\xi)|^2.
\end{aligned} \tag{9.11}$$

Because of (9.9), we have

$$\begin{aligned}
P_\alpha(\nabla)v = &-\Psi \frac{1}{2\pi} \int_0^{+\infty} \left( \Psi^{-1} S_L^{-1}(-jt, \nabla)t^{\alpha-1}e^{-j\frac{(\alpha-1)\pi}{2}} \nabla \Psi^{-1} \right) \Psi v \, dt \\
&- \Psi \frac{1}{2\pi} \int_0^{+\infty} \left( \Psi^{-1} S_L^{-1}(jt, \nabla)t^{\alpha-1}e^{j\frac{(\alpha-1)\pi}{2}} \nabla \Psi^{-1} \right) \Psi v \, dt.
\end{aligned} \tag{9.12}$$

Since $jv = j(v_1 + iv_2) = v_1 j - i(v_2 j)$ and $\Psi$ is $\mathbb{C}_j$-linear, we find

$$\Psi j \Psi^{-1}(\widehat{v_1}, \widehat{v_2})^T = (\widehat{v_1}j, \widehat{v_2}(-j))^T,$$

i.e., multiplication with $j$ on $L^2(\mathbb{R}^3, \mathbb{H})$ from the left corresponds to the multiplication with the matrix $E := \mathrm{diag}(j, -j)$ on $\widehat{X}$. As

$$Q_{-jt}(\nabla)^{-1} = (\nabla^2 + t^2)^{-1} = (-\Delta + t^2)^{-1}$$

is a scalar operator and hence commutes with any quaternion, we have

$$S_L^{-1}(-jt, \nabla) = Q_{-jt}(\nabla)^{-1}jt - \nabla Q_{-jt}(\nabla)^{-1} = (jt - \nabla)Q_{-jt}(\nabla)^{-1},$$

and in turn

$$\Psi^{-1}S_L^{-1}(-jt,\nabla)t^{\alpha-1}e^{-j\frac{(\alpha-1)\pi}{2}}\nabla\Psi^{-1}$$

$$=\Psi^{-1}\left(jt\mathcal{Q}_{-jt}(\nabla)^{-1}-\nabla\mathcal{Q}_{-jt}(\nabla)^{-1}\right)t^{\alpha-1}e^{-j\frac{(\alpha-1)\pi}{2}}\nabla\Psi^{-1}$$

$$=\left(tM_E\mathcal{Q}_{-jt}(M_G)^{-1}-M_G\mathcal{Q}_{-jt}(M_G)^{-1}\right)t^{\alpha-1}M_{\exp\left(-\frac{(\alpha-1)\pi}{2}E\right)}M_G.$$

The operator $\mathcal{Q}_{jt}(M_G)^{-1}$ is

$$\mathcal{Q}_{jt}(M_G)^{-1}=(M_G^2+t^2\mathcal{I})^{-1}=M_{(G^2+t^2\mathcal{I})^{-1}}=M_{(t^2+|\xi|^2)^{-1}\mathcal{I}}$$

with $|\xi|^2=\xi_1^2+\xi_2^2+\xi_3^2$ and the operator in the first integral of (9.12) therefore equals

$$\Psi^{-1}S_L^{-1}(-jt,\nabla)t^{\alpha-1}e^{-j\frac{(\alpha-1)\pi}{2}}\nabla\Psi^{-1}$$

$$=M_{tE(t^2+|\xi|^2)^{-1}-G(t^2+|\xi|^2)^{-1}t^{\alpha-1}}M_{\exp\left(-\frac{(\alpha-1)\pi}{2}E\right)}M_G.$$

It is hence the multiplication operator $M_{A_1(t,\xi)}$ determined by the matrix-valued function

$$A_1(t,\xi)=\frac{t^{\alpha-1}}{t^2+|\xi|^2}(tE-G(\xi))\exp\left(-\frac{(\alpha-1)\pi}{2}E\right)G(\xi)$$

$$=\frac{t^{\alpha-1}}{t^2+\xi_1^2+\xi_2^2+\xi_3^2}$$

$$\cdot\begin{pmatrix}e^{-j\frac{\alpha\pi}{2}}\xi_1(t-j\xi_1)+je^{j\frac{\alpha\pi}{2}}(\xi_2^2+\xi_3^2) & \left(e^{j\frac{\alpha\pi}{2}}\xi_1+e^{-j\frac{\alpha\pi}{2}}(\xi_1+jt)\right)(\xi_2+j\xi_3)\\ \left(je^{-j\frac{\alpha\pi}{2}}\xi_1+e^{j\frac{\alpha\pi}{2}}(-t+j\xi_1)\right)(j\xi_2+\xi_3) & e^{j\frac{\alpha\pi}{2}}(-t+j\xi_1)\xi_1-je^{-j\frac{\alpha\pi}{2}}(\xi_2^2+\xi_3^2)\end{pmatrix}.$$

Similarly, the operator in the second integral of (9.12) is

$$\Psi^{-1}S_L^{-1}(jt,\nabla)t^{\alpha-1}e^{j\frac{(\alpha-1)\pi}{2}}\nabla\Psi^{-1}$$

$$=M_{-tE(t^2+|\xi|^2)^{-1}-G(t^2+|\xi|^2)^{-1}t^{\alpha-1}}M_{\exp\left(\frac{(\alpha-1)\pi}{2}E\right)}M_G.$$

It is hence the multiplication operator $M_{A_2(t,\xi)}$ determined by the matrix-valued function

$$A_2(t,\xi)=\frac{t^\alpha}{t^2+|\xi|^2}(-tE-G(\xi))\exp\left(\frac{(\alpha-1)\pi}{2}E\right)G(\xi)$$

$$=\frac{t^{\alpha-1}}{t^2+\xi_1^2+\xi_2^2+\xi_3^2}$$

$$\cdot\begin{pmatrix}e^{j\frac{\alpha\pi}{2}}\xi_1(t+j\xi_1)-je^{-j\frac{\alpha\pi}{2}}(\xi_2^2+\xi_3^2) & -\left(e^{j\frac{\alpha\pi}{2}}(-jt+\xi_1)+e^{-j\frac{\alpha\pi}{2}}\xi_1\right)(\xi_2+j\xi_3)\\ \left(e^{j\frac{\alpha\pi}{2}}\xi_1+e^{-j\frac{\alpha\pi}{2}}(-jt+\xi_1)\right)(\xi_2-j\xi_3) & -e^{-j\frac{\alpha\pi}{2}}(t+j\xi_1)\xi_1+je^{j\frac{\alpha\pi}{2}}(\xi_2^2+\xi_3^2)\end{pmatrix}.$$

Hence, we have

$$P_\alpha(\nabla)v=\Psi^{-1}P_\alpha(M_G)\Psi v$$

with

$$P_\alpha(M_G)\widehat{v} := -\frac{1}{2\pi}\int_0^{+\infty} M_{A_1(t,\xi)}\widehat{v}\, dt - \frac{1}{2\pi}\int_0^{+\infty} M_{A_2(t,\xi)}\widehat{v}\, dt, \qquad (9.13)$$

for $\widehat{v} = \Psi v \in \Psi\,\mathcal{D}(\nabla)$. We show now that these integrals converge for any $\widehat{v} \in \Psi\,\mathcal{D}(\nabla)$. As $\Psi$ is isometric, this is equivalent to (9.8) converging for any $v \in \mathcal{D}(\nabla)$. Since all norms on a finite-dimensional vector space are equivalent, there exists a constant $C > 0$ such that

$$\|M\| \le C \max_{\ell,\kappa\in\{1,2\}} |m_{\ell,\kappa}|, \quad \forall M = \begin{pmatrix} m_{1,1} & m_{1,2} \\ m_{2,1} & m_{2,2} \end{pmatrix} \in \mathbb{C}_j^{2\times2}. \qquad (9.14)$$

The modulus of the $(1,1)$-entry of $A_1(t,\xi)$ with $t \ge 0$ is

$$\frac{t^{\alpha-1}}{t^2 + \xi_1^2 + \xi_2^2 + \xi_3^2} \left| e^{-j\frac{\alpha\pi}{2}}\xi_1(t - j\xi_1) + je^{j\frac{\alpha\pi}{2}}(\xi_2^2 + \xi_3^2) \right|$$

$$= \frac{t^{\alpha-1}}{t^2 + \xi_1^2 + \xi_2^2 + \xi_3^2}\left(|\xi_1 t| + |\xi|^2\right) \le \frac{t^{\alpha-1}}{t^2 + |\xi|^2}\left(|\xi|t + |\xi|^2\right).$$

Similarly, one sees that the $(2,2)$-entry of $A_1(t,\xi)$ satisfies this estimate. For the $(1,2)$-entry we have on the other hand

$$\frac{t^{\alpha-1}}{t^2 + \xi_1^2 + \xi_2^2 + \xi_3^2}\left|\left(je^{-j\frac{\alpha\pi}{2}}\xi_1 + e^{j\frac{\alpha\pi}{2}}(-t + j\xi_1)\right)(j\xi_2 + \xi_3)\right|$$

$$\le \frac{t^{\alpha-1}}{t^2 + \xi_1^2 + \xi_2^2 + \xi_3^2}\left(2|\xi_1||\xi_2 + j\xi_3| + t|\xi_2 + j\xi_3|\right)$$

$$\le \frac{2t^{\alpha-1}}{t^2 + |\xi|^2}\left(|\xi|^2 + t|\xi|\right).$$

Similar computations show that the $(2,1)$-entry also satisfies this estimate and hence we deduce from (9.14) that

$$\|A_1(t,\xi)\| \le 2C\frac{t^{\alpha-1}}{t^2 + |\xi|^2}\left(|\xi|t + |\xi|^2\right).$$

Analogous arguments show that this estimate is also satisfied by $\|A_2(t,\xi)\|$. For the integrals in (9.13) we hence obtain

$$\int_0^{+\infty}\|M_{A_1(t,\xi)}\widehat{v}\|_{\widehat{X}}\, dt + \int_0^{+\infty}\|M_{A_2(t,\xi)}\widehat{v}\|_{\widehat{X}}\, dt$$

$$\le 2\int_0^{+\infty} 2C\left\|\frac{t^{\alpha-1}}{t^2 + |\xi|^2}\left(|\xi|t + |\xi|^2\right)|\widehat{v}(\xi)|\right\|_{L^2(\mathbb{R}^3)}\, dt$$

$$\le 4C\int_0^1 t^{\alpha-1}\left\|\frac{|\xi|t}{t^2 + |\xi|^2}|\widehat{v}(\xi)| + \frac{|\xi|^2}{t^2 + |\xi|^2}|\widehat{v}(\xi)|\right\|_{L^2(\mathbb{R}^3)}\, dt$$

$$+ 4C\int_1^{+\infty} t^{\alpha-2}\left\|\frac{t^2}{t^2 + |\xi|^2}|\xi\widehat{v}(\xi)| + \frac{t|\xi|}{t^2 + |\xi|^2}|\xi\widehat{v}(\xi)|\right\|_{L^2(\mathbb{R}^3)}\, dt.$$

Now observe that

$$\frac{t^2}{t^2 + |\xi|^2} \le 1, \quad \frac{|\xi|^2}{t^2 + |\xi|^2} \le 1, \quad \frac{t|\xi|}{t^2 + |\xi|^2} \le \frac{1}{2} < 1.$$

Because of (9.10), the relation $\widehat{v} \in \Psi \mathcal{D}(\nabla)$ implies that $|\widehat{v}(\xi)|$ and $\||\xi|\widehat{v}(\xi)|$ both belong to $L^2(\mathbb{R}^3)$ and hence we finally find

$$\int_0^{+\infty} \|M_{A_1(t,\xi)}\widehat{v}\|_{\widehat{X}} \, dt + \int_0^{+\infty} \|M_{A_2(t,\xi)}\widehat{v}\|_{\widehat{X}} \, dt$$

$$\le 8C\|v(\xi)\|_{L^2(\mathbb{R}^3)} \int_0^1 t^{\alpha-1} \, dt + 8C\||\xi\widehat{v}(\xi)\|_{L^2(\mathbb{R}^3)} \int_1^{+\infty} t^{\alpha-2} \, dt,$$

which is finite as $\alpha \in (0,1)$. Hence, (9.13) converges for any $\widehat{v} \in \Psi \mathcal{D}(\nabla)$ and (9.8) converges in turn for any $v \in \mathcal{D}(\nabla)$. $\qquad\square$

**Theorem 9.2.2.** *The operator $P_\alpha(\nabla)$ can be extended to a closed operator on $L^2(\mathbb{R}^3, \mathbb{H})$. For $v \in \mathcal{D}(\nabla^2) = \mathcal{D}(-\Delta)$, it is moreover given by*

$$P_\alpha(\nabla)v = (-\Delta)^{\frac{\alpha}{2}-1} \left[ \frac{1}{2}(-\Delta)^{\frac{1}{2}} + \frac{1}{2}\nabla \right] \nabla v. \tag{9.15}$$

*Proof.* Let $v \in \mathcal{D}(\nabla^2) = \mathcal{D}(-\Delta)$. Because of (3.29), we have that

$$P_\alpha(\nabla)v = \frac{1}{2\pi} \int_{-\infty}^{+\infty} (-jt\mathcal{I} + \nabla)\mathcal{Q}_{c,-jt}(\nabla)^{-1}(-j)^2(-tj)^{\alpha-1}\nabla v \, dt$$

$$= -\frac{1}{2\pi} \int_0^{+\infty} (-jt\mathcal{I} + \nabla)\mathcal{Q}_{c,-jt}(\nabla)^{-1}t^{\alpha-1}e^{-j(\alpha-1)\frac{\pi}{2}}\nabla v \, dt \tag{9.16}$$

$$- \frac{1}{2\pi} \int_0^{+\infty} (jt\mathcal{I} + \nabla)\mathcal{Q}_{c,jt}(\nabla)^{-1}t^{\alpha-1}e^{j(\alpha-1)\frac{\pi}{2}}\nabla v \, dt.$$

Due to (9.7), we have moreover

$$\mathcal{Q}_{c,jt}(\nabla)^{-1} = (-t^2 + \Delta)^{-1} = \mathcal{Q}_{c,-jt}(\nabla)^{-1}$$

and hence

$$P_\alpha(\nabla)v = -\frac{1}{2\pi} \int_0^{+\infty} t^\alpha \mathcal{Q}_{c,jt}(\nabla)^{-1} j \left( e^{j(\alpha-1)\frac{\pi}{2}} - e^{-j(\alpha-1)\frac{\pi}{2}} \right) \nabla v \, dt$$

$$- \frac{1}{2\pi} \int_0^{+\infty} \nabla \mathcal{Q}_{c,jt}(\nabla)^{-1}t^{\alpha-1} \left( e^{j(\alpha-1)\frac{\pi}{2}} + e^{-j(\alpha-1)\frac{\pi}{2}} \right) \nabla v \, dt$$

$$= \frac{\sin\left((\alpha-1)\frac{\pi}{2}\right)}{\pi} \int_0^{+\infty} t^\alpha \mathcal{Q}_{c,jt}(\nabla)^{-1}\nabla v \, dt \tag{9.17}$$

$$- \frac{\cos\left((\alpha-1)\frac{\pi}{2}\right)}{\pi} \int_0^{+\infty} \nabla \mathcal{Q}_{c,jt}(\nabla)^{-1}t^{\alpha-1}\nabla v \, dt.$$

For the first integral, we obtain

$$
\begin{aligned}
&\frac{\sin\left((\alpha-1)\frac{\pi}{2}\right)}{\pi}\int_0^{+\infty} t^\alpha \mathcal{Q}_{c,jt}(\nabla)^{-1}\nabla v\, dt\\
&=\frac{\sin\left((\alpha-1)\frac{\pi}{2}\right)}{\pi}\int_0^{+\infty} t^\alpha(-t^2+\Delta)^{-1}\nabla v\, dt\\
&=\frac{\sin\left((\alpha-1)\frac{\pi}{2}\right)}{\pi}\int_0^{+\infty} \tau^{\frac{\alpha-1}{2}}(-\tau+\Delta)^{-1}\nabla v\, d\tau\\
&=\frac{1}{2}(-\Delta)^{\frac{\alpha-1}{2}}\nabla v.
\end{aligned}
\tag{9.18}
$$

The last identity follows from the integral representation of the fractional power $A^\beta$ with $\mathrm{Re}(\beta)\in(0,1)$ of a complex linear sectorial operator $A$ given in Corollary 3.1.4 of [165], namely

$$
A^\beta v = \frac{\sin(\pi\beta)}{\pi}\int_0^{+\infty}\tau^\beta\left(\tau\mathcal{I}+A^{-1}\right)^{-1}v\, d\tau,\quad v\in\mathcal{D}(A).
\tag{9.19}
$$

As $-\Delta$ is an injective sectorial operator on $L^2(\mathbb{R}^3,\mathbb{C}_j)$, its closed inverse $(-\Delta)^{-1}$ is also a sectorial operator. Its fractional power $\left((-\Delta)^{-1}\right)^{\frac{1-\alpha}{2}}$ is, because of (9.19), given by the last integral in (9.18). Since

$$
(-\Delta)^{\frac{\alpha-1}{2}} = \left((-\Delta)^{-1}\right)^{\frac{1-\alpha}{2}},
$$

we obtain the last equality. Observe that the expression $\frac{1}{2}(-\Delta)^{\frac{\alpha-1}{2}}\nabla v$ is meaningful as we chose $v\in\mathcal{D}(\nabla^2)$. Indeed, if we consider the operators in the Fourier space $\hat{X}$ as in the proof of Theorem 9.2.1, then $-\Delta$ corresponds to the multiplication operator $M_{|\xi|^2}$ generated by the scalar function $|\xi|^2$. The operator $(-\Delta)^{\frac{\alpha-1}{2}}$ is then the multiplication operator $M_{|\xi|^{\alpha-1}}$ generated by the function $(|\xi|^2)^{\frac{\alpha-1}{2}}=|\xi|^{\alpha-1}$. Hence,

$$
\begin{aligned}
\mathcal{D}(-\Delta)^{-\frac{\alpha-1}{2}} &= \left\{v\in L^2(\mathbb{R}^3,\mathbb{H}):\ \hat{v}\in\mathcal{D}(M_{|\xi|^{\alpha-1}})\right\}\\
&= \left\{v\in L^2(\mathbb{R}^3,\mathbb{H}):\ |\xi|^{\alpha-1}\hat{v}(\xi)\in\hat{X}\right\}.
\end{aligned}
$$

If $G(\xi)$ is as in (9.5), then

$$
\widehat{\nabla v}(\xi) = M_G\hat{v}(\xi) = G(\xi)\hat{v}(\xi)\in\hat{X}
$$

and because of (9.11) we have $|G(\xi)\hat{v}(\xi)| = |\xi||\hat{v}(\xi)|\in L^2(\mathbb{R})$. As $\alpha\in(0,1)$, we therefore find that

$$
|\xi|^{\alpha-1}|M_G\hat{v}(\xi)| = |\xi|^\alpha|\hat{v}(\xi)|
$$

belongs to $L^2(\mathbb{R}^3)$ and so we have $\widehat{\nabla v}\in\mathcal{D}(M_{|\xi|^{\alpha-1}})$. This is equivalent to $\nabla v\in\mathcal{D}\left((-\Delta)^{\frac{\alpha-1}{2}}\right)$.

As $v \in \mathcal{D}(\nabla^2) = \mathcal{D}(-\Delta)$, we obtain similarly that the second integral in (9.17) equals

$$
\begin{aligned}
&-\frac{\cos\left((\alpha-1)\frac{\pi}{2}\right)}{\pi} \int_0^{+\infty} \nabla \mathcal{Q}_{c,jt}(\nabla)^{-1} t^{\alpha-1} \nabla v \, dt \\
&= \frac{\sin\left((\alpha-2)\frac{\pi}{2}\right)}{\pi} \int_0^{+\infty} \nabla(-t^2 \mathcal{I}+\Delta)^{-1} t^{\alpha-1} \nabla v \, dt \\
&= \frac{\sin\left((\alpha-2)\frac{\pi}{2}\right)}{2\pi} \int_0^{+\infty} (-\tau \mathcal{I}+\Delta)^{-1} \tau^{\frac{\alpha-2}{2}} \nabla^2 v \, d\tau \\
&= \frac{1}{2}(-\Delta)^{\frac{\alpha}{2}-1} \nabla^2 v.
\end{aligned}
\tag{9.20}
$$

Again this expression is meaningful as we assumed $v \in \mathcal{D}(\nabla^2)$. This is equivalent to $|\xi|^2 \widehat{v}(\xi) \in \widehat{X}$ because $\widehat{\nabla^2 v}(\xi) = |\xi|^2 \widehat{v}(\xi)$. Since $\alpha \in (0,1)$ and $\widehat{v} \in \mathcal{D}(M_{|\xi|^2})$, the function $|\xi|^2 \widehat{v}(\xi)$ belongs to the domain of the multiplication operator $M_{|\xi|^{\alpha-2}}$ because

$$
M_{|\xi|^{\alpha-2}} |\xi|^2 \widehat{v}(\xi) = |\xi|^\alpha \widehat{v}(\xi) \in \widehat{X}.
$$

Since $(-\Delta)^{\frac{\alpha}{2}-1}$ corresponds to $M_{|\xi|^{\alpha-2}}$ on the Fourier space $\widehat{X}$, we find $\nabla^2 v$ in $\mathcal{D}\left((-\Delta)^{\frac{\alpha}{2}-1}\right)$. Altogether, we find

$$
P_\alpha(\nabla)v = (-\Delta)^{\frac{\alpha}{2}-1} \left[ \frac{1}{2}(-\Delta)^{\frac{1}{2}} + \frac{1}{2}\nabla \right] \nabla v, \quad \forall v \in \mathcal{D}(\nabla^2). \tag{9.21}
$$

Finally, we show that $P_\alpha(\nabla)$ can be extended to a closed operator. We need to show that for any sequence $v_n \in \mathcal{D}(P_\alpha(\nabla)) = \mathcal{D}(\nabla)$ that converges to 0 and for which also the sequence $P_\alpha(\nabla)v_n$ converges, we have $z := \lim_{n \to +\infty} P_\alpha(\nabla)v_n = 0$. In order to do this, we write as in (9.17)

$$
\begin{aligned}
P_\alpha(\nabla)v =\ & \frac{\sin\left((\alpha-1)\frac{\pi}{2}\right)}{\pi} \int_0^{+\infty} t^\alpha (t^2\mathcal{I}+\Delta)^{-1} \nabla v \, dt \\
&- \frac{\cos\left((\alpha-1)\frac{\pi}{2}\right)}{\pi} \int_0^{+\infty} \nabla(t^2\mathcal{I}+\Delta)^{-1} t^{\alpha-1} \nabla v \, dt.
\end{aligned}
$$

If we choose an arbitrary, but fixed $r > 0$, then the operator $(r\mathcal{I}+\Delta)^{-1}$ commutes with $(t^2\mathcal{I}+\Delta)^{-1}$ and $\nabla$ and we deduce from the above integral representation that

$$
(r\mathcal{I}+\Delta)^{-1} P_\alpha(\nabla)v = P_\alpha(\nabla)(r\mathcal{I}+\Delta)^{-1} v, \quad \forall v \in \mathcal{D}(\nabla).
$$

We show now that the mapping $v \mapsto P_\alpha(\nabla)(r\mathcal{I}+\Delta)^{-1}v$ is a bounded linear operator on $L^2(\mathbb{R}^3, \mathbb{H})$. Since $(r\mathcal{I}+\Delta)^{-1}$ maps $L^2(\mathbb{R}^3, \mathbb{H})$ to $\mathcal{D}(\Delta) = \mathcal{D}(\nabla^2)$, the composition $\nabla^2(r\mathcal{I}+\Delta)^{-1}$ of the bounded operator $(r\mathcal{I}+\Delta)^{-1}$ and the closed operator $\nabla^2$ is bounded itself. As we have seen above, $\nabla^2$ and also the bounded operator $\nabla^2(r\mathcal{I}+\Delta)^{-1}$ map $L^2(\mathbb{R}^3, \mathbb{H})$ into the domain of the closed operator $(-\Delta)^{\frac{\alpha}{2}-1}$. Hence, their composition $(-\Delta)^{-\frac{\alpha}{2}-1}\nabla^2(r\mathcal{I}+\Delta)^{-1}$ is therefore bounded. Similarly, $\nabla(r+\Delta)^{-1}$ is a bounded operator that maps $L^2(\mathbb{R}^3,\mathbb{H})$ to $\mathcal{D}((-\Delta)^{\frac{\alpha-1}{2}})$ as we have seen above, and so the composition $(-\Delta)^{\frac{\alpha-1}{2}}\nabla(r\mathcal{I}+\Delta)^{-1}$ is also bounded. Because of (9.21), the operator

$$P_\alpha(\nabla)(r\mathcal{I}+\Delta)^{-1} = \frac{1}{2}(-\Delta)^{\frac{\alpha-1}{2}}\nabla(r\mathcal{I}+\Delta)^{-1} + \frac{1}{2}(-\Delta)^{\frac{\alpha}{2}-1}\nabla^2(r\mathcal{I}+\Delta)^{-1}$$

is the linear combination of bounded operators and hence bounded itself.

If a sequence $v_n \in \mathcal{D}(P_\alpha(\nabla))$ converges to 0 and $z = \lim_{n\to+\infty} P_\alpha(\nabla)v_n$ exists in $L^2(\mathbb{R}^3,\mathbb{H})$, then

$$(r+\Delta)^{-1}z = \lim_{n\to+\infty}(r+\Delta)^{-1}P_\alpha(\nabla)v_n = \lim_{n\to+\infty}P_\alpha(\nabla)(r+\Delta)^{-1}v_n = 0.$$

But as $(r+\Delta)^{-1}$ is the inverse of a closed operator, its kernel is trivial and so $z = \lim_{n\to+\infty} P_\alpha(\nabla)v_n = 0$. Hence, $P_\alpha(\nabla)$ can be extended to a closed operator. $\square$

**Remark 9.2.1.** The identity (9.15) might seem surprising at first glance, but it is actually rather intuitive. By the spectral theorem, there exist two spectral measures $E_{(-\Delta)}$ and $E_\nabla$ on $[0,+\infty)$ (resp. $\mathbb{R}$) such that $-\Delta = \int_{[0,+\infty)} t\,dE_{-\Delta}(t)$ and $\nabla = \int_{\mathbb{R}} r\,dE_\nabla(r)$. As $\nabla^2 = -\Delta$, the spectral measure $E_{(-\Delta)}$ is furthermore the push-forward measure of $E_\nabla$ under the mapping $t \mapsto t^2$ such that

$$\int_{[0,+\infty)} f(t)\,dE_{(-\Delta)}(t) = \int_{\mathbb{R}} f\left(t^2\right)\,dE_\nabla(t)$$

for any measurable function $f$. Hence, we have for $v \in \mathcal{D}(\nabla^2)$ that

$$\begin{aligned}
P_\alpha(\nabla) &= \int_{\mathbb{R}} t^\alpha \chi_{[0,+\infty)}(t)\,dE_\nabla(t)v \\
&= \int_{\mathbb{R}} t^{\alpha-2}\frac{1}{2}(|t|+t)t\,dE_\nabla(t)v \\
&= \int_{\mathbb{R}} t^{\alpha-2}\,dE_\nabla(t)\frac{1}{2}\left(\int_{\mathbb{R}}|t|\,dE_\nabla(t) + \int_{\mathbb{R}} t\,dE_\nabla(t)\right)\int_{\mathbb{R}} t\,dE_\nabla(t)v \\
&= \int_{[0,+\infty)} t^{\frac{\alpha}{2}-1}\,dE_{(-\Delta)}(t)\frac{1}{2} \\
&\quad \cdot \left(\int_{[0,+\infty)}|t|^{\frac{1}{2}}\,dE_{(-\Delta)}(t) + \int_{\mathbb{R}} t\,dE_\nabla(t)\right)\int_{\mathbb{R}} t\,dE_\nabla(t)v \\
&= (-\Delta)^{-\frac{\alpha}{2}-1}\left[\frac{1}{2}(-\Delta)^{\frac{1}{2}} + \frac{1}{2}\nabla\right]\nabla v.
\end{aligned}$$

The vector part of $P_\alpha(\nabla)$ is, because of (9.15), given by

$$\operatorname{Vec} P_\alpha(\nabla)v = \frac{1}{2}(-\Delta)^{\frac{\alpha-1}{2}}\nabla v.$$

If we apply the divergence to this equation with sufficiently regular $v$, we find

$$\operatorname{div}\left(\operatorname{Vec} P_\alpha(\nabla)v\right) = \frac{1}{2}(-\Delta)^{\frac{\alpha-1}{2}}\Delta v = -\frac{1}{2}(-\Delta)^{\frac{\alpha+1}{2}}.$$

We can thus reformulate the fractional heat equation (9.2) with $\alpha \in (1/2, 1)$ as

$$\frac{\partial}{\partial t}v - 2\operatorname{div}\left(\operatorname{Vec} f_\beta(\nabla)v\right) = 0, \quad \beta = 2\alpha - 1.$$

## 9.3   An example with non-constant coefficients

As pointed out before, the advantage of the above procedure is that is does not only apply to the gradient to reproduce the fractional Laplacian. Rather it applies to a large class of vector operators, in particular generalized gradients with non-constant coefficients. As a first example, we consider the operator

$$T := \xi_1\frac{\partial}{\partial\xi_1}e_1 + \xi_2\frac{\partial}{\partial\xi_2}e_2 + \xi_3\frac{\partial}{\partial\xi_3}e_3$$

on the space $L^2(\mathbb{R}^3_+, \mathbb{H}, d\mu)$ of $\mathbb{H}$-valued functions on

$$\mathbb{R}^3_+ = \{\xi = (\xi_1, \xi_2, \xi_3)^T \in \mathbb{R}^3 : \quad \xi_\ell > 0\}$$

that are square integrable with respect to

$$d\mu(\xi) = \frac{1}{\xi_1\xi_2\xi_3}d\lambda(\xi),$$

where $\lambda$ denotes the Lebesgue measure on $\mathbb{R}^3$. In order to determine $\mathcal{Q}_s(T)^{-1}$ we observe that the operator given by the change of variables $J : f \mapsto f \circ \iota$ with $\iota(x) = (e^{x_1}, e^{x_2}, e^{x_3})^T$ is an isometric isomorphism between $L^2(\mathbb{R}^3, \mathbb{H}, d\lambda(x))$ and $L^2(\mathbb{R}^3_+, \mathbb{H}, d\mu(\xi))$. Moreover, $T = J^{-1}\nabla J$ such that

$$Q_s(T) = (s^2\mathcal{I} + T\overline{T}) = J^{-1}(s^2\mathcal{I} + \Delta)J$$

and in turn

$$Q_s(T)^{-1} := (s^2\mathcal{I} - T\overline{T})^{-1} = J^{-1}(s^2\mathcal{I} + \Delta)^{-1}J.$$

We therefore have for sufficiently regular $v$ with calculations analogue to those in (9.16) and (9.17) that

$$P_\alpha(T)v = \frac{\sin((\alpha-1)\pi)}{\pi}\int_0^{+\infty} t^\alpha(-t^2\mathcal{I} + T\overline{T})^{-1}T\,dt$$

$$+ \frac{\cos((\alpha-1)\pi)}{\pi}\int_0^{+\infty} t^{\alpha-1}T(-t^2\mathcal{I} + T\overline{T})^{-1}Tv\,dt.$$

Clearly, the vector part of this operator is again given by the first integral such that

$$\text{Vec}\, P_\alpha(T)v = \frac{\sin((\alpha-1)\pi)}{\pi} \int_0^{+\infty} t^\alpha (-t^2\mathcal{I} + T\overline{T})^{-1} Tv\, dt$$

$$= \frac{\sin((\alpha-1)\pi)}{\pi} \int_0^{+\infty} t^\alpha J^{-1}(-t^2\mathcal{I} + \Delta)^{-1} JTv\, dt$$

$$= J^{-1}\frac{\sin((\alpha-1)\pi)}{\pi} \int_0^{+\infty} t^\alpha (-t^2\mathcal{I} + \Delta)^{-1}\, dt\, JTv$$

$$= \frac{1}{2}J^{-1}(-\Delta)^{\frac{\alpha-1}{2}} JTv,$$

where the last equation follows from computations as in (9.21). Choosing $\beta = 2\alpha + 1$, we thus find for sufficiently regular $v$ that

$$\text{Vec}\, f_\beta(T)v(\xi)$$

$$= \frac{1}{2}J^{-1}(-\Delta)^\alpha JTv(\xi_1, \xi_2, \xi_3)$$

$$= \frac{1}{2}J^{-1}(-\Delta)^\alpha \begin{pmatrix} e^{x_1}v_{\xi_1}(e^{x_1}, e^{x_2}, e^{x_3}) \\ e^{x_2}v_{\xi_2}(e^{x_1}, e^{x_2}, e^{x_3}) \\ e^{x_3}v_{\xi_3}(e^{x_1}, e^{x_2}, e^{x_3}) \end{pmatrix}$$

$$= \frac{1}{2}J^{-1}\frac{1}{(2\pi)^3} \int_{\mathbb{R}^3} \int_{\mathbb{R}^3} -|y|^{2\alpha} e^{iz\cdot y} e^{-x\cdot y} \begin{pmatrix} e^{x_1}v_{\xi_1}(e^{x_1}, e^{x_2}, e^{x_3}) \\ e^{x_2}v_{\xi_2}(e^{x_1}, e^{x_2}, e^{x_3}) \\ e^{x_3}v_{\xi_3}(e^{x_1}, e^{x_2}, e^{x_3}) \end{pmatrix} dx\, dy$$

$$= \frac{1}{2(2\pi)^3} \int_{\mathbb{R}^3} \int_{\mathbb{R}^3} -|y|^{2\alpha} e^{i\sum_{k=1}^3 \xi_k y_k} e^{-ix\cdot y} \begin{pmatrix} e^{x_1}v_{\xi_1}(e^{x_1}, e^{x_2}, e^{x_3}) \\ e^{x_2}v_{\xi_2}(e^{x_1}, e^{x_2}, e^{x_3}) \\ e^{x_3}v_{\xi_3}(e^{x_1}, e^{x_2}, e^{x_3}) \end{pmatrix} dx\, dy.$$

The above computations are elementary and illustrate that more complicated operators than the Nabla operator can be considered with the introduced techniques. In particular, one can define and study new types of fractional evolution equations derived from generalized gradient operators with non-constant coefficients of the form

$$T = a_1(x)\frac{\partial}{\partial x_1}e_1 + a_2(x)\frac{\partial}{\partial x_2}e_2 + a_3(x)\frac{\partial}{\partial x_3}e_3. \tag{9.22}$$

The version of the $S$-functional calculus for operators with commuting components, which we applied in order to study the Nabla operator, simplifies the computations considerably. In the next chapter, we will investigate a more involved example that shows the power of our theory.

# Chapter 10

# Applications to fractional diffusion

In this chapter, we define what we call the $S$-spectrum approach to fractional diffusion processes and we give an application of our theory to a class of fractional Fourier laws with non-constant coefficients. We point out that, since we consider the Fourier law, the operator $T$ will be the vector operator

$$T = e_1 a_1(x_1)\partial_{x_1} + e_2 a_2(x_2)\partial_{x_2} + e_3 a_3(x_3)\partial_{x_3}, \tag{10.1}$$

where the operators $a_\ell(x_\ell)\partial_{x_\ell}$, $\ell = 1, 2, 3$, contain just first order derivatives. Our theory applies to more general operators but here we limit ourselves to the problems related with the heat equation. This new approach was introduced by the authors in [55].

## 10.1 New fractional diffusion problems

In order to determine fractional powers of $T$, we assume suitable conditions on the coefficients of the operator $T$ that allow us to show that the $S$-resolvent operators $S_L^{-1}(s, T)$ and $S_R^{-1}(s, T)$ satisfy suitable estimates. We choose now any $j \in \mathbb{S}$ and for $\alpha \in (0, 1)$, $v \in \mathcal{D}(T)$, we define

$$P_\alpha(T)v := \frac{1}{2\pi} \int_{-j\mathbb{R}} S_L^{-1}(s, T)\, ds_j\, s^{\alpha-1}Tv, \tag{10.2}$$

or

$$P_\alpha(T)v := \frac{1}{2\pi} \int_{-j\mathbb{R}} s^{\alpha-1}\, ds_j\, S_R^{-1}(s, T)Tv, \tag{10.3}$$

where $ds_j = ds(-j)$. These integrals do not depend on the imaginary unit $j \in \mathbb{S}$ and both integrals define the same operator. Moreover, they correspond to a

© Springer Nature Switzerland AG 2019
F. Colombo, J. Gantner, *Quaternionic Closed Operators, Fractional Powers and Fractional Diffusion Processes*, Operator Theory: Advances and Applications 274,
https://doi.org/10.1007/978-3-030-16409-6_10

modified version of Balakrishnan's formula that takes only spectral points with the positive real part into account. These modifications are necessary because $s \mapsto s^\alpha$ for $\alpha \in (0,1)$ is not defined for $s \in (-\infty, 0)$ and, unlike in the complex setting, it is not possible to choose different branches of $s^\alpha$ to avoid this problem. When we define the fractional powers of $T$, we have to take this fact into account and choose a suitable path of integration in Balakrishnan's formula. We furthermore use the notation $P_\alpha(T)$, for $\alpha \in (0,1)$, to stress that we only take spectral values $s$ with $\mathrm{Re}(s) \geq 0$ into account, i.e., only points where $s^\alpha$ is actually defined.

Now we make some considerations on the heat equation. First, we consider the case when $T$ is minus the gradient operator because in this case we expect that our new method, based on the spectral theory on the $S$-spectrum, reproduces the classical fractional heat equation that contains the fractional powers of the negative Laplace operator. So we consider the heat flux $\mathbf{q}(\nabla)v = -\nabla v$, where $v$ is the temperature, we identify $\mathbb{R}^3 \cong \{s \in \mathbb{H} : \mathrm{Re}(s) = 0\}$ and we consider the gradient operator $\nabla$ as the quaternionic operator

$$\mathbf{q}(\nabla)v = -(e_1 \partial_{x_1} + e_2 \partial_{x_2} + e_3 \partial_{x_3})v. \tag{10.4}$$

Instead of replacing the negative Laplacian $(-\Delta)^\alpha$ in the heat equation, we replace the fractional gradient in the equation

$$\partial_t v(t,x) + \mathrm{div}\, \mathbf{q}(\nabla)v(t,x) = 0.$$

The following two observations are of crucial importance in order to define the new procedure to fractional diffusion processes.

(I) Since $s^\alpha$, for $\alpha \in (0,1)$, is not defined on $(-\infty, 0)$ and in the Hilbert space $L^2(\mathbb{R}^3, \mathbb{H})$ it is $\sigma_S(\nabla) = \mathbb{R}$, we consider the projections of the fractional powers of $\nabla^\alpha$, indicated by $P_\alpha(\nabla)$, to the subspace associated with the subset $[0, +\infty)$ of the $S$-spectrum of $\nabla$, on which the function $s^\alpha$ is well defined and slice hyperholomorphic.

(II) The above procedure gives a quaternionic operator

$$P_\alpha(\nabla) = Z_0 + e_1 Z_1 + e_2 Z_2 + e_3 Z_3,$$

where $Z_\ell$, $\ell = 0, 1, 2, 3$ are real operators obtained by the functional calculus. Finally, we take the vector part of $P_\alpha(\nabla)$ defined by

$$\mathrm{Vect}(P_\alpha(\nabla)) = e_1 Z_1 + e_2 Z_2 + e_3 Z_3$$

so that we can apply to the operator $\mathrm{Vect}(P_\alpha(\nabla))$ the divergence operator.

More explicitly, we define $\nabla^\alpha$ only on the subspace associated to $[0, \infty)$ that is

$$P_\alpha(\nabla)v = \frac{1}{2\pi} \int_{-j\mathbb{R}} S_L^{-1}(s, \nabla)\, ds_j\, s^{\alpha-1} \nabla v,$$

for $v : \mathbb{R}^3 \to \mathbb{R}$ in $\mathcal{D}(\nabla)$, where the path integral is computed taking into account just the part of the $S$-spectrum with $\operatorname{Re}(\sigma_S(\nabla)) \geq 0$ since we have proven in Theorem 9.1.1 that the $S$-spectrum of $\nabla$ as an operator on $L^2(\mathbb{R}^3, \mathbb{H})$ is $\sigma_S(\nabla) = \mathbb{R}$. With this definition and the surprising expression for the left $S$-resolvent operator

$$S_L^{-1}(-jt, \nabla) = (-jt + \nabla) \underbrace{(-t^2 + \Delta)^{-1}}_{=R_{-t^2}(-\Delta)},$$

where $R_{-t^2}(-\Delta)$ is the classical resolvent operator of the Laplacian, the fractional powers $P_\alpha(\nabla)$ become

$$P_\alpha(\nabla)v = \underbrace{\frac{1}{2}(-\Delta)^{\frac{\alpha}{2}-1}\nabla^2 v}_{\operatorname{Scal}P_\alpha(\nabla)v} + \underbrace{\frac{1}{2}(-\Delta)^{\frac{\alpha-1}{2}}\nabla v}_{=\operatorname{Vec}P_\alpha(\nabla)v}.$$

We define the scalar part of the operator $P_\alpha(\nabla)$ applied to $v$ as

$$\operatorname{Scal}P_\alpha(\nabla)v := \frac{1}{2}(-\Delta)^{\frac{\alpha}{2}-1}\nabla^2 v,$$

and the vector part as

$$\operatorname{Vec}P_\alpha(\nabla)v := \frac{1}{2}(-\Delta)^{\frac{\alpha-1}{2}}\nabla v.$$

Now we observe that

$$\operatorname{div}\operatorname{Vec}P_\alpha(\nabla)v = -\frac{1}{2}(-\Delta)^{\frac{\alpha}{2}+1}v.$$

This shows that in the case of the gradient we get the same result, that is the fractional Laplacian. The fractional heat equation for $\alpha \in (1/2, 1)$

$$\partial_t v(t, x) + (-\Delta)^\alpha v(t, x) = 0$$

can hence be written as

$$\partial_t v(t, x) - 2\operatorname{div}(\operatorname{Vec}P_\beta(\nabla)v) = 0, \qquad \beta = 2\alpha - 1.$$

Before we state our new approach to fractional diffusion processes, we will make some considerations regarding the case when the components of the operator $T$ commute among themselves and when they do not commute. When we consider non homogeneous materials in $\Omega \subseteq \mathbb{R}^3$, where $\Omega$ can be bounded or unbounded, then Fourier's law becomes

$$T := \mathbf{q}(x, \partial_x) = e_1 a(x)\partial_{x_1} + e_2 b(x)\partial_{x_2} + e_3 c(x)\partial_{x_3}, \quad x = (x_1, x_2, x_3), \quad (10.5)$$

where $e_\ell$, $\ell = 1, 2, 3$ are orthogonal unit vectors in $\mathbb{R}^3$ and the coefficients $a$, $b$, $c : \Omega \to \mathbb{R}$ depend on the space variables $x = (x_1, x_2, x_3)$, and possibly on time. In this case to define the fractional powers of $T$, we need to show that the pseudo-resolvent operator

$$\mathcal{Q}_s(T) := T^2 - 2\mathrm{Re}(s)T + |s|^2 \mathcal{I} \tag{10.6}$$

is invertible. The computations can, however, be quite complicated. In the following we consider the commutative case that provides for a better understanding of our new procedure.

When $T$ has commuting components, that is $T$ is a vector operator of the form

$$T = e_1 \, a_1(x_1)\partial_{x_1} + e_2 \, a_2(x_2)\partial_{x_2} + e_3 \, a_3(x_3)\partial_{x_3},$$

where $a_1$, $a_2$, $a_3 : \Omega \to \mathbb{R}$ are suitable real valued functions that depend on the space variables $x_1$, $x_2$, $x_3$, respectively, then the $S$-spectrum can also be determined by the commutative pseudo-resolvent operator, given by the inverse of the operator

$$\mathcal{Q}_{c,s}(T) := s^2 \mathcal{I} - 2sT_0 + T\overline{T} = a_1^2(x_1)\partial_{x_1}^2 + a_2^2(x_2)\partial_{x_2}^2 + a_3^2(x_3)\partial_{x_3}^2 + s^2 \mathcal{I},$$

because $\mathcal{Q}_{c,s}(T)$ is invertible if and only if $\mathcal{Q}_s(T)$ is invertible, cf. Theorem 3.3.4. The operator $\mathcal{Q}_{c,s}(T)$ is a scalar operator if $s^2$ is a real number. Since $T$ is a vector operator, we have $T_0 = 0$, and $T\overline{T}$ does not contain the imaginary units of the quaternions. Using the non commutative expression of the pseudo-resolvent operator $\mathcal{Q}_s(T)$, we obtain

$$\begin{aligned}
\mathcal{Q}_s(T) = &-(a_1(x_1)\partial_{x_1})^2 - (a_2(x_2)\partial_{x_2})^2 - (a_3(x_3)\partial_{x_3})^2 \\
&- 2s_0(e_1 \, a_1(x_1)\partial_{x_1} + e_2 \, a_2(x_2)\partial_{x_2} + e_3 \, a_3(x_3)\partial_{x_3}) + |s|^2 \mathcal{I}.
\end{aligned}$$

We observe that according to what we need to show in the commutative case we have two possibilities. In the next subsection we explicitly write the procedure to define the $S$-spectrum approach to fractional diffusion processes.

## 10.2   The $S$-spectrum approach to fractional diffusion processes

Suppose that $\Omega \subseteq \mathbb{R}^3$ is a suitable bounded or unbounded domain and let $X$ be a two-sided Banach space. We consider the initial boundary value problem for non-homogeneous materials. We use the notation $T = \mathbf{q}(x, \partial_x)$ and we restrict ourselves to the case of homogeneous boundary conditions (for $\tau$ positive number):

$$\begin{aligned}
T(x) &= e_1 a_1(x_1)\partial_{x_1} + e_2 a_2(x_2)\partial_{x_2} + e_3 a_3(x_3)\partial_{x_3}, \quad x = (x_1, x_2, x_3) \in \Omega, \\
\partial_t v(x,t) &+ \mathrm{div}\, T(x)v(x,t) = 0, \quad (x,t) \in \Omega \times (0,\tau], \\
v(x,0) &= f(x), \quad x \in \Omega, \\
v(x,t) &= 0, \quad x \in \partial\Omega, \ t \in [0,\tau].
\end{aligned}$$

Our general procedure consists of the following steps:

(S1) We study the invertibility of the operator

$$Q_{c,s}(T) := s^2\mathcal{I} - 2sT_0 + T\overline{T}$$
$$= a_1^2(x_1)\partial_{x_1}^2 + a_2^2(x_2)\partial_{x_2}^2 + a_3^2(x_3)\partial_{x_3}^2 + s^2\mathcal{I},$$

where $\overline{T} = -T$, to determine the $S$-resolvent operators. Precisely, let $F : \Omega \to \mathbb{H}$ be a given function with a suitable regularity and denote by $Y : \Omega \to \mathbb{H}$ the unknown function of the boundary value problem:

$$\left(a_1^2(x_1)\partial_{x_1}^2 + a_2^2(x_2)\partial_{x_2}^2 + a_3^2(x_3)\partial_{x_3}^2 + s^2\mathcal{I}\right)Y(x) = F(x), \quad x \in \Omega,$$
$$Y(x) = 0, \quad x \in \partial\Omega.$$

We study under which conditions on the coefficients $a_1$, $a_2$, $a_3 : \mathbb{R}^3 \to \mathbb{R}$ the above equation has a unique solution. We can similarly use the non-commutative version of the pseudo-resolvent operator $Q_s(T)$. In the case we deal with an operator $T$ with non-commuting components, then we can only consider $Q_s(T)$ because $Q_{c,s}(T)$ is not well-defined.

(S2) From (S1) we get the unique pseudo-resolvent operator $Q_{c,s}(T)^{-1}$ and so we can define the $S$-resolvent operator

$$S_L^{-1}(s,T) = (s\mathcal{I} - \overline{T})Q_{c,s}(T)^{-1}.$$

Then we prove that any $s \in \mathbb{H} \setminus \{0\}$ with $\mathrm{Re}(s) = 0$ belongs to $\rho_S(T)$. The $S$-resolvent operators satisfy the estimates

$$\left\|S_L^{-1}(s,T)\right\| \leq \frac{\Theta}{|s|} \quad \text{and} \quad \left\|S_R^{-1}(s,T)\right\| \leq \frac{\Theta}{|s|}, \tag{10.7}$$

for $\mathrm{Re}(s) = 0$ with a constant $\Theta > 0$ that does not depend on $s$.

(S3) Using the quaternionic Balakrishnan's formula, we define $P_\alpha(T)$ as

$$P_\alpha(T)v := \frac{1}{2\pi}\int_{-j\mathbb{R}} s^{\alpha-1}\, ds_j\, S_R^{-1}(s,T)Tv, \quad \text{for } \alpha \in (0,1),$$

and $v \in \mathcal{D}(T)$. Analogously, one can use the definition of $P_\alpha(T)$ related to the left $S$-resolvent operator.

(S4) After we define the fractional powers $P_\alpha(T)$ of the vector operator $T$, we consider its vector part $\mathrm{Vec}(P_\alpha(T))$ and we obtain the fractional evolution equation

$$\partial_t v(t,x) - \mathrm{div}(\mathrm{Vec}(P_\alpha(T)v)(t,x) = 0.$$

This approach has several advantages:

(I) It modifies the Fourier law but keeps the law of conservation of energy.

(II) It is applicable to a large class of operators that includes the gradient and also operators with variable coefficients.

(III) The fractional powers of the operator $T$ are more realistic for non-homogeneous materials.

(IV) The fact that we keep the evolution equation in divergence form allows an immediate definition of the weak solution of the fractional evolution problem.

In the next section we show explicitly how the $S$-spectrum approach to fractional diffusion processes works in the Hilbert space setting.

## 10.3   Fractional Fourier's law in a Hilbert space

These results where proved in our paper [55]. In this section, we show that, under suitable conditions on the coefficients $a_\ell(x_\ell)$ for $\ell = 1, 2, 3$, we can define the fractional operator associated with the quaternionic operator

$$T = e_1 a_1(x_1)\partial_{x_1} + e_2 a_2(x_2)\partial_{x_2} + e_3 a_3(x_3)\partial_{x_3}.$$

The conditions on the coefficients $a_\ell(x_\ell)$ for $\ell = 1, 2, 3$ are determined by the Lax-Milgram lemma since we work in a quaternionic Hilbert space. We define

$$L^2 := L^2(\Omega, \mathbb{H}) := \left\{ u : \Omega \to \mathbb{H} : \int_\Omega |u(x)|^2 \, dx < +\infty \right\}$$

with the scalar product

$$\langle u, v \rangle_{L^2} := \langle u, v \rangle_{L^2(\Omega, \mathbb{H})} := \int_\Omega \overline{u(x)} v(x) \, dx,$$

where $u(x) = u_0(x) + u_1(x)e_1 + u_2(x)e_2 + u_3(x)e_3$ and $v(x) = v_0(x) + v_1(x)e_1 + v_2(x)e_2 + v_3(x)e_3$ for $x = (x_1, x_2, x_3) \in \Omega$, and $\Omega \subset \mathbb{R}^3$ is a bounded open set with smooth boundary. We furthermore introduce the quaternionic Sobolev space

$$H^1 := H^1(\Omega, \mathbb{H}) := H^1(\Omega, \mathbb{R}) \otimes \mathbb{H},$$

where $H^1(\Omega, \mathbb{R})$ denotes the Sobolov space of all square-integrable real-valued functions that have square-integrable weak partial derivatives. Hence, the quaternionic space $H^1$ consists of those functions $u(x) = u_0(x) + \sum_{\ell=1}^3 u_\ell(x)e_\ell \in L^2(\Omega, \mathbb{H})$ such that for any $\ell = 0, \ldots, 3$ the function $u_\ell$ belongs to $H^1(\Omega, \mathbb{R})$. That is, for any $\kappa = 1, 2, 3$, there exists a function $g_{\ell,\kappa}(x) \in L^2(\Omega, \mathbb{R})$ such that

$$\int_\Omega u_\ell(x)\partial_{x_\kappa} \varphi(x) \, dx = -\int_\Omega g_{\ell,\kappa}(x)\varphi(x) \, dx, \quad \forall \varphi \in C_c^\infty(\Omega, \mathbb{R}),$$

where $C_c^\infty(\Omega, \mathbb{R})$ denotes the set of real-valued infinitely differentiable functions with compact support on $\Omega$. With the quaternionic scalar product

$$\langle u, v \rangle_{H^1} := \langle u, v \rangle_{H^1(\Omega, \mathbb{H})} := \langle u, v \rangle_{L^2} + \sum_{\ell=1}^{3} \langle \partial_{x_\ell} u, \partial_{x_\ell} v \rangle_{L^2}$$

the space $H^1$ is a quaternionic Hilbert space. We define $H_0^1 := H_0^1(\Omega, \mathbb{H})$ to be the closure of $C_c^\infty(\Omega, \mathbb{H})$ in $H^1(\Omega, \mathbb{H})$. As in the scalar case, see [49, Theorem 9.17], if we suppose that $\Omega$ is of class $C^1$ and we assume that

$$u \in H^1(\Omega, \mathbb{H}) \cap C(\overline{\Omega}, \mathbb{H}),$$

then the condition $u = 0$ on $\partial\Omega$ is equivalent to $u \in H_0^1(\Omega, \mathbb{H})$. In the general case, for arbitrary functions $u \in H^1(\Omega, \mathbb{H})$, the notation $u|_{\partial\Omega}$ has to be understood in the sense of the trace operator, see [49, p. 315]. The space $H_0^1$ turns out to be the kernel of the trace operator, i.e.,

$$H_0^1 := H_0^1(\Omega, \mathbb{H}) := \left\{ u \in H^1(\Omega, \mathbb{H}) : \quad u|_{\partial\Omega} = 0 \right\}.$$

Finally, $H_0^1(\Omega, \mathbb{H})$ is a subspace of $H^1(\Omega, \mathbb{H})$ that is a quaternionic Hilbert space itself. We define

$$\|u\|_D^2 := \sum_{\ell=1}^{3} \|\partial_{x_\ell} u\|_{L^2}^2.$$

Due to the regularity of $\partial\Omega$, the Poincaré-inequality

$$\|u\|_{L^2} \leq C_\Omega \|u\|_D$$

holds and so $\|u\|_D$ defines a norm on $H_0^1(\Omega, \mathbb{H})$ that is equivalent to $\|u\|_{H^1}$. However, we want to point out that $\|u\|_D \neq \|\nabla u\|_{L^2}$ if $\nabla$ denotes the quaternionic Nabla operator $\nabla = e_1 \partial_{x_1} + e_2 \partial_{x_2} + e_3 \partial_{x_3}$ although this notation is used for Sobolev spaces of real-valued functions. From the proof of the Poincaré inequality, it is obvious that one can even choose $C_\Omega$ such that

$$\|u\|_{L^2} \leq C_\Omega \|\partial_{x_\ell} u\|_{L^2}, \quad \ell = 1, 2, 3. \tag{10.8}$$

The first result we need to show is that all purely imaginary quaternions are in the $S$-resolvent set of $T$. Moreover, we show that the $S$-resolvent operators decay sufficiently fast at infinity so that the operator defined in $(S3)$, that is,

$$P_\alpha(T)v = \frac{1}{2\pi} \int_{-j\mathbb{R}} s^{\alpha-1} \, ds_j \, S_R^{-1}(s, T)Tv, \quad v \in \mathcal{D}(T),$$

for $\alpha \in (0, 1)$, turns out to be well-defined.

**Theorem 10.3.1.** *Let $\Omega$ be a bounded domain in $\mathbb{R}^3$ with smooth boundary. Assume that the coefficients $a_\ell : \overline{\Omega} \to \mathbb{R}$, for $\ell = 1, 2, 3$, of the operator*

$$T = e_1 a_1(x_1)\partial_{x_1} + e_2 a_2(x_2)\partial_{x_2} + e_3 a_3(x_3)\partial_{x_3}$$

*belong to $C^1(\overline{\Omega}, \mathbb{R})$ and suppose that*

$$\inf_{x \in \Omega} |a_\ell(x_\ell)^2| - \frac{\sqrt{C_\Omega}}{2} \left\| \partial_{x_\ell} a_\ell(x_\ell)^2 \right\|_\infty > 0, \quad \ell = 1, 2, 3,$$

*and*

$$\frac{1}{2} - \frac{1}{2} \|\Phi\|_\infty^2 C_\Omega^2 C_a^2 > 0,$$

*where $C_\Omega$ is the Poincaré constant of $\Omega$ satisfying (10.8) and*

$$\Phi(x) := \sum_{\ell=1}^{3} e_\ell \partial_{x_\ell} a_\ell(x_\ell) \quad and \quad C_a := \sup_{\substack{x \in \Omega \\ \ell=1,2,3}} \frac{1}{|a_\ell(x_\ell)|} = \frac{1}{\inf_{\substack{x \in \Omega \\ \ell=1,2,3}} |a_\ell(x_\ell)|}.$$

*Then any $s \in \mathbb{H} \backslash \{0\}$ with $\mathrm{Re}(s) = 0$ belongs to $\rho_S(T)$ and the $S$-resolvent operators satisfy, for $s$ with $\mathrm{Re}(s) = 0$, the estimates*

$$\left\| S_L^{-1}(s, T) \right\|_{\mathcal{B}(L^2)} \leq \frac{\Theta}{|s|} \quad and \quad \left\| S_R^{-1}(s, T) \right\|_{\mathcal{B}(L^2)} \leq \frac{\Theta}{|s|} \tag{10.9}$$

*with a constant $\Theta > 0$ that does not depend on $s$. Here, $\| \cdot \|_{\mathcal{B}(L^2)}$ denotes the operator norm on the space of bounded quaternionic operators on $L^2 = L^2(\Omega, \mathbb{H})$.*

*Proof.* As a first step, we want to show that any $s \in \mathbb{H} \backslash \{0\}$ with $\mathrm{Re}(s) = 0$ belongs to $\rho_S(T)$, i.e., that

$$\mathcal{Q}_s(T) = T^2 - 2s_0 T + |s|^2 \mathcal{I}$$

has a bounded inverse, and we want to do this by applying the Lax-Milgram lemma. Since $T$ has commuting components, this operator has for $s = js_1 \in \mathbb{H}$ the form

$$\mathcal{Q}_s(T) = T^2 + s_1^2 \mathcal{I} = -(a_1(x_1)\partial_{x_1})^2 - (a_2(x_2)\partial_{x_2})^2 - (a_3(x_3)\partial_{x_3})^2 + s_1^2 \mathcal{I}.$$

We want to show that it has an inverse on $L^2(\Omega, \mathbb{H})$ that satisfies

$$\left\| \mathcal{Q}_s(T)^{-1} \right\|_{\mathcal{B}(L^2)} \leq C \frac{1}{s_1^2}$$

with a constant $C$ that is independent of $s_1$. We consider the bilinear form

$$\langle \mathcal{Q}_s(T)u, v \rangle_{L^2} = \int_\Omega \overline{\mathcal{Q}_s(T)u(x)} v(x)\, dx$$

on $H_0^1(\Omega, \mathbb{H})$. We rewrite this as

$$\langle Q_s(T)u, v\rangle_{L^2} = s_1^2 \int_\Omega \overline{u(x)}v(x)\, dx - \sum_{\ell=1}^3 \int_\Omega \overline{(a_\ell(x_\ell)\partial_{x_\ell})^2 u(x)}v(x)\, dx$$

and further

$$\int_\Omega \overline{(a_\ell(x_\ell)\partial_{x_\ell})^2 u(x)}v(x)\, dx$$
$$= \int_\Omega \left( (a_\ell(x_\ell)\partial_{x_\ell})^2 \overline{u(x)} \right) v(x)\, dx$$
$$= \int_\Omega \left( \partial_{x_\ell} a_\ell(x_\ell)\partial_{x_\ell} \overline{u(x)} \right) a_\ell(x_\ell)v(x)\, dx.$$

Integrating by parts, we find

$$\int_\Omega \overline{(a_\ell(x_\ell)\partial_{x_\ell})^2 u(x)}v(x)\, dx$$
$$= -\int_\Omega \frac{1}{2}\left( \partial_{x_\ell} a_\ell(x_\ell)^2 \right)\left( \partial_{x_\ell}\overline{u(x)} \right)v(x)\, dx$$
$$\quad - \int_\Omega a_\ell(x_\ell)^2 \left( \partial_{x_\ell}\overline{u(x)} \right)\partial_{x_\ell}v(x)\, dx$$
$$\quad + \int_{\partial\Omega} a_\ell(x_\ell)^2 \left( \partial_{x_\ell}\overline{u(x)} \right)v(x)n_\ell(x)\, dS(x),$$

where $S$ is the surface measure on $\partial\Omega$ and $n_\ell(x)$ denotes for $x \in \partial\Omega$ the $\ell$-th component of the outward pointing normal. Since $v \in H_0^1(\Omega, \mathbb{H})$, the integral over the boundary is zero and hence we altogether obtain

$$b(u, v) := s_1^2 \int_\Omega \overline{u(x)}v(x)\, dx + \sum_{\ell=1}^3 \int_\Omega \frac{1}{2}\left( \partial_{x_\ell} a_\ell(x_\ell)^2 \right)\left( \partial_{x_\ell}\overline{u(x)} \right)v(x)\, dx$$
$$\quad + \sum_{\ell=1}^3 \int_\Omega a_\ell(x_\ell)^2 \left( \partial_{x_\ell}\overline{u(x)} \right)\partial_{x_\ell}v(x)\, dx.$$

The bilinear form $b$ is continuous on $H_0^1(\Omega, \mathbb{H})$ as

$$|b(u, v)| \le s_1^2 \int_\Omega \left|\overline{u(x)}v(x)\right|\, dx + \sum_{\ell=1}^3 \frac{1}{2}\left\|\partial_{x_\ell} a_\ell^2\right\|_\infty \int_\Omega \left|\left( \partial_{x_\ell}\overline{u(x)} \right)v(x)\right|\, dx$$
$$\quad + \sum_{\ell=1}^3 \left\|a_\ell^2\right\|_\infty \int_\Omega \left( \partial_{x_\ell}\overline{u(x)} \right)\partial_{x_\ell}v(x)\, dx$$

$$\leq s_1^2\|u\|_{L^2}\|v\|_{L^2} + \sum_{\ell=1}^{3}\left\|\partial_{x_\ell}a_\ell^2\right\|_\infty \left\|\partial_{x_\ell}u\right\|_{L^2}\|v\|_{L^2}$$

$$+ \sum_{\ell=1}^{3}\left\|a_\ell^2\right\|_\infty \left\|\partial_{x_\ell}u\right\|_{L^2}\left\|\partial_{x_\ell}v\right\|_{L^2}$$

$$\leq \left(s_1^2 + \sum_{\ell=1}^{3}\left\|\partial_{x_\ell}a_\ell^2\right\|_\infty + \sum_{\ell=1}^{3}\left\|a_\ell^2\right\|_\infty\right)\|u\|_{H^1}\|v\|_{H^1},$$

where $\|\cdot\|_\infty$ denotes the supremum norm. Furthermore, observe that for any $w \in L^2$, the map $\ell_w(v) := \langle w, v\rangle_{L^2}$ is a continuous quaternionic linear functional on $H_0^1(\Omega,\mathbb{H})$ since

$$|\ell_w(v)| = |\langle w, v\rangle_{L^2}| \leq \|w\|_{L^2}\|v\|_{L^2} \leq \|w\|_{L^2}\|v\|_{H_0^1}.$$

We can consider $H_0^1(\Omega,\mathbb{H})$ also as a real Hilbert space, if we restrict the multiplication with scalars to $\mathbb{R}$ and endow it with the real scalar product $\langle u, v\rangle_{\mathbb{R}} = \mathrm{Re}\langle u, v\rangle_{H^1}$. Then $\mathrm{Re}\,b$ is a continuous $\mathbb{R}$-bilinear form on $H_0^1(\Omega,\mathbb{H})$ and $\mathrm{Re}\,\ell_w$ is for any $w \in L^2(\Omega,\mathbb{H})$ a continuous linear functional on $H_0^1(\Omega,\mathbb{H})$. What remains to show in order to apply the Lemma of Lax–Milgram is that $\mathrm{Re}\,b$ is also coercive. We have

$$\mathrm{Re}\,b(u,u) = s_1^2\|u\|_{L^2}^2 + \sum_{\ell=1}^{3}\mathrm{Re}\int_\Omega \frac{1}{2}\left(\partial_{x_\ell}a_\ell(x_\ell)^2\right)\left(\partial_{x_\ell}\overline{u(x)}\right)u(x)\,dx$$

$$+ \sum_{\ell=1}^{3}\int_\Omega a_\ell(x_\ell)^2|\partial_{x_\ell}u(x))|^2\,dx$$

$$\geq s_1^2\|u\|_{L^2}^2 - \sum_{\ell=1}^{3}\frac{1}{2}\left\|\partial_{x_\ell}a_\ell(x_\ell)^2\right\|_\infty \int_\Omega \left|\partial_{x_\ell}\overline{u(x)}\right||u(x)|\,dx$$

$$+ \sum_{\ell=1}^{3}\inf_{x\in\Omega}|a_\ell(x_\ell)^2|\int_\Omega |\partial_{x_\ell}u(x)|^2\,dx.$$

Applying the Young inequality, we find for any $\delta > 0$ that

$$\mathrm{Re}\,b(u,u) \geq s_1^2\|u\|_{L^2}^2 + \sum_{\ell=1}^{3}\inf_{x\in\Omega}|a_\ell(x_\ell)^2|\,\|\partial_{x_\ell}u\|_{L^2}^2$$

$$- \sum_{\ell=1}^{3}\frac{1}{2}\left\|\partial_{x_\ell}a_\ell(x_\ell)^2\right\|_\infty \left(\frac{\delta}{2}\|\partial_{x_\ell}u\|_{L^2}^2 + \frac{1}{2\delta}\|u\|_{L^2}^2\right).$$

Therefore, we furthermore get

$$\operatorname{Re} b(u, u) \geq \left( s_1^2 - \frac{1}{4\delta} \sum_{\ell=1}^{3} \left\| \partial_{x_\ell} a_\ell(x_\ell)^2 \right\|_\infty \right) \|u\|_{L^2}^2$$

$$+ \sum_{\ell=1}^{3} \left( \inf_{x \in \Omega} |a_\ell(x_\ell)^2| - \frac{\delta}{4} \left\| \partial_{x_\ell} a_\ell(x_\ell)^2 \right\|_\infty \right) \|\partial_{x_\ell} u\|_{L^2}^2$$

$$\geq s_1^2 \|u\|_{L^2}^2 - \frac{C_\Omega}{4\delta} \sum_{\ell=1}^{3} \left\| \partial_{x_\ell} a_\ell(x_\ell)^2 \right\|_\infty \|\partial_{x_\ell} u\|_{L^2}^2$$

$$+ \sum_{\ell=1}^{3} \left( \inf_{x \in \Omega} |a_\ell(x_\ell)^2| - \frac{\delta}{4} \left\| \partial_{x_\ell} a_\ell(x_\ell)^2 \right\|_\infty \right) \|\partial_{x_\ell} u\|_{L^2}^2 \,,$$

where $C_\Omega$ is the Poincaré constant. The optimal choice $\delta = \sqrt{C_\Omega}$ finally yields

$$\operatorname{Re} b(u, u)$$

$$\geq s_1^2 \|u\|_{L^2}^2 + \sum_{\ell=1}^{3} \left( \inf_{x \in \Omega} |a_\ell(x_\ell)^2| - \frac{\sqrt{C_\Omega}}{2} \left\| \partial_{x_\ell} a_\ell(x_\ell)^2 \right\|_\infty \right) \|\partial_{x_\ell} u\|_{L^2}^2 \qquad (10.10)$$

$$\geq \kappa(s_1^2) \|u\|_{H^1}^2$$

with the constant

$$\kappa(s_1^2) := \min \left\{ s_1^2, \ \min_{1 \leq \ell \leq 3} \left\{ \inf_{x \in \Omega} |a_\ell(x_\ell)^2| - \frac{\sqrt{C_\Omega}}{2} \left\| \partial_{x_\ell} a_\ell(x_\ell)^2 \right\|_\infty \right\} \right\}.$$

Let now $w \in L^2(\Omega, \mathbb{H})$. The Lemma of Lax–Milgram implies, due to the arguments above, the existence of a unique $u_w \in H_0^1(\Omega, \mathbb{H})$ such that

$$\operatorname{Re} b(u_w, v) = \operatorname{Re} \ell_w(v) = \operatorname{Re} \langle w, v \rangle_{L^2}, \quad \forall v \in H_0^1(\Omega, \mathbb{H}) \qquad (10.11)$$

and in turn also

$$\langle \mathcal{Q}_s(T) u_w, v \rangle_{L^2(\mathbb{R}, \mathbb{H})} = b(u_w, v) = \langle w, v \rangle_{L^2}, \quad \forall v \in H_0^1(\Omega, \mathbb{H}) \qquad (10.12)$$

because

$$b(u_w, v) = \operatorname{Re} b(u_w, v) + \sum_{\ell=1}^{3} (\operatorname{Re} b(u_w, -v e_\ell)) \, e_\ell$$

and

$$\langle w, v \rangle_{L^2} = \operatorname{Re}\langle w, v \rangle_{L^2} + \sum_{\ell=1}^{3} (\operatorname{Re}\langle w, -v e_\ell \rangle_{L^2}) \, e_\ell.$$

Furthermore, we have

$$\|u_w\|_{L^2} \leq \|u_w\|_{H^1} \leq \frac{1}{\kappa(s_1^2)} \|w\|_{L^2}.$$

The mapping $S : w \to u_w$ is therefore a bounded linear mapping on $L^2(\Omega, \mathbb{H})$ and so the operator $\mathcal{Q}_s(T)$ has a bounded inverse on $L^2(\Omega, \mathbb{H})$ with range in $H_0^1(\Omega, \mathbb{H})$. From the estimate (10.10), we furthermore conclude

$$s_1^2 \|u_w\|_{L^2}^2 \leq \operatorname{Re} b(u_w, u_w) \leq |b(u_w, u_w)| = |\langle w, u_w \rangle_{L^2}| \leq \|w\|_{L^2} \|u_w\|_{L^2}.$$

Therefore, we have

$$\left\| \mathcal{Q}_s(T)^{-1} w \right\|_{L^2} = \|u_w\|_{L^2} \leq \frac{1}{s_1^2} \|w\|_{L^2}$$

and so

$$\left\| \mathcal{Q}_s(T)^{-1} \right\|_{\mathcal{B}(L^2)} \leq \frac{1}{s_1^2}. \tag{10.13}$$

Using this estimate, we can now show that the $S$-resolvent of $T$ decays fast enough along the set of purely imaginary quaternions. For any $v \in H_0^1(\Omega, \mathbb{H})$, we have that

$$b(u_w, v) = \langle \mathcal{Q}_s(T) u_w, v \rangle_{L^2} = \left\langle T^2 u_w, v \right\rangle_{L^2} + s_1^2 \langle u_w, v \rangle_{L^2}.$$

The first term can be expressed as

$$\langle T^2 u_w, v \rangle_{L^2} = \int_\Omega \overline{(T^2 u_w)(x)} v(x) \, dx$$

$$= \sum_{\ell=1}^3 \int_\Omega a_\ell(x_\ell) \partial_{x_\ell} \overline{(T u_w)(x)} (-e_\ell) v(x) \, dx.$$

Integration by parts yields

$$\left\langle T^2 u_w, v \right\rangle_{L^2} = \sum_{\ell=1}^3 \int_\Omega \overline{(T u_w)(x)} \, e_\ell \partial_{x_\ell} \left( a_\ell(x_\ell) v(x) \right) \, dx$$

$$+ \sum_{\ell=1}^3 \int_{\partial\Omega} \overline{(T u_w)(x)} \, n_\ell(x)(-e_\ell) a_\ell(x_\ell) v(x) \, dS(x)$$

$$= \sum_{\ell=1}^3 \int_\Omega \overline{(T u_w)(x)} e_\ell \left( \partial_{x_\ell} a_\ell(x_\ell) \right) v(x) \, dx$$

$$+ \sum_{\ell=1}^3 \int_\Omega \overline{(T u_w)(x)} e_\ell a_\ell(x_\ell) \partial_{x_\ell} v(x) \, dx$$

$$= \int_\Omega \overline{(T u_w)(x)} \left( \sum_{\ell=1}^3 e_\ell \partial_{x_\ell} a_\ell(x_\ell) \right) v(x) \, dx$$

$$+ \int_\Omega \overline{(T u_w)(x)} T v(x) \, dx,$$

where the integral over the boundary vanishes as $v(x) = 0$ on $\partial\Omega$ because $v \in H_0^1(\Omega, \mathbb{H})$. We find that

$$b(u_w, v) = \int_\Omega \overline{(Tu_w)(x)} \Phi(x) v(x)\, dx + \langle Tu_w, Tv \rangle_{L^2} + s_1^2 \langle u_w, v \rangle_{L^2}.$$

with

$$\Phi(x) := \sum_{\ell=1}^{3} e_\ell \partial_{x_\ell} a_\ell(x_\ell).$$

Choosing $v = u_w$ yields

$$b(u_w, u_w) = \int_\Omega \overline{(Tu_w)(x)} \Phi(x) u_w(x)\, dx + \|Tu_w\|_{L^2}^2 + s_1^2 \|u_w\|_{L^2}^2. \qquad (10.14)$$

We hence have

$$|b(u_w, u_w)| \geq \|Tu_w\|_{L^2}^2 + s_1^2 \|u_w\|_{L^2}^2 - \int_\Omega \left| \overline{(Tu_w)(x)} \Phi(x) u_w(x) \right| dx$$

$$\geq \|Tu_w\|_{L^2}^2 + s_1^2 \|u_w\|_{L^2}^2 - \int_\Omega \left| \overline{(Tu_w)(x)} \right| \|\Phi\|_\infty |u_w(x)|\, dx$$

$$\geq \frac{1}{2} \|Tu_w\|_{L^2}^2 + s_1^2 \|u_w\|_{L^2}^2 - \frac{1}{2} \|\Phi\|_\infty^2 \|u_w\|_{L^2}^2,$$

where the last identity follows from the Young inequality. In order to estimate the term $\|u_w\|_{L^2}^2$, we write $u_w(x) = u_{w,0}(x) + \sum_{\ell=1}^{3} u_{w,\ell}(x) e_\ell$ with $u_{w,\ell}(x) \in \mathbb{R}$. Then

$$\|u_w\|_{L^2}^2 = \sum_{\ell=0}^{3} \|u_{w,\ell}\|_{L^2}^2 \leq C_\Omega^2 \sum_{\ell=0}^{3} \|u_{w,\ell}\|_D^2$$

$$= C_\Omega^2 \sum_{\ell=0}^{3} \sum_{k=1}^{3} \|\partial_{x_k} u_{w,\ell}\|_{L^2}^2 \leq C_\Omega^2 C_a^2 \sum_{\ell=0}^{3} \sum_{k=1}^{3} \|a_k \partial_{x_k} u_{w,\ell}\|_{L^2}^2$$

with

$$C_a := \sup_{\substack{x \in \Omega \\ \ell=1,2,3}} \frac{1}{|a_\ell(x_\ell)|} = \frac{1}{\inf_{\substack{x \in \Omega \\ \ell=1,2,3}} |a_\ell(x_\ell)|}.$$

Since $u_{w,\ell}$ is real-valued, we furthermore find that

$$\|Tu_{w,\ell}\|_{L^2}^2 = \int_\Omega \left| \sum_{k=1}^{3} e_k a_k(x) \partial_{x_k} u_{w,\ell}(x) \right|^2 dx$$

$$= \sum_{k=1}^{3} \int_\Omega |a_k(x) \partial_{x_k} u_{w,\ell}(x)|^2\, dx = \sum_{k=1}^{3} \|a_k \partial_{x_k} u_{w,\ell}\|_{L^2}^2$$

and so

$$\|u_w\|_{L^2}^2 \le C_\Omega^2 C_a^2 \sum_{\ell=0}^{3} \|Tu_{w,\ell}\|_{L^2}^2.$$

Altogether, we conclude that

$$\frac{1}{2}\|Tu_w\|_{L^2}^2 + s_1^2\|u_w\|_{L^2}^2 - \frac{1}{2}\|\Phi\|_\infty^2 C_\Omega^2 C_a^2 \sum_{\ell=0}^{3} \|Tu_{w,\ell}\|_{L^2}^2 \tag{10.15}$$

$$\le |b(u_w, u_w)| = |\langle w, u_w\rangle_{L^2}| \le \|w\|_{L^2}\|u_w\|_{L^2}.$$

We observe now that the operator $\mathcal{Q}_s(T)$ is a scalar operator and hence maps real-valued functions to real-valued functions so that for $r = 0, \dots, 3$

$$b(u_w, u_{w,r}) = \langle \mathcal{Q}_s(T)u_w, u_{w,r}\rangle_{L^2}$$

$$= \langle \mathcal{Q}_s(T)u_{w,0}, u_{w,r}\rangle_{L^2} + \sum_{\ell=1}^{3}(-e_\ell)\,\langle \mathcal{Q}_s(T)u_{w,\ell}, u_{w,r}\rangle_{L^2}$$

with $\langle \mathcal{Q}_s(T)u_{w,0}, u_{w,r}\rangle_{L^2} \in \mathbb{R}$ for $\ell = 1, 2, 3$. If $w(x) = w_0(x) + \sum_{\ell=1}^{3} w_\ell(x)e_\ell$ with $w_\ell(x) \in \mathbb{R}$, we conclude from

$$\langle \mathcal{Q}_s(T)u_{w,0}, u_{w,r}\rangle_{L^2} + \sum_{\ell=1}^{3}(-e_\ell)\,\langle \mathcal{Q}_s(T)u_{w,\ell}, u_{w,r}\rangle_{L^2} = b(u_w, u_{w,r})$$

$$= \langle w, u_{w,r}\rangle_{L^2} = \langle w_0, u_{w,r}\rangle_{L^2} + \sum_{\ell=1}^{3}(-e_\ell)\langle w_\ell, u_{w,r}\rangle_{L^2}$$

that for $\ell, r = 0, \dots, 4$

$$b(u_{w,\ell}, u_{w,r}) = \langle \mathcal{Q}_s(T)u_{w,r}, u_{w,r}\rangle_{L^2} = \langle w_\ell, u_{w,r}\rangle_{L^2}$$

and in particular for $\ell = 0, \dots, 4$

$$b(u_{w,\ell}, u_{w,\ell}) = \langle \mathcal{Q}_s(T)u_{w,\ell}, u_{w,\ell}\rangle_{L^2} = \langle w_\ell, u_{w,\ell}\rangle_{L^2}.$$

Repeating the above arguments, we find that (10.15) also holds for $u_{w,\ell}$ instead of $u_w$. However, since $u_{w,\ell}$ is real-valued and has only one component, this estimate then reads as

$$\frac{1}{2}\|Tu_{w,\ell}\|_{L^2}^2 + s_1^2\|u_{w,\ell}\|_{L^2}^2 - \frac{1}{2}\|\Phi\|_\infty^2 C_\Omega^2 C_a^2\|Tu_{w,\ell}\|_{L^2}^2 \tag{10.16}$$

$$\le |b(u_{w,\ell}, u_{w,\ell})| = |\langle w_\ell, u_{w,\ell}\rangle_{L^2}|$$

$$\le \|w_\ell\|_{L^2}\|u_{w,\ell}\|_{L^2} \le \|w\|_{L^2}\|u_w\|_{L^2}.$$

If we set

$$K := \frac{1}{2} - \frac{1}{2}\|\Phi\|_\infty^2 C_\Omega^2 C_a^2 > 0,$$

then (10.16) turns into

$$K\|Tu_{w,\ell}\|_{L^2}^2 + s_1^2\|u_{w,\ell}\|_{L^2}^2 \le \|w\|_{L^2}\|u_w\|_{L^2},$$

which implies in particular

$$\|Tu_{w,\ell}\|_{L^2}^2 \le \frac{1}{K}\|w\|_{L^2}\|u_w\|_{L^2}.$$

From (10.15), we finally conclude that

$$
\frac{1}{2}\|Tu_w\|_{L^2}^2 + s_1^2\|u_w\|_{L^2}^2 \le \frac{1}{2}\|\Phi\|_\infty^2 C_\Omega^2 C_a^2 \sum_{\ell=0}^{3}\|Tu_{w,\ell}\|_{L^2}^2 + \|w\|_{L^2}\|u_w\|_{L^2}
$$
$$
\le \left(1 + \frac{1}{2}\|\Phi\|_\infty^2 C_\Omega^2 C_a^2 \frac{4}{K}\right)\|w\|_{L^2}\|u_w\|_{L^2}
$$

(10.17)

so that, after setting

$$\tau := \frac{1}{2}\left(1 + \frac{1}{2}\|\Phi\|_\infty^2 C_\Omega^2 C_a^2 \frac{4}{K}\right)^{-1} > 0,$$

we have

$$\tau\|Tu_w\|_{L^2}^2 \le \|w\|_{L^2}\|u_w\|_{L^2}.$$

Since $u_w = \mathcal{Q}_s(T)^{-1}w$, we find, because of (10.13), that $\|u_w\|_{L^2} \le \frac{1}{s_1^2}\|w\|_{L^2}$ so that in turn

$$\tau\|Tu_w\|_{L^2}^2 \le \frac{1}{s_1^2}\|w\|_{L^2}^2.$$

Hence, we have

$$\left\|T\mathcal{Q}_s(T)^{-1}w\right\|_{L^2} = \|Tu_w\|_{L^2} \le \frac{1}{\sqrt{\tau}s_1}\|w\|_{L^2}$$

and so

$$\left\|T\mathcal{Q}_s(T)^{-1}\right\| \le \frac{1}{\sqrt{\tau}s_1}.$$

If we set

$$\Theta := 2\max\left\{1, \frac{1}{\sqrt{\tau}}\right\},$$

then the above estimate and (10.13) yield

$$\left\|S_R^{-1}(s,T)\right\| = \left\|(T - \bar{s}\mathcal{I})\mathcal{Q}_s(T)^{-1}\right\|$$
$$\le \left\|T\mathcal{Q}_s(T)^{-1}\right\| + \left\|\bar{s}\mathcal{Q}_s(T)^{-1}\right\| \le \frac{\Theta}{s_1}$$

and

$$\left\|S_L^{-1}(s,T)\right\| = \left\|TQ_s(T)^{-1} - Q_s(T)^{-1}\bar{s}\right\|$$
$$\leq \left\|TQ_s(T)^{-1}\right\| + \left\|Q_s(T)^{-1}\bar{s}\right\| \leq \frac{\Theta}{s_1},$$

for any $s = js_1 \in \mathbb{H}$.                                                    □

As a consequence of the previous results, we are now in the position to prove the following crucial result.

**Theorem 10.3.2.** *Let the coefficients of $T$ be as in Theorem 10.3.1 and let $\alpha \in (0,1)$. For any $v \in \mathcal{D}(T)$, the integral*

$$P_\alpha(T)v := \frac{1}{2\pi} \int_{-j\mathbb{R}} s^{\alpha-1} \, ds_j \, S_R^{-1}(s,T)Tv.$$

*converges absolutely in $L^2(\Omega, \mathbb{H})$.*

*Proof.* The right $S$-resolvent equation implies

$$S_R^{-1}(s,T)Tv = sS_R^{-1}(s,T)v - v, \quad \forall v \in \mathcal{D}(T)$$

and so

$$\frac{1}{2\pi} \int_{-j\mathbb{R}} \left\|s^{\alpha-1} \, ds_j \, S_R^{-1}(s,T)Tv\right\|$$

$$\leq \frac{1}{2\pi} \int_{-\infty}^{-1} |t|^{\alpha-1} \left\|S_R^{-1}(-jt,T)\right\| \|Tv\| \, dt$$

$$+ \frac{1}{2\pi} \int_{-1}^{1} |t|^{\alpha-1} \left\|(-jt)S_R^{-1}(-jt,T)v - v\right\| \, dt$$

$$+ \frac{1}{2\pi} \int_{1}^{+\infty} t^{\alpha-1} \left\|S_R^{-1}(jt,T)\right\| \|Tv\| \, dt.$$

As $\alpha \in (0,1)$, the estimate (10.9) now yields

$$\frac{1}{2\pi} \int_{-j\mathbb{R}} \left\|s^{\alpha-1} \, ds_j \, S_R^{-1}(s,T)Tv\right\|$$

$$\leq \frac{1}{2\pi} \int_{1}^{+\infty} t^{\alpha-1} \frac{\Theta}{t} \|Tv\| \, dt + \frac{1}{2\pi} \int_{-1}^{1} |t|^{\alpha-1} \left(|t|\frac{\Theta}{|t|} + 1\right) \|v\| \, dt$$

$$+ \frac{1}{2\pi} \int_{1}^{+\infty} t^{\alpha-1} \frac{\Theta}{t} \|Tv\| \, dt < +\infty.$$                    □

We conclude this section by observing that the above result can be proved in different function spaces, not only in the Hilbert setting, using different techniques. Once the above result is established, this theory will open the way to the study of the corresponding fractional evolution problem, a field that is currently under investigation.

## 10.4 Concluding Remarks

During the referee process of this book, there was an intensive study of some problems associated with the fractional powers of vector operators that continues to this day. This book has laid the foundations for the treatment of several problems that involve fractional powers of vector operators. A systematic treatment of all these problems will be considered in a future monograph. Here we just recall the conditions to generate the fractional powers of the noncommutative Fourier's law investigated in the paper [74]. Precisely, let $\Omega$ be a bounded domain in $\mathbb{R}^3$ and let $\tau > 0$. Denote by $v$ the temperature of the material contained in $\Omega$. Suppose that the heat flux, in the non-homogeneous material contained in $\Omega$, is given by the local operator:

$$T(x) = a(\underline{x})\partial_x e_1 + b(\underline{x})\partial_y e_2 + c(\underline{x})\partial_z e_3, \quad \underline{x} = (x, y, z) \in \Omega. \tag{10.18}$$

We consider the following evolution problem

$$\begin{cases} \partial_t v(\underline{x}, t) + \operatorname{div} T(\underline{x}) v(\underline{x}, t) = 0, & (\underline{x}, t) \in \Omega \times (0, \tau], \\ v(\underline{x}, 0) = f(\underline{x}), & \underline{x} \in \Omega, \\ v(\underline{x}, t) = 0, & \underline{x} \in \partial\Omega, \ t \in [0, \tau], \end{cases} \tag{10.19}$$

where $f$ is a given datum.

We wish to determine the non-local Fourier's law, associated with $T$, by defining the fractional powers $P_\alpha(T)$ of $T$ using the integral formula (10.2) with homogeneous Dirichlet boundary conditions.

**Remark 10.4.1.** In this way we obtain the associated fractional evolution problem, replacing operator $T$ by the vector part of $P_\alpha(T)$ in the system (10.19).

The main result is the following theorem and for its proof see [74].

**Theorem 10.4.1.** *Let $\Omega$ be a bounded $C^1$-domain in $\mathbb{R}^3$, let $T = a(\underline{x})\partial_x e_1 + b(\underline{x})\partial_y e_2 + c(\underline{x})\partial_z e_3$ with $a, b, c \in C^1(\overline{\Omega})$ and set*

$$F_{(a,b,c)} := e_1 \partial_x(a) + e_2 \partial_y(b) + e_3 \partial_z(c).$$

*Let $a, b, c \geq m > 0$, and assume that*

$$\min\{\inf_{\underline{x} \in \Omega} a^2, \inf_{\underline{x} \in \Omega} b^2, \inf_{\underline{x} \in \Omega} c^2\} - (2\max\{\sup_{\underline{x} \in \Omega} a^2, \sup_{\underline{x} \in \Omega} b^2, \sup_{\underline{x} \in \Omega} c^2\})^{1/2} C_\Omega \|F_{(a,b,c)}\|_{L^\infty} > 0$$
$$\tag{10.20}$$

*and*

$$1 - 2\|F_{(a,b,c)}\|_{L^\infty}\left(1 + 4C_\Omega^2 \max\{\sup_{\underline{x} \in \Omega}(1/a^2), \sup_{\underline{x} \in \Omega}(1/b^2), \sup_{\underline{x} \in \Omega}(1/c^2)\}\right) > 0, \tag{10.21}$$

*where $C_\Omega$ is the Poincaré constant of $\Omega$ and*

$$\|F_{(a,b,c)}\|_{L^\infty} := \sup_{\underline{x} \in \Omega}(|\partial_x(a)| + |\partial_y(b)| + |\partial_z(c)|).$$

*Then for any $\alpha \in (0,1)$ and for any $v \in \mathcal{D}(T)$, the integral (10.2) converges absolutely in $L^2(\Omega, \mathbb{H})$.*

In the non-commutative case, there are some difficulties that we explain shortly in the following. The sesquilinear form $\mathcal{B}(u,v)$, associated with the invertibility of the operator $\mathcal{Q}_s(T) := T^2 - 2s_0 T + |s|^2 \mathcal{I}$, with homogeneous Dirichlet boundary conditions, has to be considered with care.

**Remark 10.4.2.** We point out some challenges that appear in the application of the Lax–Milgram lemma according to the dimension $d = 2, 3$ of $\Omega$.

(I) The quadratic form $\mathcal{B}(u,v)$ associated with the operator $\mathcal{Q}_s(T)$ is in general degenerate on $H_0^1(\Omega, \mathbb{H})$.

(II) In dimension $d = 3$ , when $\Omega$ is a $\mathcal{C}^1$ bounded set in $\mathbb{R}^3$ and $a \neq 0$, $b \neq 0$, $c \neq 0$, it turns out that $\mathcal{B}(u,v)$ is continuous and coercive under suitable conditions on the coefficients on $\mathcal{Y} \times \mathcal{Y}$, where

$$\mathcal{Y} := \{H_0^1(\Omega, \mathbb{H}) : \quad v_0 = v_1 = v_2 = v_3\}$$

is a closed subspace of $H_0^1(\Omega, \mathbb{H})$ and the $S$-resolvent operators satisfy suitable growth conditions which ensure the existence of the fractional powers.

(III) In dimension $d = 2$, when $\Omega$ is a $\mathcal{C}^1$ bounded set in $\mathbb{R}^2$ and $a \neq 0$, $b \neq 0$, it turns out that $\mathcal{B}(u,v)$ is continuous and coercive under suitable conditions on the coefficients on $\mathcal{X} \times \mathcal{X}$, where

$$\mathcal{X} := \{v \in H_0^1(\Omega, \mathbb{H}) : \quad v_0 = v_2 \text{ and } v_1 = v_3\}$$

is a closed subspace of $H_0^1(\Omega, \mathbb{H})$ and the $S$-resolvent operators satisfy suitable growth conditions which ensure the existence of the fractional powers.

(IV) If we consider the quadratic form in dimension $d = 3$, that is when $\Omega$ is a $\mathcal{C}^1$ bounded set in $\mathbb{R}^3$ and $a \neq 0$, $b \neq 0$, $c = 0$, then the quadratic form is not coercive because $c = 0$. This case cannot be treated using Lax–Milgram, but a suitable method for degenerate equations has to be used.

(V) From the physical point of view, the case for $a \neq 0$, $b \neq 0$, $c = 0$, in dimension $d = 3$ is the case in which the conductivity is the direction $z$ goes to zero.

(VI) The proofs for the continuity and coercivity are similar in any dimension; the estimate for the $S$-resolvent operators have some differences according to the fact that we work in $\mathcal{Y}$ or in $\mathcal{X}$.

# Chapter 11

# Historical notes and References

Several years ago, motivated by the paper [37] of G. Birkhoff and J. von Neumann and the book [4], one of the authors and I. Sabadini started to look for an appropriate notion of spectrum for quaternionic linear operators. They had realized that the notion of spectrum of a quaternionic linear operator had not been understood and that, as a consequence, quaternionic spectral theory could not be fully developed. It was clear that the existing notions of left spectrum and right spectrum of a quaternionic linear operator were insufficient to construct a quaternionic spectral theory. The main reason was that the left spectrum and the right spectrum mimic the definition of eigenvalues in the complex case, but they do not shed light on the true nature of the quaternionic spectrum. After more than 10 years of research and 70 years after the paper [37] that motivated quaternionic operator theory, in 2006, F. Colombo and I. Sabadini understood that the $S$-spectrum was the correct notion of spectrum to develop quaternionic spectral theory. This notion works also for n-tuples of not necessarily commuting operators, so it opens the way to a very general new spectral theory. For more details on the discovery of the $S$-spectrum and of the $S$-functional calculus, see Chapter 1 in [57].

The discovery of the $S$-spectrum of a quaternionic linear operator was the starting point for the development of the quaternionic spectral theory. The present monograph and the book [57] contain a systematic treatment of quaternionic operator theory and some of its applications. The monograph [93], which was published in 2011, contains some results of the $S$-functional calculus for quaternionic linear operators and for $n$-tuples of noncommuting operators that were hitherto understood. In the same monograph, one also finds the basic results of the function theory of slice hyperholomorphic functions, which is the theory of hyperholomorphic functions associated with the $S$-spectrum. Later on, several functional calculi based on the $S$-spectrum had been introduced. Some of those are the quaternionic versions of functional calculi for complex linear operators like the $S$-functional calculus, the quaternionic Phillips functional calculus or the quaternionic $H^\infty$-

© Springer Nature Switzerland AG 2019

F. Colombo, J. Gantner, *Quaternionic Closed Operators, Fractional Powers and Fractional Diffusion Processes*, Operator Theory: Advances and Applications 274, https://doi.org/10.1007/978-3-030-16409-6_11

functional calculus, but the $F$-functional calculus or the $W$-functional calculus do not have a complex analogue. The above mentioned functional calculi naturally extend to the case of $n$-tuples of operators. These results are spread in several papers published throughout the years, so we provide a list of references for both the spectral theory based on the $S$-spectrum and the related slice hyperholomorphic function theory. For more details we refer the reader to the comments at the end the chapters of the book [57].

The theory of slice hyperholomorphic functions and the spectral theory on the $S$-spectrum are nowadays very well developed and the main monographs on these topics are [18], [57], [93], [100], [129], [139].

## 11.1   Theory of slice hyperholomorphic functions

There are three possible ways to define slice hyperholomorphic functions. One can define them

- (i) as functions that are left (resp. right) holomorphic on each complex plane $\mathbb{C}_j$ as in [141],

- (ii) as slice functions with components that satisfy the Cauchy–Riemann equations (this definition comes from the Fueter mapping theorem [116], [117] for the quaternionic case or the Sce mapping theorem for Clifford algebra-valued functions [195]), or

- (iii) as functions in the kernel of the global differential operator of slice hyperholomorphic functions, which was introduced in [62].

The second definition is the most appropriate for operator theory and it is the one that we used. The interest in slice hyperholomorphic functions, which were defined in [141], arose in 2006 also because of their applications to operator theory. Similar functions were, however, already used much earlier by Fueter who considered functions of the form

$$f(q) = f_0(u + iv) + jf_1(u + iv), \quad q = u + jv,$$

where $f_0, f_1$ are the real-valued components of an analytic function $F(z) = f_0(z) + \iota f_1(z)$ that depends on a complex variable $z = z_0 + iz_1$, in order to define what he called *hyperanalytic functions* [116]. These hyperanalytic functions are nothing but intrinsic slice hyperholomorphic functions. In [117], the author generates Fueter regular functions by applying the Laplace operator to such functions. The relation $\breve{f} = \Delta f$ between Fueter regular function $\breve{f}$ and the slice hyperholomorphic function $f$ is nowadays called the Fueter mapping theorem. In [195], Sce extended this theorem to functions with values in Clifford algebras of odd dimension. The extension to Clifford algebras of even dimensions needs arguments based on Fourier multipliers. In [186], Qian used the even and odd condition (2.4) in order to define entire slice hyperholomorphic functions and he generalized the theorem of Sce.

In [141], slice hyperholomorphic functions were defined as those functions that satisfied the properties shown in Lemma 2.1.5; that is they are those functions whose restrictions to complex planes $\mathbb{C}_j$ are left (resp. right) holomorphic. As was shown in Proposition 2.1.11 on axially symmetric slice domains, this definition is equivalent to Definition 2.1.2. Precisely, one can show that such functions satisfy the Structure Formula [76] (more often called Representation Formula) when they are defined on an axially symmetric slice domain. Considering only functions on axially symmetric slice domains is, however, not sufficient for developing a rich theory of quaternionic linear operators. For operator theory it is important to consider functions that are defined on axially symmetric open sets that are not necessarily slice domains, so for this reason we use Definition 2.1.2 for slice hyperholomorphicity.

The last approach to slice hyperholomorphic functions uses the global differential operator

$$G(q) := |\underline{q}|^2 \frac{\partial}{\partial q_0} + \underline{q} \sum_{j=1}^{3} q_j \frac{\partial}{\partial q_j},$$

which was introduced in [62].

If $U \subseteq \mathbb{H}$ is an open set and $f : U \to \mathbb{H}$ is a slice hyperholomorphic function then

$$G(q)f(q) = 0.$$

Defining slice hyperholomorphic functions as those functions that are in the kernel of the operators $G$, we have a possible way to define slice hyperholomorphic functions in several variables. Here the theory is far from being developed because we have a system of non constant differential operators and the power series expansion disappears. We recall that in the paper [102] there are some results associated with the theory of slice hyperholomorphic functions in several variables but the global operator is not used there, see also [2], [3] and [150]. Regarding the global operator see also [92], [105], [155].

**Function theory**

Slice hyperholomorphic functions are not only defined over the quaternions but they are also defined for Clifford algebra-valued functions. In the quaternionic setting, slice hyperholomorphic functions are also called slice regular functions and the theory has been developed by several authors. Some of the most important contributions were published in [38], [60], [66], [67], [106], [118], [119], [136], [137], [138], [140], [141], [142], [143], [144], [145], [146], [147] [160], [188], [197], [198] [199].

Slice hyperholomorphic functions with values in a Clifford algebra are also called slice monogenic functions. The main results of this theory are contained in the papers [76], [78], [94], [95], [96], [97], [98], [99] [157], [189], [206]. We point out that the results proved for Clifford algebra-valued functions hold in particular for the quaternionic setting but the converse is, in general, not true.

The Fueter–Sce mapping theorem and the inverse Fueter–Sce mapping theorem provide a relation between slice hyperholomorphic functions and the classical theory of monogenic functions, see [63], [85], [86], [87], [88], [89], [90], [91], [104], [109], [176]. Another relation of this type, which is provided by the Radon transform and the dual Radon transform, can be found in the paper [71]. In [73], the Cauchy transform has been treated in the slice hyperholomorphic setting. Finally, the theory of slice hyperholomorphic functions has been extended to the setting of functions with values in a real alternative ∗-algebra [33], [151], [152], [153], [154], [156].

## Approximation theorems

Important approximation theorems and related topics for slice hyperholomorphic functions are studied in the papers [118], [119], [120], [121], [122], [123], [124], [125], [126], [127], [128]. The monograph [129] contains a systematic treatment of slice hyperholomorphic approximation theory.

## Function spaces

Several complex function spaces have been extended to the slice hyperholomorphic setting. The quaternionic Hardy space $H_2(\Omega)$, where $\Omega$ is either the quaternionic unit ball $\mathbb{B}$ or the half space $\mathbb{H}^+$ of quaternions with positive real part, was introduced and studied in [13], [21], [22], [25], [34]. The Hardy spaces $H^p(\mathbb{B})$ for arbitrary $0 < p < +\infty$ were studied in [194]. The fractional Hardy spaces were studied in [29]. Carleson measures for Hardy and Bergman spaces in the quaternionic unit ball were studied in [193]. The BMO- and VMO-spaces of slice hyperholomorphic functions were treated in [135]. The slice hyperholomorphic Bergman spaces are studied studied in [61], [64], [65], the slice hyperholomorphic Fock space is considered in [31] and weighted Bergman spaces, the Bloch, Besov and Dirichlet spaces of slice hyperholomorphic functions on the unit ball $\mathbb{B}$ were introduced in [50]. Inner product spaces and Krein spaces in the quaternionic setting are studied in [26]. A class of quaternionic positive definite functions and their derivatives is studied in [27]. A quaternionic analogue of the Segal-Bargmann transform is studied in [108].

## Slice hyperholomorphic Schur Analysis

In recent years, a slice hyperholomorphic version of Schur analysis has been developed in [1], [2], [8], [9], [13], [16], [17], [21], [22], [23], [24], [25], [32]. The Boolean convolution in the quaternionic setting has been studied in [10]. The monograph [18] is the first book that contains a systematic study of Schur analysis in the slice hyperholomorphic setting. It is based on some of the previous papers, but it also contains several new results. Recent results on Schur analysis, related topics and quaternionic polynomials can be found in the papers [39], [40],

[41], [42], [43], [44], [45]. An overview of classical Schur analysis can be found, for example, in [7].

## 11.2 Spectral theory on the $S$-spectrum

The spectral theory based on the notion of $S$-spectrum is the study of some functional calculi, the spectral theorem, spectral operators and the characteristic operator functions. Precisely, we have: the $S$-functional calculus, the Phillips functional based on groups, the $H^\infty$-functional calculus, $F$-functional calculus, and the $W$-functional calculus. The $F$- and the $W$-functional calculi are monogenic functional calculi in the spirit of the monogenic functional calculus introduced by A. McIntosh and his collaborators, see [172], [173], [178], [179] and the book [171]. The monogenic functional calculus is based on the more classical theory of monogenic functions, that is functions that are in the kernel of the Dirac operator, see [46] [103], [162].

The spectral theorem on the $S$-spectrum, the spectral integration theory and spectral operators in the quaternionic setting are fundamental tools of this theory.

### The $S$-functional calculus

The $S$-functional calculus for bounded operators has been developed in [11], [58], [68], [70], [77], [82], [83], [133]. In the case we consider intrinsic functions, the $S$-functional calculus can be defined for one-sided quaternionic Banach space as it was shown in [131]. In the paper [131], the author has also developed the theory of spectral operators in quaternionic Banach spaces, see also [134].

The $S$-functional calculus can be defined also for $n$-tuples of noncommuting operators using slice hyperholomorphic functions with values in a Clifford-algebra (also called slice monogenic functions), see [78], [101]. The commutative version of the $S$-functional calculus, that is the $S$-functional calculus for operators with commuting components, is studied in [80]. The $S$-functional calculus can also be defined for unbounded operators. We refer the reader to the papers [69], [101], where we reduce the case of unbounded operators, with suitable transformations, to the case of bounded operators. The direct approach has been studied in the more recent paper [130], while the $S$-resolvent equation is in [52]. The $S$-functional calculus was the starting point for the development of various quaternionic functional calculi, see [75] for a friendly introduction.

### The Phillips functional calculus and semigroups

In the classical setting, the Phillips functional can be found, for example, in [110], [185]. The Phillips functional calculus for generators of strongly continuous groups, based on the quaternionic version of the Laplace-Stieltjes transform, was introduced in [12]. Groups and semigroups of quaternionic linear operators were considered in [19], [79], [158].

## The $H^\infty$-functional calculus and fractional powers

In the paper [30], the authors introduce the $H^\infty$-functional calculus based on the $S$-spectrum; this is the quaternionic analogue of the calculus introduced by McIntosh [170] (see also [6]). In [30] it is also considered the $H^\infty$-functional calculus for $n$-tuples of noncommuting operators.

In a more general version of the $H^\infty$-functional calculus, fractional powers of quaternionic linear operators are treated in [53], [54]. The authors also show how the fractional powers of quaternionic linear operators define new fractional diffusion and evolution processes. Fractional powers of vector operators and their applications to fractional Fourier's law in a Hilbert space have been studied in [55]. For a more direct approach to fractional powers of quaternionic operators, including the Kato formula, see the paper [52]. See also [134] where these arguments are treated.

## The $F$-functional calculus

The Fueter–Sce mapping theorem in integral form, introduced in [88], is an integral transform that maps slice hyperholomorphic functions into Fueter regular functions (or into monogenic functions). By formally replacing the scalar variable in this integral transform by a quaternionic operator $T$ (or by a paravector operator), we obtain a functional calculus for Fueter regular functions (or for monogenic functions) that is based on the theory of slice hyperholomorphic functions. The $F$-functional calculus was introduced and studied in the following papers [20], [56], [81], [84], [88].

The $F$-functional calculus for paravector operators defines a monogenic functional calculus for $n$-tuples of commuting operators. Because of the structure of the Fueter–Sce mapping theorem in integral form, the $F$-functional calculus depends on the dimension of the Clifford algebra. The Fueter–Sce–Qian mapping theorem, was proved by Michele Sce [195] for the case when $n$ is odd and by Tao Qian [186] for the case when $n$ is even. Later, the theorem of Fueter–Sce–Qian was generalized to the case in which a slice hyperholomorphic function $f$ was multiplied by a monogenic homogeneous polynomial of degree $k$, see [176], [182], [183], and to the case in which the function $f$ was defined on an open set $U$ not necessarily chosen in the upper complex plane, see [186], [187]. A more recent result can be found in [184]. The definition of the $F$-functional calculus can be extended to the case unbounded quaternionic operators and also to the case of $n$-tuples of unbounded operators. As it is well known in the case of unbounded operators, the notion of commutativity is more delicate and one has to pay attention to the domains of the operators, see [81].

## The $W$-functional calculus

Using the notion of slice hyperholomorphic functions, it is possible to define a transform that maps slice hyperholomorphic functions into Fueter regular functions (or monogenic functions) of plane wave type. This transform is different from the Fueter–Sce mapping theorem in integral form. With such integral transform we can define the $W$-functional calculus. This calculus has been introduced in [72] for monogenic functions and it was inspired by the paper [201].

## The spectral theorem on the $S$-spectrum

The spectral theorem based on the notion of $S$-spectrum for quaternionic bounded and unbounded normal operators has been proved in 2015 and published in the paper [14] that appeared in 2016. As in the complex case, the spectral theorem for unitary operators can be deduced by the quaternionic version of Herglotz's theorem, proved in [17]. The spectral theorem for unitary operators based on Herglotz's theorem was proved in [15]. The simple case of compact normal operators was shown in [149]. For quaternionic matrices, the spectral theorem based on the right spectrum was proved in [114]. In the finite dimensional case, the right spectrum equals the $S$-spectrum. The starting point to prove the spectral theorem is to define spectral integrals in the quaternionic setting. In fact, inspired by the complex setting, see Chapter 4 of the book [200], spectral integrals were extended to the quaternionic setting. Most of the proofs of the properties of spectral integrals are easily adapted from the classical case, i.e., when $\mathcal{H}$ is a complex Hilbert space, but some facts require additional arguments which have to be studied carefully. The main ingredients to prove the spectral theorem for bounded normal operators are

- the Riesz representation theorem for the dual of $\mathcal{C}(X, \mathbb{R})$, where $X$ is a compact Hausdorff space,

- the Riesz representation theorem for quaternionic Hilbert spaces,

- the Teichmüller-decomposition of a normal bounded operator $T = A + JB$,

- the continuous functional calculus based on the $S$-spectrum.

To prove the spectral theorem for unbounded normal operators it was necessary to introduce the notion of spectral integrals that depend on the imaginary operator $J$. Precisely, the main ingredients can be summarized in the following points:

- the spectral theorem for bounded normal operators,

- the spectral integrals in the quaternionic setting depending on the imaginary operator $J$,

- suitable transformations (in the spirit of von Neumann) that reduce the case of unbounded operators to the case of bounded operators.

Finally, using spectral integrals, we define a functional calculus for unbounded normal operators. For more details see the book [57]. The continuous functional calculus was first introduced in [148], while in the book [57], the continuous functional calculus was studied for Teichmüller decomposition of a normal bounded operator $T = A + JB$.

Perturbation of normal quaternionic operators were studied in [51], the Schatten class and the Berezin transform for quaternionic operators have been studied in [59]. Beyond the spectral theorem, there is the theory of the characteristic operator functions that has been initiated in [28].

Several papers have appeared in the literature that claimed to introduce a spectral theorem for normal operators on a quaternionic Hilbert space (see [202], [115], [112], [203]). However, in all of the aforementioned papers, a precise notion of spectrum is not made clear. We will now enter into a discussion concerning the papers of Teichmüller [202] and Viswanath [203]. Teichmüller's paper [202] is the first to claim a spectral theorem for normal operators and appeared in 1936. Despite not making the notion of spectrum clear, [202] does have a number of valid and important observations such as the decomposition $T = A + JB$. Finally, the spectral resolution in [202] takes the form

$$N = \int_{-\infty}^{\infty} \int_{0}^{\infty} (\lambda' + T_0 \lambda'') dQ_{\lambda''} dP_{\lambda'} \tag{11.1}$$

where $N$ is a normal operator while $T_0$ is what he calls an "Imaginäroperator" (the german word for "imaginary operator") on $\overline{\operatorname{ran} B}$, i.e., $T_0 T_0^* = \mathcal{I}_{\operatorname{ran} B}$ and $T_0^* = -T_0$. In 1971 the paper [203] of Viswanath also claimed to have a spectral theorem for normal operators on a quaternionic Hilbert space. It is worth noting that [202] is not quoted in Viswanath's paper [203]. The approach of [203] is very different from [202] in so far as the symplectic image of a normal operator is used and the spectral theorem is allegedly deduced from the classical spectral theorem and some kind of lifting argument. Viswanath's spectral resolution takes the form

$$T = \int_{\mathbb{C}^+} \lambda dE, \tag{11.2}$$

where $T$ is a normal operator and $E$ is a projection-valued measure.

One of the main applications of the quaternionic spectral theorem is in quaternionic quantum mechanics. In the old papers related to quaternionic quantum mechanics [112], [115], [166] the authors use the notion of right spectrum. To explore the equivalent formulations of complex and quaternionic quantum mechanics, see [132]. For recent applications of the spectral theory on the $S$-spectrum to quantum mechanics see [180], [181]. We also mention the papers [175] on the coherent state transforms to Clifford analysis and [174] on the S-spectrum and associated continuous frames on quaternionic Hilbert spaces.

**Spectral integration and spectral operators**

There existed different approaches to this spectral integration in the literature, but these approaches required the introduction of a left multiplication on the Hilbert space (even though this multiplication was sometimes only assumed to be defined for quaternions in one complex plane and not for all $q \in \mathbb{H}$). This left multiplication was in general only partially determined by the a priori given mathematical structures. It had to be extended randomly and the necessary procedure did not generalize to the Banach space setting.

In [131], an approach to spectral integration of intrinsic slice functions on a quaternionic right Banach space has been developed. This integration is done with respect to a spectral system instead of a spectral measure $E$, a concept that specifies ideas of [203]. A spectral system is a couple $(E, J)$ that consists of a spectral measure $E$ on the axially symmetric Borel sets of $\mathbb{H}$ and an imaginary operator $J$ satisfying $E(\mathbb{H} \setminus \mathbb{R}) = -J^2$. It has a clear and intuitive interpretation in terms of the right linear structure of the space: the spectral measure $E$ associates projections with sets of spectral spheres $[s]$ and the imaginary operator $J$ defines how to multiply the different values in the spheres onto the vectors in the corresponding subspaces. Additionally, it is compatible with the complex theory. The spectral system $(E, J)$ of a normal operator $T$ is fully determined by $T$ and it is exactly the structure that is necessary to prove the spectral theorem as it is done in [14]. In order to fit to existing approaches to spectral integration, the operator $J$ has been, however, usually extended randomly in order to define a left multiplication.

In [131] and in [134], we consider spectral measures that are defined on axially symmetric subsets of $\mathbb{H}$ instead of subsets of a complex halfplane $\mathbb{C}_j^+$. Both approaches are equivalent for intrinsic slice functions: we can identify any axially symmetric set with its intersection with one complex halfplane $\mathbb{C}_j^+$ in order to obtain a bijective relation between these two types of sets. Quaternionic linear spectral operators are studied in [131] and in [134], where the complex linear theory in [111] is generalized.

# 11.3  The monographs on operators and functions

For the convenience of the reader, we summarize in the following the contents of monographs [18], [57], [93], [100], [129], [139] and the long self-contained papers [28] and [131] that treat specific arguments.

**Slice hyperholomorphic Schur analysis**

The monograph [18] is the first treatise of Schur analysis in the slice hyperholomorphic setting. The book is divided into three parts: I. Classical Schur analysis, II. Quaternionic analysis and III. Quaternionic Schur analysis.

I. Classical Schur analysis.

Chapter 1. Some history, Krein spaces, Pontryagin spaces, and negative squares, the Wiener algebra, the Nehari extension problem, the Carathéodory–Toeplitz extension problem and various classes of functions and realization theorems.

Chapter 2. Rational functions and minimal realizations, minimal factorization, rational functions J-unitary on the imaginary line, rational functions J-unitary on the circle.

Chapter 3. The Schur algorithm, interpolation problems, first order discrete systems, the Schur algorithm and reproducing kernel spaces.

II. Quaternionic analysis.

Chapter 4. Some preliminaries on quaternions, polynomials with quaternionic coefficients, matrices with quaternionic entries, matrix equations.

Chapter 5. Quaternionic locally convex linear spaces, quaternionic inner product spaces, quaternionic Hilbert spaces, partial majorants, majorant topologies and inner product spaces, quaternionic Hilbert spaces, weak topology, Quaternionic Pontryagin spaces, Quaternionic Krein spaces, Positive definite functions and reproducing kernel quaternionic Hilbert spaces, Negative squares and reproducing kernel quaternionic Pontryagin spaces.

Chapter 6. Elements of slice hyperholomorphic functions: The scalar case. The Hardy space of the unit ball, Blaschke products (unit ball case), The Wiener algebra, The Hardy space of the open half-space, Blaschke products (half-space case).

Chapter 7. Slice hyperholomorphic operator-valued functions. Definition and main properties, S-spectrum and S-resolvent operator, The functional calculus, Two results on slice hyperholomorphic extension, Slice hyperholomorphic kernels, The space $H^2_{\mathcal{H}}(B)$ and slice backward-shift invariant subspaces.

III. Quaternionic Schur analysis.

Chapter 8. Reproducing kernel spaces and realizations. The various classes of functions, The Potapov–Ginzburg transform, Schur and generalized Schur functions of the ball, Contractive multipliers, inner multipliers and Beurling–Lax theorem, A theorem on convergence of Schur multipliers, The structure theorem, Carathéodory and generalized Carathéodory functions, Schur and generalized Schur functions of the half-space, Herglotz and generalized Herglotz functions.

Chapter 9. Rational slice hyperholomorphic functions. Definition and first properties, Minimal realizations, Realizations of unitary rational functions, Rational slice hyperholomorphic functions, Linear fractional transformation, Backward-shift operators.

Chapter 10. First applications: scalar interpolation and first order discrete systems. The Schur algorithm, The reproducing kernel method, Carathéodory–Fejér interpolation, Boundary interpolation, First order discrete linear systems, Discrete systems: the rational case.

Chapter 11. Interpolation: operator-valued case. Formulation of the interpolation problems, The problem $\mathbf{IP}(H^2_{\mathcal{H}}(B))$: the non-degenerate case, Left-tangential interpolation in $\mathcal{S}(\mathcal{H}_1; \mathcal{H}_2; \mathbb{B})$, Interpolation in $\mathcal{S}(\mathcal{H}_1; \mathcal{H}_2; \mathbb{B})$. The non degenerate case, Interpolation: The case of a finite number of interpolating conditions, Leechs theorem, Interpolation in $\mathcal{S}(\mathcal{H}_1; \mathcal{H}_2; \mathbb{B})$. Nondegenerate case: Sufficiency.

## Spectral theory on the S-spectrum for quaternionic operators

The book [57] can be considered the first part of quaternionic operator theory that has been further developed in the present monograph.

Chapter 1. Introduction. What is quaternionic spectral theory. Some historical remarks on the S-spectrum. The discovery of the $S$-spectrum. Why such a long time to understand the S-spectrum.

Chapter 2. Slice hyperholomorphic functions. Slice hyperholomorphic functions. The Fueter mapping theorem in integral form. Vector-valued slice hyperholomorphic functions. Comments and remarks.

Chapter 3. The S-spectrum and the S-functional calculus. The S-spectrum and the S-resolvent operators. Definition of the S-functional calculus. Comments and remarks. The left spectrum $\sigma_L(T)$ and the left resolvent operator. Power series expansions and the S-resolvent equation.

Chapter 4. Properties of the S-functional calculus for bounded operators. Algebraic properties and Riesz-projectors. The Spectral Mapping Theorem and the Composition Rule. Convergence in the S-resolvent sense. The Taylor formula for the S-functional calculus. Bounded operators with commuting components. Perturbations of the SC-resolvent operators. Some examples. Comments and remarks. The S-functional calculus for $n$-tuples of operators. The W-functional calculus for quaternionic operators.

Chapter 5. The S-functional calculus for unbounded operators. The S-spectrum and the S-resolvent operators. Definition of the S-functional calculus. Comments and remarks.

Chapter 6. The $H^\infty$ functional calculus. The rational functional calculus. The S-functional calculus for operators of type $\omega$. The $H^\infty$ functional calculus. Boundedness of the $H^\infty$ functional calculus. Comments and remarks. Comments on fractional diffusion processes.

Chapter 7. The F-functional calculus for bounded operators. The F-resolvent operators and the F-functional calculus. Bounded perturbations of the F-resolvent. The F-resolvent equations. The Riesz projectors for the F-functional calculus. The Cauchy–Fueter functional calculus. Comments and remarks. The F-functional calculus for $n$-tuples of operators. The inverse Fueter–Sce mapping theorem.

Chapter 8. The F-functional calculus for unbounded operators. Relations between F-resolvent operators. The F-functional calculus for unbounded operators. Comments and remarks. F-functional calculus for $n$-tuples of unbounded operators.

Chapter 9. Quaternionic operators on a Hilbert space. Preliminary results. The S-spectrum of some classes of operators. The splitting of a normal operator and consequences. The continuous functional calculus. Comments and remarks.

Chapter 10. Spectral integrals. Spectral integrals for bounded measurable functions. Spectral integrals for unbounded measurable functions. Comments and remarks.

Chapter 11. The spectral theorem for bounded normal operators. Construction of the spectral measure. The spectral theorem and some consequences. Comments and remarks.

Chapter 12. The spectral theorem for unbounded normal operators. Some transformations of operators. The spectral theorem for unbounded normal operators. Some consequences of the spectral theorem. Comments and remarks.

Chapter 13. Spectral theorem for unitary operators. Herglotzs theorem in the quaternionic setting. Preliminaries for the spectral resolution. Further properties of quaternionic Riesz projectors. The spectral resolution. Comments and remarks.

Chapter 14. Spectral integration in the quaternionic setting. Spectral integrals of real-valued slice functions. Imaginary operators. Spectral systems and spectral integrals of intrinsic slice functions. On the different approaches to spectral integration.

Chapter 15. Bounded quaternionic spectral operators. The spectral decomposition of a spectral operator. Canonical reduction and $S$-functional calculus.

## Noncommutative Functional Calculus

The book [93] is the first monograph on slice hyperholomorphic functions and the relater functional calculi. The content of the book is the following.

Chapter 1. Introduction to the theory.

Chapter 2. Slice monogenic functions. Clifford algebras. Slice monogenic functions: definition and properties. Power series. Cauchy integral formula I. Zeros of slice monogenic functions. The slice monogenic product. Slice monogenic Cauchy kernel. Cauchy integral formula II. Duality Theorems. Topological duality theorems and the chapter is concluded with some notes.

Chapter 3. Functional calculus for n-tuples of operators. The S-resolvent operator and the S-spectrum. Properties of the S-spectrum. The $S$-functional calculus. Algebraic rules. The spectral mapping and the S-spectral radius theorems. The $S$-functional calculus for unbounded operators and algebraic properties.

Chapter 4. Quaternionic functional calculus. Definition of slice regular functions. Properties of slice regular functions. Representation Formula for slice regular functions. The slice regular Cauchy kernel. The Cauchy integral formula II. Linear bounded quaternionic operators. The S-resolvent operator series. The S-spectrum and the S-resolvent operators. Examples of S-spectra. The quaternionic functional calculus. Algebraic properties of the quaternionic functional calculus. The S-spectral radius. The S-spectral mapping and the composition theorems. Bounded perturbations of the S-resolvent operator. Linear closed quaternionic operators. The functional calculus for unbounded operators. An application: uniformly continuous quaternionic semigroups.

## Entire Slice Regular Functions

In the book [100], the theory of entire slice hyperholomorphic functions is considered.

Chapter 1. Introduction.

Chapter 2. Slice regular functions: algebra. Definition and main results. The Representation Formula. Multiplication of slice regular functions. Quaternionic intrinsic functions. Composition of power series.

Chapter 3. Slice regular functions: analysis. Some integral formulas. Riemann mapping theorem. Zeros of slice regular functions. Modulus of a slice regular function and Ehrenpreis–Malgrange lemma. Cartan theorem.

Chapter 4. Slice regular infinite products. Infinite products of quaternions. Infinite products of functions. Weierstrass theorem. Blaschke products. Growth of entire slice regular functions. Growth scale. Jensen theorem. Carathéodory theorem. Growth of the *-product of entire slice regular functions. Almost universal entire functions. Entire slice regular functions of exponential type.

## Quaternionic Approximation

In the book [129], the authors systematically treat the approximation theory for slice hyperholomorphic functions, but there are also considerations for other settings.

Chapter 1. Preliminaries on hypercomplex analysis. An introduction to slice regular functions. Slice regular functions, an alternative definition. Further properties of slice regular functions. Preliminary results on quaternionic polynomials. Slice monogenic functions. Other hypercomplex function theories.

Chapter 2. Approximation of continuous Functions. Weierstrass–Stone type results. Carleman type results. Notes on Müntz–Szász type Results.

Chapter 3. Approximation in compact balls by Bernstein and convolution type operators. Approximation by $q$-Bernstein polynomials, $q \geq 1$, in compact balls. Approximation by convolution operators in compact balls and in Cassini cells. Approximation by quaternionic polynomial convolutions. Approximation by nonpolynomial quaternion convolutions. Approximation by convolution operators of a paravector variable.

Chapter 4. Approximation of slice regular functions in compact sets. Runge type results. Mergelyan type results. Riemann mappings for axially symmetric sets. Approximation in some particular compact sets. Arakelian type results. Arakelian's theorem for slice monogenic functions. Approximation by Faber type polynomials. Biholomorphic bijections between complements of compact axially symmetric sets. Quaternionic Faber polynomials and Faber series. Approximation by Polynomials in Bergman spaces. Approximation in Bergman spaces of first kind. Approximation in Bergman spaces of second kind.

Chapter 5. Overconvergence, equiconvergence and universality properties. Overconvergence of Chebyshev and Legendre Polynomials. Walsh equiconvergence type results. The interpolation problem. Walsh equiconvergence theorem. Universality properties of power series and entire functions. Almost universal power series. Almost universal entire functions.

Chapter 6. Inequalities for quaternionic polynomials. The complex case. Bernstein's inequality. Erdös-.Lax's inequality. Turán's inequality. Notes on Turan type inequalities. Approximation of nullsolutions of generalized Cauchy–Riemann operators. Runge type results. Approximation of axially monogenic functions. Polynomial approximation of monogenic quasi-conformal maps. Approximation in $L^2$. Complete orthogonal systems of monogenic polynomials.

## Regular Functions of a Quaternionic Variable

In the book [139], the case of quaternionic slice hyperholomorphic functions is treated.

Chapter 1. Definitions and Basic Results. Regular Functions, Affine Representation. Extension Results, Algebraic Structure.

Chapter 2. Regular Power Series. Convergence of Power Series. Series Expansion and Analyticity. Zeros.

Chapter 3. Basic Properties of the Zeros. Algebraic Properties of the Zeros. Topological Properties of the Zeros. On the Roots of Quaternions. Factorization of Polynomials. Multiplicity. Division Algorithm and Bezout Theorem. Grobner Bases for Quaternionic Polynomials.

Chapter 4. Infinite Products. Infinite Products of Quaternions. The Quaternionic Logarithm. Infinite Products of Functions Defined on $\mathbb{H}$. Convergence of an Infinite Product. Convergence-Producing Regular Factors. Weierstrass Factorization Theorem.

Chapter 5. Singularities. Regular Reciprocal and Quotients. Laurent Series and Expansion. Classification of Singularities. Poles and Quotients. Casorati–Weierstrass Theorem.

Chapter 6. Integral Representations. Cauchy Theorem and Morera Theorem. Cauchy Integral Formula. Pompeiu Formula. Derivatives Using the Cauchy Formula. Coefficients of the Laurent Series Expansion. Argument Principle.

Chapter 7. Maximum Modulus Theorem and Applications. Maximum and Minimum Modulus. Open Mapping Theorem. Real Parts of Regular Functions. Phragmen–Lindelof Principle. An Ehrenpreis–Malgrange Lemma.

Chapter 8. Spherical Series and Differential Spherical Series and Expansions. Integral Formulas and Cauchy Estimates. Symmetric Analyticity. Differentiating Regular Functions. Rank of the Differential.

Chapter 9. Fractional Transformations and the Unit Ball. Transformations of the Quaternionic Space. Regular Fractional Transformations. Transformations of the Quaternionic Riemann Sphere. Schwarz Lemma and Transformations of the Unit Ball. Rigidity and a Boundary Schwarz Lemma. Borel–Carathéodory Theorem. Bohr Theorem.

Chapter 10. Generalizations and Applications. Slice Regularity in Algebras Other than $\mathbb{H}$. The Case of Octonions. The Case of $\mathbb{R}_3$. The Monogenic Case. Quaternionic Functional Calculus. Orthogonal Complex Structures Induced by Regular Functions.

## De Branges spaces and characteristic operator function: the quaternionic case

In the long paper [28], the authors systematically investigate the characteristic operator functions and related topics in the quaternionic setting.

Chapter 1. Introduction. The complex numbers setting. The quaternionic setting.

Chapter 2. Quaternions. Toeplitz matrices. Hankel matrices. Functional analysis.

Chapter 3. Slice hyperholomorphic functions. The S-resolvent operators and the S-spectrum. The map $\chi$ and applications. Slice hyperholomorphic weights: half-plane. Slice hyperholomorphic weights: unit ball.

Chapter 4. Rational functions. Rational slice hyperholomorphic functions. Symmetries.

Chapter 5. Operator models. Rota's model in the quaternionic setting. Other operator models.

Chapter 6. Structure theorems for $H(A, B)$ spaces. $H(A, B)$ spaces. The structure theorem: Half-space case. The unit ball case. A theorem on the zeros of a polynomial.

Chapter 7. $J$-contractive functions. $J$-contractive functions in the unit ball. $J$-contractive functions in the right half-space. The case of entire functions.

Chapter 8. The characteristic operator function. The problem with close to self-adjoint operators in the quaternionic case. Properties of the characteristic operator function. Examples. Inverse problems.

Chapter 9. $L(F)$ spaces. $L(F)$ spaces associated to analytic weights. Linear fractional transformations. Canonical differential systems.

Chapter 10. The matrizant. The characteristic spectral functions. Canonical differential systems.

## Operator Theory on One-Sided Quaternionic Linear Spaces: Intrinsic S-Functional Calculus and Spectral Operators

The long paper [131] contains fundamental results on the operator theory on one-sided quaternionic linear spaces, the intrinsic S-functional calculus and the spectral operators on quaternionic Banach spaces.

Chapter 1. Introduction.

Chapter 2. Preliminaries: Slice hyperholomorphic functions. The S-functional calculus. The spectral theorem for normal operators.

Chapter 3. Intrinsic S-functional calculus on one-sided Banach spaces.

Chapter 4. Spectral integration in the quaternionic setting. On the different approaches to spectral integration.

Chapter 5. Bounded quaternionic spectral operators.

Chapter 6. Canonical reduction and intrinsic S-functional calculus for quaternionic spectral operators.

Chapter 7. Concluding Remarks.

# Chapter 12

# Appendix: Principles of functional Analysis

The principles of functional analysis do not depend on the quaternionic structure, so with minor changes these can be proved also in quaternionic functional analysis. For the convenience of the reader, we collect such results in this appendix. Some of the results were already proved in [110], so we quote those here.

**Theorem 12.0.1** (The open mapping theorem). *Let $X$ and $W$ be two right quaternionic Banach spaces, and let $T$ be a right linear continuous quaternionic operator from $X$ onto $W$. Then the image of every open set is open.*

*Proof.* Let $X$ and $W$ be two right quaternionic Banach spaces and let $T : X \to W$ be a right linear continuous map such that $TX = W$. It is enough to prove the statement for a neighborhood of 0, more precisely for balls. We denote by $B_X(r)$ the ball in $X$ of radius $r > 0$ and centered at the origin. We prove that the closure $\overline{TB_X(r)}$ of the image of any ball $B_X(r)$ centered at 0 in $X$ contains a neighborhood of 0 in $W$. Moreover, since $TB_X(r) = rTB_X(1)$, we only need to show that $TB_X(r)$ is a neighborhood of the origin for some positive $r$.

We will make use of the notation $B_X(r) - B_X(r)$ to denote the set of elements of the form $u - v$ where $u, v \in B_X(r)$.

Observe that the function $u - v$ is continuous in $u$ and $v$. Also, notice that there exists an open ball $B_X(r')$, for suitable $r' > 0$, such that $B_X(r') - B_X(r') \subseteq B_X(r)$.

For every $v \in X$, we have that $v/n \to 0$ as $n \to \infty$ so $v \in nB_X(r')$ for a suitable $n \in \mathbb{N}$. So

$$X = \bigcup_{n=1}^{\infty} nB_X(r') \quad \text{and} \quad W = TX = \bigcup_{n=1}^{\infty} nTB_X(r').$$

By the Baire category theorem, one of the sets $\overline{nTB_X(r')}$ contains a nonempty

© Springer Nature Switzerland AG 2019
F. Colombo, J. Gantner, *Quaternionic Closed Operators, Fractional Powers and Fractional Diffusion Processes*, Operator Theory: Advances and Applications 274,
https://doi.org/10.1007/978-3-030-16409-6_12

open set. The map $w \mapsto nw$ is a homeomorphism in $W$ and $\overline{TB_X(r')}$ contains a nonempty open set denoted by $\mathcal{B}$, so

$$\overline{TB_X(r)} \supseteq \overline{TB_X(r')} - \overline{TB_X(r')} \supseteq \overline{TB_X(r')} - \overline{TB_X(r')} \supseteq \mathcal{B} - \mathcal{B}.$$

The map $w \mapsto u - w$ is a homeomorphism; this implies that the set $u - B_X(r)$ is open. Since the set $\mathcal{B} - \mathcal{B} = \bigcup_{u \in \mathcal{B}} (u - \mathcal{B})$ is the union of open sets, it is open and it contains the origin and so it is a neighborhood of the origin. Thus, we have proved that the closure of the image of the neighborhood of the origin contains a neighborhood of the origin.

For any $\varepsilon > 0$, consider the two spheres $B_X(\varepsilon)$ and $B_W(\varepsilon)$ centered at the origin of $X$ and $W$, respectively.

Choose an arbitrary positive real number $\varepsilon_0$ and let $\varepsilon_\ell > 0$ be a sequence such that $\sum_{\ell \in \mathbb{N}} \varepsilon_\ell < \varepsilon_0$.

Then, according to what we have proved above, there exists a sequence $\{\theta_\ell\}_{\ell \in \mathbb{N} \cup \{0\}}$ with $\theta_\ell > 0$ and $\theta_\ell \to 0$ such that

$$\overline{TB_X(\varepsilon_\ell)} \supset B_W(\theta_\ell), \quad \ell \in \mathbb{N} \cup \{0\}. \tag{12.1}$$

Now take $w \in B_W(\theta_0)$. We show that there exists $v \in B_X(2\varepsilon_{\theta_0})$ such that $Tv = w$. From (12.1), for $\ell = 0$ there exists a $v_0 \in B_X(\varepsilon_0)$ such that

$$\|w - Tv_0\| < \theta_1.$$

Since $w - Tv_0 \in B_W(\theta_1)$ again from (12.1) with $\ell = 1$, there is $v_1 \in B_X(\varepsilon_1)$ with

$$\|w - Tv_0 - Tv_1\| < \theta_2.$$

So we construct a sequence $\{v_n\}_{n \in \cup\{0\}}$ such that $v_n \in B_X(\varepsilon_n)$ and

$$\left\| w - T\sum_{\ell=0}^{n} v_\ell \right\| < \theta_{n+1}, \quad n \in \mathbb{N} \cup \{0\}. \tag{12.2}$$

Let us denote $p_m = \sum_{\ell=0}^{m} v_\ell$. So for $m > n$,

$$\|p_m - p_n\| = \|v_{n+1} + \ldots + v_m\| < \varepsilon_{n+1} + \ldots + \varepsilon_m,$$

which shows that $p_m$ is a Cauchy sequence and that the series $\sum_{\ell=0}^{\infty} v_\ell$ converges at a point $v$ with

$$\|v\| \leq \sum_{\ell=0}^{\infty} \varepsilon_\ell = 2\varepsilon_0.$$

Now recall that $T$ is continuous and from (12.2) we have $w = Tv$.

This means that an arbitrary sphere $B_X(2\varepsilon_0)$, about the origin in $X$, maps onto the set $TB_X(2\varepsilon_0)$ which contains the sphere $B_W(\theta_0)$ about the origin in $W$. So if $\mathcal{X}$ is a neighborhood of the origin in $X$, then $T\mathcal{X}$ contains a neighborhood of the origin of $W$. Since $T$ is linear then the above procedure works for every neighborhood of every point. $\qquad\square$

**Theorem 12.0.2** (The Banach continuous inverse theorem). *Let $X$ and $W$ be two right quaternionic Banach spaces and let $T$ be a right linear continuous quaternionic operator that is one-to-one from $X$ onto $W$. Then $T$ has a right linear continuous inverse.*

*Proof.* Let $X$ and $W$ be two right quaternionic Banach spaces and $T$ be a right linear continuous and one-to-one operator such that $TX = W$. By Theorem 12.0.1 $T$ maps open sets onto open sets, so if we write $T$ as $(T^{-1})^{-1}$, it is immediate that $T^{-1}$ is continuous. Now take $w_1$, $w_2 \in W$ and $v_1$, $v_2 \in X$ such that $Tv_1 = w_1$, $Tv_2 = w_2$ and $p \in \mathbb{H}$. Then,

$$T(v_1 + v_1) = Tv_1 + Tv_2 = w_1 + w_2, \quad T(v_1 p) = T(v_1)p = w_1 p$$

so that

$$T^{-1}(w_1 + w_2) = v_1 + v_2$$

and

$$T^{-1}(w_1 p) = v_1 p,$$

so $T^{-1}$ is right linear quaternionic operator. □

**Definition 12.0.3.** Let $X$ and $W$ be two right quaternionic Banach spaces. Suppose that $T$ is a right linear quaternionic operator whose domain $\mathcal{D}(T)$ is a (right) linear manifold contained in $X$ and whose range belongs to $W$. The graph of $T$ consists of all point $(v, Tv)$, with $v \in \mathcal{D}(T)$, in the product space $X \times W$.

**Definition 12.0.4.** We say that $T$ is a closed operator if its graph is closed in $X \times W$.

**Remark 12.0.1.** Equivalently, we can say that $T$ is closed if $v_n \in \mathcal{D}(T)$, $v_n \to v$, and $Tv_n \to y$ imply that $v \in \mathcal{D}(T)$ and $Tv = y$.

**Theorem 12.0.5** (The closed graph theorem). *Let $X$ and $W$ be two right quaternionic Banach spaces. Let $T : X \to W$ be a right linear closed quaternionic operator. Then $T$ is continuous.*

*Proof.* Since $X$ and $W$ are two right quaternionic Banach spaces, we have that $X \times W$ with the norm $\|(v, w)\|_{X \times W} = \|v\|_X + \|w\|_W$ is a right quaternionic Banach space. The graph of $T$ denoted by

$$\mathcal{G}(T) = \{(v, Tv) : \quad v \in \mathcal{D}(T)\}$$

is a closed linear manifold in the product space $X \times W$ so it is a right quaternionic Banach space. The projection

$$P_X : \mathcal{G}(T) \mapsto X, \quad P_X(v, Tv) = v$$

is one-to-one and onto, linear and continuous, so by Theorem 12.0.2, its inverse $P_X^{-1}$ is continuous. Now consider the projection

$$P_W : \mathcal{G}(T) \mapsto W, \quad P_W(v, Tv) = Tv,$$

since $T = P_W P_X^{-1}$, so we get the statement. □

**Theorem 12.0.6** (The Hahn–Banach theorem). *Let $X_0$ be a right subspace of a right quaternionic Banach space $X$ on $\mathbb{H}$. Suppose that $p$ is a norm on $X$ and let $\phi$ be a linear and continuous functional on $X_0$ such that*

$$|\langle \phi, v \rangle| \leq p(v), \quad \forall v \in X_0. \tag{12.3}$$

*Then it is possible to extend $\phi$ to a linear and continuous functional $\Phi$ on $X$ satisfying the estimate (12.3) for all $v \in X$.*

*Proof.* Note that, for any quaternion $q$, we have $q = q_0 + q_1 i + q_2 j + q_3 k = z_1(q) + z_2(q)j$, where $z_1, z_2 \in \mathbb{C}_i = \mathbb{R} + \mathbb{R}i$ and $qj = -z_2(q) + z_1(q)j$, so $q = z_1(q) - z_1(qj)j$. The functional $\phi$ can be written as $\phi = \phi_0 + \phi_1 i + \phi_2 j + \phi_3 k = \psi_1(\phi) + \psi_2(\phi)j$, with $\psi_1(\phi) = \phi_0 + \phi_1 i$ and $\psi_2(\phi) = \phi_2 + \phi_3 i$ which are complex functionals. It is immediate that

$$\langle \phi, v \rangle = \langle \psi_1, v \rangle - \langle \psi_1, vj \rangle j, \quad \forall v \in X_0,$$

where $\psi_1$ is a $\mathbb{C}$–linear functional. So we can apply the complex version of the Hahn–Banach theorem to deduce the existence of a functional $\tilde{\psi}_1$ that extends $\psi_1$ to the whole of $X$. The functional $\Psi$, given by

$$\langle \Psi, v \rangle = \langle \tilde{\psi}_1, v \rangle - \langle \tilde{\psi}_1, vj \rangle j,$$

is defined on $X$ and it is the extension that satisfies estimate (12.3) for all $v \in X$.  $\square$

The following result is an immediate consequence of the quaternionic version of the Hahn–Banach theorem.

**Corollary 12.0.7.** *Let $X$ be a right quaternionic Banach space and let $v \in X$. If $\langle \phi, v \rangle = 0$ for every linear and continuous functional $\phi$ in $X^*$, then $v = 0$.*

We can reformulate this with the following corollary.

**Corollary 12.0.8.** *The dual space of a quaternionic right Banach space separates points.*

In the following paragraphs, we have restated the quaternionic version of the results that we have previously used in this book. The proofs in the complex case, found in [110], are very similar.

**Theorem 12.0.9** (Uniform boundedness principle). *Let $X$ and $W$ be two right quaternionic Banach spaces and let $\{T_\alpha\}_{\alpha \in A}$ be bounded linear maps from $X$ to $W$. Suppose that*

$$\sup_{\alpha \in A} \|T_\alpha v\| < \infty, \quad v \in X.$$

*Then,*

$$\sup_{\alpha \in A} \|T_\alpha\| < \infty.$$

*Proof.* For the proof see Theorem 11 in [110, p. 52].  $\square$

Also, the following extension theorem is in [110].

**Theorem 12.0.10** (Extension by continuity). *Let $X$ and $W$ be two-sided quaternionic Banach spaces. Let $F : D \subset X \to W$ be a uniformly continuous operator and suppose that $D$ is dense in $X$. Then $F$ has a unique continuous extension $\widetilde{F} : X \to W$ which is uniformly continuous.*

**Lemma 12.0.11** (Corollary of Ascoli–Arzelá theorem). *Let $G_1$ be a compact subset of a topological group $G$ and let $K$ be a bounded subset of the space of continuous functions $C(G_1)$. Then $K$ is conditionally compact if and only if for every $\varepsilon > 0$ there is a neighborhood $U$ of the identity in $G$ such that $|f(t) - f(s)| < \varepsilon$ for every $f \in K$ and every pair $s$, $t$ in $S$ with $t \in U$.*

*Proof.* It is Corollary 9 [110, p. 267] and its proof can be obtained with the same arguments. $\qquad\square$

**Definition 12.0.12.** We say that a quaternionic topological vector space $X$ has the fixed point property if for every continuous mapping $T : X \to X$, there exists $u \in X$ such that $u = T(u)$.

**Lemma 12.0.13.** *Let $K$ be a compact convex subset of a locally convex linear quaternionic space $X$ and let $T : K \to K$ be continuous. If $K$ contains at least two points, then there exists a proper closed convex subset $K_1 \subset K$ such that $T(K_1) \subseteq K_1$.*

**Theorem 12.0.14** (Schauder–Tychonoff). *A compact convex subset of a locally convex quaternionic linear space has the fixed point property.*

*Proof.* The proof is based on Zorn lemma and on Lemma 12.0.13, see [110]. $\qquad\square$

**Definition 12.0.15.** Let $X_0$ be a subset of $X$ and let $\mathrm{span}(X_0)$ be the subspace of $X$ spanned by $X_0$. We say that $X_0$ is a fundamental set if $\mathrm{span}(X_0) = X$.

The above definition is useful to state the following result.

**Theorem 12.0.16.** *Let $X$ be a quaternionic Banach space and let $A_m$ be a sequence of linear bounded quaternionic operators on $X$ to itself. Then the limit $Av = \lim_{m \to \infty} A_m v$ exists for every $v \in X$ if and only if*

(a) *the limit $Av$ exists for every fundamental set,*

(b) *for each $v \in X$ we have $\sup_{m \in \mathbb{N}} \|A_m v\| < \infty$.*

*When the limit $Av$ exists for each $v \in X$, the operator $A$ is bounded and*

$$\|A\| \leq \liminf_{m \to \infty} \|A_m\| \leq \sup_{m \in \mathbb{N}} \|A_m\| < \infty.$$

*Proof.* It mimics the proof of Theorem II.3.6 in [110] for complex Banach spaces. $\qquad\square$

**Lemma 12.0.17.** *Let F and G belong to $L^1(\mathbb{R}, \mathbb{H})$ with respect to the Lebesgue measure. Then the convolution*

$$(F * G)(t) := \int_0^t F(t - \tau)G(\tau)d\tau$$

*is defined for almost all t, is a function in $L^1(\mathbb{R}, \mathbb{H})$, thus*

$$\|(F * G)\|_{L^1} \le \|F\|_{L^1}\|G\|_{L^1}.$$

(a) *If $F \in L^1(\mathbb{R}, \mathbb{H})$ and there exists a positive constant M such that $|G(t)| \le M$, then*

$$\|(F * G)\|_{L^1} \le M\|F\|_{L^1}.$$

(b) *Let F and G be defined for $t \ge 0$ and let them be Lebesgue integrable over every finite interval. Then $(F * G)(t)$ is Lebesgue integrable over every finite interval.*

*Proof.* It follows the proof of Lemma 24 in [110, p.634].                    □

**Theorem 12.0.18.** *Let V and W be quaternionic two-sided Banach spaces and let T be a closed linear quaternionic operator on a domain $\mathcal{D}$ and with range W. Let $(S, \mu)$ be a measure quaternionic space and let $\mathcal{F}$ be a $\mu$-integrable function with values in $\mathcal{D}$. Suppose that $T\mathcal{F}$ is a $\mu$-integrable function then we have*

(a) $\int_S \mathcal{F}(\tau)\mu(d\tau) \in \mathcal{D}$, *and*

(b) $T\int_S \mathcal{F}(\tau)\mu(d\tau) = \int_S T\mathcal{F}(\tau)\mu(d\tau)$.

*Proof.* It follows with obvious modifications from the proof of the Theorem 20 in [110, p. 153].                    □

# Bibliography

[1] K. Abu-Ghanem, D. Alpay, F. Colombo, D. P. Kimsey, I. Sabadini, Boundary interpolation for slice hyperholomorphic Schur functions, *Integral Equations Operator Theory* **82** (2015), 223–248.

[2] K. Abu-Ghanem, D. Alpay, F. Colombo, I. Sabadini, Gleason's problem and Schur multipliers in the multivariable quaternionic setting, *J. Math. Anal. Appl.* **425** (2015), 1083–1096.

[3] K. Abu-Ghanem, D. Alpay, F. Colombo, I. Lewkowicz, I. Sabadini, Herglotz functions of several quaternionic variables, *J. Math. Anal. Appl.* **466** (2018), 169–182.

[4] S. Adler, *Quaternionic Quantum Field Theory*, Oxford University Press, 1995.

[5] S. Agrawal, S. H. Kulkarni, An analogue of the Riesz-representation theorem, *Novi Sad J. Math.* **30** (2000), 143–154.

[6] D. Albrecht, X. Duong, A. McIntosh, Operator theory and harmonic analysis, in: *Instructional Workshop on Analysis and Geometry, Part III (Canberra, 1995)*, pp. 77–136, Proc. Centre Math. Appl. Austral. Nat. Univ. **34**, Austral. Nat. Univ., Canberra, 1996.

[7] D. Alpay, *The Schur algorithm, reproducing kernel spaces and system theory*, American Mathematical Society, Providence, RI, 2001. Translated from the 1998 French original by Stephen S. Wilson, Panoramas et Synthèses.

[8] D. Alpay, V. Bolotnikov, F. Colombo, I. Sabadini, Self-mappings of the quaternionic unit ball: multiplier properties, the Schwarz-Pick inequality, and the Nevanlinna-Pick interpolation problem, *Indiana Univ. Math. J.* **64** (2015), 151–180.

[9] D. Alpay, V. Bolotnikov, F. Colombo, I. Sabadini, Interpolation problems for certain classes of slice hyperholomorphic functios, *Integral Equations Operator Theory* **86** (2016), 165–183.

[10] D. Alpay, M. Bozejko, F. Colombo, D.P. Kimsey, I. Sabadini, Boolean convolution in the quaternionic setting, *Linear Algebra Appl.* **506** (2016), 382–412.

© Springer Nature Switzerland AG 2019

F. Colombo, J. Gantner, *Quaternionic Closed Operators, Fractional Powers and Fractional Diffusion Processes*, Operator Theory: Advances and Applications 274,
https://doi.org/10.1007/978-3-030-16409-6

[11] D. Alpay, F. Colombo, J. Gantner, I. Sabadini, A new resolvent equation for the $S$-functional calculus, *J. Geom. Anal.* **25** (2015), 1939–1968.

[12] D. Alpay, F. Colombo, J. Gantner, D.P. Kimsey, Functions of the infinitesimal generator of a strongly continuous quaternionic group, *Anal. Appl. (Singap.)* **15** (2017), 279–311.

[13] D. Alpay, F. Colombo, I. Lewkowicz, I. Sabadini, Realizations of slice hyperholomorphic generalized contractive and positive functions, *Milan J. Math.* **83** (2015), 91–144.

[14] D. Alpay, F. Colombo, D.P. Kimsey, The spectral theorem for for quaternionic unbounded normal operators based on the $S$-spectrum, *J. Math. Phys.* **57** no. 2 (2016), 023503, 27 pp.

[15] D. Alpay, F. Colombo, D.P. Kimsey, I. Sabadini. The spectral theorem for unitary operators based on the $S$-spectrum, *Milan J. Math.* **84** (2016), 41–61.

[16] D. Alpay, F. Colombo, D.P. Kimsey, I. Sabadini, Wiener algebra for quaternions, *Mediterr. J. Math.* **13** (2016), 2463–2482.

[17] D. Alpay, F. Colombo, D.P. Kimsey, I. Sabadini, An extension of Herglotz's theorem to the quaternions, *J. Math. Anal. Appl.* **421** (2015), 754–778.

[18] D. Alpay, F. Colombo, I. Sabadini, *Slice hyperholomorphic Schur analysis*, Operator Theory: Advances and Applications **256**, Birkhäuser/Springer, Cham, 2016. xii+362 pp.

[19] D. Alpay, F. Colombo, I. Sabadini, Perturbation of the generator of a quaternionic evolution operator, *Anal. Appl. (Singap.)* **13** (2015), 347–370.

[20] D. Alpay, F. Colombo, I. Sabadini, On some notions of convergence for n-tuples of operators, *Math. Methods Appl. Sci.* **37** (2014), 2363–2371.

[21] D. Alpay, F. Colombo, I. Sabadini, Schur functions and their realizations in the slice hyperholomorphic setting, *Integral Equations Operator Theory* **72** (2012), 253–289.

[22] D. Alpay, F. Colombo, I. Sabadini, Pontryagin De Branges Rovnyak spaces of slice hyperholomorphic functions, *J. Anal. Math.* **121** (2013), 87–125.

[23] D. Alpay, F. Colombo, I. Sabadini, Krein–Langer factorization and related topics in the slice hyperholomorphic setting, *J. Geom. Anal.* **24** (2014), 843–872.

[24] D. Alpay, F. Colombo, I. Sabadini, Generalized quaternionic Schur functions in the ball and half-space and Krein-Langer factorization, in: *Hypercomplex Analysis: New Perspectives and Applications*, pp. 19–41, Trends in Mathematics, Birkhäuser, 2014.

[25] D. Alpay, F. Colombo, I. Sabadini, Quaternionic Hardy spaces in the open unit ball and in the half space and Blaschke products, *Journal of Physics: Conference Series* **597** (2015), 012009.

[26] D. Alpay, F. Colombo, I. Sabadini, Inner product spaces and Krein spaces in the quaternionic setting in: *Recent Advances in Inverse Scattering, Schur Analysis and Stochastic Processes*, pp. 33–65, Operator Theory: Advances and Applications **244**, Birkhäuser, 2015.

[27] D. Alpay, F. Colombo, I. Sabadini, On a class of quaternionic positive definite functions and their derivatives, *J. Math. Phys.* **58** (2017), 033501, 15 pp.

[28] D. Alpay, F. Colombo, I. Sabadini, *Characteristic operator function and canonical differential expressions: The quaternionic case*, Preprint, 2017. Collezione dei Quaderni di Dipartimento, numero QDD 229. Inserito negli Archivi Digitali di Dipartimento in data 09-01-2018.

[29] D. Alpay, F. Colombo, I. Sabadini, On slice hyperholomorphic fractional Hardy spaces, *Math. Nachr.* **290** (2017), 2725–2739.

[30] D. Alpay, F. Colombo, T. Qian, I. Sabadini, The $H^\infty$ functional calculus based on the $S$-spectrum for quaternionic operators and for $n$-tuples of non-commuting operators, *J. Funct. Anal.* **271** (2016), 1544–1584.

[31] D. Alpay, F. Colombo, I. Sabadini, G. Salomon, Fock space in the slice hyperholomorphic setting, in: *Hypercomplex Analysis: New Perspectives and Applications*, pp. 43–59, Trends in Mathematics, Birkhäuser, 2014.

[32] D. Alpay, I. Sabadini, Beurling-Lax type theorems in the complex and quaternionic setting, *Linear Algebra Appl.* **530** (2017), 15–46.

[33] A. Altavilla, Some properties for quaternionic slice regular functions on domains without real points, *Complex Var. Elliptic Equ.* **60** (2015), 59–77.

[34] N. Arcozzi, G. Sarfatti, Invariant metrics for the quaternionic Hardy space, *J. Geom. Anal.* **25** (2015), 2028–2059.

[35] W.G. Bade, An operational calculus for operators with spectrum in a strip, *Pacific J. Math.* **3** (1953), 257–290.

[36] A.V. Balakrishnan, Fractional powers of closed operators and the semigroups generated by them, *Pacific J. Math.* **10** ( 1960), 419–437.

[37] G. Birkhoff, J. von Neumann, The logic of quantum mechanics, *Ann. of Math.* **37** (1936), 823-843.

[38] C. Bisi, C. Stoppato, The Schwarz-Pick lemma for slice regular functions, *Indiana Univ. Math. J.* **61** (2012), 297–317.

[39] V. Bolotnikov, Zeros, factorizations and least common multiples of quaternion polynomials, *J. Algebra Appl.* **16** (2017), 1750181, 23 pp.

[40] V. Bolotnikov, Confluent Vandermonde matrices, divided differences, and Lagrange-Hermite interpolation over quaternions, *Comm. Algebra* **45** (2017), 575–599.

[41] V. Bolotnikov, On the Sylvester equation over quaternions, in: *Noncommutative Analysis, Operator Theory and Applications*, pp. 43–75, Oper. Theory Adv. Appl. **252**, Linear Oper. Linear Syst., Birkhüser/Springer, Cham, 2016.

[42] V. Bolotnikov, Confluent Vandermonde matrices and divided differences over quaternions, *C.R. Math. Acad. Sci. Paris* **353** (2015), 391–395.

[43] V. Bolotnikov, Polynomial interpolation over quaternions, *J. Math. Anal. Appl.* **421** (2015), 567–590.

[44] V. Bolotnikov, Pick matricies and quaternionic power series, *Integral Equations Operator Theory* **80** (2014), 293–302.

[45] V. Bolotnikov, Lagrange interpolation problem for quaternion polynomials, *C.R. Math. Acad. Sci. Paris* **352** (2014), 577–581.

[46] F. Brackx, R. Delanghe, F. Sommen, *Clifford Analysis*, Pitman Res. Notes in Math. **76**, 1982.

[47] N. Bourbaki, *Algebra I. Chapters 1–3*, Elements of Mathematics, Springer, Berlin, 1998. Reprint of the 1989 English translation.

[48] C. Bucur, E. Valdinoci, *Nonlocal diffusion and applications*, Lecture Notes of the Unione Matematica Italiana **20**, Springer, Cham; Unione Matematica Italiana, Bologna, 2016.

[49] H. Brezis, *Functional analysis, Sobolev spaces and partial differential equations*, Universitext, Springer, New York, 2011. xiv+599 pp.

[50] C.M.P. Castillo Villalba, F. Colombo, J. Gantner, J.O. González-Cervantes, Bloch, Besov and Dirichlet Spaces of Slice Hyperholomorphic Functions, *Complex Anal. Oper. theory* **9** (2015), 479–517.

[51] P. Cerejeiras, F. Colombo, U. Kähler, I. Sabadini, Perturbation of normal quaternionic operators, to appear in *Trans. Amer. Math. Soc.*, arXiv:1710.10730.

[52] F. Colombo, J. Gantner, Fractional powers of quaternionic operators and Kato's formula using slice hyperholomorphicity, *Trans. Amer. Math. Soc.* **370** (2018), 1045–1100.

[53] F. Colombo, J. Gantner, An introduction to fractional powers of quaternionic operators and new fractional diffusion processes, in: Advances in Complex Analysis and Operator Theory (F. Colombo, I. Sabadini, D.C. Struppa, M.B. Vajiac, eds.), Trends in Mathematics, pp. 101–134, Springer Basel, 2017.

[54] F. Colombo, J. Gantner, An application of the $S$-functional calculus to fractional diffusion processes, *Milan J. Math.* **86** (2018), 225–303.

[55] F. Colombo, J. Gantner, Fractional powers of vector operators and fractional Fourier's law in a Hilbert space, *J. Phys. A* **51** no. 30 (2018), 305201, 25 pp.

[56] F. Colombo, J. Gantner, Formulations of the F-functional calculus and some consequences, *Proc. Roy. Soc. Edinburgh Sect. A* **146** (2016), 509–545.

[57] F. Colombo, J. Gantner, D.P. Kimsey, *Spectral theory on the S-spectrum for quaternionic operators*, Operator Theory: Advances and Applications, vol. 270, Birkhäuser, 2019, ISBN 978-3-030-03074-2.

[58] F. Colombo, J. Gantner, On power series expansions of the S-resolvent operator and the Taylor formula, *J. Geom. Phys.* **110** (2016), 154–175.

[59] F. Colombo, J. Gantner, T. Jansens, The Schatten class and the Berezin transform for quaternionic operators, *Math. Methods Appl. Sci.* **39** (2016), 5582–5606.

[60] F. Colombo, J. Gantner, T. Jansens, Berezin transform of slice hyperholomorphic functions, *Complex Var. Elliptic Equ.* **62** (2017), 1204–1220.

[61] F. Colombo, J. O. González-Cervantes, M. E. Luna-Elizarraras, I. Sabadini, M. Shapiro, On two approaches to the Bergman theory for slice regular functions, in: *Advances in hypercomplex analysis*, Springer INdAM Ser., vol. 1, pp. 39–54, Springer, Milan, 2013.

[62] F. Colombo, J. O. González-Cervantes, I. Sabadini, A nonconstant coefficients differential operator associated to slice monogenic functions, *Trans. Amer. Math. Soc.* **365** (2013), 303–318.

[63] F. Colombo, J. O. González-Cervantes , I. Sabadini, The Bergman-Sce transform for slice monogenic functions, *Math. Methods Appl. Sci.* **34** (2011), 1896–1909.

[64] F. Colombo, J. O. González-Cervantes, I. Sabadini, On slice biregular functions and isomorphisms of Bergman spaces, *Complex Var. Elliptic Equ.* **57** (2012), 825–839.

[65] F. Colombo, J. O. González-Cervantes, I. Sabadini, The C-property for slice regular functions and applications to the Bergman space, *Complex Var. Elliptic Equ.*, **58** (2013), 1355–1372.

[66] F. Colombo, G. Gentili, I. Sabadini, A Cauchy kernel for slice regular functions, *Ann. Global Anal. Geom.* **37** (2010), 361–378.

[67] F. Colombo, G. Gentili, I. Sabadini, D.C. Struppa, Extension results for slice regular functions of a quaternionic variable, *Adv. Math.* **222** (2009), 1793–1808.

[68] F. Colombo, G. Gentili, I. Sabadini, D.C. Struppa A functional calculus in a non commutative setting, *Electron. Res. Announc. Math. Sci.* **14** (2007), 60–68.

[69] F. Colombo, G. Gentili, I. Sabadini, D.C. Struppa, Non commutative functional calculus: unbounded operators, *J. Geom. Phys.* **60** (2010), 251–259.

[70] F. Colombo, G. Gentili, I. Sabadini, D.C. Struppa, Non commutative functional calculus: bounded operators, *Complex Anal. Oper. Theory* **4** (2010), 821–843.

[71] F. Colombo, R. Lavicka, I. Sabadini, V. Soucek, The Radon transform between monogenic and generalized slice monogenic functions, *Math. Ann.* **363** (2015), 733–752.

[72] F. Colombo, R. Lavicka, I. Sabadini, V. Soucek, Monogenic plane waves and the W-functional calculus, *Math. Methods Appl. Sci.* **39** (2016), 412–424.

[73] F. Colombo, S. Mongodi, The Cauchy transform in the slice hyperholomorphic setting and related topics, *J. Geom. Phys.* **137** (2019), 162–183.

[74] F. Colombo, S. Mongodi, M. Peloso, S. Pinton *Fractional powers of the noncommutative Fourier's law by the S-spectrum approach, Math. Methods Appl. Sci.* (2019), online first: doi.org/10.1002/mma.5466

[75] F. Colombo, D. P. Kimsey, A Panorama on Quaternionic Spectral Theory and Related Functional Calculi, in: *Modern Trends in Hypercomplex Analysis* (S. Bernstein, U. Kähler, I. Sabadini, F. Sommen, eds.), pp. 111–142, Trends in Mathematics, Springer International Publishing, Cham, 2016.

[76] F. Colombo, I. Sabadini, A structure formula for slice monogenic functions and some of its consequences, in: *Hypercomplex Analysis* (I. Sabadini, M. Shapiro, F. Sommen, eds.), pp. 101–114, Trends in Math., Birkhäuser, Basel, 2009.

[77] F. Colombo, I. Sabadini, On some notions of spectra for quaternionic operators and for n-tuples of operators, *C.R. Math. Acad. Sci. Paris* **350** (2012), 399–402.

[78] F. Colombo, I. Sabadini, The Cauchy formula with s-monogenic kernel and a functional calculus for noncommuting operators, *J. Math. Anal. Appl.* **373** (2011), 655–679.

[79] F. Colombo, I. Sabadini, The quaternionic evolution operator, *Adv. Math.* **227** (2011), 1772–1805.

[80] F. Colombo, I. Sabadini, The $\mathcal{F}$-spectrum and the $\mathcal{SC}$-functional calculus, *Proc. Roy. Soc. Edinburgh Sect. A* **142** (2012), 479–500.

[81] F. Colombo, I. Sabadini, The F-functional calculus for unbounded operators, *J. Geom. Phys.* **86** (2014), 392–407.

[82] F. Colombo, I. Sabadini, On some properties of the quaternionic functional calculus, *J. Geom. Anal.* **19** (2009), 601-627.

[83] F. Colombo, I. Sabadini, On the formulations of the quaternionic functional calculus, *J. Geom. Phys.* **60** (2010), 1490–1508.

[84] F. Colombo, I. Sabadini, *Bounded perturbations of the resolvent operators associated to the $\mathcal{F}$-spectrum*, in: *Hypercomplex Analysis and Applications* (I. Sabadini, F. Sommen, eds.), pp. 13–28, Trends in Mathematics, Birkhäuser, 2011.

[85] F. Colombo, I. Sabadini, F. Sommen, The inverse Fueter mapping theorem, *Commun. Pure Appl. Anal.* **10** (2011), 1165–1181.

[86] F. Colombo, I. Sabadini, F. Sommen, The inverse Fueter mapping theorem in integral form using spherical monogenics, *Israel J. Math.* **194** (2013), 485–505.

[87] F. Colombo, I. Sabadini, F. Sommen, The Fueter primitive of biaxially monogenic functions, (con I. Sabadini, F. Sommen), *Commun. Pure Appl. Anal.* **13** (2014), 657–672.

[88] F. Colombo, I. Sabadini, F. Sommen, The Fueter mapping theorem in integral form and the $\mathcal{F}$-functional calculus, *Math. Methods Appl. Sci.* **33** (2010), 2050–2066.

[89] F. Colombo, I. Sabadini, F. Sommen, The inverse Fueter mapping theorem in integral form using spherical monogenics, *Israel J. Math.* **194** (2013), 485–505.

[90] F. Colombo, I. Sabadini, F. Sommen, The Fueter primitive of biaxially monogenic functions, *Commun. Pure Appl. Anal.* **13** (2014), 657–672.

[91] F. Colombo, D. Peña-Peña, I. Sabadini, F. Sommen, A new integral formula for the inverse Fueter mapping theorem, *J. Math. Anal. Appl.* **417** (2014), 112–122.

[92] F. Colombo, F. Sommen, Distributions and the global operator of slice monogenic functions, *Complex Anal. Oper. Theory* **8** (2014), 1257–1268.

[93] F. Colombo, I. Sabadini, D.C. Struppa, *Noncommutative functional calculus. Theory and applications of slice Hyperholomorphic functions*, Progress in Mathematics. vol. 289, Birkhäuser/Springer Basel AG, Basel, 2011.

[94] F. Colombo, I. Sabadini, D.C. Struppa, Sheaves of slice regular functions, *Math. Nachr.* **285** (2012), 949–958.

[95] F. Colombo, I. Sabadini, D.C. Struppa, The Pompeiu formula for slice hyperholomorphic functions, *Michigan Math. J.* **60** (2011), 163–170.

[96] F. Colombo, I. Sabadini, D.C. Struppa, The Runge theorem for slice hyperholomorphic functions, *Proc. Amer. Math. Soc.* **139** (2011), 1787–1803.

[97]  F. Colombo, I. Sabadini, D.C. Struppa, An extension theorem for slice mono-
      genic functions and some of its consequences, *Israel J. Math.* **177** (2010),
      369–389.

[98]  F. Colombo, I. Sabadini, D.C. Struppa, Duality theorems for slice hyper-
      holomorphic functions, *J. Reine Angew. Math.* **645** (2010), 85–105.

[99]  F. Colombo, I. Sabadini, D.C. Struppa, Slice monogenic functions, *Israel J.
      Math.* **171** (2009), 385–403.

[100] F. Colombo, I. Sabadini, D. C. Struppa, *Entire Slice Regular Functions*,
      Operator Theory: Advances and Applications, vol. 256, Springer Basel, 2017.

[101] F. Colombo, I. Sabadini, D.C. Struppa, A new functional calculus for non
      commuting operators, *J. Funct. Anal.* **254** (2008), 2255–2274.

[102] F. Colombo, I. Sabadini, D.C. Struppa, Algebraic properties of the module
      of slice regular functions in several quaternionic variables, *Indiana Univ.
      Math. J.* **61** (2012), 1581–1602.

[103] F. Colombo, I. Sabadini, F. Sommen, D.C. Struppa, *Analysis of Dirac Sys-
      tems and Computational Algebra*, Progress in Mathematical Physics, vol. 39,
      Birkhäuser, Boston, 2004.

[104] A.K. Common and F. Sommen, Axial monogenic functions from holomorphic
      functions, *J. Math. Anal. Appl.* **179** (1993), 610–629.

[105] L. Cnudde, H. De Bie, G. Ren, Algebraic approach to slice monogenic func-
      tions, *Complex Anal. Oper. Theory* **9** (2015), 1065–1087.

[106] C. Della Rocchetta, G. Gentili, G. Sarfatti, The Bohr theorem for slice reg-
      ular functions, *Math. Nachr.* **285** (2012), 2093–2105.

[107] H. Dietrich On the Pettis measurability theorem, *Bull. Austral. Math. Soc.*
      **50** (1994), 109–116.

[108] K. Diki, A. Ghanmi, A quaternionic analogue of the Segal-Bargmann trans-
      form, *Complex Anal. Oper. Theory* **11** (2017), 457–473.

[109] B. Dong, K.I. Kou, T. Qian, I. Sabadini, On the inversion of Fueter's theo-
      rem, *J. Geom. Phys.* **108** (2016), 102–116.

[110] N. Dunford, J. Schwartz. *Linear Operators, part I: General Theory*, J. Wiley
      and Sons (1988).

[111] N. Dunford, J. Schwartz. *Linear Operators, part III: General Theory*, J.
      Wiley and Sons (1988).

[112] G. Emch, Mécanique quantique quaternionienne et relativité restreinte, I,
      *Helv. Phys. Acta* **36** (1963), 739–769.

[113] K.J. Engel, R. Nagel, *One-Parameter Semigroups for Linear Evolution Equa-
      tions*. Graduate Texts in Mathematics. Springer, 2000.

[114] D.R. Farenick, B.A.F. Pidkowich, The spectral theorem in quaternions, *Linear Algebra Appl.* **371** (2003), 75–102.

[115] D. Finkelstein, J.M. Jauch, S. Schiminovich, D. Speiser, Foundations of quaternion quantum mechanics, *J. Mathematical Phys.* **3** (1962), 207–220.

[116] R. Fueter, Analytische Funktionen einer Quaternionenvariablen (German), *Comment. Math. Helv.* **4** no. 1 (1932), 9–20.

[117] R. Fueter, Die Funktionentheorie der Differentialgleichung $\Delta u = 0$ and $\Delta\Delta u = 0$ mit vier reellen Variablen (German), *Comment. Math. Helv.* **7** no. 1 (1934), 307–330.

[118] S. Gal, O.J. González-Cervantes, I. Sabadini, On some geometric properties of slice regular functions of a quaternion variable, *Complex Var. Elliptic Equ.* **60** (2015), 1431–1455.

[119] S.G: Gal, O.J. González-Cervantes, I. Sabadini, Univalence results for slice regular functions of a quaternion variable, *Complex Var. Elliptic Equ.* **60** (2015), 1346–1365.

[120] S.G. Gal, I. Sabadini, Approximation in compact balls by convolution operators of quaternion and paravector variable, *Bull. Belg. Math. Soc. Simon Stevin* **20** (2013), 481–501.

[121] S.G. Gal, I. Sabadini, Carleman type approximation theorem in the quaternionic setting and applications, *Bull. Belg. Math. Soc. Simon Stevin* **21** (2014), 231–240.

[122] S.G. Gal, I. Sabadini, Walsh equiconvergence theorems in the quaternionic setting, *Complex Var. Elliptic Equ.* **59** (2014), 1589–1607.

[123] S.G. Gal, I. Sabadini, On Bernstein and Erdos-Lax's inequalities for quaternionic polynomials, *C.R. Math. Acad. Sci. Paris* **353** (2015), 5–9.

[124] S.G. Gal, I. Sabadini, Arakelian's approximation theorem of Runge type in the hypercomplex setting, *Indag. Math. (N.S.)* **26** (2015), 337–345.

[125] S.G. Gal, I. Sabadini, Universality properties of the quaternionic power series and entire functions, *Math. Nachr.* **288** (2015), 917–924.

[126] S.G. Gal, I. Sabadini, Approximation by polynomials on quaternionic compact sets, *Math. Methods Appl. Sci.* **38** (2015), 3063–3074.

[127] S.G. Gal, I. Sabadini, Faber polynomials on quaternionic compact sets, *Complex Anal. Oper. Theory* **11** (2017), 1205–1220.

[128] S.G. Gal, I. Sabadini, Overconvergence of Chebyshev and Legendre expansions in quaternionic ellipsoids, *Adv. Appl. Clifford Algebr.* **27** (2017), 125–133.

[129] S.G. Gal, I. Sabadini, *Quaternionic Approximation*, Frontiers in Mathematics, Birkhäuser, 2019. DOI 10.1007/978-3-030-10666-9.

[130] J. Gantner, A direct approach to the $S$-functional calculus for closed operators, *J. Operator Theory* **77** (2017), 287–331.

[131] J. Gantner, Operator Theory on One-Sided Quaternionic Linear Spaces: Intrinsic S-Functional Calculus and Spectral Operators, to appear in: *Mem. Amer. Math. Soc.*, arXiv:1803.10524.

[132] J. Gantner, On the equivalence of complex and quaternionic quantum mechanics, *Quantum Stud. Math. Found.* **5** (2018), 357–390.

[133] J. Gantner, *Slice hyperholomorphic functions and the quaternionic functional calculus*, Master Thesis, 2014, Vienna University of Technology.

[134] J. Gantner, *Contributions to quaternionic operator theory and applications*, PhD Thesis, 2017, Politecnico di Milano.

[135] J. Gantner, J.O. González-Cervantes, T. Janssens, BMO- and VMO-spaces of slice hyperholomorphic functions, *Math. Nachr.* **290** (2017), 2259–2279.

[136] G. Gentili, S. Salamon, C. Stoppato, Twistor transforms of quaternionic functions and orthogonal complex structures, *J. Eur. Math. Soc. (JEMS)* **16** (2014), 2323–2353.

[137] G. Gentili, G. Sarfatti, Landau-Toeplitz theorems for slice regular functions over quaternions, *Pacific J. Math.* **265** (2013), 381–404.

[138] G. Gentili, C. Stoppato, D.C. Struppa, A Phragmen-Lindelof principle for slice regular functions, *Bull. Belg. Math. Soc. Simon Stevin* **18** (2011), 749–759.

[139] G. Gentili, C. Stoppato, D.C. Struppa, *Regular functions of a quaternionic variable*, Springer Monographs in Mathematics, Springer, Heidelberg, 2013.

[140] G. Gentili, C. Stoppato, The open mapping theorem for regular quaternionic functions, *Ann. Sc. Norm. Super. Pisa Cl. Sci.* **8** (2009), 805–815.

[141] G. Gentili, D.C. Struppa, A new approach to Cullen-regular functions of a quaternionic variable, *C.R. Math. Acad. Sci. Paris* **342** (2006), 741–744.

[142] G. Gentili, D.C. Struppa, A new theory of regular functions of a quaternionic variable, *Adv. Math.* **216** (2007), 279–301.

[143] G. Gentili, D.C. Struppa, On the multiplicity of zeroes of polynomials with quaternionic coefficients, *Milan J. Math.* **76** (2008), 15–25.

[144] G. Gentili, C. Stoppato, The zero sets of slice regular functions and the open mapping theorem, in: *Hypercomplex Analysis and Applications* (I. Sabadini, F. Sommen, eds.), pp. 95–107, Trends in Math., Birkhäuser/Springer Basel AG, Basel, 2011.

[145] G. Gentili, D. C. Struppa, Lower bounds for polynomials of a quaternionic variable, *Proc. Amer. Math. Soc.* **140** (2012), 1659–1668.

[146] G. Gentili, C. Stoppato, Power series and analyticity over the quaternions, *Math. Ann.* **352** (2012), 113–131.

[147] G. Gentili, I. Vignozzi, The Weierstrass factorization theorem for slice regular functions over the quaternions, *Ann. Global Anal. Geom.* **40** (2011), 435–466.

[148] R. Ghiloni, V. Moretti, A. Perotti, Continuous slice functional calculus in quaternionic Hilbert spaces, *Rev. Math. Phys.* **25** (2013), 1350006, 83 pp.

[149] R. Ghiloni, V. Moretti, A. Perotti, Spectral properties of compact normal quaternionic operators, in: *Hypercomplex Analysis: New Perspectives and Applications* pp. 133–143, Trends in Mathematics, Birkhäuser, 2014.

[150] R. Ghiloni, A. Perotti, Slice regular functions of several Clifford variables, in: *Proceedings of i.c.n.p.a.a. 2012 – Workshop "Clifford algebras, Clifford analysis and their applications"*, pp. 734–738, AIP Conf. Proc., vol. 1493, 2012.

[151] R. Ghiloni, A. Perotti, A new approach to slice regularity on real algebras, in: *Hypercomplex analysis and applications*, pp. 109–123, Trends in Math., Birkhäuser/Springer Basel AG, Basel, 2011.

[152] R. Ghiloni, A. Perotti, Slice regular functions on real alternative algebras, *Adv. Math.* **226** (2011), 1662–1691.

[153] R. Ghiloni, A. Perotti, Zeros of regular functions of quaternionic and octonionic variable: a division lemma and the camshaft effect, *Ann. Mat. Pura Appl.* **190** (2011), 539–551.

[154] R. Ghiloni, A. Perotti, Volume Cauchy formulas for slice functions on real associative C*-algebras, *Complex Var. Elliptic Equ.* **58** (2013), 1701–1714.

[155] R. Ghiloni, A. Perotti, Global differential equations for slice regular functions, *Math. Nachr.* **287** (2014), 561–573.

[156] R. Ghiloni, A. Perotti, Power and spherical series over real alternative C*-algebras, *Indiana Univ. Math. J.* **63** (2014), 495–532.

[157] R. Ghiloni, A. Perotti, Lagrange polynomials over Clifford numbers, *J. Algebra Appl.* **14** no. 5 (2015), 1550069, 11 pp.

[158] R. Ghiloni, V. Recupero, Semigroups over real alternative *-algebras: generation theorems and spherical sectorial operators, *Trans. Amer. Math. Soc.* **368** (2016), 2645–2678.

[159] R. Ghiloni, V. Recupero, Slice regular semigroups, *Trans. Amer. Math. Soc.* **370** (2018), 4993–5032.

[160] J.O. González-Cervantes, I. Sabadini, On some splitting properties of slice regular functions, *Complex Var. Elliptic Equ.* **62** (2017), 1393–1409.

[161] I.S. Gradshteyn, and I.M. Ryzhik, *Table of integrals, series, and products,* Elsevier/Academic Press, Amsterdam, 2007.

[162] K. Gürlebeck, K. Habetha, and W. Sprößig, *Holomorphic functions in the plane and n-dimensional space,* Birkhäuser Verlag, Basel, 2008. Translated from the 2006 German original.

[163] K. Gürlebeck, W. Sprössig, *Quaternionic analysis and elliptic boundary value problems,* International Series of Numerical Mathematics, vol. 89, Birkhäuser Verlag, Basel, 1990. 253 pp.

[164] A. Guzman, Growth properties of semigroups generated by fractional powers of certain linear operators, *J. Functional Analysis* **23** no. 4 (1976) 331–352.

[165] M. Haase, *The functional calculus for sectorial operators,* Operator Theory: Advances and Applications, vol. 169, Birkhäuser Verlag, Basel, 2006.

[166] L.P. Horwitz, L.C. Biedenharn, Quaternion quantum mechanics: Second quantization and gauge fields, *Annals of Physics* **157** (1984), 4327488.

[167] T. Kato, Note on fractional powers of linear operators, *Proc. Japan Acad.* **36** (1960), 94–96.

[168] T. Kato, Fractional powers of dissipative operators, *J. Math. Soc. Japan* **13** (1961), 246–274.

[169] H. Komatsu, Fractional powers of operators, *Pacific J. Math.* **19** ( 1966), 285–346.

[170] A. McIntosh, Operators which have an $H^\infty$ functional calculus, in: *Miniconference on operator theory and partial differential equations (North Ryde, 1986),* pp. 210–231, Proc. Centre Math. Anal. Austral. Nat. Univ., vol. 14, Austral. Nat. Univ., Canberra, 1986.

[171] B. Jefferies, *Spectral properties of noncommuting operators,* Lecture Notes in Mathematics, vol. 1843, Springer-Verlag, Berlin, 2004.

[172] B. Jefferies, A. McIntosh, The Weyl calculus and Clifford analysis, *Bull. Austral. Math. Soc.* **57** (1998), 329–341.

[173] B. Jefferies, A. McIntosh, J. Picton-Warlow, The monogenic functional calculus, *Studia Math.* **136** (1999), 99–119.

[174] M. Khokulan, K. Thirulogasanthar, B. Muraleetharan, S-spectrum and associated continuous frames on quaternionic Hilbert spaces, *J. Geom. Phys.* **96** (2015), 107–122.

[175] W.D. Kirwin, J. Mourao, J.P. Nunes, T. Qian, Extending coherent state transforms to Clifford analysis, *J. Math. Phys.* **57** (2016), 103505, 10 pp.

[176] K.I. Kou, T. Qian, F. Sommen, Generalizations of Fueter's theorem, *Meth. Appl. Anal.* **9** (2002), 273–290.

[177] P.D. Lax, *Functional analysis*, Pure and Applied Mathematics (New York), Wiley-Interscience, John Wiley & Sons, New York, 2002.

[178] C. Li, A. McIntosh, T. Qian, Clifford algebras, Fourier transforms and singular convolution operators on Lipschitz surfaces, *Rev. Mat. Iberoamericana* **10** (1994), 665–721.

[179] A. McIntosh, A. Pryde, A functional calculus for several commuting operators, *Indiana U. Math. J.* **36** (1987), 421–439.

[180] B. Muraleetharan, I. Sabadini, K. Thirulogasanthar, *S*-Spectrum and the quaternionic Cayley transform of an operator, *J. Geom. Phys.*, **124** (2018), 442–455

[181] B. Muraleetharan, K. Thirulogasanthar, I. Sabadini, A Representation of Weyl-Heisenberg Lie Algebra in the Quaternionic Setting, *Ann. Physics* **385** (2017), 180–213.

[182] D. Peña-Peña, *Cauchy-Kowalevski extensions, Fueter's theorems and boundary values of special systems in Clifford analysis*, PhD Dissertation, Gent, 2008.

[183] D. Peña-Peña, T. Qian, F. Sommen, An alternative proof of Fueter's theorem, *Complex Var. Elliptic Equ.* **51** (2006), 913–922.

[184] D. Peña-Peña, I. Sabadini, F. Sommen, Fueter's theorem for monogenic functions in biaxial symmetric domains, *Results Math.*, **72** (2017), 1747–1758.

[185] R.S. Phillips, Spectral theory for semi-groups of linear operators, *Trans. Amer. Math. Soc.* **71** ( 1951), 393–415.

[186] T. Qian, Generalization of Fueter's result to $\mathbb{R}^{n+1}$, *Atti Accad. Naz. Lincei Cl. Sci. Fis. Mat. Natur. Rend. Lincei (9) Mat. Appl.* **8** no. 2 (1997), 111–117.

[187] T. Qian, Singular integrals on star-shaped Lipschitz surfaces in the quaternionic space, *Math. Ann.* **310** (1998), 601–630.

[188] G. Ren, X. Wang, Caratheodory theorems for slice regular functions, *Complex Anal. Oper. Theory* **9** (2015), 1229–1243.

[189] G. Ren, Z. Xu, Schwarz's Lemma for Slice Clifford Analysis, *Adv. Applied Cliff. Alg.* **25** (2015), 965–976.

[190] L. Rodman, *Topics in Quaternion Linear Algebra*, Princeton University Press, Princeton, NJ, 2014.

[191] W. Rudin, *Functional Analysis*, McGraw-Hill Series in Higher Mathematics, McGraw-Hill Book Co., New York–Düsseldorf–Johannesburg, 1973.

[192] W. Rudin, *Real and complex analysis*, McGraw-Hill, New York, 3rd ed., 1987.

[193] I. Sabadini, A. Saracco, Carleson measures for Hardy and Bergman spaces in the quaternionic unit ball, *J. Lond. Math. Soc.* **95** (2017), 853–874.

[194] G. Sarfatti, *Elements of function theory in the unit ball of quaternions*, PhD thesis, Universitá di Firenze, 2013.

[195] M. Sce, Osservazioni sulle serie di potenze nei moduli quadratici, *Atti Acc. Lincei Rend. Fisica* **23** (1957), 220–225.

[196] C. Sharma, T.J. Coulson, Spectral theory for unitary operators on a quaternionic Hilbert space, *J. Math. Phys.* **28(9)** (1987), 1941–1946

[197] C. Stoppato A new series expansion for slice regular functions, *Adv. Math.* **231** (2012), 1401–1416.

[198] C. Stoppato, Singularities of slice regular functions, *Math. Nachr.* **285** (2012), 1274–1293.

[199] C. Stoppato, Regular Moebius transformations of the space of quaternions, *Ann. Global Anal. Geom.* **39** (2011), 387–401.

[200] K. Schmüdgen, *Unbounded self-adjoint operators on Hilbert space*, Graduate Texts in Mathematics, vol. 265, Springer, Heidelberg, 2012.

[201] F. Sommen, Clifford Analysis and integral geometry, in: *Clifford algebras and their applications in mathematical physics (Montpellier, 1989)*, pp. 293–311, Fund. Theories Phys., vol. 47, Kluwer Acad. Publ., Dordrecht, 1992.

[202] O. Teichmüller, Operatoren im Wachsschen Raum, *J. Reine Angew. Math.* **174** (1936), 73–124.

[203] K. Viswanath, Normal operations on quaternionic Hilbert spaces, *Trans. Amer. Math. Soc.* **162** (1971), 337–350.

[204] J. Watanabe, On some properties of fractional powers of linear operators, *Proc. Japan Acad.* **37** (1961), 273–275.

[205] K. Yosida, Fractional powers of infinitesimal generators and the analyticity of the semi-groups generated by them, *Proc. Japan Acad.* **36** (1960), 86–89.

[206] Y. Yang, T. Qian, Zeroes of slice monogenic functions, *Math. Methods Appl. Sci.* **34** (2011), 1398–1405.

# Index

Printed in the United States
By Bookmasters